Biological and Medical Physics, Biomedical Engineering

This series is intended to be comprehensive, covering a broad range of topics important to the study of the physical, chemical and biological sciences. Its goal is to provide scientists and engineers with textbooks, monographs, and reference works to address the growing need for information. The fields of biological and medical physics and biomedical engineering are broad, multidisciplinary and dynamic. They lie at the crossroads of frontier research in physics, biology, chemistry, and medicine.

Books in the series emphasize established and emergent areas of science including molecular, membrane, and mathematical biophysics; photosynthetic energy harvesting and conversion; information processing; physical principles of genetics; sensory communications; automata networks, neural networks, and cellular automata. Equally important is coverage of applied aspects of biological and medical physics and biomedical engineering such as molecular electronic components and devices, biosensors, medicine, imaging, physical principles of renewable energy production, advanced prostheses, and environmental control and engineering.

Sarat Kumar Swain · Anuradha Biswal
Editors

Chitosan Nanocomposites

Bionanomechanical Applications

Springer

Editors
Sarat Kumar Swain
Department of Chemistry
Veer Surendra Sai University of Technology
Burla, Sambalpur, Odisha, India

Anuradha Biswal
Department of Chemistry
Veer Surendra Sai University of Technology
Burla, Sambalpur, Odisha, India

ISSN 1618-7210 ISSN 2197-5647 (electronic)
Biological and Medical Physics, Biomedical Engineering
ISBN 978-981-19-9645-0 ISBN 978-981-19-9646-7 (eBook)
https://doi.org/10.1007/978-981-19-9646-7

This Springer imprint is published by the registered company Springer Nature Singapore Pte Ltd.
The registered company address is: 152 Beach Road, #21-01/04 Gateway East, Singapore 189721, Singapore

The editors dedicate this book to ...

Mr. Rabindra Nath Swain
(Father of Prof. Sarat Kumar Swain)
&

The loving memories of Late Mr. Himanshu Sekhar Biswal
(Father of Ms. Anuradha Biswal)

Preface

Chitosan, a biomaterial usually extracted from the wastes obtained from sea food processing industries, have slowly taken the centre stage, both scientifically and commercially. Starting from health supplements to medical devices, food additives to cosmetics, chitosan has found its applicability in different facets of life. Following its discovery in 1811 to the present day, this cheap and readily available biopolymer has found its way from laboratories to industries, owing to its wonderful properties like biodegradability, biocompatibility, mucoadhesivity, non-toxicity, antimicrobial nature, optical properties and many more. Even after multitudes of scientific research, still there exists some grey areas related to chitosan, namely structure, mechanical properties, solubility, stability and processibility which opens up new windows for further improvements. This task is taken upon by the field of material science which suggested overcoming these limitations by the means of chemical and physical modifications. Chitosan has been composited in various ways by using materials to impart desirable properties to the overall material. Several materials like polymers, nanomaterials, natural bioactive materials, amino acids have been utilized to get robust materials with accentuated properties. Out of these, combining chitosan with different nanomaterials to form chitosan-based nanocomposites have emerged as a viable option to widen the applicability of chitosan by enhancing its properties by manifolds. Last few decades have witnessed the incorporation of a wide variety of nanomaterials like metal/metal oxide nanoparticles, carbonaceous nanomaterials, polymeric nanomaterials, peptide nanostructures, layered nanomaterials, and ceramics for designing of chitosan-based nanocomposites which overcomes the drawbacks faced by the pristine biopolymer. These hybrid materials have been employed in various applications like sensing, biomedical, theranostics, agriculture, food additives, environmental remediation, packaging and other industrial applications. Out of these, health care sector has seen an exponential rise in the use of chitosan-based nanocomposites for applications like drug delivery, wound healing, biosensing, cancer therapy, gene delivery, and tissue engineering. But, the present world requires the mechanization of these biomedical applications to ensure advanced outcomes in health care sector. Chitosan-based nanocomposites are being increasingly used for different bionanomechanical applications. Studying the release mechanics, entrapment mechanism in nanoscale and other related parameters of these nanocomposites have become the prime interest of material scientists. Amalgamation of different bionanomechanical applications of chitosan-based nanocomposites under a unified literature is the main driving force behind the present book.

The prime objective of this book "Chitosan Nanocomposites: Bionanomechanical Applications" is to provide a broad coverage on chitosan and its potential uses in tissue engineering, dentistry, wound healing, targeted drug delivery, nanomedicines and many more. This book is divided into seventeen chapters, each of which focuses on one of the important issues stated above while also highlighting recent developments made by eminent academics and researchers. The chapters discussing the importance of chitosan, extraction and isolation techniques, chemical and biological aspects, modifications, nanocomposite formation and its potential applications in different bionanomechanical applications and its future perspectives stands as the backbone of the present book. The book presents the journey of chitosan from crustaceans wastes to its advanced applications in different industries, as an illustrative story to invite the interest of readers towards this beautiful biomaterial and its magical properties. Different aspects of chitosan based nanocomposites have been schematically presented by representative figures, schemes and tables. The book aims on providing students, academicians, research community and industrialists new insights on chitosan nanocomposites in bionanomechanical applications as well as present ideas for further scope for research and improvement in the field. The present book contains seventeen chapters in various topics with contribution of authors with affiliation from thirteen different countries.

We, the editors, would like to extend our sincere gratitude to the Springer Nature and the eminent handling editor, Dr. Loyola D'Silva for giving us an opportunity to present our ideas through this book as part of the book series, "Biological and Medical Physics, Biomedical Engineering". We express our thanks to Dr. Saranya Kalidoss and Dr. Saranya Devi Balasubraminan for their timely help and support throughout the editing process. We would like to express our deep sense of gratitude to the research groups of Dr. Butnariu, Dr. Chatterji, Dr. Kassab, Dr. Staden, Dr. Szopa, Dr. Palanisamy, Dr. Nie, Dr. Li, Dr. Sarkar, Dr. Altuntas, Dr. Hajam, Dr. Yasri, Dr. Rajivgandhi and Dr. Damiri for contributing their valuable works to make this book a success. We would also like to acknowledge the untiring efforts of all members of our research laboratories, namely Ms. Swapnita Patra, Ms. Krishna Manjari Sahu, Mr. Susobhan Swain, Mr. Shuvendu Shuvankar Purohit and Mr. Sachit Kumar Das, for the successful completion of the book. If our efforts, in partnership with our contributing authors and publishers, are successful in satisfying the requirements of academics, researchers, and students in the fields of engineering, material science, biomedical sciences, as well as those in the pharmaceutical discipline, we will be glad.

Burla, Sambalpur, Odisha, India

Prof. Sarat Kumar Swain
Ms. Anuradha Biswal

Contents

Editors and Contributors

About the Editors

Prof. (Dr.) Sarat Kumar Swain is presently working as a Professor of Chemistry at Veer Surendra Sai University of Technology (VSSUT), Burla, Sambalpur, Odisha, India. He has worked as the Post-Doctoral Fellow in the Department of Polymer Engineering, University of Akron, OH, USA, after receiving his doctoral degree from Utkal University, Bhubaneswar, India. He has also worked as a visiting researcher at Indian Association for Cultivation for Science, Kolkata and Jawaharlal Nehru Centre for Advanced and Scientific Research, Bangalore with awarding INSA SR Fellowship and JNCASR Fellowship respectively. He has more than 25 years of experience in teaching and research in the area of Organic Chemistry, Materials Chemistry, Nanotechnology and Polymer Chemistry at UG, PG and PhD levels. Prof. Swain has successfully supervised seventeen PhD scholars and three post-doctoral fellows along with more than fifty master theses as the output of his research. His research interests include, reinforced polymer nanocomposites, hybrid nanomaterials, advanced materials like nanoclay, graphene, CNF, CNT and CQD for improving of fire-retardant, gas barrier, thermal, chemical and mechanical resistance properties of the materials. He has designed various kinds of nanostructured materials for waste water treatment, packaging, anti-corrosion performance, superconductor properties, sensing and biomedical applications. Prof. Swain has published more than 150 research articles in different international SCI/Scopus indexed journals and contributed chapters for different books. So far, he has published six books in different publishers such as Springer, Elsevier and RSC, and invented of two patents (One USA and One Indian) to his credit. He has received several awards such as BOYSCAST Fellowship, DAE-Young Scientist Research Award from Government of India along with prestigious "Samanta Chandra Sekhar Award" of Government of Odisha for various academic and scientific achievements. Besides his academic credits, he has also successfully rendered various administrative responsibilities such as Dean (Academic Affairs), Dean (PG Studies and Research), Head of the Department of Chemistry etc. during his professional career.

Ms. Anuradha Biswal is currently working as a research scholar in the Veer Surendra Sai University of Technology, Burla, Sambalpur, India after completing her post graduate degree from Orissa University of Agriculture and Technology, Bhubaneswar, India. She is pursuing her doctoral degree under the direct supervision of Prof. Sarat Kumar Swain. During her Ph.D., she has published six research articles in different SCII/SCOPUS indexed international journals and contributed six book chapters in various edited books of international rapute. Her area of research includes the designing of polymer-based biomaterials for wound dressing applications.

Contributors

Derya Bal Altuntaş Faculty of Engineering and Architecture, Department of Bioengineering, Recep Tayyip Erdogan University, Rize, Turkey

Swetha Andra Department of Chemistry, Rajalakshmi Institute of Technology, Chennai, Tamil Nadu, India

Sema Aslan Faculty of Science, Department of Chemistry, Mugla Sitki Kocman University, Mugla, Turkey

Satheeshkumar Balu Department of Oral Pathology, Saveetha Dental College and Hospitals, Saveetha Institute of Medical and Technical Sciences, Chennai, Tamil Nadu, India

Mohammed Berrada Laboratory of Biomolecules and Organic Synthesis (BIOSYNTHO), Department of Chemistry, Faculty of SciencesBen M'Sick, University Hassan II of Casablanca, Casablanca, Morocco

Debasrita Bharatiya Department of Chemistry, Veer Surendra Sai University of Technology, Burla, Sambalpur, Odisha, India

Katarzyna Bialik-Wąs Faculty of Chemical Engineering and Technology, Institute of Organic Chemistry and Technology, Cracow University of Technology, Kraków, Poland

Anuradha Biswal Department of Chemistry, Veer Surendra Sai University of Technology, Sambalpur, Odisha, India

Monica Butnariu Chemistry and Biochemistry Discipline, University of Life Sciences "King Mihai I" From Timisoara, Timis, Romania

Tanushri Chatterji School of Bioscience, Institute of Management Studies (University Courses Campus), Ghaziabad, UP, India

Feng Cheng MIIT Key Laboratory of Critical Materials Technology for New Energy Conversion and Storage, School of Chemistry and Chemical Engineering, Harbin Institute of Technology, Harbin, PR China

Catalina Cioates Negut Laboratory of Electrochemistry and PATLAB, National Institute of Research for Electrochemistry and Condensed Matter, Bucharest, Romania

Jiliang Dai College of Light Industry and Textile, Qiqihar University, Qiqihar, Heilongjiang, PR China

Fouad Damiri Laboratory of Biomolecules and Organic Synthesis (BIOSYNTHO), Department of Chemistry, Faculty of SciencesBen M'Sick, University Hassan II of Casablanca, Casablanca, Morocco

Hamza Daoudi Materials Science, Energy and Nano-Engineering (MSN) Department, Mohammed VI Polytechnic University (UM6P), Benguerir, Morocco

Peng Ding College of Life Sciences, Xinyang Normal University, Xinyang, China

Halina Ekiert Chair and Department of Pharmaceutical Botany, Medical College, Jagiellonian University, Kraków, Poland

Mounir El Achaby Materials Science, Energy and Nano-Engineering (MSN) Department, Mohammed VI Polytechnic University (UM6P), Benguerir, Morocco

Chirâa El Idrissi El Hassani Laboratory of Materials, Catalysis, and Valorization of Natural Resources, URAC 24, Faculty of Science and Techniques of Mohammedia, University Hassan II, Hassan, Casablanca, Morocco

Damaris-Cristina Gheorghe Laboratory of Electrochemistry and PATLAB, National Institute of Research for Electrochemistry and Condensed Matter, Bucharest, Romania

B. H. Jaswanth Gowda Department of Pharmaceutics, Yenepoya Pharmacy College & Research Centre, Yenepoya (Deemed to Be University), Mangalore, Karnataka, India

Younis Ahmad Hajam Department of Life Sciences and Allied Health Sciences, Sant Baba Bhag Singh University, Jalandhar, Punjab, India

Ruxandra -Maria Ilie-Mihai Laboratory of Electrochemistry and PATLAB, National Institute of Research for Electrochemistry and Condensed Matter, Bucharest, Romania

Ceyhun Işık Faculty of Science, Department of Chemistry, Mugla Sitki Kocman University, Mugla, Turkey

Karolina Jafernik Chair and Department of Pharmaceutical Botany, Medical College, Jagiellonian University, Kraków, Poland

Zineb Kassab Materials Science, Energy and Nano-Engineering (MSN) Department, Mohammed VI Polytechnic University (UM6P), Benguerir, Morocco

Lovepreet Kaur Department of Life Sciences and Allied Health Sciences, Sant Baba Bhag Singh University, Jalandhar, Punjab, India

Namrata Khanna Department of Biochemistry, M A Rangoonwala College of Dental Sciences and Research Centre, Pune, Maharashtra, India

Nivedita Khanna Scientist-Biopharmaceutical Services, SGS Canada Inc, Mississauga, ON, Canada

Praveen Kumar Department of Mechanical Engineering, Sant Longowal Institute of Engineering and Technology, Longowal, Sangrur, Punjab, India

Rajesh Kumar Department of Biosciences, Himachal Pradesh University, Shimla, India

Hongbin Li MIIT Key Laboratory of Critical Materials Technology for New Energy Conversion and Storage, School of Chemistry and Chemical Engineering, Harbin Institute of Technology, Harbin, PR China;
College of Light Industry and Textile, Qiqihar University, Qiqihar, Heilongjiang, PR China

Jingyu Li College of Life Sciences, Xinyang Normal University, Xinyang, China

Vellaichamy Mageshwaran Microbial Technology Laboratory, ICAR-National Bureau of Agriculturally Important Microorganisms, Uttar Pradesh, Kushmaur, Mau Nath Bhanjan, India

Magdalena Anna Malinowska Faculty of Chemical Engineering and Technology, Institute of Organic Chemistry and Technology, Cracow University of Technology, Kraków, Poland

Atilla Eren Mamuk Faculty of Science, Department of Physics, Mugla Sitki Kocman University, Mugla, Turkey

Muthuchamy Maruthupandy Laboratorio de Nanocelulosa y Biomateriales, Departamento de Ingeniería Química, Biotecnología y Materiales, Universidad de Chile, Santiago, Chile

Yalavarthi Nagaraju Microbial Technology Laboratory, ICAR-National Bureau of Agriculturally Important Microorganisms, Uttar Pradesh, Kushmaur, Mau Nath Bhanjan, India

Neelam Department of Life Sciences and Allied Health Sciences, Sant Baba Bhag Singh University, Jalandhar, Punjab, India

Lei Nie College of Life Sciences, Xinyang Normal University, Xinyang, China; Université Libre de Bruxelles (ULB), École Polytechnique de Bruxelles, Brussels, Belgium

Oseweuba Valentine Okoro College of Life Sciences, Xinyang Normal University, Xinyang, China

Biswajit Parhi Department of Chemistry, Veer Surendra Sai University of Technology, Burla, Sambalpur, Odisha, India

Shuvendu S. Purohit Department of Chemistry, Veer Surendra Sai University of Technology, Burla, Sambalpur, Odisha, India

Franck Quero Laboratorio de Nanocelulosa y Biomateriales, Departamento de Ingeniería Química, Biotecnología y Materiales, Universidad de Chile, Santiago, Chile

Seema Rai Department of Zoology, Guru Ghasidas Vishwavidayalaya (A Central University), Bilaspur, Chhattisgarh, India

Govindan Rajivgandhi Laboratorio de Nanocelulosa y Biomateriales, Departamento de Ingeniería Química, Biotecnología y Materiales, Universidad de Chile, Santiago, Chile

Mohd Salim Reshi Department of Zoology, School of Biosciences and Biotechnology, Baba Ghulam Shah Badshah University, Rajouri, Jammu and Kashmir, India

Satish Rojekar Department of Pharmacological Sciences, Icahn School of Medicine at Mount Sinai, New York, NY, USA

Dhiman Santra Department of Chemistry, University of Kalyani, Kalyani, West Bengal, India

Mitali Sarkar Department of Chemistry, University of Kalyani, Kalyani, West Bengal, India

Marta Sharafan Faculty of Chemical Engineering and Technology, Institute of Organic Chemistry and Technology, Cracow University of Technology, Kraków, Poland

Amin Shavandi College of Life Sciences, Xinyang Normal University, Xinyang, China

Elżbieta Sikora Faculty of Chemical Engineering and Technology, Institute of Organic Chemistry and Technology, Cracow University of Technology, Kraków, Poland

Suruchi Singh Life Sciences, Tata Consultancy Services Limited, Thane (W), Mumbai, Maharashtra, India

Palanisamy Sivasubramanian Department of Mechanical Engineering, Dilkap Research Institute of Engineering and Management Studies, At Village Mamdapur, Tal- Karjat, Dist Raigad, India

Raluca-Ioana Stefan-van Staden Laboratory of Electrochemistry and PATLAB, National Institute of Research for Electrochemistry and Condensed Matter, Bucharest, Romania

Sarat Kumar Swain Department of Chemistry, Veer Surendra Sai University of Technology, Burla, Sambalpur, Odisha, India

Agnieszka Szopa Chair and Department of Pharmaceutical Botany, Medical College, Jagiellonian University, Kraków, Poland

Qianqian Wei College of Life Sciences, Xinyang Normal University, Xinyang, China

Viroj Wiwanitkit Dr. D. Y. Patil Medical College, Hospital and Research Centre, D. Y. Patil Vidyapeeth, Pune, India;
Department of Eastern Medicine, Government College University Faisalabad, Faisalabad, Pakistan;
Hainan Medical University, Haikou, China;
Faculty of Medicine, University of Nis, Nis, Serbia;
Joseph Ayobalola University, Ikeji-Arakeji, Nigeria

Sora Yasri KM Center, Bangkok, Thailand

Chenxu Zhao College of Life Sciences, Xinyang Normal University, Xinyang, China

Abbreviations

3D	Three dimensional
5-ALA	5-aminolevulinic acid
AA	Alginic acid
ACE	Angiotensin-1-converting enzyme
ACOP	Acetaminophen
AD	Alzheimer's Disease
AgNP/AgNPs	Silver Nanoparticles
AgNS	Silver nanospheres
ALP	Alkaline phosphatase
AMPK	Adenosine monophosphate protein kinase
AND	Andrographolide
ANE	Anemoside B4
AOT	Sodium bis(2-ethylhexyl) sulfosuccinate
ASA	Aspirin
AuNP	Gold Nanoparticles
AuNS	Gold nanospheres
BB	Bacillus cereus
BBB	Blood brain barrier
BG	Bioactive glass
b-GP	b-Sodium glycerophosphate
BMI	Body mass index
BMP-7	Bone morphogenetic protein-7
BMSC	Bone marrow stromal cells
BN	Boron nitride
BTE	Bone-tissue engineering
C/CH/CS	Chitosan
CAS	Chemical Abstracts Service
CB	Candida bombicola
CDKAL1	Cdk5 Regulatory related protein 1-like 1
Ce-BG	Cerium-Doped Bioactive Glass
CEC	Corneal endothelial cell

CG	Chitosan-l-glutamic acid
ChNPs	Chitosan nanoparticles, chitosan-based nanoparticles, chitosan nanostructures
CJMSCs	Conjunctiva mesenchymal stem cells
CMC	Carboxymethyl cellulose
CMC	Carboxymethyl chitosan
C-NMR	Carbon-nuclear magnetic resonance
CNPs	Chitosan nanoparticles
CNT	Carbon nanotubes
Col	Collagen
COS	Chito-oligosaccharide/Chitosan oligomers
CPC	Calcium phosphate cement
CPPs	Cell-penetrating peptides
CPs	Conductive polymers
CrI	Crystallinity index
CS/CpG ODNs	Cytosine-phosphate guanine oligodeoxynucleotides
CS/CS-DMMA	Chitosan/dimethylmaleic anhydride
CSNPs	Chitosan nanoparticles
CS-SF	Chitosan–silk–fibroin
CTA	Computed tomography angiography
CTAB	Cetyltrimethylammonium bromide
CuCNPs	Copper-loaded chitosan nanoparticles
Cur	Curcumin
CV	Cyclic voltammetry
CVD	Cardiovascular diseases
DA	Degree of Acetylation/Dicarboxylic acid
DCS	Decellularized scaffolds
DD	Degree of deacetylation
DFUs	Diabetic foot ulcers
DM	Diabetes mellitus
DNA	Deoxyribonucleic acid
DOM	Domperidone
DPP4	Dipeptidyl peptidase-4
DPP-4	Dipeptidyl peptidase-four
DPPH	2,2-diphenyl-1-picryl-hydrazyl-hydrate radical essay
DPV	Differential pulse voltammetry
E. coli	Escherichia coli
ECM	Extra cellular matrix
EDX	Energy dispersive X-ray spectroscopy
EF	Euryale ferox
EGF	Epidermal growth factor
eHAp/eHAP	Erbium ion doped hydroxyapatite
EIS	Electrochemical impedance spectroscopy
EnSCs	Endometrial stem cells
EPO	Erythropoietin

ESP	Electrospinning technique
FDA	Food and drug administration
Fe_3O_4	Iron oxide
FeNPs	Fe_3O_4 nanoparticles
FESEM	Field emission scanning electron microscopy
FTIR	Fourier transform infrared spectroscopy
GADA	Glutamic acid decarboxylase
GAGs	Glycosaminoglycans
GAuCS	Graphene/gold nanoparticle/chitosan
GDM	Gestational Diabetes Mellitus
GE	Gelatin
Gen	Gentamicin
GFAP	Glial fibrillary acidic protein
GLP-1	Glucagon-like peptide 1
GMA	Glycidyl methacrylate
GNPs	Gold nanoparticles
GO	Graphene oxide
GON	Graphene oxide nanoparticles
GTA	Glutaraldehyde
HA/HAD/HAL	Hyaluronic acid
HA/HAP	Hydroxyapatite
HAD	Hot air drying
HBCS	Hydroxybutyl Chitosans
HHEX	Hematopoietically Expressed Homebox
HLA	Human leucocyte complex
HMW	High molecular weight
H-NMR	Hydrogen-nuclear magnetic resonance
HPLC	High Performance Liquid Chromatography
HRTE	MHigh resolution transmission electron microscopy
IA2A	Insulinoma-affiliation protein antibodies
IAA	Anti-insulin
IFO	Ifosfamide
IGF2BP2	Insulin-like boom aspect 2 mRNA-binding protein 2
JAZF1	Juxtaposed with every other zinc finger gene 1
kDa	Kilodalton
LA	Lipoic Acid
LB	Langmuir–Blodgett
LBL/LbL	Layer-by-layer
LCs	Liquid crystals
LMW	Low molecular weight
LN	Lithium niobate
LNKN	Lithium sodium potassium niobate
LOD	Limit of detection
LPSSD	Low-pressure steam drying
M. Wt./Mw/MW	Molecular weight

MAP	Element mapping analysis
MAP-2	Microtubule-associated protein 2
MCS	Maleilated chitosan
MDR	Multidrug resistance
MET	Metronidazole benzoate
MIC	Minimum inhibitory concentration
micro-CT	Micro-computed tomography
MMT	Montmorilonite
MOs	Metal oxides
Mox	Moxifloxacin
MPC	5-methyl pyrrolidinone chitosan
MSCs	Mesenchymal stem cells
MSF	Methacrylated silk fibroin
MTT	3-(4, 5-di-methylthiazol-2-yl)-2, 589 5-diphenyltetrazolium bromide
MWCNTs	Multi-walled carbon nanotubes
MX	Meloxicam
N,O-CMC	N,O-carboxymethyl chitosan
N-2-HACC	N-2- hydroxypropyl trimethyl ammonium chloride chitosan
NB	New bone
NBMs	Nano Biomaterials
NFC	Nanofibrillated Cellulose
NFN	Nanofibernet
N-G	Nitrogen-doped graphene
NGO	Nano graphene oxide
nHAP/n-HAp	Nano-hydroxyapatite
NIR	Near infrared
NLC	Nanostructured lipid carriers
NMs	Nanomaterials
NPs	Nanoparticles, nanostructures
NT	Neurotensin
OHNPs	Hydroxides nanoparticles
ORN	Ornidazole
OVX	ovariectomy-induced osteoporosis
PAG	Polyaniline/graphene
PAN	Poly acrylonitrile
PB	Pseudomonas aeruginosa
PCL	Polycaprolactone
PDA	Polydopamine
PDLCs	Polymer dispersed liquid crystals
PEDOT	Poly(3,4-ethylenedioxythiophene)
PEG	Polyethylene glycol
PEM	Polyelectrolyte multilayer
PEO	Polyethene oxide
PEPCK	Phosphoenolpyruvate carboxykinase

PHB	Polyhydroxy butyrate
PHDLC	Polymer hydrogel dispersed liquid crystals
PLA	Polylactic acid
PLGA	Poly(D, L-lactic-co-glycolic acid)
PLLA	Poly(l-lactic acid)
PMAA	Poly(methacrylic acid)
PNIPAM	Poly(N-Isopropylacrylamide)
PPARG	Peroxisome Proliferator Activated Receptor Gamma
PpIX	Protoporphyrin IX
PRP	Platelet-rich plasma
Psi	Porous silicon
PSS	Polystyrene sulfonate
Pt	Platinum
PTP	Protein tyrosine phosphate
PU	Polyurethane
PVA	Polyvinyl alcohol
PVDF	Poly(vinylidene fluoride)
PVPI	polyvinylpyrrolidoneiodine
PZT	Lead zirconate titanate
QAS-PEI	Quarternary ammonium polyethyleneimine
QC	Quaternized chitosan
QCS	Quaternized Chitosan
QCSG	Glycidyl Methacrylate Functionalized Quaternized Chitosan
QCSP	Quaternized Chitosan-Graft-Polyaniline
QNHC	Quaternary ammonium chitosan
rhAm	Recombinant human amelogenin
RNA	Ribose Nucleic Acid
ROS	Reactive oxygen species
S aureus	Staphylococcus aureus
SA	Salicylaldehyde
SA-CS-Col	Sodium algenate-Chitosan-Collagen
SA-CS-Col-GO	Sodium algenate-Chitosan-Collagen-Graphene Oxide
SBMA	Sulfobetaine methacrylate
SCASA	Supercritical-CO_2-assisted solubilization and atomization
SCHI	Succinate-modified chitosan
SCI	Spinalcord injury
S-COS	Sulfated-chitosan oligomers
SEM	Scanning Electron Microscope
SeNPs	Chitosan-selenium nanoparticles
SF	Silk fibroin
SL30 A8	Solute provider own circle of relatives 30 member 8
SS	Stainless steel
ssDNA	Single-stranded DNA
STPP	Sodium tripolyphosphate
SUM	Sumatriptan

SWNTs	Single walled carbon nanotubes
T1DM	Type 1 Diabetes Mellitus
T2DM	Type 2 Diabetes Mellitus
T3DM	Type 3 Diabetes Mellitus
TC	Thiolated chitosan
TCF7L2	Transcription aspect like 7-like 2
TCNPs	Thiolated chitosan-based nanoparticles
TCS	Thiolated chitosan
TE	Tissue Engineering
tg-ChNPs-ChPRP	Tigecycline Loaded Chitosan Nanoparticles
TGF-β1	Transforming growth factor-β1
TIPS	Thermally Induced Phase Separation
TMC	N-trimethyl chitosan
TMPyP	Tetrakis(1-methylpyridinium-4-yl)porphyrin
TPP	Tripolyphosphate
VAC	Vaccarin
WH	Whitlockite
wjhMSC-MT	Wharton's jelly Mesenchymal stem cells micro tissue
Wt	Weight
XRD	X-ray diffraction
Zn	Zinc
α-LA	α-lipoic acid
β-TCP	β-tricalcium phosphate

Chapter 1
Chitosan: A Smart Biomaterial

Anuradha Biswal and Sarat Kumar Swain

Abstract Chitosan is the second most abundant biopolymer after cellulose for which it has been given a special class of attention by researchers for its versatile properties. Its biocompatibility, antibacterial behaviour, antifungal properties, and non-toxicity play a vital role in choosing chitosan as a smart material for different applications. Structurally, chitosan is the deacetylated form of chitin with achieving better functionalities due to the presence of free amino groups. For that reason, it is a multifunctional polymer by which it achieves maximum compatibility during its fabrication. This introductory chapter of the present book reveals the history of chitin and chitosan including the source and isolation of chitin and gives an overall idea of chitosan as a smart material. The physicochemical properties of chitosan are an essential point for designing chitosan-based devices. The processibility of chitosan along with various applications is highlighted in this chapter. It is found as a prospective biomaterial having potential applications in agriculture, paper, textile, and food industries; wastewater treatment; cosmetics; environmental remediation; and various biomedical areas like wound healing, tissue engineering, and gene therapy. The outcome of the present chapter may explore a special interest for the readers towards the active concentration and involvement in different chapters of the present book.

1.1 Introduction

With the onset of the green revolution, the search for eco-friendly and renewable alternatives to be used in different applications is the main aim of the scientific community. Production of cheap and sustainable materials for different day-to-day applications is the need of the hour. For this purpose, biomaterials have emerged as a great choice because of their abundance, biodegradability, and biocompatibility. Recent research scenario is directed towards designing materials and devices with a proper selection of a wide range of available biomaterials. Biomaterials can be most accurately

A. Biswal · S. K. Swain (✉)
Department of Chemistry, Veer Surendra Sai University of Technology, Sambalpur, Odisha, India
e-mail: skswain_chem@vssut.ac.in

S. K. Swain and A. Biswal (eds.), *Chitosan Nanocomposites*,
Biological and Medical Physics, Biomedical Engineering,
https://doi.org/10.1007/978-981-19-9646-7_1

defined as "any material or combination of materials of synthetic or natural origin, that can be used for a prolonged period of time, having the ability to replace any tissue, organ partially or completely which in turn improves the quality of life of an individual" [1]. Four categories of materials fall under this definition, namely metals, polymers, composites, and ceramics. However, the biodegradability and biocompatibility of synthetic polymers limit their usage over natural polymers. The natural polymers like cellulose, starch, chitin, hyaluronate, alginates, etc., have gained increasing interest in many biological and biomedical applications due to their close structural resemblance with biological macromolecules. This makes their biorecognition easier and therefore are conveniently degraded into non-toxic residues to be removed by natural processes from the bioenvironment [2]. Naturally found polysaccharides have now dominated many fields like biomedical applications, packaging, textile, paper industry, agriculture, and environmental protection.

In this regard, chitin and chitosan have emerged as rare gems to be utilized in almost every facet of life. Its rising importance may be credited to the source of its extraction, i.e., marine waste products, biodegradability, antibacterial and anti-inflammatory properties, and mucoadhesivity which makes them suitable for application in many biomedical fields (Fig. 1.1) [3]. A huge number of scientific literatures have been published to establish the importance of chitin and chitosan in different applications. From Scopus, about 17,000 citations can be found on chitosan, its chemistry, and related applications [4]. This impressive citation statistics reflect the wide investigation of chitosan for different applications. Being a smart biomaterial, chitosan is highly employed for different biomedical applications. Either in a virgin state, functionalized state, or in combination with other polymers, chitosan shows potential antimicrobial and antioxidant activities [5]. Chitosan is composed of β-1,4 connected D-glucosamine and N-acetyl-D-glucosamine units that are distributed randomly in the polymer. It is the partially deacetylated form of chitin. It is cationic in nature as compared to other biopolymers which are anionic or neutral in nature. This factor is behind the unparalleled functionalities shown by chitosan over other polymers. This also aids the formation of electrostatic complexes with other negatively charged natural or synthetic polymers [6]. The physicochemical attributes of chitosan such as molecular weight, degree of crystallinity, and degree of deacetylation strongly affect its biological properties [7]. Chitosan has been employed and tested for a plethora of applications like packaging, drug delivery [8], tissue engineering [9], wound dressings, water treatment [10], etc. in the past few years.

This particular chapter is dedicated to explore the past and present of this unique biomaterial. The evolution of chitosan, sources, and extraction techniques have been elaborately discussed along with the inclusion of illustrative schemes. The present chapter also sheds light on some important properties of chitosan along with its wide range of applicability in different industrial and biomedical fields. The chapter will help students, researchers, and industrialists in understanding the different outstanding properties of chitosan and aid in the discovery of novel applications.

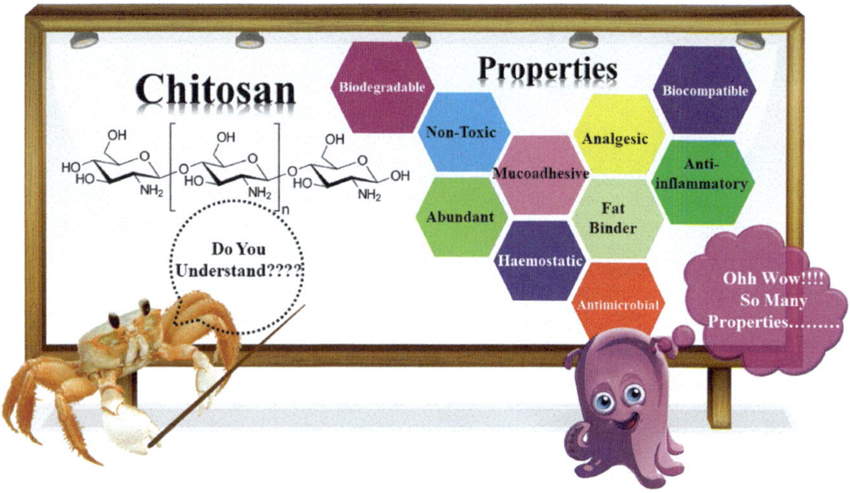

Fig. 1.1 Unique properties of chitosan

1.2 Chitin/Chitosan: A Smart Biomaterial

Chitin is a polysaccharide of natural origin that is characterized as a white, inelastic, hard nitrogen containing compound that is usually obtained as a by-product of food processing industries. It is the second most abundant material after cellulose. It constitutes the major component of the exoskeletons of crustaceans like crabs, shrimps, lobsters, molluscs, insects, and fungi [11].

Chitosan, on the other hand, is referred to as the resulting product obtained from the partial deacetylation of chitin. Structurally chitosan is a copolymeric structure containing β-(1 → 4)-2-amino-D-glucose units and β-(1 → 4)-2-acetamido-D-glucose units with the former exceeding by 80% [12]. This is mainly comprised of two sugar moieties, i.e., N-acetylglucosamine and glucosamine. The ratio of both sugar components is dependent on the treatment with alkali. Each monomeric unit of chitosan contains one free primary amine group along with two hydroxyl groups. It is analogous to the structure of cellulose [13]. Chitosan is considered unique in comparison to other polysaccharides due to the presence of about 6.89% nitrogen content, which creates commercial interest in the biopolymer [14].

Both chitin and chitosan show the phenomena of polymorphism in which different forms of the same compound is resulted from differences in polarities and packing orientation of adjacent polymeric chains in subsequent sheets. The presence of highly reactive amino and hydroxyl groups results in affinity towards inter- and intramolecular hydrogen bonding to yield linear aggregates along with rigid crystalline structures. Spectroscopic techniques have detected three crystalline forms of chitosan, i.e., α, β, and γ forms of chitosan. Among these, the α-allomorphic form is more dominant with adjacent chains arranged in opposite directions. The subsequent sections

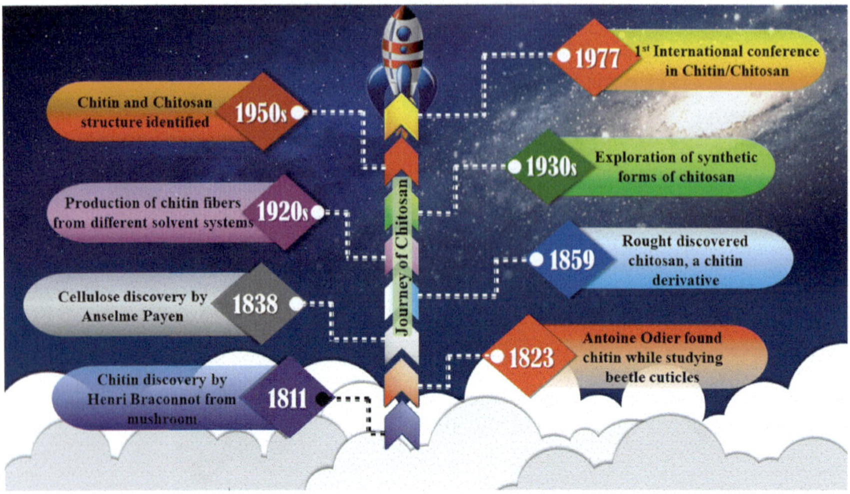

Fig. 1.2 Timeline of chitosan evolution

discuss the history of chitosan discovery, isolation, and properties of chitosan in detail.

1.2.1 History of Chitin and Chitosan

The history of chitin and chitosan can be traced back to 1811 in France, when the chemist Henri Braconnot denominated the insoluble residue obtained from the extraction of fungi with alcohol, water, and dilute alkali as fungine/fongine [15]. The detailed timeline of the discovery of chitosan till date is represented schematically in Fig. 1.2. He discovered the natural polysaccharide from fungi 30 years prior to the discovery of cellulose. It was further named as chitine by Odier in 1823. In 1843, Lassaigne detected the presence of nitrogen in chitine. However, in 1859, Charles Rouget treated the then reported chitine with concentrated caustic potash under reflux conditions to obtain a new compound named "*chitine modifiée*". The newly discovered material was found to be soluble in organic acids [16]. The term "chitosan" was finally coined by Felix Hoppe-Seyler in 1894 which referred to a modified form of chitin, obtained from treated shells of scorpions, crabs, and spiders, that is easily solubilized in acid [17]. This was followed by an era of controversy and confusion. The confusion rose due to severely disputing results obtained by different laboratories due to a lack of a clear understanding of the structure as well as the nomenclature of different polysaccharides investigated during that time [18]. After the differentiation between chitin and cellulose by using, X-ray diffraction techniques, chitin as well as chitosan grabbed a lot of attention due to the study of natural fibres. This led to the formation of chitosan films and fibres which were patented by George W. Rigby in mid-1930s [19]. This period witnessed an exponential rise in the use of chitosan in different industrial applications like the paper industry [20], photography [21], textile [22], and adhesives [23]. In the 1960s, chitosan was popularized due to its combination ability with red blood cells. With such an upsurge in the application of chitosan, the 1st international conference dedicated to chitosan was held in 1977. Till date there are about 2000 established applications of chitin/chitosan and their derivatives [24]. These polysaccharides have opened up a new window for researchers to explore numerous applications attributing to their ease of processibility and modification.

1.2.2 Sources and Biosynthesis of Chitin

Chitin, being the second most abundant biopolymer, is readily available in nature which aids its increasing demand to a great extent. It usually is found as ordered macro-sized fibrils present in the exoskeletons of crustaceans, molluscs, cuticles of insects, fungi, etc. (Fig. 1.3) [25]. The main sources of chitin have been summarized as follows [26, 27]:

- Crustacean shells (shrimps, lobsters, crayfish, krill, and barnacles).
- Molluscs (cuttlefish, octopus, clams, snails, squids, and oysters).
- Algae (diatoms, green algae, and brown algae).
- Fungi cell wall (Ascomycetes, Phycomycetes, and Basidiomycetes like Mucor rouxii, Aspergillus niger, Penicillium notatum, etc.).
- Insects (silkworms, housefly, spiders, ants, brachiopods, beetles, and scorpions).

Recently, chitosan employed for industrial application is primarily isolated from crustacean shells, i.e., shells of prawns, shrimps, and crabs that can be conveniently

Fig. 1.3 Various sources of chitin

collected from wastes of food industries. Furthermore, cocoon breeding in silk industries, fermentation of fungal specimens, and protein extraction from insects have also become a leading contributor of chitin. Fish scales do not give impressive yields and therefore are discarded as they contain chitin of only 1 wt% of their total weight [28]. According to records, 10^{12}–10^{14} tons of chitosan are extracted from crustacean wastes every year [29]. Many leading companies like Heppe Medical Chitosan GmbH, Chinova Bioworks, G.T.C. Bio-corporation, and Golden-Shell Biochemical contribute a huge part in the production of chitosan from shrimp shell wastes for different applications including food, medical, waste management, textile, drug, and many more [30].

In order to have a complete understanding regarding chitin, chitosan, choice of extraction routes, and their related properties, knowledge about the biosynthesis mechanism of chitin in different organisms is a must. The biosynthesis pathway of chitin is same for all organisms, starting from insects to fungi and crustacean shells to algae. The biosynthesis process can be broadly divided into five stages, i.e., production of N-acetylglucosamine-6-phosphate from sugar molecules like glycogen, glucose through hexosamine pathway [31], generation of amino sugar uridine diphosphate N-acetylglucosamine-(UDP-N-acetylglucosamine), and the polymerization reaction of UDP-N-acetylglucosamine into chitosan, catalyzed by chitin synthase. This is followed by the deposition of synthesized chitin along the surface of the cell membrane and subsequent release into the extracellular space. In the final stage, the chitin gets self-assembled to form nanofibrils [32–34]. The chains then condense together to form microfibrils and get oriented in the ECM, i.e., cuticles, cell walls, and peritrophic matrices [35]. Chitin composition depends on species, gender, age, season, habitat, and other related environmental conditions of the source from which it is extracted [36]. X-ray diffraction studies differentiate chitin into three allomorphic forms, i.e., α, β, and γ-chitin. These microfibrils are also differentiated depending on the number of chains, orientation, unit size, and degree of hydration. The α-chitin is the highly abundant and stable crystalline form that is found in exoskeletons of arthropods like crabs, lobsters, krill, and insect cuticles [33]. The β-chitin is isolated from squid pens and diatoms, where the microfibrils are oriented in the parallel direction, rendering flexibility to the structure [37]. The γ-chitin exhibits random orientation of the microfibrils [38]. Chitin is found in a combined state with other components like minerals, proteins, lipids, and carbohydrates in the crustacean shells. Hence, isolating chitin and subsequent synthesis of chitosan are essential for necessary applications.

1.2.3 Isolation of Chitin

Isolation of chitin is a very laborious process as compared to cellulose. Naturally found chitin is extracted in a combined state with different biomolecules, as discussed in the earlier section. Removal of these impurities becomes very essential to obtain the pristine product for onward applications [39]. Classical extraction methods are very

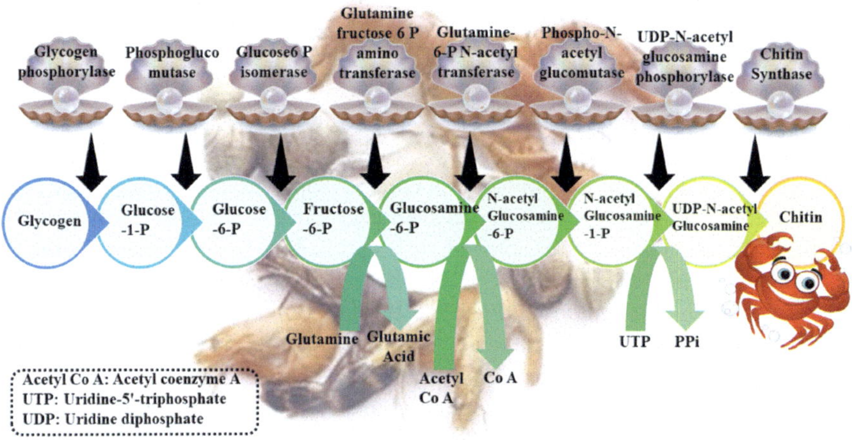

Fig. 1.4 Extraction process of chitin from crustacean wastes

time-consuming and have a severe impact on the environment due to the necessity of energy, huge quantities of alkali and mineral acids, and other hazardous chemicals [40–42].

1.2.3.1 Steps involved in Isolation Process

The isolation of chitosan from biomass involves a series of steps as follows:

- Pre-treatment of biomass.
- Deproteination.
- Demineralization.
- Decolouration.
- Post-treatment processing.

These steps are described elaborately in the next section (Fig. 1.4). The order of these series of steps can be altered or some steps can be skipped depending on the source and type of biomass obtained. The degree of deacetylation of chitin and the degree of crystallinity are highly dependent on the source and can be changed during the purification stage.

1.2.3.2 Pre-treatment Step

This step comprises all the processing that is necessary to prepare the collected biomass for the extraction of chitin. This includes scrapping to eliminate soft tissues, boiling, and subsequent pressing. For many organisms, boiling may be included in the hygienization step. The biomass is then dried and size reduction is done for further processing.

1.2.3.3 Demineralization

The high mineral content of the biomass is removed in the demineralization step. As the crustacean exoskeletons consist of higher $CaCO_3$ content (>50% (w/w)), demineralization becomes advisable [43, 44]. This is done by two general methods, i.e., biological and chemical demineralization. The chemical demineralization technique involves the utilization of acids like H_2SO_4, HNO_3, CH_3COOH, HCOOH, and HCl [39]. Whereas, the biological demineralization technique involves the use of bacteria or bioenzymes like Alcalase® [41]. The acids generated by these species interact with the calcium carbonate and subsequently form a precipitate which can be easily eliminated by simple washing.

1.2.3.4 Deproteination

Chitin chains remain entangled in a matrix made up of protein, which requires removal during purification to be used in food or biomedical applications as some proteins are allergenic [45]. The chitin chains are connected to the proteins by several hydrogen bridge bonds as well as covalent interaction of the free amino groups present in the skeletal structure [46]. Hence, high processing is done to obtain complete deproteination. In the chemical deproteination method, NaOH is commonly selected as a reagent. The efficacy of this process is highly dependent on processing temperature, the ratio of alkali to biomass as well as the concentration of the alkali used [47]. However, during the deproteination stage, chitin can show some alterations like hydrolysis and partial deacetylation of the biopolymeric chain [45]. Biological deproteination can also be done using proteases and isolated proteases (purified or crude) [48]. As the impact of biological deproteination is less harsh as compared to the chemical counterpart, certain smaller amino acids and peptide fragments are left attached to the chitin chain after the process [45] along with very less alteration in the degree of deacetylation [49].

1.2.3.5 Decolouration and Post-Treatment Processing

This step is adopted to remove the coloured pigments that are obtained for some organisms (for example, pink colour pigments in the case of biomass obtained from crustacean shells). For this, a mild reaction with potassium permanganate or hydrogen peroxide is done [50, 51]. This can also be achieved by carrying out extraction using solvents like chloroform, acetone, or ethanol [52]. The post-treatment processing comprises of drying, neutralization, and milling to obtain the final chitin of the required size and texture.

1.2.4 Conversion of Chitin to Chitosan

The final isolated chitin obtained after several stages of extraction has very limited application due to the presence of rigid crystalline structures, acetylated groups, and most importantly lack of solubility in aqueous solutions [53]. Removal of these acetylated groups, even partially, can imbibe desirable characteristics like biocompatibility, biodegradability, and solubility. This partially deacetylated product (degree of deacetylation of more than 50%) is termed as "chitosan". Usually, two methods, namely enzymatic and chemical methods, are adopted for successful deacetylation of chitin to get chitosan, a material that is highly utilized in various applications [54, 55].

1.2.4.1 Enzymatic Deacetylation Technique

Enzymatic deacetylation refers to a rather eco-friendly process of chitosan production from chitin by using an enzyme called chitin deacetylase. But, the complex process of selecting, breeding, cultivation, and extraction of good quality of bacteria secreting the concerned enzyme for this method restrains the choice of this method [56]. The careful selection of bacterial strains led to the production of viable chitosan for successful utilization in different applications. Bacterial strains are highly preferred over the current use of fungal strain as it is easier to breed and cultivate during fermentation [57].

1.2.4.2 Chemical Deacetylation Technique

It is the most prominently used method for mass industrial production of chitosan. The raw materials are subjected to decalcification, followed by deproteination and decolouration to get chitin (Fig. 1.5). This is then reacted with 40–50% sodium hydroxide solution to obtain the deacetylated product, chitosan. The degree of deacetylation can be manipulated by varying the alkali concentration, chitin/alkali ratio, and reaction temperature. This particular method is mostly chosen due to its simplicity of the method, ease of implementation, high efficacy level, and control over the production process [58, 59]. The limitations associated with this technique are the environmental impacts and huge energy requirements.

1.2.5 Physicochemical Properties of Chitosan

The physicochemical properties of chitosan establish the effect of structural attributes with its potential application in different fields [60, 61]. The various parameters like time, temperature, and reagent concentration used during the extraction process have

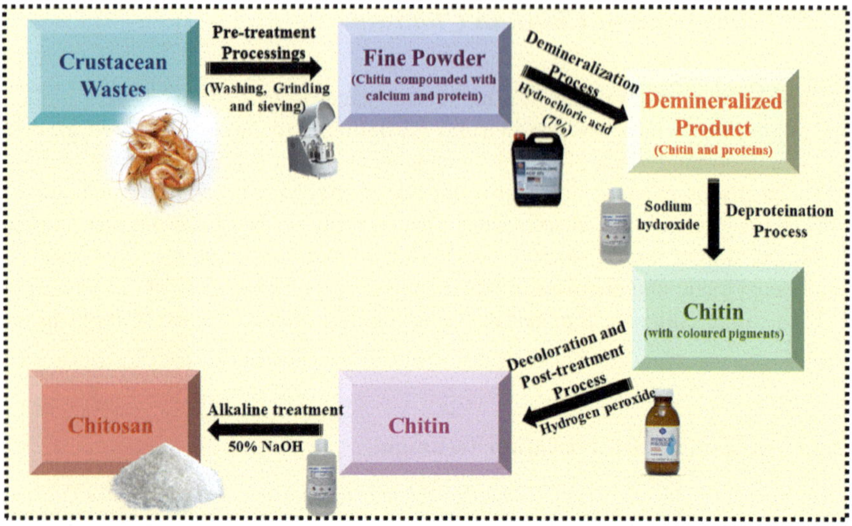

Fig. 1.5 Conversion of chitin to chitosan

a deep impact on the physicochemical properties of chitosan. There are quite a few physicochemical properties of chitosan like degree of deacetylation, average molar mass, solubility, water retention, viscosity, crystallinity, nitrogen and ash content, and presence of water. Out of these, the degree of deacetylation, solubility, viscosity, and average molecular mass are the most essential criteria that decide the applicability of chitosan. The percentage of protein and heavy metal content along with endotoxin levels determine its utilization as a biomaterial [62].

1.2.5.1 Average Molecular Mass

Determination of the final chitosan product is considered as one of the most essential analyses as it determines the biological properties exhibited by chitosan. Extensive efforts have established an increase in the biological activity of chitosan with a decrease in molecular mass [63–65]. Molar mass detection holds much importance in medical applications of chitosan where the structure-activity relationship is vital. It predicts the chitosan-assisted biochemical processes that take part in the cells [64–66]. It also influences the crystalline size along with the morphological character exhibited by the final chitosan product.

1.2.5.2 Degree of Deacetylation

The degree of deacetylation is a measure of the acetyl content of chitosan. As discussed earlier, the amino group imparts activity to chitosan. To increase the degree of deacetylation, chitosan is repeatedly subjected to treatment with an alkali. The degree of deacetylation determines the solubility and crystalline nature of chitosan. An increase in the degree of deacetylation is accompanied by an increase in charge density in the polymeric chains and affects the flexibility. It also plays a vital role in cell migration and adhesion. It is observed that a lower degree of deacetylation supports cell growth and adhesion [67]. However, it does not have any effect on the cytocompatibility of chitosan.

1.2.5.3 Solubility

Chitosan is known to be soluble in weakly acidic solutions above pH value 7. It shows solubility in certain organic and inorganic acids like lactic acid, phosphoric acid, hydrochloric acid, acetic acid, etc. after a long period of stirring. Further, its solubility is dependent on the strength of the acidic medium and the pKa value.

1.2.5.4 Viscosity

The viscosity of the chitosan solution is directly proportional to the degree of deacetylation and concentration, but is inversely proportional to temperature. Commercially available chitosan has different viscosities. Higher molecular mass and linear structure enable it to be applied as a viscosity improving agent. The viscosity of chitosan decreases with an increase in shearing rate and hence is termed as pseudoplastic [68, 69].

1.3 Processibility of Chitosan

Based on the application, chitosan can be easily processed into different physical forms like powder [70], membranes/films [71], flakes [72], composites [73], and fibres [74]. Furthermore, chitosan can be manipulated at the nanoscale to form different nano-dimensional materials like nanoparticles [75], nanofibres [76], and nanocomposites [77], as illustrated in Fig. 1.6. This section provides a brief discussion regarding various physical forms of chitosan.

1.3.1 Chitosan Flakes and Powders

Chitosan flakes are considered to be a easiest to manage form of chitosan. They have been used in pristine or combined form for a plethora of applications, especially in environmental remediation. Pietrelli et al. studied the Cr^{3+} removal efficacy from wastewater systems using chitosan flakes [78]. The flakes showed higher order Cr^{3+}

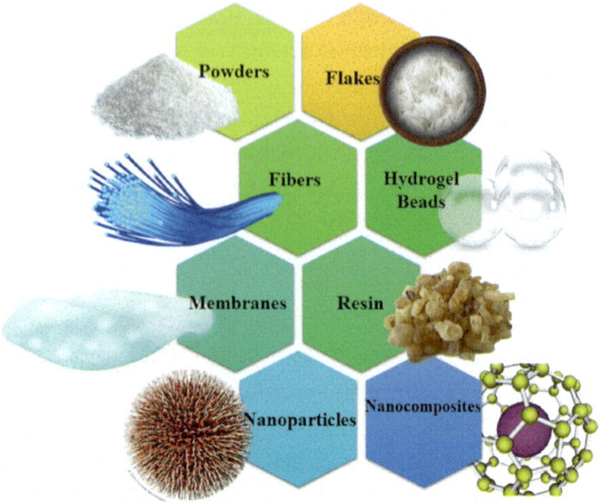

Fig. 1.6 Chitosan processed into various forms

loading capacities and recovery percentages. They have also been reported for the elimination of impurities from palm oil effluents [79]. Chitosan has been used in powdered forms in many fields like the removal of heavy metals [80], dye removal [70], and many more. Along with good adsorption ability, chitosan powders can act as an excellent coagulating and flocculating agent.

1.3.2 Chitosan Hydrogel Beads

Chitosan-based hydrogels act as a lucrative option for many applications because of their flexibility, 3D structure, encapsulation efficiency, stimuli responsiveness, and adsorptive nature. Starting from biomedical to agricultural industries have witnessed a rise in the application of chitosan-based gels. Further, the ease of removal of material also provides an additional benefit that makes the recovery of material easier. These hydrogel beads are very easily prepared and can be synthesized on a nanoscale (2–3 nm). The gel beads are used for catalysis [81], removal of dyes and heavy metals [82], wound healing [83], drug delivery [84], biosensing, and many more.

1.3.3 Chitosan Fibres and Resins

Chitosan fibres and resins were first reported in 1926 but their production for commercial purposes was a costlier affair. A higher degree of deacetylation of chitin is a necessity for the production of chitosan fibres. The treatment of chitin with alkali yields chitosan fibres with excellent strength analogous to viscous fibres. These fibres exhibit high mechanical strength, tensile strength, and biocompatibility which makes them suitable for various biomedical applications like wound healing, tissue engineering, drug delivery, etc. [85]. Based on the crystallinity, structure, and average crystallite size, chitosan fibres can be divided into different types [86]. Chitosan is susceptible to various alterations which enable the addition of different cationic groups via the quaternization modification method. This leads to the formation of anion exchange resins, as a strong alternative to the existing anion exchange resin models [87]. Chitosan-based resins are widely employed in heavy metal removal [88] and dye removal [89].

1.3.4 Chitosan Membranes

Chitosan, due to its film-forming affinity, stability in alkaline medium, eco-friendly nature, biodegradability, outstanding mechanical strength, and ability to withstand high temperatures, is a viable candidate for membrane preparation [90]. Electrospinning has evolved as an effective technique for the production of nonwoven, porous, fibrous membranes for many applications like biosensing, catalysis, tissue engineering, etc. These synthesized membranes are also good candidates to be used in water purification [91], wound healing [92], and removal of heavy metals [93].

1.3.5 Chitosan Nanoparticles

Chitosan nanoparticles were initially reported by Ohya and the group in 1994. They synthesized chitosan nanoparticles by emulsification and crosslinking to deliver 5-fluorouracil, an anticancer drug, intravenously. Chitosan nanoparticles possess the inherent properties of chitosan biopolymer and the characteristics of nanoparticles like size, interface and surface effects, and quantum effects [94]. Till date there have been five synthetic techniques, namely emulsification solvent diffusion, microemulsion, ionotropic gelation, reverse micellar method, and polyelectrolyte complexation, used for the preparation of chitosan nanoparticles [95]. Out of these methods, polyelectrolyte complexation and ionic gelation are the most popular. These are very convenient and avoid the use of organic solvents and high shearing force [96]. Chitosan nanoparticles possess many interesting characters like mucoadhesivity, ability of transmucosal permeability, and antibacterial and biological properties which enable them to be used in various biomedical applications [97].

1.3.6 Chitosan-Based Composites and Nanocomposites

Virgin chitosan materials may have superior biological properties but a lack of mechanical attributes limits their utilization in various applications. Hence, combining it with different materials to form hybrid materials with enhanced properties yields chitosan-based composite materials. This method not only improves the mechanical strength but also accentuates the biological and chemical properties of chitosan. When chitosan is combined with a nano-dimensional material (size between 1 and 100 nm), the resultant product is said to be chitosan-based nanocomposites. Recently, chitosan nanocomposites have revolutionized the field of material science. Here, both the biopolymer as well as nanomaterials incorporated have an impact on the enhancement of particular properties of the overall material. The nanofiller (dispersed material in chitosan matrix) exhibits visible improvement in thermal, mechanical, and physical properties along with enhancing the inherent properties like surface area and other physicochemical attributes [98]. Different chitosan nanocomposites formed by incorporating chitosan with clay, polymeric, carbonaceous, and metal/metal oxide-based nanomaterials have led to novel materials with extensive applications in environmental remediation [99], biomedical fields [100], food industry [98], packaging industry [101], and many more.

1.4 Different Applications of Chitosan

Chitosan has been proved to be associated with a wide range of applications which is attributed to its multiple functionalities such as abundance, versatility, low cost, biodegradability, biocompatibility, digestibility, and non-toxic nature. Chitosan and its derivatives have been used to answer numerous biomedical and environmental problems [102]. The different applications of chitosan and its derivatives (Fig. 1.7) have been discussed in detail in this section.

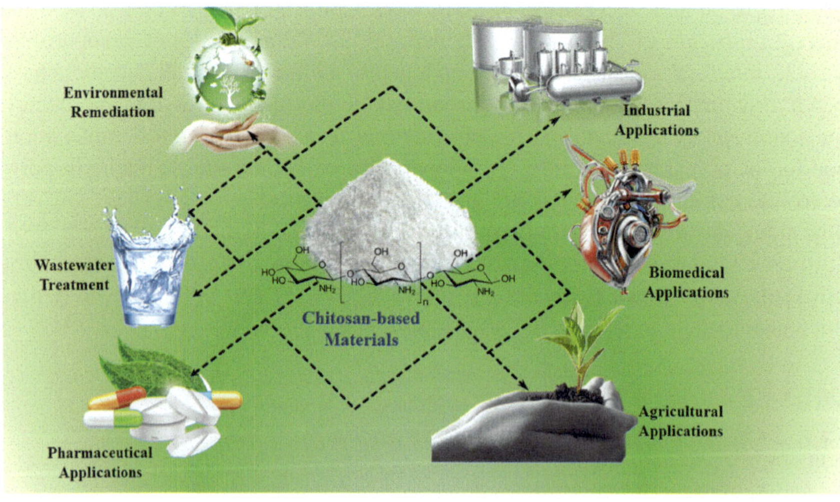

Fig. 1.7 Different applications of chitosan-based materials

1.4.1 *Agriculture and Industrial Applications*

1.4.1.1 Agriculture

Chitosan has been extensively used in agriculture in the past several years. It is used as additives in fertilizers, regulators of plant growth, antiviral agents, and preservatives for agro-based products [103]. Apart from these, chitosan is also used in other sectors as in soil quality promoter, plant protection, enhancement in the production of secondary metabolites, and as a defence activator. The lamination of chitosan films on seeds as a single-layer coating can act as an antimicrobial barrier and protects the seed from fungal growth as well as improves the disease resistance capacity. Chandra et al. reported that chitosan-assisted camellia shows immunity towards vesicular fusarium wilt [104]. Further chitosan on breaking down by microbial action acts as a source of nutrients for the promotion of plant growth and increases actinomycetes in soil [105]. Additionally, chitosan and its associated derivatives help in building the immunity of plants against diseases and climatic changes, as well as lesser dependence on harsh fertilizers and pesticides, thus lowering soil pollution.

1.4.1.2 Textile Industry

Chitosan has found multiple applications in the textile industry. Chitosan and derivatives have been employed as a coating material to yield novel characteristics to the textile [106, 107]. The main application of chitosan in the textile industry is in the form of chitosan fibres. Chitosan fibres exhibit remarkable properties like biodegradability, biocompatibility, antibacterial properties, and non-toxicity. Derivatives of chitosan imbibe soil repellent and antistatic behaviour to the textiles. In medical textiles, chitosan-based materials are also used as sutures, fibres, and threads. Chitosan is also used for decolouration of pigments or dye removal in textiles. Functionalized chitosan is also used to give an antimicrobial finish to the cotton [108]. The use of chitosan in cloth-making process inhibits bacterial and fungal growth in textiles. These chitosan-based fibres are usually synthesized by electrospinning and wet spinning method. Li et al. employed the wet spinning method by taking LiOH/urea as a solvent for the dissolution of chitosan and manufactured fibres with an improved mechanical strength of about 21.6% [109]. Chitosan is also used to provide an anti-wrinkle and antibacterial finish to the textile as well as employed for the enhancement of dyes.

1.4.1.3 Food Industry

In the food industry, edible chitosan (degree of deacetylation > 83%) has mass-scale utilization as dietary food additive. Chitosan is used as a supplement for obesity control because of its capacity to combine with fats [110]. Upon binding with fats, chitosan forms a mass that cannot be absorbed by the body and gets easily eliminated by the human body. Chitosan fibres also exhibit hypocholesterolemic properties [14, 111]. Being unable to be digested by the human body with zero calorific value, chitosan is taken as a weight loss supplement. It offers an eco-friendly alternative to be used in the food packaging industry as it exhibits enhanced shelf life and aroma barrier attributes [112, 113]. Chitosan is also employed as food preservatives for meat, vegetables, and other food articles which provides a barrier against spoilage due to fungal growth and loss of nutrition [114, 115]. It is also utilized for the preservation of beverages, encapsulation, and clarification of bioactive compounds [116].

1.4.1.4 Paper Industry

The paper industry has witnessed a rise in the use of chitosan and derivatives of chitosan in past decades. Chitosan has been successfully employed for the strengthening of papers obtained via recycling. It is also used in the production of eco-friendly food wrapping materials, packaging materials, cardboard packaging, and toilet papers. With a very close resemblance with cellulose structure, extensive use of chitosan in the paper industry has led to a lowering in the use of chemical additives. Papers manufactured with the usage of chitosan is reported to be resistant to moisture and have smooth surface [117]. Another application of chitosan can be as a flocculating agent in the wastewater from pulp mills [118].

1.4.2 Wastewater Treatment

Chitosan is increasingly used for removing effluents from the water like grease, oils, fine particulates, and heavy metal pollutants from wastewater [119]. Use of chitosan in water treatment processes dates to over 30 years. It is used for clearance of water spills, as when treated with oil spills, chitosan binds the oil mass which makes it easy to remove [120]. Chitosan, having a better affinity towards residual oil, hydrophilicity, biodegradability, flocculating, and adsorption capacity, has replaced traditionally used coagulants like aluminium, polyaluminium chloride, and polyacrylamide. Derivatives of chitin and chitosan have been successfully used in the treatment of anionic wastewater streams. N-benzyl sulfonate derivatives of chitosan have been used for metal ions removal in an acidic medium [28]. As they show a chelating effect and adsorption towards organic particles and heavy metal ions (Hg^{2+}, Cd^{2+}, Pb^{2+}, Co^{2+}, etc.), derivatives of chitosan can easily capture these heavy metal ions from wastewater and result in subsequent removal. Chitosan is also used in the removal of traces of radioactive pollutants like Pu, $_{60}Co$, etc. [121].

1.4.3 *Pharmaceutical Industry*

1.4.3.1 Cosmetics

Chitosan, being a natural gum, has shown vast use in different applications in the cosmetics industry. It is used in skin and hair treatments as well as in other cosmetic items like nail colours, lotions, and creams. Chitosan upon application on the skin surface creates a moisturizing and protective shield due to its film-forming ability. Hence, it gives a smoother finish to the skin and provides a shielding effect from harsh chemicals present in detergents and other environmental conditions. This is the driving idea behind using chitosan in sunscreens and several other lotions. It has higher order hydrating characteristics even in comparison to hyaluronic acid. Due to this, chitosan is used actively in lip care as it prevents drying and chapping of lips. Furthermore, chitosan also ensures softer lips with greater adhesion of coloured pigments. It is used in deodorants [120], hair styling polymer, facial masks [122], shampoos, and mouthwash [123].

1.4.3.2 Drug Delivery

Chitosan is a non-toxic biopolymer that shows excellent biological properties like biodegradability, biocompatibility, encapsulation and film-forming ability, and mucoadhesivity, and can be easily absorbable in the body [124]. This makes chitosan a potential candidate to be used in various pharmaceutical applications. It is considered as a viable drug carrier [125]. The most interesting advantage of chitosan in drug delivery application is that after the release of the encapsulated drugs, chitosan-based carrier is degraded by the lysozymes present in the human body. These degraded products are completely reabsorbed by the body, which makes chitosan significant in drug delivery applications. The deliverable drugs are usually blended with chitosan via coating, dissolution, or adsorption techniques to get tablets, microspheres, microcapsules, gels, and films for sustained release of the encapsulated drugs. Chitosan-based delivery systems have been used for the effective release of growth factors, antibiotics, proteins, peptides, vaccines, and anti-inflammatory drugs. It is also used in different delivery routes like nasal, oral, transdermal, and ocular drug delivery [126].

1.4.4 *Biomedical Application*

Chitosan shows unprecedented applicability in various biomedical applications like wound healing, tissue engineering, gene therapy, phototherapy, anticancer treatment, and many more. This is credited to its mechanical stability, optical clarity, gas permeability, immunological activity, wettability, and antibacterial behaviour [107]. This particular section discusses some of the biomedical applications of chitosan and its derivatives.

1.4.4.1 Wound Healing

Healing of a wound is a complex and dynamic method that involves several parameters and six steps. The process is comprised of inflammation, cell proliferation, angiogenesis, synthesis of a matrix, deposition of collagen, and reepithelialisation. The presence of NAG subunit (N-acetylglucosamine) in chitosan holds utmost importance in the field of skin tissue repair. Chitosan not only promotes the growth of cells because of its high positive surface charge but also supports blood clotting and thrombosis. Additionally, the presence of free amino groups forms a complex with the acidic groups present in the blood cells. Chitosan also regulates the functioning of macrophages, collagenase, tumour necrosis factor, and interleukins secretion to promote wound healing [127]. Chitosan and its derivatives have shown excellent efficacy as wound dressings [128]. Hence, chitosan exhibits enhanced wound healing activity and is considered a suitable candidate to be used as wound dressing material.

1.4.4.2 Tissue Engineering

Chitosan has multiple properties like non-toxicity, hemocompatibility, antibacterial and analgesic activities, and several other biological properties that support cell migration and repair of tissues. It has unparalleled stability in human body environment and the ability to promote angiogenesis which renders it suitable for the development of a matrix for the process of tissue regeneration [129]. Chitosan also exhibits remarkable water retention, oxygen permeability, and mechanical attributes that are significant for the tissue repair and regeneration process. Chitosan has been used as potential scaffolds in skin, cornea, cartilage, liver tissue as well as blood vessel tissue engineering.

1.4.4.3 Gene Therapy

Chitosan exhibits the capability of binding with nucleic acids via electrostatic interactions and gets endocytosed in the cell avoiding further dissociation of the formed DNA-chitosan complex [63]. It also has a vital role in the protection of the DNA molecules from the lysosomal activity for effective transfection of the cell as well as in membranal adhesion. The DNA-chitosan hybrids are formed as nanospheres and complexes. This effect has many advantages like (1) embedment of lysosomotropic agents which aids the protection of DNA complex from degradation by lysosomes, (2) attachment of ligands with the DNA-chitosan complex for receptor-assisted endocytosis, (3) additional encapsulation of bioactives for multiple plasmids, (4) polyplexes lyophilization for prolonged storage without considerable activity loss, and (5) higher shielding from degradation by serum nuclease [130]. This renders chitosan-based materials apt for gene delivery applications.

1.4.5 Environmental Remediation

The presence of free amino functional moieties in chitosan enables it to interact with negative surfaces [120]. This effect makes it easier for chitosan to form complexes with anionic polyelectrolytes which are insoluble in water [131]. Owing to this, chitosan is used as an effective adsorbent of phenolic compounds, dyes, pesticides, and herbicides from water bodies [132]. In addition to this, due to its distinct structure chitosan shows a high affinity towards many dyes like acid, naphthol, and sulphur-based dyes [133]. Chitosan is also used as a coagulating agent as well as in the treatment of activated sludge eliminated from biological waste treatment. The coagulated by-products obtained from food industries consist of 30–75% proteins that have great applicability as animal feeds. The use of chitosan ensures successful recovery of this protein-rich content from wastewater from food processing industries. Moreover, it is also used for coagulating amino acids from wastewater procured from seafood processing industries.

1.5 Remark

Although chitosan is accepted as the most promising biomaterial for various applications, improving its strength, stability, and processibility is a big challenge in research. To overcome these limitations of chitosan, compatible fillers may suitably be chosen for designing composite materials. When fillers are nano-dimensional, the dispersibility and interference become highest giving various remarkable properties to chitosan-based nanocomposites. The aim of converting chitosan from a smart biomaterial to sustainable chitosan-based nanocomposites covers the journey from laboratory to industry. Based on the idea of the present chapter, the book is designed towards bionanomechanical applications of chitosan nanocomposites

Conflict of Interest The authors declare that there is no conflict of interest in publishing this article.

References

1. C.P. Stumpf, Biomaterials in dental ceramics, topics in mining, metallurgy and materials engineering (2013)
2. S. Thomas, N. Ninan, S. Mohan, E. Francis (eds.), *Natural polymers, Biopolymers, Biomaterials, and Their Composites, Blends, and IPNs*. CRC Press (2012)
3. A. Muxika, A. Etxabide, J. Uranga, P. Guerrero, K. De La Caba, Chitosan as a bioactive polymer: Processing, properties and applications. Int. J. Biol. Macromol. **105**, 1358–1368 (2017)
4. M. Friedman, V.K. Juneja, Review of antimicrobial and antioxidative activities of chitosans in food. J. Food Prot. **73**(9), 1737–1761 (2010)
5. Y. Qin, P. Li, Antimicrobial chitosan conjugates: Current synthetic strategies and potential applications. Int. J. Mol. Sci. **21**(2), 499 (2020)
6. J. Venkatesan, S.K. Kim, Chitosan composites for bone tissue engineering—An overview. Mar. Drugs **8**(8), 2252–2266 (2010)
7. S. Bagheri-Khoulenjani, S.M. Taghizadeh, H. Mirzadeh, An investigation on the short-term biodegradability of chitosan with various molecular weights and degrees of deacetylation. Carbohyd. Polym. **78**(4), 773–778 (2009)
8. S. Peers, A. Montembault, C. Ladavière, Chitosan hydrogels for sustained drug delivery. J. Control. Release **326**, 150–163 (2020)
9. M.M. Islam, M. Shahruzzaman, S. Biswas, M.N. Sakib, T.U. Rashid, Chitosan based bioactive materials in tissue engineering applications—A review. Bioactive materials **5**(1), 164–183 (2020)
10. L. Das, P. Das, A. Bhowal, C. Bhattachariee, Synthesis of hybrid hydrogel nano-polymer composite using graphene oxide, Chitosan and PVA and its application in waste water treatment. Environ. Technol. Innov. **18**, 100664 (2020)
11. V.S. Yeul, S.S. Rayalu, Unprecedented chitin and chitosan: A chemical overview. J. Polym. Environ. **21**(2), 606–614 (2013)
12. H.G. Yoon, H.Y. Kim, Y.H. Lim, H.K. Kim, D.H. Shin, B.S. Hong, H.Y. Cho, Thermostable chitosanase from Bacillus sp. strain CK4: Cloning and expression of the gene and characterization of the enzyme. Appl. Environ. Microbiol. **66**(9), 3727–3734 (2000)
13. M. Ibrahim, O. Osman, A.A. Mahmoud, Spectroscopic analyses of cellulose and chitosan: FTIR and modeling approach. J. Comput. Theor. Nanosci. **8**(1), 117–123 (2011)
14. W. Xia, P. Liu, J. Zhang, J. Chen, Biological activities of chitosan and chitooligosaccharides. Food Hydrocolloids **25**(2), 170–179 (2011)
15. H. Braconnot, Sur la nature des champibnons. Ann Chi **79**, 265–304 (1881)
16. C. Rouget, Des substances amylacées dans les tissus des animaux, spécialement des Articulés (chitine). Comp. Rend **48**, 792–795 (1859)
17. F. Hoppe-Seyler, Ueber chitin und cellulose. Ber. Dtsch. Chem. Ges. **27**(3), 3329–3331 (1894)
18. M.V. Tracey, Chitin. Pure Appl. Chem. **7**, 1–14 (1957)
19. G.W. Rigby, U.S. Patent No. 2,033,787 (U.S. Patent and Trademark Office, Washington, DC, 1936)
20. T.E. Larson, Bacteria, corrosion and red water. J. Am. Water Works Assoc. **31**(7), 1186–1196 (1939)
21. M. Marasco, Antistatic photographic film. US Patent2 (1938)
22. L.B. Arnold, Fibrous product. US Patent 2,142,986 (1939)
23. R.W. Maxwell, Adhesive. US Patent 2,182,524 (1939)
24. N. Morin-Crini, E. Lichtfouse, G. Torri, G. Crini, Fundamentals and applications of chitosan, in *Sustainable Agriculture Reviews*, vol. 35. (Springer, Cham, 2019), pp. 49–123
25. M. Kaya, E. Lelešius, R. Nagrockaitė, I. Sargin, G. Arslan, A. Mol, B. Bitim, Differentiations of chitin content and surface morphologies of chitins extracted from male and female grasshopper species. PLoS ONE **10**(1), e0115531 (2015)

26. S.M. Joseph, S. Krishnamoorthy, R. Paranthaman, J.A. Moses, C. Anandharamakrishnan, A review on source-specific chemistry, functionality, and applications of chitin and chitosan. Carbohydr. Polym. Technol. Appl. **2**, 100036 (2021)
27. S. Crognale, C. Russo, M. Petruccioli, A. D'annibale, Chitosan production by fungi: Current state of knowledge, future opportunities and constraints. Fermentation **8**(2), 76 (2022)
28. P.K. Dutta, J. Dutta, V.S. Tripathi, Chitin and chitosan: Chemistry, properties and applications (2004)
29. G.S. Dhillon, S. Kaur, S.K. Brar, M. Verma, Green synthesis approach: extraction of chitosan from fungus mycelia. Crit. Rev. Biotechnol. **33**(4), 379–403 (2013)
30. T. Philibert, B.H. Lee, N. Fabien, Current status and new perspectives on chitin and chitosan as functional biopolymers. Appl. Biochem. Biotechnol. **181**(4), 1314–1337 (2017)
31. J. Hou, B.E. Aydemir, A.G. Dumanli, Understanding the structural diversity of chitins as a versatile biomaterial. Phil. Trans. R. Soc. A **379**(2206), 20200331 (2021)
32. V.P. Santos, N.S. Marques, P.C. Maia, M.A.B.D. Lima, L.D.O. Franco, G.M.D. Campos-Takaki, Seafood waste as attractive source of chitin and chitosan production and their applications. Int. J. Mol. Sci. **21**(12), 4290 (2020)
33. D. Elieh-Ali-Komi, M.R. Hamblin, Chitin and chitosan: production and application of versatile biomedical nanomaterials. Int. J. Adv. Res. **4**(3), 411 (2016)
34. X. Liu, A.M. Cooper, J. Zhang, K.Y. Zhu, Biosynthesis, modifications and degradation of chitin in the formation and turnover of peritrophic matrix in insects. J. Insect Physiol. **114**, 109–115 (2019)
35. H. Merzendorfer, The cellular basis of chitin synthesis in fungi and insects: common principles and differences. Eur. J. Cell Biol. **90**(9), 759–769 (2011)
36. L. Bastiaens, L. Soetemans, E. D'Hondt, K. Elst, Sources of chitin and chitosan and their isolation. Chitin Chitosan Prop. Appl. 1–34 (2019)
37. T. Hahn, E. Tafi, A. Paul, R. Salvia, P. Falabella, S. Zibek, Current state of chitin purification and chitosan production from insects. J. Chem. Technol. Biotechnol. **95**(11), 2775–2795 (2020)
38. C. Casadidio, D.V. Peregrina, M.R. Gigliobianco, S. Deng, R. Censi, P. Di Martino, Chitin and chitosans: Characteristics, eco-friendly processes, and applications in cosmetic science. Mar. Drugs **17**(6), 369 (2019)
39. A. Percot, C. Viton, A. Domard, Optimization of chitin extraction from shrimp shells. Biomacromol **4**(1), 12–18 (2003)
40. K. Kurita, Chitin and chitosan: functional biopolymers from marine crustaceans. Mar. Biotechnol. **8**(3), 203–226 (2006)
41. S. Kaur, G.S. Dhillon, Recent trends in biological extraction of chitin from marine shell wastes: A review. Crit. Rev. Biotechnol. **35**(1), 44–61 (2015)
42. J. Li, J.F. Revol, R.H. Marchessault, Effect of degree of deacetylation of chitin on the properties of chitin crystallites. J. Appl. Polym. Sci. **65**(2), 373–380 (1997)
43. A. Sombke, G.D. Edgecombe, Morphology and evolution of Myriapoda. Arthropod Struct. Dev. **43**(1), 3–4 (2014)
44. M. Kaya, K. Sofi, I. Sargin, M. Mujtaba, Changes in physicochemical properties of chitin at developmental stages (larvae, pupa and adult) of Vespa crabro (wasp). Carbohyd. Polym. **145**, 64–70 (2016)
45. I. Younes, M. Rinaudo, Chitin and chitosan preparation from marine sources. Structure, properties and applications. Mar. Drugs**13**(3), 1133–1174 (2015)
46. R.A. Muzzarelli, Chitin nanostructures in living organisms, in Chitin (Springer, Dordrecht, 2011), pp. 1–34
47. F.A. Al Sagheer, M.A. Al-Sughayer, S. Muslim, M.Z. Elsabee, Extraction and characterization of chitin and chitosan from marine sources in Arabian Gulf. Carbohyd. Polym. **77**(2), 410–419 (2009)
48. W. Arbia, L. Arbia, L. Adour, A. Amrane, Chitin extraction from crustacean shells using biological methods—A review. Food Technol. Biotechnol. **51**(1), 12–25 (2013)

49. L. Manni, O. Ghorbel-Bellaaj, K. Jellouli, I. Younes, M. Nasri, Extraction and characterization of chitin, chitosan, and protein hydrolysates prepared from shrimp waste by treatment with crude protease from Bacillus cereus SV1. Appl. Biochem. Biotechnol. **162**(2), 345–357 (2010)
50. A. Tolaimate, J. Desbrieres, M. Rhazi, A. Alagui, Contribution to the preparation of chitins and chitosans with controlled physico-chemical properties. Polymer **44**(26), 7939–7952 (2003)
51. A. Waśko, P. Bulak, M. Polak-Berecka, K. Nowak, C. Polakowski, A. Bieganowski, The first report of the physicochemical structure of chitin isolated from Hermetia illucens. Int. J. Biol. Macromol. **92**, 316–320 (2016)
52. M. Kaya, S. Erdogan, A. Mol, T. Baran, Comparison of chitin structures isolated from seven Orthoptera species. Int. J. Biol. Macromol. **72**, 797–805 (2015)
53. D. Sivaramakrishna, B. Bhuvanachandra, M.K. Mallakuntla, S.N. Das, B. Ramakrishna, A.R. Podile, Pretreatment with KOH and KOH-urea enhanced hydrolysis of α-chitin by an endo-chitinase from Enterobacter cloacae subsp. cloacae. Carbohydr. Polym. **235**, 115952 (2020)
54. M. Yadav, P. Goswami, K. Paritosh, M. Kumar, N. Pareek, V. Vivekanand, Seafood waste: a source for preparation of commercially employable chitin/chitosan materials. Bioresour. Bioprocess. **6**(1), 1–20 (2019)
55. L.C. Wang, L.Q. Di, J.S. Li, L.H. Hu, J.M. Cheng, H. Wu, Elaboration in type, primary structure, and bioactivity of polysaccharides derived from mollusks. Crit. Rev. Food Sci. Nutr. **59**(7), 1091–1114 (2019)
56. J. Cai, J. Yang, Y. Du, L. Fan, Y. Qiu, J. Li, J.F. Kennedy, Enzymatic preparation of chitosan from the waste Aspergillus niger mycelium of citric acid production plant. Carbohyd. Polym. **64**(2), 151–157 (2006)
57. K. Kaur, V. Dattajirao, V. Shrivastava, U. Bhardwaj, *Isolation and characterization of chitosan-producing bacteria from beaches of Chennai* (Enzyme research, India, 2012), p.2012
58. H. El Knidri, R. Belaabed, A. Addaou, A. Laajeb, A. Lahsini, Extraction, chemical modification and characterization of chitin and chitosan. Int. J. Biol. Macromol. **120**, 1181–1189 (2018)
59. S. Kumari, P. Rath, A.S.H. Kumar, T.N. Tiwari, Extraction and characterization of chitin and chitosan from fishery waste by chemical method. Environ. Technol. Innov. **3**, 77–85 (2015)
60. M.N.R. Kumar, A review of chitin and chitosan applications. React. Funct. Polym. **46**(1), 1–27 (2000)
61. M. Rinaudo, Chitin and chitosan: Properties and applications. Prog. Polym. Sci. **31**(7), 603–632 (2006)
62. M.A. Barbosa, I.C. Gonçalves, P.M.D. Moreno, R.M. Gonçalves, S.G. Santos, A.P. Pêgo, I.F. Amaral, 2.13 Chitosan (2017)
63. I. Aranaz, M. Mengíbar, R. Harris, I. Paños, B. Miralles, N. Acosta, Á. Heras, Functional characterization of chitin and chitosan. Curr. Chem. Biol. **3**(2), 203–230 (2009)
64. P.K. Dutta, M.N.V. Ravikumar, J. Dutta, Chitin and chitosan for versatile applications. J. Macromol. Sci. Part C Polym. Rev. **42**(3), 307–354 (2002)
65. M. Lavertu, S. Methot, N. Tran-Khanh, M.D. Buschmann, High efficiency gene transfer using chitosan/DNA nanoparticles with specific combinations of molecular weight and degree of deacetylation. Biomaterials **27**(27), 4815–4824 (2006)
66. H.K. No, S.P. Meyers, Preparation and characterization of chitin and chitosan—A review. J. Aquat. Food Prod. Technol. **4**(2), 27–52 (1995)
67. Z. Zakaria, Z. Izzah, M. Jawaid, A. Hassan, Effect of degree of deacetylation of chitosan on thermal stability and compatibility of chitosan-polyamide blend. BioResources **7**(4), 5568–5580 (2012)
68. EL-HAFIAN, E. A., Elgannoudi, E. S., Mainal, A., & Yahaya, A. H. B., Characterization of chitosan in acetic acid: Rheological and thermal studies. Turk. J. Chem. **34**(1), 47–56 (2010)
69. J. Shaji, V. Jain, S. Lodha, Chitosan: A novel pharmaceutical excipient. Int. J. Pharm. Appl. Sci **1**, 11–28 (2010)
70. M. Sadeghi-Kiakhani, M. Arami, K. Gharanjig, Preparation of chitosan-ethyl acrylate as a biopolymer adsorbent for basic dyes removal from colored solutions. J. Environ. Chem. Eng. **1**(3), 406–415 (2013)

71. L. Cui, S. Gao, X. Song, L. Huang, H. Dong, J. Liu, S. Yu, Preparation and characterization of chitosan membranes. RSC Adv. **8**(50), 28433–28439 (2018)
72. K.J. Adarsh, G. Madhu, A comparative study on metal adsorption properties of different forms of chitosan. Int. J. Innov. Res. Sci. Eng. Technol. **3**(2), 9609–9617 (2014)
73. S. Hussain, M. Kamran, S.A. Khan, K. Shaheen, Z. Shah, H. Suo, U. Ghani, Adsorption, kinetics and thermodynamics studies of methyl orange dye sequestration through chitosan composites films. Int. J. Biol. Macromol. **168**, 383–394 (2021)
74. S.W. Xiong, Y. Yu, P. Wang, M. Liu, S.H. Chen, X.Z. Yin, H. Wang, Growth of $AgBr/Ag_3PO_4$ heterojunction on chitosan fibers for degrading organic pollutants. Adv. Fiber Mater. **2**(5), 246–255 (2020)
75. M. Hadidi, S. Pouramin, F. Adinepour, S. Haghani, S.M. Jafari, Chitosan nanoparticles loaded with clove essential oil: Characterization, antioxidant and antibacterial activities. Carbohyd. Polym. **236**, 116075 (2020)
76. F. Tao, Y. Cheng, X. Shi, H. Zheng, Y. Du, W. Xiang, H. Deng, Applications of chitin and chitosan nanofibers in bone regenerative engineering. Carbohyd. Polym. **230**, 115658 (2020)
77. S. Roy, H.C. Kim, L. Zhai, J. Kim, Preparation and characterization of synthetic melanin-like nanoparticles reinforced chitosan nanocomposite films. Carbohyd. Polym. **231**, 115729 (2020)
78. L. Pietrelli, I. Francolini, A. Piozzi, M. Sighicelli, I. Silvestro, M. Vocciante, Chromium (III) removal from wastewater by chitosan flakes. Appl. Sci. **10**(6), 1925 (2020)
79. A.L. Ahmad, S. Sumathi, B.H. Hameed, Adsorption of residue oil from palm oil mill effluent using powder and flake chitosan: equilibrium and kinetic studies. Water Res. **39**(12), 2483–2494 (2005)
80. M.A. Hassan, T.P. Li, Z.Z. Noor, Coagulation and flocculation treatment of wastewater in textile industry using chitosan. J. Chem. Nat. Resour. Eng. **4**(1), 43–53 (2009)
81. P.K. Sahu, P.K. Sahu, S.K. Gupta, D.D. Agarwal, Chitosan: An efficient, reusable, and biodegradable catalyst for green synthesis of heterocycles. Ind. Eng. Chem. Res. **53**(6), 2085–2091 (2014)
82. G. Crini, G. Torri, E. Lichtfouse, G.Z. Kyzas, L.D. Wilson, N. Morin-Crini, Dye removal by biosorption using cross-linked chitosan-based hydrogels. Environ. Chem. Lett. **17**(4), 1645–1666 (2019)
83. A.M. Craciun, L.M. Tartau, M. Pinteala, L. Marin, Nitrosalicyl-imine-chitosan hydrogels based drug delivery systems for long term sustained release in local therapy. J. Colloid Interface Sci. **536**, 196–207 (2019)
84. V.T. Nooshabadi, M. Khanmohamadi, E. Valipour, S. Mahdipour, A. Salati, Z.V. Malekshahi, J. Ai, Impact of exosome-loaded chitosan hydrogel in wound repair and layered dermal reconstitution in mice animal model. J. Biomed. Mater. Res., Part A **108**(11), 2138–2149 (2020)
85. B. Sibaja, E. Culbertson, P. Marshall, R. Boy, R.M. Broughton, A.A. Solano, M.L. Auad, Preparation of alginate–chitosan fibers with potential biomedical applications. Carbohyd. Polym. **134**, 598–608 (2015)
86. C.K. Pillai, W. Paul, C.P. Sharma, Chitin and chitosan polymers: Chemistry, solubility and fiber formation. Prog. Polym. Sci. **34**(7), 641–678 (2009)
87. K.Z. Elwakeel, Removal of Cr (VI) from alkaline aqueous solutions using chemically modified magnetic chitosan resins. Desalination **250**(1), 105–112 (2010)
88. L. Zhou, C. Shang, Z. Liu, G. Huang, A.A. Adesina, Selective adsorption of uranium (VI) from aqueous solutions using the ion-imprinted magnetic chitosan resins. J. Colloid Interface Sci. **366**(1), 165–172 (2012)
89. K.Z. Elwakeel, M.A. Abd El-Ghaffar, S.M. El-kousy, H.G. El-Shorbagy, Synthesis of new ammonium chitosan derivatives and their application for dye removal from aqueous media. Chem. Eng. J. **203**, 458–468 (2012)
90. S.S. Silva, J.F. Mano, R.L. Reis, Ionic liquids in the processing and chemical modification of chitin and chitosan for biomedical applications. Green Chem. **19**(5), 1208–1220 (2017)

91. Q. Huang, G. Li, M. Chen, S. Dong, Graphene oxide functionalized O-(carboxymethyl)-chitosan membranes: fabrication using dialysis and applications in water purification. Colloids Surf., A **554**, 27–33 (2018)
92. Z. Liu, W. Wei, P.L. Tremblay, T. Zhang, Electrostimulation of fibroblast proliferation by an electrospun poly (lactide-co-glycolide)/polydopamine/chitosan membrane in a humid environment. Colloids Surf., B **220**, 112902 (2022)
93. D. Yang, L. Li, B. Chen, S. Shi, J. Nie, G. Ma, Functionalized chitosan electrospun nanofiber membranes for heavy-metal removal. Polymer **163**, 74–85 (2019)
94. A. Ingle, A. Gade, S. Pierrat, C. Sonnichsen, M. Rai, Mycosynthesis of silver nanoparticles using the fungus Fusarium acuminatum and its activity against some human pathogenic bacteria. Curr. Nanosci. **4**(2), 141–144 (2008)
95. W. Tiyaboonchai, Chitosan nanoparticles: a promising system for drug delivery. Naresuan Univ. J. Sci. Technol. (NUJST) **11**(3), 51–66 (2013)
96. A.K. Sailaja, P. Amareshwar, P. Chakravarty, Different techniques used for the preparation of nanoparticles using natural polymers and their application. Int. J. Pharm. Pharm. Sci. **3**(2), 45–50 (2011)
97. N.A. Peppas, Y. Huang, Nanoscale technology of mucoadhesive interactions. Adv. Drug Deliv. Rev. **56**(11), 1675–1687 (2004)
98. S. Kumar, F. Ye, S. Dobretsov, J. Dutta, Chitosan nanocomposite coatings for food, paints, and water treatment applications. Appl. Sci. **9**(12), 2409 (2019)
99. P.F.P. Barbosa, L.R. Cumba, R.D.A. Andrade, do Carmo, D. R., Chemical modifications of cyclodextrin and chitosan for biological and environmental applications: Metals and organic pollutants adsorption and removal. J. Polym. Environ. **27**(6), 1352–1366 (2019)
100. M. Xie, K. Huang, F. Yang, R. Wang, L. Han, H. Yu, F. Wu, Chitosan nanocomposite films based on halloysite nanotubes modification for potential biomedical applications. Int. J. Biol. Macromol. **151**, 1116–1125 (2020)
101. N. Oladzadabbasabadi, A.M. Nafchi, F. Ariffin, M.J.O. Wijekoon, A.A. Al-Hassan, M.A. Dheyab, M. Ghasemlou, Recent advances in extraction, modification, and application of chitosan in packaging industry. Carbohyd. Polym. **277**, 118876 (2022)
102. S.K. Kim, N. Rajapakse, Enzymatic production and biological activities of chitosan oligosaccharides (COS): A review. Carbohyd. Polym. **62**(4), 357–368 (2005)
103. M.E. Badawy, E.I. Rabea, A biopolymer chitosan and its derivatives as promising antimicrobial agents against plant pathogens and their applications in crop protection. Int. J. Carbohydr. Chem. (2011)
104. S. Chandra, N. Chakraborty, K. Panda, K. Acharya, Chitosan-induced immunity in Camellia sinensis (L.) O. Kuntze against blister blight disease is mediated by nitric-oxide. Plant Physiol. Biochem. **115**, 298–307 (2017)
105. A.F. Rodríguez, D.C. Menéndez, E.O. Delgado, O.L. Díaz, J.C. Pino, Evaluation of chitosan as an inhibitor of soil-borne pathogens and as an elicitor of defence markers and resistance in tobacco plants. Span. J. Agric. Res. **5**(4), 533–541 (2007)
106. M.R. Kumar, R. Muzzarelli, C. Muzzarelli, H. Sashiwa, A.J. Domb, Chitosan chemistry and pharmaceutical perspectives. Chem. Rev. **104**(12), 6017–6084 (2004)
107. S.K. Kim, (ed.), *Chitin and Chitosan Derivatives: Advances in Drug Discovery and Developments*. CRC Press (2013)
108. X. Fu, Y. Shen, X. Jiang, D. Huang, Y. Yan, Chitosan derivatives with dual-antibacterial functional groups for antimicrobial finishing of cotton fabrics. Carbohyd. Polym. **85**(1), 221–227 (2011)
109. Y. Li, P. Zhuang, Y. Zhang, Z. Wang, Q. Hu, A new approach for preparing chitosan fibers from a LiOH/urea solvent system. Mater. Lett. **84**, 73–76 (2012)
110. R. Ramya, J. Venkatesan, S.K. Kim, P.N. Sudha, Biomedical applications of chitosan: An overview. J. Biomater. Tissue Eng. **2**(2), 100–111 (2012)
111. T. van der Gronde, A. Hartog, C. van Hees, H. Pellikaan, T. Pieters, Systematic review of the mechanisms and evidence behind the hypocholesterolaemic effects of HPMC, pectin and chitosan in animal trials. Food Chem. **199**, 746–759 (2016)

112. H. Wang, J. Qian, F. Ding, Emerging chitosan-based films for food packaging applications. J. Agric. Food Chem. **66**(2), 395–413 (2018)
113. M. Mujtaba, R.E. Morsi, G. Kerch, M.Z. Elsabee, M. Kaya, J. Labidi, K.M. Khawar, Current advancements in chitosan-based film production for food technology; A review. Int. J. Biol. Macromol. **121**, 889–904 (2019)
114. G. Kerch, Chitosan films and coatings prevent losses of fresh fruit nutritional quality: A review. Trends Food Sci. Technol. **46**(2), 159–166 (2015)
115. C. Duan, X. Meng, J. Meng, M.I.H. Khan, L. Dai, A. Khan, Y. Ni, Chitosan as a preservative for fruits and vegetables: A review on chemistry and antimicrobial properties. J. Bioresour. Bioprod. **4**(1), 11–21 (2019)
116. M.A.M. Rocha, M.A. Coimbra, C. Nunes, Applications of chitosan and their derivatives in beverages: A critical review. Curr. Opin. Food Sci. **15**, 61–69 (2017)
117. E. Bobu, R. Nicu, M. Lupei, F.L. Ciolacu, J. Desbrieres, Synthesis and characterization of n-alkyl chitosan for papermaking applications. Cellul. Chem. Technol. **45**(9), 619 (2011)
118. J.P. Wang, Y.Z. Chen, S.J. Yuan, G.P. Sheng, H.Q. Yu, Synthesis and characterization of a novel cationic chitosan-based flocculant with a high water-solubility for pulp mill wastewater treatment. Water Res. **43**(20), 5267–5275 (2009)
119. W.J. Hennen, *Chitosan: Natural Fat Blocker*. Woodland Pub (1996)
120. L.C. Yo, *Preparation and Characterization of Water-Soluble Chitosan Gel for Skin Hydration* (Doctoral dissertation, Universiti Sains Malaysia, 2008)
121. R. Yang, H. Li, M. Huang, H. Yang, A. Li, A review on chitosan-based flocculants and their applications in water treatment. Water Res. **95**, 59–89 (2016)
122. J. Hu, W. Liu, B. Liu, U.S. Patent No. 7,780,979 (U.S. Patent and Trademark Office, Washington, DC, 2010)
123. E.M. Costa, S. Silva, A.R. Madureira, A. Cardelle-Cobas, F.K. Tavaria, M.M. Pintado, A comprehensive study into the impact of a chitosan mouthwash upon oral microorganism's biofilm formation in vitro. Carbohyd. Polym. **101**, 1081–1086 (2014)
124. Y. Habibi, L.A. Lucia, *Polysaccharide Building Blocks: A Sustainable Approach to the Development of Renewable Biomaterials* (John Wiley & Sons, 2012)
125. M. Dash, F. Chiellini, R.M. Ottenbrite, E. Chiellini, Chitosan—A versatile semi-synthetic polymer in biomedical applications. Prog. Polym. Sci. **36**(8), 981–1014 (2011)
126. B.E. Benediktsdóttir, Ó. Baldursson, M. Másson, Challenges in evaluation of chitosan and trimethylated chitosan (TMC) as mucosal permeation enhancers: From synthesis to in vitro application. J. Control. Release **173**, 18–31 (2014)
127. Y. Rosen, P. Gurman, N. Elman, (eds.), *Drug Delivery: An Integrated Clinical and Engineering Approach*. CRC Press (2017)
128. V. Patrulea, V. Ostafe, G. Borchard, O. Jordan, Chitosan as a starting material for wound healing applications. Eur. J. Pharm. Biopharm. **97**, 417–426 (2015)
129. P.R. Sivashankari, M. Prabaharan, Prospects of chitosan-based scaffolds for growth factor release in tissue engineering. Int. J. Biol. Macromol. **93**, 1382–1389 (2016)
130. M.R. Tehrani, S. Safari, S. Jafari, M. Amini, M.H. Ghahremani, N. Ostad, Gene delivery using N-diethylmethyl chitosan to Hela cell line. J. Control Release **1**, 34–38 (2012)
131. S. Dumitriu, *Polysaccharides: Structural Diversity and Functional Versatility* (CRC Press, 2004)
132. S. Sayed, A. Jardine, Chitosan derivatives as important biorefinery intermediates. Quaternary tetraalkylammonium chitosan derivatives utilized in anion exchange chromatography for perchlorate removal. Int. J. Mol. Sci. **16**(5), 9064–9077 (2015)
133. N.M. Alves, J.F. Mano, Chitosan derivatives obtained by chemical modifications for biomedical and environmental applications. Int. J. Biol. Macromol. **43**(5), 401–414 (2008)

Chapter 2
Biological and Chemical Aspects of Chitosan

Monica Butnariu

Abstract Chitosan is a polysaccharide that is made from chitin—a matter forming from the operation of treating shrimp shells, as well as other crustaceans, with NaOH (sodium hydroxide). Chitosan has profited from a substantial number of trading and equally biomedical uses. It can be used in agriculture, for seed treatment, and as a biopesticide, helping plants fight fungal infections. In winemaking, it is used to prevent the wine from spoiling. On an industrial scale, it is also used in water filtration processes. Being considered a good haemostatic, chitosan is useful in medicine for the creation of bandages designed to reduce bleeding as well as an antibacterial agent, being used especially in the treatment of gum disease or bleeding. It is also anti-carcinogenic. Somewhat controversially, chitosan has been claimed for its role in limiting fat absorption, which makes it useful for dieting, with no evidence yet to contradict this special property of chitosan. Thus, chitosan has been increasingly used to support weight loss efforts due to its valuables that reduce fat absorption in the intestines. Chitin and chitosan (chitinous substances) are found in abundance in nature and are renewable sources, which have attracted interest in developing new applications based on these simple substances. Moreover, chitinous biopolymers are biocompatible, biodegradable, non-toxic, valuables that have allowed the use of their derivatives in medicine.

2.1 Background

In 1811, the director of the biological garden in Nancy, France, Prof. H. Braconnot, separated a fibrous matter or chitinous substances (CSs) from a specific type of mushroom. Then he noticed that this matter does not dissolve in aqueous solutions of acids, such as H_2SO_4 (sulfuric acid). A decade later, the same matter was found in certain insects and in 1823 Ojear named it chitin (from the Greek khiton meaning

M. Butnariu (✉)
Chemistry and Biochemistry Discipline, University of Life Sciences "King Mihai I" From Timisoara, Calea Aradului 119, 300645 Timis, Romania
e-mail: monicabutnariu@yahoo.com

S. K. Swain and A. Biswal (eds.), *Chitosan Nanocomposites*,
Biological and Medical Physics, Biomedical Engineering,
https://doi.org/10.1007/978-981-19-9646-7_2

envelope). In 1859, Prof. C. Rouget subjected chitin to alkaline treatment and noticed that the resulting matter dissolves in acids; but only in 1894, this matter was called chitosan by Hoppe-Seyler. Meanwhile, in 1878, Ledderhose proposed that chitin be made from glucosamine and CH_2COOH (acetic acid).

In the 1930s and 1940s, these glucosamine biopolymers aroused the interest of the Eastern world, especially in medicine and in the field of water purification. In 1970, interest in these renewed bio-macromolecules was renewed at the first conference on chitin and chitosan organized in the USA in 1977. Muzzarelli's studies in 1980 led to a clearer understanding and characterization of these materials [1]. Today, it is known that chitin and chitosan (CSs) are found in abundance in nature and are renewable sources, a fact that attracted interest in the development of new applications based on these simple matters. Moreover, CSs are fibrous compounds derived from the shells of crustaceans such as shrimp, lobsters, and crabs. The CSss biopolymers are biocompatible, biodegradable, and non-toxic valuables that allowed the use of their derivatives in medicine [2].

Chitin $[(C_8H_{13}O_5N)n]$ is a polymer composed of N-acetylglucosamine fraction joined by p-(1 → 4) covalent bonds. N-acetylglucosamine is a derivative of glucose. Structurally, chitin is comparable to cellulose $[(C_6H_{10}O_5)_n]$, which consists of glucose fragments and is also linked by β-(1 → 4) linkages, excluding that a hydroxyl assembly on a $(C_6H_{10}O_5)_n$ monomer is superseded by an acetyl amine assembly in a chitin monomer. Functionally, chitin is most reasonable to the protein keratin, which is used as a constitutional constituent in many organisms. Chitin is the second most abundant biopolymer in the world, after $(C_6H_{10}O_5)_n$. The hydrogen bonds between the monomers in chitin make it very strong. Pure chitin is translucent and flexible. However, in many animals, chitin is combined with other molecules to form a composite material. For example, in molluscs and crustaceans, it combines with calcium carbonate ($CaCO_3$) to form hard and often colored shells. In insects, chitin is often stacked into crystals that produce iridescent colors used for biomimicry, communication, and to attract mates [3]. CSs are complementary compounds: to get chitosan, there must be chitin. The latter can also be formed by the combination of pitch, conchiolin, aragonite, and $CaCO_3$. It is the second most important polymer after $(C_6H_{10}O_5)_n$; In addition, it is biocompatible, biodegradable, and non-toxic.

2.2 The Skeletal Structure of Chitosan

Information about the valuables and skeleton of $(C_6H_{10}O_5)_n$ (the most important representative of *polysaccharides*, which is the main constitutional constituent of plants) is presented in the specialized literature in an accessible form. However, there is much less information about the skeleton of CSs [4]. However, it is the basis of the skeletal system, which supports the skeleton of the cells that form the tissues in the cuticle of insects, crustacean shells, and the cell wall of bacteria and fungi. The fact that CSs skeletons in insect and crustacean organisms are inherent in hardness is associated with the formation of a special chitin-carbonate complex. It appears as a result of the deposition of the matter we are interested in on $CaCO_3$, which acts as a kind of inorganic matrix. One of the major differences between CSs biopolymers and semi-synthetic $(C_6H_{10}O_5)_n$ derivatives is the nitrogen content which can have a direct effect in some applications such as the chelating agent in the treatment of waste water [5].

The higher the nitrogen content in CSs biopolymers, especially in chitosan, the more advantageous it provides numerous nucleophilic locations (positions) on the biopolymer chain, for example, compared to $(C_6H_{10}O_5)_n$. There is a certain analogy between the skeleton of $(C_6H_{10}O_5)_n$ and chitin. However, unlike the first, in chitin, the acetamide assembly is the substitute for the second carbon atom of the elementary unit. In $(C_6H_{10}O_5)_n$, the hydroxyl plays the same role. Native (i.e. natural) chitin macromolecules usually contain a certain number of units with free primary amino clusters. This matter is added to enhance the flavor and aroma of food, to improve the appearance, or is used as a preservative. There are also nutritional supplements that contain it. The composition of chitin is such that this matter has medicinal valuables [6]. Its benefits are considered to be the following (Fig. 2.1).

Chitin is a very useful matter. What it is and what are its medicinal valuables, it would be good to remember. It occurs very often in nature. So much so that it ranks second in prevalence (the first belongs to $(C_6H_{10}O_5)_n$). A number of scientists even believe that humanity will switch to an exclusively chitinous diet in the near future.

For example, Sam Hudson, Professor of polymer chemistry, recently reported that researchers are now on the verge of discovering a "new world" in which the number of products that can be made from chitin will be endless. These locations offer possibilities for an easier and selective synthetic transformation of these biopolymers

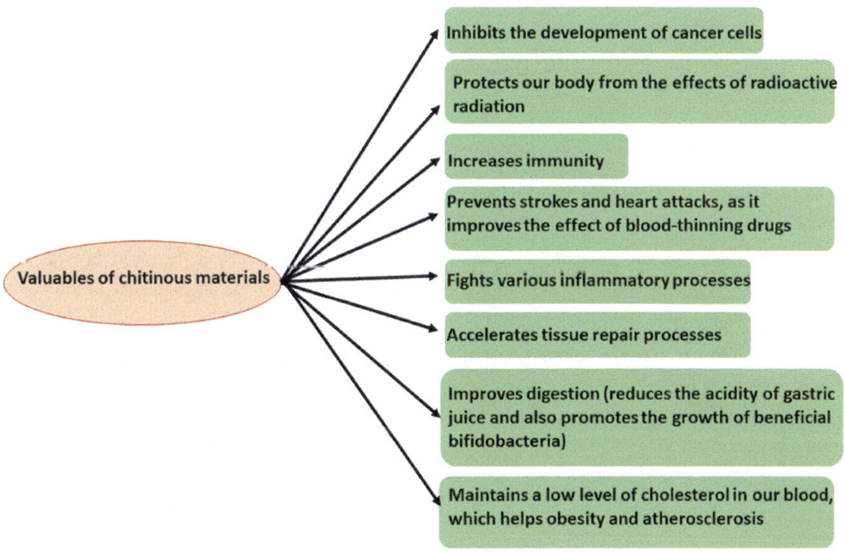

Fig. 2.1 Different properties of chitinous materials

to obtain the various derivatives, which would be needed for the development of the desired applications [7].

The annual production of chitin in nature is estimated at 10^9–10^{10} tons, and at the present time, it is too little used and too little commercialized compared to the $(C_6H_{10}O_5)_n$ industry. Chitin is the second most widespread polyglucide in nature, surpassed only by $(C_6H_{10}O_5)_n$, being present in the exoskeletons of crustaceans (crab, shrimp, lobster, etc.). Chitin is a hard and flexible material (or hard when combined with $CaCO_3$). It is synthesized by special glands (chitin glands) of many animal and fungal species. It plays an essential constitutional role in several eukaryotic taxa, particularly macroscopic fungi and yeasts, mycorrhizae and lichens, and arthropods. As such, it is considered to be one of the synapomorphies of the *Opisthochonts* assembly [8]. This constitutes, in particular, the external cuticles of insects and crustacean shells or in the shell of cephalopods (Nautilus, for example) and many molluscs (it is also one of the constituents from the mother to the of-perla) and pseudonacres that constitute the interior some clams and pearls of pearl oysters or freshwater pearl mussels. However, in bivalves, it is present in a highly variable proportion depending on the species. It is also found in many microorganisms, for example, in the cuticle that protects the eggs of rotifers [9] (Fig. 2.2).

The chemical skeleton of chitin consists of linear and repetitive units of 2-acetamide-2-dioxy-D-glucopyranose linked by ß-(1 → 4) linkages. In nature, chitin

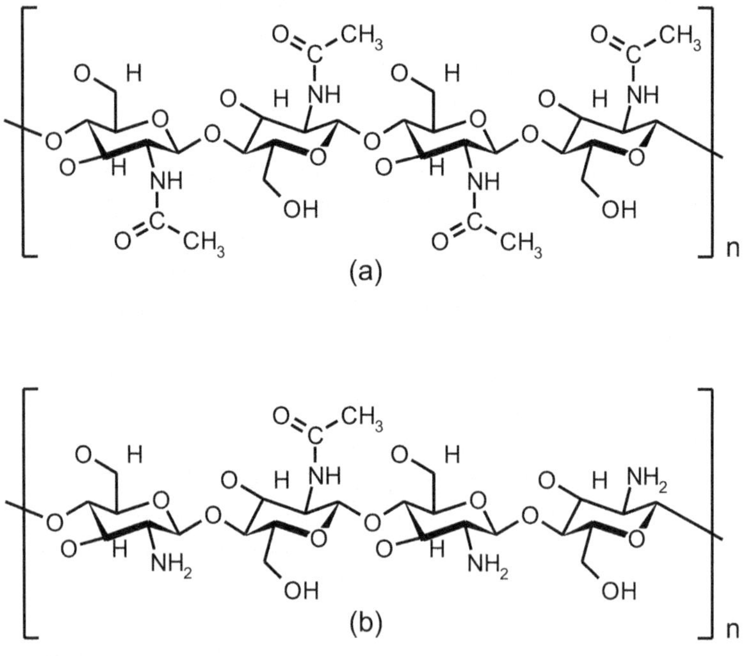

Fig. 2.2 Chemical skeleton of chitin (**a**) and chitosan (**b**)

Table 2.1 Approximate chitin content in various living species

Species	Mass % $(C_8H_{13}O_5N)n$/body weight
Mushroom (Fungi)	5–20%
Invertebrate animals (Worms)	20–38%
Devil fish (Octopus)	3–20%
Predatory arachnids (Scorpions)	30
Arachnids (Spiders)	38
Small insects (Kitchen bugs)	35
Family dytiscidae (Water beetle)	37
Silkworm	44
Snail crab species	69
Edible crab	70

performs a protective effect and reference function, providing resistance to crustaceans, fungi, and bacteria [10]. In this sense, it is reasonable to $(C_6H_{10}O_5)_n$, which is the supporting material of the plant cell wall. But chitin is more reactive, according to the materials of the recent researches. When heated and treated with concentrated alkali, it turns into chitosan. This biopolymer can dissolve in dilute acid solutions, as well as bind and react with other chemicals [11]. In addition to crustaceans, it is also found in various insects, worms, fungi, and mushrooms, in varying proportions from species to species and from region to region (Table 2.1).

Thus, chemists sometimes refer to chitosan as a "builder" that can be used to create different biopolymers. In order to obtain chitin in its pure form, those containing organic matter remove proteins, calcium, and other minerals, turning them into a soluble form. The result is a chitinous crumb. Observations made under an electron microscope show that certain pollutants (heavy metals such as cadmium, copper, chromium, zinc, and lead, for example) can interfere with the biosynthesis of chitin and degrade bivalve nacre (a phenomenon studied in the anodont swan, family Unionidae) [12]. This could suggest that certain parasites affecting oysters and their shells could be facilitated by contamination of the shells by certain heavy metals (actually observed contamination). In nature, chitin has the role of a constitutional polyglucide, providing adhesion between the fiber beds of the overlapping laminae. For example, in the case of mollusk shells, the shell matrix is composed of two constitutional units (schematic representation in Fig. 2.3). The first is a high molecular weight chitin-protein complex arranged in the form of beds and layers. The second unit called the mineralization matrix is a polypeptide with a pronounced affinity for calcium. During the growth of the carapace, the chitin-protein complex remains trapped between the layers of the mineralization matrices, offering the possibility of adhesion between the two adjacent matrices [13].

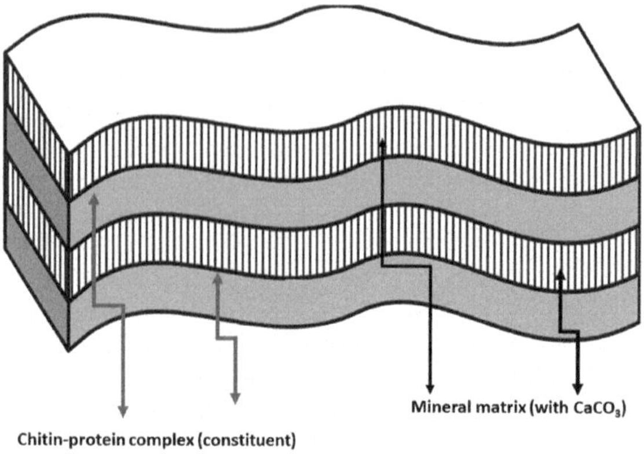

Fig. 2.3 Schematic representation of the outer shell of molluscs

This tight assembly provides excellent constitutional rigidity and mechanical strength to the shells.

Chitosan (Table 2.2b) is deacetylated chitin, a constitutional modification of chitin often carried out by alkaline hydrolysis. The process itself used during hydrolysis determines the deacetylation of chitin, and consequently, the commercially available chitin samples show between 70 and 100% degrees of deacetylation.

Besides the degree of deacetylation for a given chitosan sample, the molecular weight of the macromolecule, which can vary between 150,000 and 600,000 daltons, is also characteristic. Chitin and chitosan are of high commercial interest due to their rich nitrogen content (6.89%), which allows their use as chelating agents. In addition, both chitin and chitosan are biodegradable, biocompatible, non-toxic, non-allergic, and renewable biomaterials and find their application in fields such as medicine, perfumes and cosmetics, food industry, and agriculture [14].

One of the major differences between chitinous biopolymers and semi-synthetic $(C_6H_{10}O_5)_n$ derivatives is the nitrogen content which can have a direct effect on some applications such as chelating agents in wastewater treatment. The higher the nitrogen content in chitinous biopolymers, especially in chitosan, the more advantageous it provides numerous nucleophilic locations on the biopolymer chain, for example, compared to $(C_6H_{10}O_5)_n$.

These locations offer possibilities for an easier and more selective synthetic transformation of these biopolymers to obtain the various derivatives, which would be needed for the development of the desired applications. Chitin is very hydrophobic and insoluble in water, acids, bases, and common organic solvents. The organic solvents in which it is soluble are hexafluoroisopropanol, hexafluoroacetone, chloroalcohols in the presence of mineral acid solutions, and dimethyl acetamide in the presence of 5% Lithium chloride (LiCl). On the other hand, chitosan, due to the presence of the primary amino assembly in most of the sugar units that make up the biopolymer skeleton, dissolves in dilute organic acids, but is insoluble in water, above pH 6–7 and in common organic solvents [15].

Table 2.2 Examples of applications of chitosan derivatives

No.	Cross-linking	Controlled release systems	Other applications
1	Alkyl	Erodible Hydrogel for Controlled Drug Release	Foam emulsion stabilizer
2	Aldehyde	Hydrogels for controlled drug release	Biological adhesive for soft tissues
3	Palmitic acid	Hydrogel for the controlled release of Rhodamine B	…
4	Carboxymethyl	Injectable gel for sustained release of morphine, and slow release of gentamicin	Reparative agent in the reduction of periodontal pockets
5	Fructose	…	Films for fruit preservation
6	2-hydroxyethyl methacrylate	Blood-compatible films for controlled glucose release	Gel or solution for preventing post-surgical erythonal adhesion
7	Lactic-glycolic acid	Hydrogels for controlled gastric release	Matrices for the attachment of hepatocytes
8	Pluronic	Hydrogels for nasal delivery of antiasthmatic proteins	…
9	Ethylenediaminetetra CH_3COOH	Bioadhesive hydrogels for nasal, ocular, and vaginal release	…
10	Poly(ethylene glycol)	Bioadhesive hydrogels for the release of peptides at the gastric level	…
11	Poly(ethylene glycol) galactosyl	Gelled particles for the release of insulin or N-phenyl-1-naphthylamine	Constituent of hydro- and hydroalcoholic gels for topical use

The solubility of chitinous matters is usually associated with the crystallinity of the sample. Higher crystallinity suggests greater or increased molecular interactions between biopolymer chains. A chitinous chemical matter can dissolve only under the conditions in which these interactions are canceled [16]. The intra- and intermolecular hydrogen bonds of the biopolymer chains are the major cause of these interactions and have an important role in the low solubility of these matters. However, chemical modifications of chitosan result in derivatives that are soluble in water in a wider pH range, including in strongly basic environments. The modifications consist in the introduction of ionic clusters or substituents in the biopolymer skeleton, which dissolve in polar solvents such as water through polar-polar interactions and determine the solubility of the macromolecule [17].

The skeleton of chitin and $(C_6H_{10}O_5)_n$ leads to physico-chemical proximity of the two biomolecules, which allows them to perform reasonable functions in living systems. Like cellulose $(C_6H_{10}O_5)_n$ molecules, chitin $(C_8H_{13}O_5N)_n$ molecules have a high rigidity and a pronounced tendency to form intermolecular associations with the formation of highly ordered skeletons. Types of such skeletons are represented by (-chitin) formations, which differ in the degree of order and mutual orientation of individual biopolymer chains [18].

CSs biosynthesis takes place in special cellular organs (chitosomes) with the participation of the enzyme chitin synthase in a cyclic sequence; followed by the transfer of N-acetyl-D-glucosamine residues from uridine diphosphate-N-acetyl-D-glucosamine to the growing biopolymer chain. Chitosan, whose presence is particularly characteristic of the cell walls of certain fungi, is formed by the enzymatic N-deacetylation of chitin. In nature, chitin is in a complex with other polyglucides and minerals and is covalently bound to a protein [19]. To isolate chitin, its insolubility is taken into account. So, the shells of crabs or lobsters, which contain up to 25% chitin, are demineralized with hydrochloric acid, Protein solutions, and in hot alkali, and chitin bleaching is done with H_2O_2. Milder isolation conditions are demineralization with complexes and treatment with oxidizing agents at neutral pH. The chitin obtained in this way has a mass of several millions. Chitin is the second most common natural biopolymer after $(C_6H_{10}O_5)_n$. Available sources of chitin are wastes from marine invertebrate fisheries and the mycelium of lower fungi. Chitosan is more promising, in solution it can form salts, giving solutions with high viscosity.

Chitosan yields powerful compounds; with proteins, anionic polyglucides form chelated complexes with metals, etc., which is the basis of its use for the removal of proteins from wastewater in food production (meat, fish, dairy industry, cheese making), making chelating ion exchangers, immobilization living cells in biotechnology, in the manufacture of honey, and preparation of finishing paper and textile fibers [20]. Certain N-acyl derivatives of chitosan are good gelling agents; when chitosan is acylated with dicarionic acid derivatives, cross-linked gels are obtained.

In nature, chitin has a protective and supportive function, providing resistance to crustaceans, fungi, and bacteria. In this sense, it is reasonable to $(C_6H_{10}O_5)_n$, which is the supporting material of the plant cell wall. But chitin is more reactive. When heated and treated with concentrated alkali, it turns into chitosan. This biopolymer can dissolve in dilute acid solutions and can bind and react with other chemicals. Thus, chemists sometimes refer to chitosan as a "builder" that can be used to create different biopolymers. To obtain pure chitin, proteins, calcium, and other minerals are removed from the organic matter that contain it, turning them into a soluble form [21]. Chitin is formed by a series of glycosidic bonds between substituted glucose molecules. Chitin is different from $(C_6H_{10}O_5)_n$ because of the substitution that occurs on the glucose molecule. Instead of a hydroxyl (OH) assembly, the glucose molecules in chitin have an amyl assembly attached which consists of carbon and nitrogen.

2.3 Chitin Isolation and Chitosan Synthesis

In biology, chitin $(C_8H_{13}NO_5)$n is a polyglucide, an organic matter resembling $(C_6H_{10}O_5)_n$, which forms the exoskeleton of insects or other arthropods (crustaceans or arachnids). It is a modified sugar containing nitrogen, polyglucide, from the assembly of acetylglucosamine (N-acetyl-glucosamine) linked together by β-1,4 covalent bonds (reasonable to the bonds between glucose clusters in $(C_6H_{10}O_5)_n$ molecules). Thus, chitin can be described as $(C_6H_{10}O_5)_n$ with one hydroxyl assembly of each monomer superseded by an acetylamine assembly. This formula allows for tighter bonds between the hydrogen atoms of adjacent biopolymers, resulting in a stronger material [22]. The chemical name of the molecule is poly N-acetyl-D-glucosamine, β-(1,4)-2-Acetamido-2-deoxy-D-glucose, or simply N-acetyl-D-glucosamine β-(1,4) N-acetyl- D-glucosamine. In its pure form, chitin is translucent, pliable, resilient, and hard. In arthropods, it is often modified, being incorporated into a hard protein matrix that forms the exoskeleton. In its pure form, it has the texture and consistency of leather but combined with $CaCO_3$, it becomes much more resistant [23].

The difference between the modified and unmodified forms can be seen by comparing a (unmodified) caterpillar to a (modified) beetle. Although it is translucent, different sizes of the parallel layers of microfibers can create constitutional colors, due to the phenomenon of interference. Chitin is one of the constituents of the cuticle in insects, spiders, or crustaceans and has a protective role. Combined with $CaCO_3$, it becomes rigid and forms the exoskeleton of crustaceans, for example, snails. The chitin found in the cuticle of arthropods is alpha chitin, in which the chains of chitin molecules arealigned in an antiparallel manner by hydrogen bonds. In brachiopods, cephalopods, and annelids, beta chitin is found where the chains of molecules are aligned in a parallel manner [24].

Long molecular chains are associated in units of one biomolecule (beta (β) chitin), units of two biomolecules (alpha (α) chitin), or units of three biomolecules (gamma chitin). Gamma chitin [γ-*chitin*] is rarer and found in brachiopods. In fungi, chitin is an essential constituent of the side wall and protects fungal cells from the environment. Chitin actively participates in the rigidity of the fungal wall [25].

The chitin biopolymer of biological origin is synthesized by an enzymatic activity called chitin synthesis. In *Saccharomyces cerevisiae,* several activities of chitin synthesis occur in different intracellular locations and at certain stages of budding of this unicellular fungus. Within natural resources of commercial interest, chitin does not exist as an independent biopolymer, but rather in the form of conglomerates with other biomaterials, mainly proteins, lipids, and inorganic salts [26].

Chitin is a polysaccharide made up of acetylglucosamine units (exactly: 2-acetamido-2-deoxy-D-glucopyranose or N-acetyl-D-glucosamine for short, abbreviated: GlcNAc). Acetylglucosamine units are linked by β-1,4-glycosidic linkages—this is the same type of linkage as the glucose molecules in $(C_6H_{10}O_5)_n$. Therefore, chitin can be understood as a variant of $(C_6H_{10}O_5)_n$ in which the hydroxyl clusters in the 2-position of the monomeric units have been superseded by acetamido clusters. This allows for stronger hydrogen bonding between adjacent polymers, making chitin harder and more stable than $(C_6H_{10}O_5)_n$. However, natural chitin is usually not a uniform polymer but a mixture of random copolymers of D-glucosamine (GlcN) and N-acetyl-D-glucosamine (GlcNAc), i.e. not all amino clusters are acetylated [27].

The degree of acetylation determines its valuables in addition to the degree of polymerization (chain length) and chain folding. The transition to chitosan, which has significantly fewer (ideally no) acetyl clusters, is therefore fluid. If the degree of acetylation is greater than 50%, it is usually called chitin; if it is smaller, it is usually called chitosan.

The skeleton of α-chitin. Hydrogen atoms are not shown. For oxygen atom no. 6, there are two positions—shown semi-transparently—which are statistically filled chitin occurs naturally in at least two conformations: arthropod chitin occurs mainly as α-chitin, in molluscs as β-chitin. A mixture of α- and β-chitin, which occurs in beetle and cephalopod larvae, is sometimes referred to as γ-chitin. Chitin is colorless. The familiar brown color (as well as firmness) of insect shells is caused by sclerotin, a constitutional protein. Chitin is mostly insoluble in weakly ionic aqueous organic solvents and tolerable to health; in strongly ionic solvents, the "solubility" is based on a depolymerization [28].

"Soluble chitin" is mostly a matter of chitin hydrochloride, some of which is even soluble in water. In nature, chitin forms complex skeletons that form in a multi-step process. Chitin molecules are synthesized by transglycosylating, membrane-bound enzymes known as chitin synthetase EC 2.4.1.16, which use uridine diphosphate-N-acetylglucosamine (UDPGlcNAc) as a substrate. In fungi, for example, this happens in special vesicles called chitosomes. Chitin molecules or chito oligomers are secreted into the extracellular space. Various changes take place outside the cells that affect the valuables. One of them is partial hydrolysis by chitinases EC 3.2.1.14. In addition to hydrolase, chitinases also have transglycosidase activity, so that chitin can be linked to glucans. Another modification is partial deacetylation by special deacetylases. Some unmodified chitin molecules crystallize and are partially covalently linked to other chitin molecules via proteins [29].

The resulting supramolecular skeleton matures through further cross-linking and the incorporation of various matters. There is much less information about the skeleton of chitin. However, it is the basis of the skeletal system, which supports the skeleton of the cells that form the tissues in the cuticle of insects, crustacean shells, and the cell wall of bacteria and fungi. The fact that chitinous skeletons in insect and crustacean organisms with inherent hardness are associated with the formation of a special chitin-carbonate complex. It appears as a result of the deposition of the matter we are interested in on $CaCO_3$, which acts as a kind of inorganic matrix [30]. The chitin isolation process begins in the marine food industry (Scheme 2.1).

One of the secondary products of this process, such as crab shells and shrimps, are first crushed to a powdery consistency to create a larger surface for the heterogeneous processes that will follow. An initial treatment of the carp with 5% sodium hydroxide dissolves several proteins leaving chitin, lipids, and calcium salt (mainly in the form of $CaCO_3$). By treating with 30% hydrochloric acid, lipids are hydrolyzed, calcium salts are decomposed (demineralization), and other minor inorganic constituents [31].

Thus, the obtained chitin can be hydrolyzed using 50% sodium hydroxide at high temperatures to obtain chitosan. Alternatively, if isolation of chitin is not desired, the sequence based on acid treatment can be reversed to directly produce chitosan.

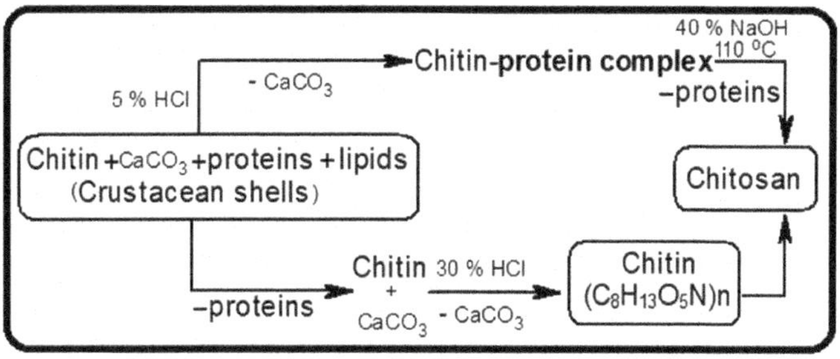

Scheme 2.1 Chitin isolation and chitosan synthesis—schematic process

In this method, the crushed shells are first treated with 5% hydrochloric acid to remove calcium salts, a process often followed by the removal of proteins and lipids by treatment with 40% sodium hydroxide at a higher temperature [32]. During the treatment with a basic medium, the simultaneous hydrolysis of the acetamide clusters in chitin takes place, the result being the formation of chitosan. Chitosan is obtained when the chitin molecule has been completely deacetylated. Chitosan, on the other hand, is left with one acetyl assembly per unit to replicate. To obtain chitosan, it is necessary to obtain chitin first. Then it is deacetylated (the acetyl molecule it has in the skeleton is removed), so that only the amino assembly remains. The process begins by obtaining the raw material, which is the exoskeleton of crustaceans, especially shrimp and prawns [33].

A washing treatment is performed to remove all impurities such as salt and mineral residues that may be embedded in the species' exoskeleton. The material is dried well and then ground to a flake form of about 1 mm. The depigmentation process follows. This procedure is optional and is done with acetone (an organic solvent in which chitosan is insoluble), xylene, ethanol, or hydrogen peroxide [34]. The previous process is followed by the decarbonization process; in which HCl is used. Once this process is finished, deproteinization is continued, which is done in a basic medium using NaOH. Wash with plenty of water and finally filter. The compound obtained is chitin. It is treated with 50% NaOH at a temperature of about 110 °C for 3 h. This process allows the removal of the acetyl assembly from the chitin skeleton so that chitosan can be obtained. To be packaged, dehydration and grinding are carried out until the particle size is 250 μm. The first has a direct impact on the secondary skeleton of the biopolymer chain and can also influence the solubility of the biopolymer in organic or aqueous solvents. It can also affect the chemical reactivity of the sample in homogeneous processes [35]. According to a selective nomenclature, chitinous matters that do not dissolve in dilute organic acids (for example 1–2% CH_2COOH) are called chitin, a biopolymer with a low degree of deacetylation. On the other hand, chitinous matters that dissolve in dilute aqueous acids are called chitosan. Solubility in aqueous acid solutions is achieved by deacetylation to the extent of 60%. However, at the degree of deacetylation between 50–60%, the distribution of the remaining acetyl clusters, grafted along the biopolymer chain, influences the solubility of the sample [36].

A distribution of acetyl clusters on the biopolymer skeleton results in homogeneous processing conditions and gives the biopolymers solubility in aqueous solutions of weak acids. On the other hand, under heterogeneous processing conditions, biopolymers are formed with distinct blocks of acetylated sugar residues and are not soluble in reasonable solvents [37]. The molecular weight of chitosan obtained at the end of the production process depends on the process parameters, time, temperature, and concentration of HCl and NaOH. The process parameters used in the production of chitosan are drastic and the process is accompanied by the splitting of the chitin skeleton. The degradation of the chitinous chain can be extended. In one preparation, a sample of chitin with a molecular weight of 1.03×10^6 kDa produced chitosan with a weight of 1×10^5 kDa. The estimation of the molecular mass of chitosan can be achieved by various techniques such as light scattering spectroscopy, viscometry, and gel permeation chromatography [38]. However, the charged nature of chitosan, its tendency to form free aggregates, and the differences in the degree of deacetylation for different chitosan samples require careful implementation of the constants in Eq. (2.1).

$$[\eta] = \mathrm{K} \cdot \mathrm{M}_{\mathrm{n}}^{\mathrm{a}} \tag{2.1}$$

Thus, the degree of deacetylation and the molecular weight of chitosan can be modified to obtain different physico-mechanical valuables.

2.4 Valuables of Chitosan

2.4.1 Physical–Chemical Valuables

The physical valuables of chitinous matters are governed by two factors: the degree of deacetylation and the molecular mass [39]. These features are as follows:

- Chitosan is a compound insoluble in water.
- Its approximate molar weight is 1.26×10^5 g/mol of biopolymer, obtained by the viscometer method.
- It has chemical valuables that make it suitable for various biomedical applications.
- It is a linear polyamide.
- It has amino clusters -NH_2 and reactive hydroxyl clusters -OH.
- It has chelating valuables for many transition metal ions.
- With lactic acid and CH_2COOH, very tight chitosan films could be formed in which, by infrared (IR) spectrum, no variation in the chemical skeleton of chitosan was observed. However, when formic acid was used, variations in the skeleton could be observed [40].

2.4.1.1 The Degree of N-Acetylation and Its Estimation

Chitosan is differentiated either from the point of view of the degree of acetylation, which corresponds to the number of N-acetylamine clusters or from the point of view of the degree of deacetylation, which corresponds to the D-glucosamine clusters.

The degree of acetylation has a major influence on the physico-chemical valuables (molecular mass, viscosity, and solubility), thus being one of the most important parameters.

It is very important to determine the degree of acetylation as simply and as correctly as possible. A multitude of methods for its estimation have been reported, such as elemental analysis, hydrolysis of acetamide clusters, titration of free amine clusters, spectroscopy (IR, UV, CD, NMR), enzymatic degradation, pyrolysis gas chromatography, and thermal analysis. The best technique for determining the degree of acetylation for rapid characterization appears to be IR spectroscopy.

Confirmation of the chemical skeleton is highlighted by functionalizing of chitosan (medium molecular weight chitosan) by treatment with N-chloroacetyl-sulfonamide derivatives derived from sulfadiazine, sulfamerazine, sulfamethoxy-diazine, sulfadime-toxin, sulfamethoxazole, and sulfisoxazole, which is confirmed by the appearance in the IR spectrum of both bands absorption characteristics of chitosan, respectively, the glucosamine unit, as well as the sulfonamide constituent [41].

Thus, in the spectrum of chitosan and its functionalized derivatives, the characteristic bands of the amide assembly were identified, due to the vibrations of dC = O (amide I) in the range 1628–1655 cm^{-1} and d_{NH} (amide II) in the range 1564–1597 cm^{-1}. At the same time, in the 3320–3400 cm^{-1} region, a broad band attributed to the valence vibration of the alcoholic OH clusters was identified. The sulfonamide constituent is present in the spectrum through the characteristic absorption bands of the sulfonamide assembly, dSO_2-N in the ranges 1255–1261 cm^{-1} and 1162–1167 $cm\text{-}^{1}$. The aromatic nucleus is present in the form of multiple bands, the most intense of which are those in the region 1538–1547 cm^{-1} and 833–898 cm^{-1} [42].

2.4.1.2 Molecular Mass

The definition of the average molecular mass of polyglucides as well as the understanding of its consequences on the physico-chemical behavior presented a real challenge for chemists for a long time. In the case of chitin and its derivatives, the knowledge of these data is of great importance in industrial applications and in numerous research fields. Although the primary skeleton of chitosan consists of acetylated β-1,4-D-glucosamine units, the name chitosan is actually a collective term that defines chitins with various degrees of deacetylation and which differ in terms of crystallinity, optical characteristics, the degree of acetylation, the content of impurities, and the molecular mass.

The methods of obtaining and the origin are the main factors responsible for the above differences. Depending on the further applications of chitosan, high-accuracy molecular mass estimation methods are needed [43].

Current methods are based on viscometric measurements. Lately, chromatographic estimation methods, also known as permeable gel chromatography or filtering gel chromatography, are used. They are used both to determine the molecular mass and its distribution for various materials: synthetic polymers, biopolymers, or natural polymers [44].

Another recently used method with very good results is IR spectroscopy (which measures the refractive index).

2.4.1.3 Solubility

Chitosan is insoluble in water at an almost neutral pH, in concentrated acids, with the exception of H_2SO_4, as well as in organic acids.

Being a polyamine, it is soluble in dilute acid solutions, forming the respective quaternary ammonium salt with organic acids; it dissolves in hydrochloric acid and solutions of organic acids such as formic, acetic, oxalic, and lactic acid. Solubility depends on the concentration and type of acid [45]. The solubility of chitosan depends on both its degree of dissociation and its mode of preparation.

2.4.1.4 Crystal Skeleton

Chitosan has a crystalline skeleton, and its polymorphism depending on its physical state. Its varied skeletons include an anhydrous form, a hydrated form, and various salts, recently analyzed by X-ray diffraction. The polymer chains in the hydrated forms are four in number, which pass through the unit cell. The two adjacent chains along the b-axis are crystallographically independent, arranged in an antiparallel manner, and connected by two rows of $N_2 \ldots O_6$ hydrogen bonds [46]. In anhydrous chitosan, the two adjacent polymer chains are found along the a-axis.

2.4.2 Technical Valuables

2.4.2.1 Flocculation

As mentioned in the introduction, chitosan represents a promising alternative in obtaining synthetic polycations, then used as flocculation agents. Water treatment offers many possibilities, from the removal of humic acid from drinking water to various techniques for treating wastewater or dewatering sludge. However, the skeleton-function relationship has rarely been studied from the point of view of flocculation in the case of chitosan [47].

However, an increase in the efficiency of chitosan, as a flocculating agent, could be correlated with the decrease in pH and the number of fragments of acetylated units and with a significant increase in density.

2.4.2.2 Gelation

Until now, no ionic and non-toxic cross-linking agent has been found that can confer a better reproducibility of chitosan gels like the calcium ions used for the gelation of alginates. However, it was observed that chitosan-based hydrogels, cross-linked with molybdate polyoxyanions, have a transparent, thermo-resistant skeleton and are able to increase their dimensions several times compared to the initial ones after swelling [48].

The gelation kinetics as well as the final elastic valuables of these gels were influenced by the fragment of acetylated units, and more precisely the final valuables of the hydrogels were determined by the interchain interactions between these acetylated clusters.

2.4.3 Biomedical Valuables

Probably the most promising uses of chitin and chitosan refer to their use in biomedical, biotechnological, and pharmaceutical applications. For such applications, chitosan must present a high purity and the chemical composition and distribution of the molecular chains must be well determined. Chitosan has a biodegradable skeleton and therefore its degradation in the human body must be very well controlled. This is possible through rigorous control of crystallinity [49]. Also when it is used in tissue engineering or tissue reconstruction, a particularly important quality is biocompatibility.

2.4.3.1 Adsorption Capacity of Drugs

Valuables of chitosan such as bioadherence, biodegradability, and low toxicity are very important in controlled drug release applications. "In vitro" studies have shown the effect of chitosan's chemical composition, molecular mass, and molecular mass distribution on adsorption capacity [50].

It was thus possible to demonstrate that the most efficient adsorption occurred in the case of chitosan with a low fragment of acetylated units but with a high molecular mass.

2.4.3.2 Genetic Engineering

Gene transfer systems of nonviral origin have recently attracted attention as an alternative in the transport of DNA plasmids to cell nuclei in genetic engineering. A number of polycations, including synthetic polycations, polylysines, lipid cations, and more recently chitosan, have been successfully used as condensing agents in genetic engineering [51]. As in the case of systems with controlled drug release, the fragment of acetylated clusters, the molecular weight, and the degree of polydispersity decisively influence the ability of chitosan to be used in such applications.

2.4.3.3 Capability to Form Capsules

Since most living cells and tissues have a negative exchange surface, chitosan, used as a polycation, has a very good adhesion. Chitosan can also form complexes with a series of polyanions such as DNA, alginates, and mucin [52]. Chitosan with a low fraction of acetylated clusters shows much better adhesion, and changing the pH to higher values positively influences adhesion. Surface treatments are often used for better bioadherence.

2.5 Chitosan Modifications: Chitosan Derivatives of Chitosan

Chitosan, as well as other polyglucides, can be chemically modified in various ways to improve their valuables in order to meet certain specific requirements. However, there are numerous uses of chitosan and its derivatives (Table 2.2).

CSs are less susceptible to potential reactants than $(C_6H_{10}O_5)_n$, probably due to the characteristic crystalline supramolecular skeleton, with strong intermolecular bonds. The chemical reactions in which these two polyglucides take part are usually carried out with difficulty due to their limited solubility and reactivity. Thus, most of the time, the modification reactions of chitin and chitosan take place in heterogeneous conditions [53].

2.5.1 Hydrolysis of the Main Chain

2.5.1.1 Chemical Hydrolysis

The glycosidic bonds of chitosan are relatively stable to the action of alkalis, being, instead, cleaved by acids. Total hydrolysis with hot hydrochloric acid of chitosan results in D-glucosamine. Under mild conditions, through acid hydrolysis, mixtures of oligosaccharides can result. Hydrochloric acid is primarily used in the case of partial hydrolyzes, and the products are mainly oligomers with a degree of polymerization of 2–4 [54]. Hydrofluoric acid hydrolysis of chitosan results in obtaining chitooligosaccharides in the form of glucopyranosyl α-fluoride at the reduced end. Ultrasonic treatments facilitate chemical degradation.

2.5.1.2 Enzymatic Hydrolysis

Both CSs and chitosan are degraded in nature by a multitude of microorganisms. Most chitinases in microorganisms hydrolyze glucosamine N-acetyl-β (1 → 4) bonds randomly. Unlike chitinases, EC 3.2.1.132 or chitosanases (catalyzing the endo-hydrolysis of β1-4-glycosidic linkages between GlcNAc and GlcN) have been studied less. However, it is known that chitosan is susceptible to *E.C.* 3.5.1.52 or N-*glycanases*, (catalyzes the hydrolysis of a glycan), E.C. 3.1.1.3 (triacylglycerol hydrolases) or lipases (catalyze the hydrolysis of triacylglycerols [TAGs] to glycerol and fatty acids [FAs]), and EC 3.4.21 or/and proteases (enzymatic degradation of proteins/peptides by hydrolytic cleavage of the peptide bond in an exergonic reaction).

It was observed that by degrading chitosan with the help of chitosanase obtained from *Bacillus* sp. corresponding dimers and pentamers result without the formation of monomers, and the cellulase from *T. viride* hydrolyzes chitosan giving products that vary from hexamers to octamers, separated in the crystalline form [55].

2.5.1.3 Deacetylation and Adsorption of Metals

Chitosan, as well as chitin, show significant affinities for metal cations. Chitosan has a high capacity to collect heavy metal cations, being one of the best absorbents found in nature, and it absorbs a wide variety of metals (Cu, Hg, Cd, Fe, Ni, Zn, Pb, Ag) and highly toxic organo-mercury compounds [56]. The adsorption capacity of metals is greatly influenced by the degree of deacetylation (by the amount of free amino clusters) and hydrophilicity.

2.5.1.4 Acetylation

N-acetylation of chitosan can be easily controlled, when it occurs in aqueous solutions of CH_2COOH or in the state of highly swollen gel, in pyridine. In this case, a degree of N-acetylation of 50% was reached, the products being soluble in water.

Furthermore, no appreciable degradation was observed during chitosan acetylation. Introduction of acetyl clusters to the amino function is effective using a highly swollen gel of chitosan in pyridine [C_5H_5N] with acetyl chloride [CH_3COCl] [57].

The reactions can be controlled with an amount of acid chlorides to give two series of acylated products with different degrees of substitution.

2.5.1.5 Acylation

Chitosan can be acylated with carboxylic acid chlorides, with a long aliphatic chain (hexanoyl, nonanoyl, dedecanoyl, tetradecanoyl chlorides), in a mixture of pyridine and chloroform to form derivatives with degrees of acylation. Due to the presence of the free amino assembly and the toxicity low, chitosan is an efficient carrier for biologically active species. To carry out coupling reactions to the amino assembly of chitosan, in organic solvents, it can be acylated, to obtain derivatives with increased solubility in common polar solvents. Thus, the 6-O-triphenylmethyl-chitosan derivative, obtained by acylation, can be conjugated with the pentapeptide Tyr-Ile-Gly-Ser-Arg (YIGSR), to obtain a product with significant antitumor activity demonstrated against lung metastases with B16BL6 murine melanoma cells in C57BL/6 mice. The pentapeptide [$C_{21}H_{35}N_6O_{10}$] having a laminin sequence, which is known to play a role in the metastasis of tumor cells, itself inhibits the formation of metastases in vitro; however, its digestion by proteases in vivo requires immobilization on appropriate polymeric supports for promising results [58].

2.5.1.6 N-phthaloylation of Chitosan

Phthaloylation

The functionality of the amino clusters can be protected by the phthaloyl assembly. Phthaloylation can be a practical way of solubilization in organic solvents, since the phthaloyl assembly removes the hydrogen from the amino assembly, preventing the formation of the hydrogen bond. Thus, the N-phthaloyl assembly can be indispensable both for protection and solubilization [59]. N-phthaloylation of total deacetylated chitosan can be obtained by treating a suspension of chitosan in dimethyl formamide [C_3H_7NO] with an excess of phthalic anhydride [$C_8H_4O_3$], at 120–130 °C.

Modification of phthaloyl-chitosan

The phthaloyl assembly can be removed by reaction with hydrazine to regenerate the free amino assembly. The main chain of chitosan is considered to be sufficiently rigid to facilitate the formation of mesophases, suggested by the liquid crystal valuables of chitosan in acidic aqueous solutions. Both N-phthaloyl-chitosan and the acetylated derivative 3,6-di-O-acetyl-N-phthaloyl-chitosan form lyotropic liquid crystal phases in Dimethyl sulfoxide; [$(CH_3)_2SO$] abbreviation DMSO [60]. The thermal behavior of the gel phases in the lyotropic liquid crystal of the above derivative reveals that the sol-gel transition is reversible with significant temporal hysteresis.

Phthaloyl-chitosan-based branching

One of the most important modifications of N-phthaloyl-chitosan is the introduction of saccharide branches in chitosan to obtain synthetic branched polyglucides. These polyglucides have been shown to have unique physical valuables and biological activities. They are very hygroscopic and show high degrees of water absorption and retention [61]. Also, they seem to be susceptible to lysozymes, enzymatically degradable, and show increased antimicrobial activity.

Coughing

The improved solubility observed in chitosan derivatives with block clusters indicates that the introduction of such clusters effectively destroys the tight arrangement of the polyglucide molecules in the crystalline solid state. Since the radical of p-toluene sulfonic acid (tosyl) is a block assembly and its ester is highly reactive, it makes it a good candidate for substitution in order to obtain a soluble precursor that gives controllable reactive modifications [62].

2.5.1.7 Formation of Schiff Bases

The free amino assembly of chitosan in reaction with an aldehyde leads to the formation of the corresponding Schiff base. Such a reaction can take place in a mixture of CH_2COOH and methanol, sometimes reaching high degrees of substitution, close to 1. During the reaction, the solution gels due to the low solubility of the resulting Schiff base derivatives. Aldehydes can be used to protect the amino clusters of chitosan, allowing only the hydroxyl clusters to be modified [63]. Thus, through a series of reactions, fully acylated chitin can be obtained.

2.5.1.8 Reductive Alkylation

The best way to selectively introduce alkyl clusters or substituted alkyl clusters to the amino clusters of chitosan is reductive alkylation. Thus, chitosan is treated with aldehyde or ketone resulting in an imine (Schiff base) which can be easily transformed into N-alkyl derivative by reduction with sodium borohydride or sodium cyanoborohydride. In the presence of reducing sugars (glucose, galactose, lactose, and cellobiose) reductive alkylation occurs easily, with the introduction of saccharide clusters in chitosan [64]. The reductive alkylation of chitosan with salicylaldehyde or phthalaldehyde acid should also be considered here in order to increase the adsorption capacity for metals, obtaining a substantial improvement in chelation.

2.5.1.9 O-carboxymethylation

The introduction of the carboxymethyl assembly in chitin or chitosan leads to the formation of anionic derivatives. Carboxymethylation of chitin takes place by treating alkali-chitin with monochloro CH_2COOH, the reaction being greatly facilitated by the use of a suitable organic solvent. It is assumed that the carboxymethylation reaction occurs preferentially at C-6, as observed from the results of the constitutional analysis of the monomeric unit obtained by hydrolysis [65].

Chitosan can be carboxymethylated under reasonable conditions, but in this case, the carboxymethyl clusters are introduced both to the hydroxyl and to the amino clusters with the formation of N, O-carboxymethyl chitosan. During the reaction, N-deacetylation takes place along with the formation of polymers, having both carboxyl and amino clusters.

2.5.1.10 Radical Copolymerization

This copolymerization reaction, which has recently seen progress, introduces new side chains to make possible the existence of several different molecular arrangements, allowing the emergence of new types of hybrid materials, composed of natural polyglucides and synthetic polymers. The valuables of the radical copolymerization products can be easily controlled by the characteristics of the side chains (molecular skeleton, length, and number) [66]. Radical copolymers based on chitosan and chitin could have new applications in certain fields such as water treatment, adsorption of metal cations, medicine, and agriculture [67].

Poly(ethylene glycol) [(PEG) $C_{2n}H_{4n+2}O_{n+1}$] is a unique polymer, which presents multiple characteristics, including solubility in water and organic solvents, forms complexes with metal cations, and has low toxicity and good biocompatibility. That is why the most numerous studies have been done on the grafting of chitosan with PEG, from the point of view of the various possible applications. When chitosan is grafted with polyaniline, products are obtained from which films or fibers can be made, the valuables of which vary depending on the percentage of aniline.

Chitin graft copolymerization with methyl acrylate, methyl methacrylate, and vinyl monomers (acrylamide, acrylic acid) was experimented with Ce (IV) and styrene by gamma irradiation. The products obtained by copolymerization with vinyl monomers demonstrated a much improved swelling and solubilization behavior, having a much higher hygroscopicity compared to chitin, and in the case of copolymerization with methyl acrylate and methyl methacrylate, it was observed that at a grafting percentage of over 400%, the swelling became very pronounced, giving rise to transparent films [68].

2.5.1.11 Chitosan Derivatives Soluble in Water at Neutral pH

Many of the applications of any chemical matter, natural or synthetic, require chemical processability. Thus, chitosan, a white powder, is difficult to handle due to solubility problems in neutral water, bases, and organic solvents. The pKa value of the primary amino clusters in chitosan is 6.5. So, even if chitosan and its derivatives are soluble at a pH lower than 6, most of its applications in basic or neutral environments cannot be realized [69]. On the other hand, the acidic solutions in which chitosan is soluble are not compatible with many applications, such as those in cosmetics, medicine, and food. There are two approaches in the literature about improving the solubility of chitosan at neutral pH [70].

The first is the chemical derivatization of chitosan (for example with substituents containing a quaternary ammonium assembly, by carboxymethylation or sulfation) so that the added substituent is hydrophilic. The second method uses chitosan with a diacetylation of 50% prepared by homogeneous processing of chitin. Under homogeneous processing conditions, the obtained chitosan remains in the solution after neutralization and there is no need for derivatization [71]. However, some applications of chitosan use its derivatized forms, and in order to improve solubility, the introduction of ionic clusters into the biopolymer skeleton is required.

2.5.1.12 Quaternary Derivatives of Chitinous Matters

The simplest derivative of this class is the trimethylammonium salt of chitosan. Repeated treatment of chitosan in N-methyl-2-pyrrolidone (NMP) containing sodium iodide and methyl iodide in the presence of sodium hydroxide results in the trimethylammonium salt of chitosan with a high degree of substitution. For reasons of stability, an ion exchange is required (iodide with chloride). The resulting product is soluble in water at neutral pH. Daly and Guerrini developed a method using the commercial Quat-188 [(3-chloro-2-hydroxy)-propyltrimethylammonium chloride] salt in the preparation of derivatives [72].

The second approach produces the glycidyl reagent in situ and reduces the concerns associated with using epoxies on a large scale. The derived Chit-Quat has an excellent solubility in water. The process of introducing quaternary ammonium substituents into the chitosan skeleton is called quaternization, its agent being Quat-188. The triethylaminoethyl derivative of chitin [triethylaminoethyl–chitin (TEAE–chitin)] has a good solubility in water at neutral pH, being prepared by activating the C-6 primary hydroxy assembly in chitin at low temperature to minimize deacetylation [73]. Activated chitin was subjected to the reaction with Diethylaminoethyl Chloride [$C_6H_{15}Cl_2N$], followed by quaternization using ethyl iodide [74]. Interestingly, the diethylaminoethylation (DEAE)-Chitin intermediate was soluble at neutral pH.

2.5.1.13 Carboxyalkylated Derivatives of Chitosan

Through the carboxyalkylation process, acidic clusters are introduced into the biopolymer skeleton. But due to the presence of free amino clusters in chitosan, these derivatives show amphotericity. So they have an isoelectric point seen in other amphoteric molecules such as polypeptides, where the biopolymer has no net electrical charge. At the isoelectric pH, the biopolymer remains undissolved. Solubility in water is obtained at pH values below or above the isoelectric point. Both chitosan derivatives, N-carboxyalkyl and O-carboxyalkyl, were prepared by using different reaction conditions to obtain N vs. selectivity, more correctly O selectivity [75].

A third route of selective synthesis with the formation of N-carboxyalkylates uses carboxyaldehydes in a reductive amination sequence. The reaction is carried out in homogeneous conditions, if the aldehyde used is soluble in water, allowing to increase the degree of substitution that must be distributed along the skeleton of the biopolymer [76]. Sequential substitutions lead to the formation of bis-carboxymethyl derivatives by using glyoxylic acid. By their nature, the N-sulfonated derivatives of chitosan are amphoteric and can be prepared under heterogeneous reaction conditions by using 2-sulfobenzoic acid anhydride.

2.5.1.14 Other Derivatives

Other non-ionic, polar derivatives of chitosan were also prepared which showed solubility in water at neutral pH. The polar substituent used is either of carbohydrate origin (with some hydrophilic hydroxy clusters) or of a polar biopolymer (e.g. polyacrylic acid) from the chitosan skeleton. Products with ionizable chain amino acid substituents (amino clusters and carboxylic acid) also showed solubility in water at neutral pH. Commercial production of chitin and chitosan is mainly based on waste from marine industries (e.g. fishing, shellfish harvesting, etc.). Different raw material sources result in different qualities of chitin and chitosan, resulting in production and quality fluctuations due to seasonal fishing variations. In addition, chitosan derived from fungal sources offers superior reporting valuables such as homogeneous polymer length and higher solubility compared to chitosan from marine sources [77]. To provide uniform chitosan, extraction of chitin from fungal species has become a stable alternative production. The production of chitin and chitosan from mushrooms can be easily and reliably achieved using ultrasonic extraction and deacetylation technology.

2.6 Conclusions and Recommendations

Chitin and chitosan are complementary compounds: to obtain chitosan, chitin must be present. Certain enzymes (chitinases) in the digestive tract of certain predators digest it (birds and mammals for example). These enzymes can also be produced by the intestinal flora of the host, for example, the bacterial intestinal flora of the snail *Helix pomatia* which can thus digest the fungi. Finding chitin in an organism does not always mean that it produced it on its own; it can come from its food or from an intestinal flora (endoflora) that secretes it. The chitin of dead insects or crustaceans is degraded more or less rapidly in ecosystems by communities of microorganisms and bacteria. Research confirms that chitin has extremely low toxicity. Side effects are possible in people with shellfish allergies.

Since chitosan is much easier to process than the chitin from which it comes, there are numerous systems described in the literature, used mainly in the biomedical field. However, it should be noted that the stability of these chitosan-based materials is lower, due to their more hydrophilic nature and sensitivity to pH. Numerous techniques (chemical modifications) are used to control the chemical and mechanical valuables of chitosan-based systems. Biodegradable polymers such as chitosan must be cross-linked to reach the performance of modulating their general valuables and giving them good stability over time. Certain reagents such as glutaric aldehyde, tripolyphosphate, ethylene glycol, diglycylether, and diisocyanate were used to cross-link chitosan. There are studies that have demonstrated that cross-linking agents are more or less cytotoxic, and thus can drastically decrease the biocompatibility of release systems. Thus, it is desirable that the cross-linker used in biomedical applications presents low toxicity and forms biocompatible and stable cross-linked products.

References

1. A.A. Bell, J.C. Hubbard, L. Liu, R.M. Davis, K.V. Subbarao, Effects of Chitin and Chitosan on the incidence and severity of fusarium yellows of celery. Plant Dis. **82**(3), 322–328 (1998). https://doi.org/10.1094/PDIS.1998.82.3.322
2. M. Habibizadeh, S. Nadri, A. Fattahi, K. Rostamizadeh, P. Mohammadi, S. Andalib, M. Hamidi, N. Forouzideh, Surface modification of neurotrophin-3 loaded PCL/chitosan nanofiber/net by alginate hydrogel microlayer for enhanced biocompatibility in neural tissue engineering. J. Biomed. Mater. Res. A **109**(11), 2237–2254 (2021). https://doi.org/10.1002/jbm.a.37208
3. T. Nagata, S. Shinya, T. Ohnuma, T. Fukamizo, Multi-functionality of a tryptophan residue conserved in substrate-binding groove of GH19 Chitinases. Sci. Rep. **11**(1), 2494 (2021). https://doi.org/10.1038/s41598-021-81903-3
4. S. Zhang, J. He, S. Xiong, Q. Xiao, Y. Xiao, F. Ding, H. Ji, Z. Yang, Z. Li, Construction and nanostructure of chitosan/nanocellulose hybrid aerogels. Biomacromol **22**(8), 3216–3222 (2021). https://doi.org/10.1021/acs.biomac.1c00266
5. H. Jayasantha Kumari, P. Krishnamoorthy, T.K. Arumugam, S. Radhakrishnan, D. Vasudevan, An efficient removal of crystal violet dye from waste water by adsorption onto TLAC/Chitosan composite: A novel low cost adsorbent. Int. J. Biol. Macromol. **96**, 324–333 (2017). https://doi.org/10.1016/j.ijbiomac.2016.11.077
6. M. Zhang, F. Zhang, C. Li, H. An, T. Wan, P. Zhang, Application of Chitosan and its derivative polymers in clinical medicine and agriculture. Polymers (Basel) **14**(5), 958 (2022). https://doi.org/10.3390/polym14050958
7. M. Rajabi, J. Cabral, S. Saunderson, M.A. Ali, Green synthesis of chitooligosaccharide-PEGDA derivatives through aza-Michael reaction for biomedical applications. Carbohydr Polym. **295**, 119884 (2022). https://doi.org/10.1016/j.carbpol.2022.119884

8. E.P. Milan, V.C.A. Martins, M.M. Horn, A.M.G. Plepis, Influence of blend ratio and mangosteen extract in chitosan/collagen gels and scaffolds: Rheological and release studies. Carbohydr Polym. **292**, 119647 (2022). https://doi.org/10.1016/j.carbpol.2022.119647
9. Ö. Varol Avcılar, E.E. Onbaşılar, A. Kocakaya, Effects of coated hatching eggs obtained from old broiler breeders with chitosan on embryonic growth, hatching results and chick quality. J. Anim. Physiol. Anim. Nutr. (Berl). **105**(5), 946–951 (2021). https://doi.org/10.1111/jpn.13506
10. F. Khan, D. Oh, P. Chandika, D.M. Jo, N.I. Bamunarachchi, W.K. Jung, Y.M. Kim, Inhibitory activities of phloroglucinol-chitosan nanoparticles on mono- and dual-species biofilms of *Candida albicans* and bacteria. Colloids Surf. B Biointerfaces. **211**, 112307 (2022). https://doi.org/10.1016/j.colsurfb.2021.112307
11. J. Wu, L. Zhang, Dissolution behavior and conformation change of chitosan in concentrated chitosan hydrochloric acid solution and comparison with dilute and semidilute solutions. Int. J. Biol. Macromol. **121**, 1101–1108 (2019). https://doi.org/10.1016/j.ijbiomac.2018.10.128
12. D. Hao, Y. Liang, Adsorption of Cu^{2+}, Cd^{2+} and Pb^{2+} in wastewater by modified chitosan hydrogel. Environ. Technol. **43**(6), 876–884 (2022). https://doi.org/10.1080/09593330.2020.1807612
13. P. Phuong, N.C. Minh, H.N. Cuong, N. Van Minh, N.T. Han, N. Van Hoa, H. Yen, T.S. Trung, Recovery of protein hydrolysate and chitosan from black tiger shrimp (*Penaeus monodon*) heads: approaching a zero waste process. J. Food Sci. Technol. **54**(7), 1850–1856 (2017). https://doi.org/10.1007/s13197-017-2616-6
14. M. Stasińska-Jakubas, B. Hawrylak-Nowak, Protective, biostimulating, and eliciting effects of chitosan and its derivatives on crop plants. Molecules **27**(9), 2801 (2022). https://doi.org/10.3390/molecules27092801
15. M. Li, H. Liu, J. Liu, Y. Pei, X. Zheng, K. Tang, F. Wang, Hydrophobic and self-recoverable cellulose nanofibrils/N-alkylated chitosan/poly(vinyl alcohol) sponge for selective and versatile oil/water separation. Int. J. Biol. Macromol. **192**, 169–179 (2021). https://doi.org/10.1016/j.ijbiomac.2021.09.189
16. A.J. Worthen, K.S. Irving, Y. Lapitsky, Supramolecular strategy effects on chitosan bead stability in acidic media: A comparative study. Gels **5**(1), 11 (2019). https://doi.org/10.3390/gels5010011
17. A.S. Kritchenkov, A.V. Kletskov, A.R. Egorov, A.G. Tskhovrebov, A.V. Kurliuk, N.V. Zhaliazniak, T.V. Shakola, V.N. Khrustalev, New water-soluble chitin derivative with high antibacterial properties for potential application in active food coatings. Food Chem. **343**, 128696 (2021). https://doi.org/10.1016/j.foodchem.2020.128696
18. O. Owoseni, Y. Su, S. Raghavan, A. Bose, V.T. John, Hydrophobically modified chitosan biopolymer connects halloysite nanotubes at the oil-water interface as complementary pair for stabilizing oil droplets. J. Colloid Interface Sci. **620**, 135–143 (2022). https://doi.org/10.1016/j.jcis.2022.03.142
19. I. Benucci, C. Lombardelli, I. Cacciotti, M. Esti, Papain covalently immobilized on chitosan-clay nanocomposite films: Application in synthetic and real white wine. Nanomaterials (Basel) **10**(9), 1622 (2020). https://doi.org/10.3390/nano10091622
20. N. Bian, X. Yang, X. Zhang, F. Zhang, Q. Hou, J. Pei, A complex of oxidised chitosan and silver ions grafted to cotton fibres with bacteriostatic properties. Carbohydr Polym. **262**, 117714 (2021). https://doi.org/10.1016/j.carbpol.2021.117714
21. D. Sikorski, K. Gzyra-Jagieła, Z. Draczyński, The kinetics of chitosan degradation in organic acid solutions. Mar. Drugs. **19**(5), 236 (2021). https://doi.org/10.3390/md19050236
22. Y. Wang, J. Ma, T. Qiu, M. Tang, X. Zhang, W. Dong, In vitro and in vivo combinatorial anticancer effects of oxaliplatin- and resveratrol-loaded N, O-carboxymethyl chitosan nanoparticles against colorectal cancer. Eur J Pharm Sci. **163**, 105864 (2021). https://doi.org/10.1016/j.ejps.2021.105864
23. E. Zhang, R. Xing, S. Liu, K. Li, Y. Qin, H. Yu, P. Li, Comparison in docetaxel-loaded nanoparticles based on three different carboxymethyl chitosans. Int. J. Biol. Macromol. **101**, 1012–1018 (2017). https://doi.org/10.1016/j.ijbiomac.2017.03.195

24. W. Wang, Q. Meng, Q. Li, J. Liu, M. Zhou, Z. Jin, K. Zhao, Chitosan derivatives and their application in biomedicine. Int. J. Mol. Sci. **21**(2), 487 (2020). https://doi.org/10.3390/ijms21020487
25. I. Badreddine, C. Lafitte, L. Heux, N. Skandalis, Z. Spanou, Y. Martinez, M.T. Esquerré-Tugayé, V. Bulone, B. Dumas, A. Bottin, Cell wall chitosaccharides are essential components and exposed patterns of the phytopathogenic oomycete *Aphanomyces euteiches*. Eukaryot Cell **7**(11), 1980–1993 (2008). https://doi.org/10.1128/EC.00091-08
26. M.M. Afroz, M.N.H. Kashem, K.M.P.S. Piash, N. Islam, Saccharomyces cerevisiae as an untapped source of fungal chitosan for antimicrobial action. Appl. Biochem. Biotechnol. **193**(11), 3765–3786 (2021). https://doi.org/10.1007/s12010-021-03639-0
27. S. Rathinam, S. Ólafsdóttir, S. Jónsdóttir, M.A.Hjálmarsdóttir, M. Másson, Selective synthesis of N,N,N-trimethylated chitosan derivatives at different degree of substitution and investigation of structure-activity relationship for activity against P. aeruginosa and MRSA. Int. J. Biol. Macromol. **160**, 548–557 (2020). https://doi.org/10.1016/j.ijbiomac.2020.05.109
28. N. Yan, X.F. Wan, X.S. Chai, Simultaneous determination of the degree of deacetylation and substitution on carboxymethyl chitosan by headspace gas chromatography. J. Agric. Food Chem. **67**(31), 8700–8705 (2019). https://doi.org/10.1021/acs.jafc.9b02974
29. S. Nishida, M. Shibano, H. Kamitakahara, T. Takano, Basic study for acyl chitosan isothiocyanates synthesis by model experiments using glucosamine derivatives. Int. J. Biol. Macromol. **132**, 17–23 (2019). https://doi.org/10.1016/j.ijbiomac.2019.03.114
30. S. Kumari, S.H. Kumar Annamareddy, S. Abanti, R.P. Kumar, Physicochemical properties and characterization of chitosan synthesized from fish scales, crab and shrimp shells. Int. J. Biol. Macromol. **104**(Pt B), 1697–1705 (2017). https://doi.org/10.1016/j.ijbiomac.2017.04.119
31. M. Psarianos, S. Ojha, R. Schneider, O.K. Schlüter, Chitin isolation and chitosan production from house crickets (*Acheta domesticus*) by environmentally friendly methods. Molecules **27**(15), 5005 (2022). https://doi.org/10.3390/molecules27155005
32. N. Nwe, T. Furuike, H. Tamura, Isolation and characterization of chitin and chitosan from marine origin. Adv. Food Nutr. Res. **72**, 1–15 (2014). https://doi.org/10.1016/B978-0-12-800269-8.00001-4
33. H.A. Said Al Hoqani, N. Al-Shaqsi, M.A.Hossain, M.A. Al Sibani, Isolation and optimization of the method for industrial production of chitin and chitosan from Omani shrimp shell. Carbohydr. Res. **492**, 108001 (2020). https://doi.org/10.1016/j.carres.2020.108001
34. R.M. Abdel-Rahman, R. Hrdina, A.M. Abdel-Mohsen, M.M. Fouda, A.Y. Soliman, F.K. Mohamed, K. Mohsin, T.D. Pinto, Chitin and chitosan from Brazilian Atlantic coast: Isolation, characterization and antibacterial activity. Int. J. Biol. Macromol. **80**, 107–120 (2015). https://doi.org/10.1016/j.ijbiomac.2015.06.027
35. T.S. Trung, L.H. Tram, N. Van Tan, N. Van Hoa, N.C. Minh, P.T. Loc, W.F. Stevens, Improved method for production of chitin and chitosan from shrimp shells. Carbohydr. Res. **489**, 107913 (2020). https://doi.org/10.1016/j.carres.2020.107913
36. J. Synowiecki, N.A. Al-Khateeb, Production, properties, and some new applications of chitin and its derivatives. Crit. Rev. Food Sci. Nutr. **43**(2), 145–171 (2003). https://doi.org/10.1080/10408690390826473
37. H. Zhang, S. Yun, L. Song, Y. Zhang, Y. Zhao, The preparation and characterization of chitin and chitosan under large-scale submerged fermentation level using shrimp by-products as substrate. Int. J. Biol. Macromol. **96**, 334–339 (2017). https://doi.org/10.1016/j.ijbiomac.2016.12.017
38. F. Samadi, Z. Es'haghi, Determination of phthalate esters in cosmetics and baby care products by a biosorbent based on lawsone capped chitosan and followed by liquid chromatography. J. Chromatogr. Sci. **60**(3), 287–297 (2022). https://doi.org/10.1093/chromsci/bmab062
39. J. Brokešová, M. Slámová, P. Zámostný, M. Kuentz, J. Koktan, L. Krejčík, B. Vraníková, P. Svačinová, Z. Šklubalová, Mechanistic study of dissolution enhancement by interactive mixtures of chitosan with meloxicam as model. Eur. J. Pharm. Sci. **169**, 106087 (2022). https://doi.org/10.1016/j.ejps.2021.106087
40. P. Deng, J. Chen, L. Yao, P. Zhang, J. Zhou, Thymine-modified chitosan with broad-spectrum antimicrobial activities for wound healing. Carbohydr. Polym. **257**, 117630 (2021). https://doi.org/10.1016/j.carbpol.2021.117630]

41. W. Zhu, T. Chen, R. He, Y. Ding, T. Duan, B. Xiao, Understanding the interfacial interactions of bioinspired chitosan-calcite nanocomposites by first principles molecular dynamics simulations and experimental FT-IR spectroscopy. Carbohydr. Polym. **223**, 115054 (2019). https://doi.org/10.1016/j.carbpol.2019.115054
42. B. Gieroba, A. Sroka-Bartnicka, P. Kazimierczak, G. Kalisz, A. Lewalska-Graczyk, V. Vivcharenko, R. Nowakowski, I.S. Pieta, A. Przekora, Surface chemical and morphological analysis of chitosan/1,3-β-d-glucan polysaccharide films cross-linked at 90 °C. Int. J. Mol. Sci. **23**(11), 5953 (2022). https://doi.org/10.3390/ijms23115953
43. D. Ren, Z. Shang, M. Zhang, S. Xu, Z. Xu, The effect of chitosan molecular weight on CO_2-triggered switching between emulsification and demulsification. Soft Matter **17**(41), 9332–9338 (2021). https://doi.org/10.1039/d1sm01036k
44. G.E. Chaudhry, C.S. Thirukanthan, K.M. NurIslamiah, Y.Y. Sung, T.S.M. Sifzizul, A.W.M. Effendy, Characterization and cytotoxicity of low-molecular-weight chitosan and chito-oligosaccharides derived from tilapia fish scales. J. Adv. Pharm. Technol. Res. **12**(4), 373–377 (2021). https://doi.org/10.4103/japtr.japtr_117_21
45. J. Long, E. Xu, X. Li, Z. Wu, F. Wang, X. Xu, Z. Jin, A. Jiao, X. Zhan, Effect of chitosan molecular weight on the formation of chitosan-pullulanase soluble complexes and their application in the immobilization of pullulanase onto Fe_3O_4-κ-carrageenan nanoparticles. Food Chem. **202**, 49–58 (2016). https://doi.org/10.1016/j.foodchem.2016.01.119
46. R. Kumar, K.H. Prakash, P. Cheang, L. Gower, K.A. Khor, Chitosan-mediated crystallization and assembly of hydroxyapatite nanoparticles into hybrid nanostructured films. J. R. Soc. Interface **5**(21), 427–439 (2008). https://doi.org/10.1098/rsif.2007.1141
47. C. Dong, W. Chen, C. Liu, Flocculation of algal cells by amphoteric chitosan-based flocculant. Bioresour. Technol. **170**, 239–247 (2014). https://doi.org/10.1016/j.biortech.2014.07.108
48. K. Kadiya, M. Sharma, S. Ghosh, Effect of the chitosan second layer on the gelation and controlled digestion of Citrem-chitosan bilayer emulsions. Food Funct. **13**(5), 2515–2533 (2022). https://doi.org/10.1039/d1fo02409d
49. W. Feng, Z. Wang, Biomedical applications of chitosan-graphene oxide nanocomposites. iScience **25**(1), 103629 (2021). https://doi.org/10.1016/j.isci.2021.103629
50. A.G. Díaz, D.A. Quinteros, J.M. Llabot, S.D. Palma, D.A. Allemandi, G. Ghersi, V. Zylberman, F.A. Goldbaum, S.M. Estein, Spray dried microspheres based on chitosan: A promising new carrier for intranasal administration of polymeric antigen BLSOmp31 for prevention of ovine brucellosis. Mater. Sci. Eng. C Mater. Biol. Appl. **62**, 489–496 (2016). https://doi.org/10.1016/j.msec.2016.01.084
51. C. Jiang, S. Xu, S. Zhang, L. Jia, Chitosan functionalized magnetic particle-assisted detection of genetically modified soybeans based on polymerase chain reaction and capillary electrophoresis. Anal. Biochem. **420**(1), 20–25 (2012). https://doi.org/10.1016/j.ab.2011.09.004
52. H. Lu, S. Ren, X. Li, J. Guo, G. Dong, J. Li, L. Gao, Poly(ethylene glycol)/chitosan/sodium glycerophosphate gel replaced the joint capsule with slow-release lubricant after joint surgery. J. Biomater. Sci. Polym. Ed. **29**(11), 1331–1343 (2018). https://doi.org/10.1080/09205063.2018.1459351
53. E.Y. Wardhono, M.P. Pinem, I. Kustiningsih, M. Effendy, D. Clausse, K. Saleh, E. Guénin, Heterogeneous deacetylation reaction of chitin under low-frequency ultrasonic irradiation. Carbohydr. Polym. **267**, 118180 (2021). https://doi.org/10.1016/j.carbpol.2021.118180
54. E. Guibal, C. Milot, O. Eterradossi, C. Gauffier, A. Domard, Study of molybdate ion sorption on chitosan gel beads by different spectrometric analyses. Int. J. Biol. Macromol. **24**(1), 49–59 (1999). https://doi.org/10.1016/s0141-8130(98)00067-1
55. S.L. Wang, C.P. Liu, T.W. Liang, Fermented and enzymatic production of chitin/chitosan oligosaccharides by extracellular chitinases from Bacillus cereus TKU027. Carbohydr. Polym. **90**(3), 1305–1313 (2012). https://doi.org/10.1016/j.carbpol.2012.06.077
56. W. Shen, Y. Fang, M. Azeem, Y. Gao, X. Li, P. Zhao, A. Ali, M. Li, R. Li, Chitosan crosslinked with polyamine-co-melamine for adsorption of Hg^{2+}: Application in purification of polluted water. Int. J. Biol. Macromol. **181**, 778–785 (2021). https://doi.org/10.1016/j.ijbiomac.2021.03.166

57. I.M. Lipatova, A.A. Yusova, N.V. Losev, E.A. Indeikin, Gelation in solutions of low deacetylated chitosan initiated by high shear stresses. Int. J. Biol. Macromol. **139**, 550–557 (2019). https://doi.org/10.1016/j.ijbiomac.2019.07.164
58. B. Huang, Z. Zhang, N. Ding, Y. Zhuang, G. Zhang, P. Fei, Preparation of acylated chitosan with caffeic acid in non-enzymatic and enzymatic systems: Characterization and application in pork preservation. Int. J. Biol. Macromol. **194**, 246–253 (2022). https://doi.org/10.1016/j.ijbiomac.2021.11.193
59. K. Kurita, H. Ikeda, Y. Yoshida, M. Shimojoh, M. Harata, Chemoselective protection of the amino groups of chitosan by controlled phthaloylation: Facile preparation of a precursor useful for chemical modifications. Biomacromol **3**(1), 1–4 (2002). https://doi.org/10.1021/bm0101163
60. L. Liu, Y. Wang, X. Shen, Y. Fang, Preparation of chitosan-g-polycaprolactone copolymers through ring-opening polymerization of epsilon-caprolactone onto phthaloyl-protected chitosan. Biopolymers **78**(4), 163–170 (2005). https://doi.org/10.1002/bip.20261
61. C. Chen, S. Tao, X. Qiu, X. Ren, S. Hu, Long-alkane-chain modified N-phthaloyl chitosan membranes with controlled permeability. Carbohydr. Polym. **91**(1), 269–276 (2013). https://doi.org/10.1016/j.carbpol.2012.08.042
62. B. Harini, S. Rajeshkumar, A. Roy, Biomedical application of chitosan and piper longum-assisted nano zinc oxide-based dental varnish. Appl. Biochem. Biotechnol. **194**(3), 1303–1309 (2022). https://doi.org/10.1007/s12010-021-03712-8
63. A. Pestov, Y. Privar, A. Slobodyuk, A. Boroda, S. Bratskaya, Chitosan cross-linking with acetaldehyde acetals. Biomimetics (Basel) **7**(1), 10 (2022). https://doi.org/10.3390/biomimetics7010010
64. P. Sahariah, B.E. Benediktssdóttir, M.Á. Hjálmarsdóttir, O.E. Sigurjonsson, K.K. Sørensen, M.B. Thygesen, K.J. Jensen, M. Másson, Impact of chain length on antibacterial activity and hemocompatibility of quaternary N-alkyl and n, n-dialkyl chitosan derivatives. Biomacromol **16**(5), 1449–1460 (2015). https://doi.org/10.1021/acs.biomac.5b00163
65. A.P. Zhu, N. Fang, Adhesion dynamics, morphology, and organization of 3T3 fibroblast on chitosan and its derivative: the effect of O-carboxymethylation. Biomacromol **6**(5), 2607–2614 (2005). https://doi.org/10.1021/bm050328q
66. W. Al-Mughrabi, A.O. Al-Dossary, A. Abdel-Naby, Free radical copolymerization of diallylamine and itaconic acid for the synthesis of chitosan base superabsorbent. Polymers (Basel) **14**(9), 1707 (2022). https://doi.org/10.3390/polym14091707
67. S. Fang, G. Wang, P. Li, R. Xing, S. Liu, Y. Qin, H. Yu, X. Chen, K. Li, Synthesis of chitosan derivative graft acrylic acid superabsorbent polymers and its application as water retaining agent. Int. J. Biol. Macromol. **115**, 754–761 (2018). https://doi.org/10.1016/j.ijbiomac.2018.04.072
68. J. Bi, C. Tian, G.L. Zhang, H. Hao, H.M. Hou, Novel procyanidins-loaded chitosan-graft-polyvinyl alcohol film with sustained antibacterial activity for food packaging. Food Chem. **365**, 130534 (2021). https://doi.org/10.1016/j.foodchem.2021.130534
69. D. Zhu, H. Cheng, J. Li, W. Zhang, Y. Shen, S. Chen, Z. Ge, S. Chen, Enhanced water-solubility and antibacterial activity of novel chitosan derivatives modified with quaternary phosphonium salt. Mater. Sci. Eng. C Mater. Biol. Appl. **61**, 79–84 (2016). https://doi.org/10.1016/j.msec.2015.12.024
70. Y. Shi, L. Liu, M. Yin, Z. Zhao, Y. Liang, K. Sun, Y. Li, Mucus- and pH-mediated controlled release of core-shell chitosan nanoparticles in the gastrointestinal tract for diabetes treatment [published online ahead of print. J. Drug Target 1–9 (2022). https://doi.org/10.1080/1061186X.2022.2104296
71. Y.A. Skorik, C.A. Gomes, M.T. Vasconcelos, Y.G. Yatluk, N-(2-Carboxyethyl)chitosans: regioselective synthesis, characterisation and protolytic equilibria. Carbohydr. Res. **338**(3), 271–276 (2003). https://doi.org/10.1016/s0008-6215(02)00432-9
72. W. Sajomsang, P. Gonil, S. Tantayanon, Antibacterial activity of quaternary ammonium chitosan containing mono or disaccharide moieties: Preparation and characterization. Int. J. Biol. Macromol. **44**(5), 419–427 (2009). https://doi.org/10.1016/j.ijbiomac.2009.03.003

73. S. Cao, P. Gao, W. Xia, S. Liu, B. Wang, A novel chitosanase from penicillium oxalicum M2 for chitooligosaccharide production: Purification Identification and Characterization. Mol. Biotechnol. **64**(9), 947–957 (2022). https://doi.org/10.1007/s12033-022-00461-9
74. S. Keely, A. Rullay, C. Wilson, A. Carmichael, S. Carrington, A. Corfield, D.M. Haddleton, D.J. Brayden, In vitro and ex vivo intestinal tissue models to measure mucoadhesion of poly (methacrylate) and N-trimethylated chitosan polymers. Pharm. Res. **22**(1), 38–49 (2005). https://doi.org/10.1007/s11095-004-9007-1
75. H. Jing, X. Du, L. Mo, H. Wang, Self-coacervation of carboxymethyl chitosan as a pH-responsive encapsulation and delivery strategy. Int. J. Biol. Macromol. **192**, 1169–1177 (2021). https://doi.org/10.1016/j.ijbiomac.2021.10.072
76. A. Ortega, A. Sánchez, G. Burillo, Binary graft of poly(*N*-vinylcaprolactam) and poly(acrylic acid) onto chitosan hydrogels using ionizing radiation for the retention and controlled release of therapeutic compounds. Polymers (Basel) **13**(16), 2641 (2021). https://doi.org/10.3390/polym13162641
77. L. Ding, Z. Liu, S. Song, M.A. Abubaker, L. Chen, J. Shi, Z. Fan, J. Zhang, Structural characteristics and rheological properties of hydroxypropyl trimethyl ammonium chloride chitosan. Int. J. Biol. Macromol. **216**, 312–321 (2022). https://doi.org/10.1016/j.ijbiomac.2022.06.175

Chapter 3
A Medico-Biological Approach to Chitosan Bionanocomposites

Namrata Khanna, Tanushri Chatterji, Suruchi Singh, and Nivedita Khanna

Abstract Significance of biopolymers like chitosan and its nanosized derivatives and the procedure of their implementation in various disciplines, particularly in medicine, is continuously increasing. The distinctive biological, physicochemical, physiological and ecological properties like biodegradability, biocompatibility, reduced toxicity, environmental stability, enhanced metabolic activity, cost-effectiveness, capability to chelate ions and elevated absorption make chitosan a promising prospect for use in various biomedical applications. Hydroxyl and amino functional groups make chitosan a propitious agent for several biomedical processes like precision carrier of drugs, bone or tissue engineering, healing of wounds, alternative medicine, haemostasis, dental procedures, cosmetic treatment and in various biotechniques. Chitosan and derivatives are also involved in mucosal delivery of vaccines. Nanoparticles conjugated with chitosan have immense scope for development of innovative drugs and vaccines. Based on recent research, the current chapter elaborates the techniques employed for preparation of chitosan-based nanocomposites.

N. Khanna
Department of Biochemistry, M A Rangoonwala College of Dental Sciences and Research Centre, 2390-B, K.B. Hidayatullah Road, Azam Campus, Camp, Pune 411001, Maharashtra, India

T. Chatterji (✉)
School of Bioscience, Institute of Management Studies (University Courses Campus), Adhyatmik Nagar, NH-09, Ghaziabad 201015, UP, India
e-mail: tanushri@gmail.com

S. Singh
Life Sciences, Tata Consultancy Services Limited, Hiranandani Estate, Thane (W), Mumbai 400607, Maharashtra, India

N. Khanna
Scientist-Biopharmaceutical Services, SGS Canada Inc, 6490 Vipond Drive, Mississauga, ON L5T 1WB, Canada

S. K. Swain and A. Biswal (eds.), *Chitosan Nanocomposites*,
Biological and Medical Physics, Biomedical Engineering,
https://doi.org/10.1007/978-981-19-9646-7_3

3.1 Introduction

Chitosan is a biodegradable polysaccharide, synthesized by deacetylation from chitin in an alkaline environment or in presence of chitin deacetylase through hydrolysis. Nanomaterials derived from chitosan have allured biomedical researchers worldwide by virtue of their distinctive non-toxic, biocompatible, biodegradable, antineoplastic, immunoenhancing and antimicrobial properties. Chitosan and products of its enzymatic hydrolysis interact with cells but do not cause any adverse reaction in the body. Chitosan reduces absorption of cholesterol, blocks free radical chain reactions and is a potent microbicidal agent, acting against several fungi, bacteria and yeasts. Chitosan, by itself or with polymers and metallic nanostructured composite materials, has a widespread use in several biomedical procedures, as a microbicidal agent in wound dressing; as a nanostructured carrier for delivery of drugs to malignant tissue, scarcely affecting the normal tissue; for delivery of genes; in dental procedures; for haemodialysis and for immobilization of enzymes.

Chitosan reactive groups are primary amino group on carbon 2 and primary, secondary hydroxyl groups on carbons 6 and 3 (Fig. 3.1). Other functional groups are glycosidic bonds and the acetamide moiety. All functional groups are capable of modification, forming polymers of different properties and behaviour. Several derivatives of chitosan have been developed to further improve chitosan's features or introduce innovative features. For example, water solubility of chitosan is enhanced by several processes, specially depolymerization, deacetylation or quaternization [1]. Modified chitosan exhibits innovative activities, for instance, 6-O-sulphated chitosan enhances differentiation of neurons and phosphorylated chitosan inhibits corrosion [2, 3].

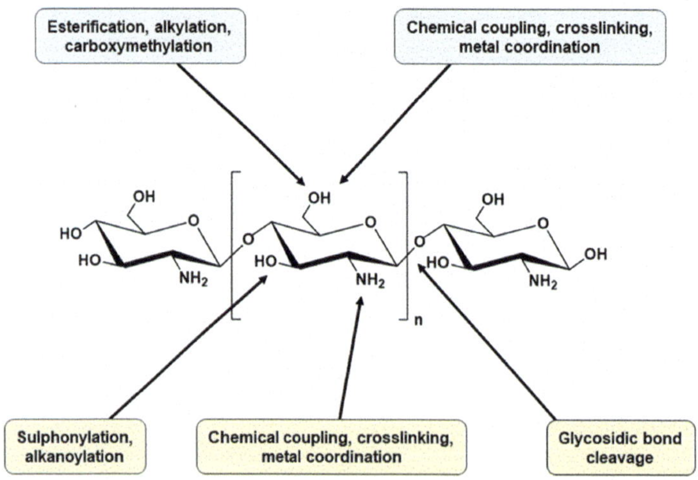

Fig. 3.1 Chemically modifiable functional groups in chitosan, used for preparing chitosan-based nanocomposites

Many diverse viewpoints regarding the antibacterial effect and methodology employed by chitosan-based nanomaterials are described in the current chapter. Chitosan and its nanostructured derivatives increase permeability of drugs through the blood–brain barrier, hence are effective carriers of drugs to the brain. Chitosan-based nanomaterials are used for uptake and transfer of drugs to target tissues and their utilization in many other biomedical procedures are also reviewed.

3.2 Unique Biological Properties of Chitosan Employed for Different Biomedical Applications

Chitosan basically is a linear homopolysaccharide. It consists of repeating units of N-acetyl glucosamine, which are linked through β-(1, 4) glycosidic bonds. It is obtained through deacetylation of chitin, which is an unbranched glucose-derived homopolymer. Chitin is present in crustaceans, in exoskeleton of many insects and in cell wall of a few bacteria and fungi [4]. Chitosan exhibits different qualities depending on origin of chitin and its depolymerization and deacetylation. For instance, chitin extracted from three different crustaceans, *Paralomis granulosa, Lithodes antarcticus and Palinurus vulgaris,* vary in physiochemical properties and crystallinity, hence the resulting chitosan exhibits different characteristics [5]. Chitosan exhibits unique properties, which makes it a significant tool for several biomedical applications. A few exceptional characteristics are discussed in the following sections and represented in Fig. 3.2.

Fig. 3.2 Chitosan with distinctive properties has a significant role in biomedical applications

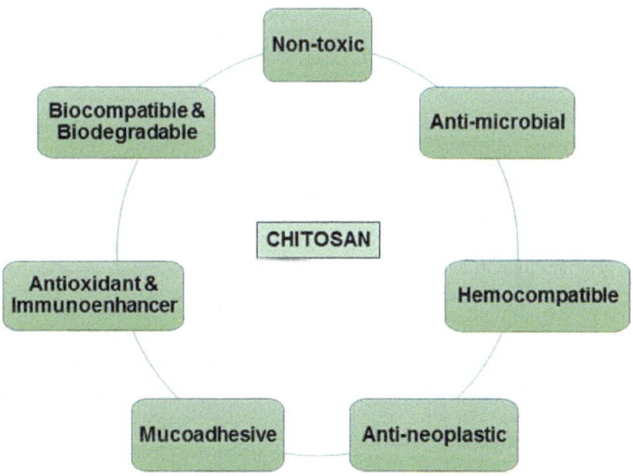

3.2.1 Non-toxicity

Chitosan has a remarkable property of preventing excessive inflammation and averting immune responses. Investigations reveal that chitosan varying in molecular weight and level of deacetylation exhibits reduced toxicity, which does not change even after uptake of nanoparticles (NPs) or in chitosan functionalization with succinate [6, 7].

3.2.2 Antimicrobial Activity

Several hypotheses explain the antimicrobial action of chitosan. The most accepted one states that positive amino group of glucosamine group in chitosan interacts with negative charge on bacterial surface, causing leakage of cellular material and resulting in bacterial death [8]. Another study on E.coli suggests that chitosan binds with DNA, preventing transcription and subsequent mRNA entry into the nucleus [9]. Chitosan molecules have chelating characteristics and bind to essential metal ions to prevent bacterial growth. For instance, amino group reacts with calcium and magnesium ions on bacterial surface and prevents induction of toxic materials, inhibiting bacterial cell growth [10]. Chitosan may also prevent entry of oxygen and essential materials within the bacterial cell, resulting in cell death.

Antimicrobial activity of chitosan increases when pH and degree of acetylation are low. As molecular weight decreases, the antibacterial action of gram-negative bacteria increases and action for gram-positive bacteria decreases. Antimicrobial action is more pronounced as chitosan and derivatives interact with the cell wall to induce a higher bacterial mortality [11]. Table 3.1 summarizes the factors affecting antimicrobial activity of chitosan. This interaction depends on hydrophilic character of the bacterial wall, which explains chitosan's non-toxicity to human cells [12]. Environmental temperature also affects chitosan's antimicrobial activity. Temperature influences viscosity and molecular weight of chitosan molecule [13].

3.2.3 Hemocompatibility

Chitosan can speed up wound healing by interacting with platelets, hence it has been employed in various investigations involving coagulation [15]. Chitosan is a potent material for wound dressing owing to the unique properties of chemoattraction, analgesic character, antimicrobial properties, enhanced granulation tissue, re-epithelialization, role in activating macrophages and neutrophils, regulated scar development and contraction [16].

Table 3.1 Factors affecting microbicidal activities of chitosan [14]

Factors	Inferences
Molecular weight	Chitosan composites with higher molecular weight pile up on bacterial surface, preventing entry of nutrients and finally killing the bacteria Chitosan composites with lower molecular weight enter the bacterial cell thereby binding with DNA and preventing transcription
Extent of deacetylation	A high positive charge on chitosan molecule defines the extent of deacetylation
pH	Antimicrobial activity increases at acidic pH, when constituent amino groups ionize and the positive charge on chitosan binds with negatively charged bacterial surface
Temperature	Temperature variation affects antimicrobial activity of chitosan, owing to changes in viscosity and molecular weight
Fungi	High molecular weight and low extent of acetylation decrease anti-fungal action of chitosan
Gram-positive bacteria	Cell wall consists of teichoic acid and peptidoglycan, responsible for functioning of various enzymes and gives stability to cell wall
Gram-negative bacteria	Cell wall consists of peptidoglycan which gives a high negative charge and binds easily to positively charged chitosan. Hence chitosan has a stronger antimicrobial action on gram-negative bacteria

3.2.4 *Antineoplastic Activity*

Recent studies involving both in vivo and in vitro systems reveal antineoplastic character of chitosan. Activity of chitosan and derivatives results in increased secretion of interleukin (IL)-1 and 2, which causes cytolytic T lymphocytes to mature and infiltrate [17].

3.2.5 *Mucoadhesivity*

Positively charged chitosan molecules have a unique property of binding to surfaces. This property aids not only in formulation of strategies which mediate delivery of favourable molecules by employing mucosal pathways but is also useful in adsorbing molecules which cannot bind to the negatively charged mucus [18]. Chitosan linked to polymers enhances their capability to adhere to surfaces through opening of tight epithelial junctions [19].

3.2.6 Antioxidant Activity

Antioxidants are instrumental in maintenance of health and wellness. They are actively involved in intercepting reactive oxygen species (ROS), responsible for breakdown of membrane proteins, lipids and DNA [20]. A few in vitro studies reveal that ROS can be removed by action of chitosan and its products. Removal of free radicals is better in presence of chitosan molecules with lower molecular weight [21]. According to a recent investigation, the antioxidant property of chitosan depends on activity of amino functional groups and carboxyl functional groups, which stabilize free radicals [22].

3.2.7 Immunoenhancing Effects

Chitosan has the ability to generate numerous anti- and pro-inflammatory cytokines [23]. It is also capable of stimulating macrophages and natural killer (NK) cells [24].

3.2.8 Biodegradability and Biocompatibility

Enzymes like lysozyme, chitinases and bacteria inhabiting the colon degrade chitosan into non-toxic products like N-acetyl glucose and glucosamine, both are easily removed and do not induce an immune reaction. This makes chitosan an excellent choice for various biomedical applications. Chitosan is highly biocompatible and forms hydrogels via crosslinking methods. It has structural and functional similarities with glycosaminoglycans (GAGs), which constitute the extracellular matrix. Hence, chitosan is often employed for devising biodegradable and biocompatible delivery systems [25].

3.3 Synthesis of Chitosan-Based Bionanocomposites

Chitosan-based bionanocomposites were first developed in a study by Ohya et al. (1994) as a chemotherapeutic agent for the delivery of 5-fluorouracil [26]. Subsequent studies have conducted extensive investigations with chitosan-based nanocomposites and devised various methodologies considering various factors like stability, size, drug-bearing capacity and retention time. The basic strategies employed to form chitosan nanocomposites involve precipitation, emulsification, covalent crosslinking or ionic interactions, evaporation or their combinations. Chitosan nanoparticles carry many lone-pair electrons and can bind easily with empty orbitals. They are extensively employed for delivery of drugs and genes [27], for use as biosensors [28], and in optical imaging [29, 30].

These activities of chitosan biocomposites depend on the size of particle and uniform character of chitosan microspheres. Antigen distribution, adsorption and related immune responses are contingent on particle size of nanocomposite. Chitosan-based nanocomposites are synthesized by emulsion crosslinking, spray drying, solvent evaporation, ionic crosslinking, flocculation or precipitation and chitosan coating.

3.3.1 Emulsion Crosslinking

The nanocomposites synthesized according to this method involve covalent crosslinking and employ chitosan along with a crosslinking agent. Aldehyde group of the crosslinking agent and positive amino group of chitosan react to constitute the biocomposite. The emulsion consists of chitosan in aqueous phase and phase of Span 80 as toluene, stabilizer and glutaraldehyde as crosslinking agent. Both the phases are thoroughly mixed to facilitate crosslinking and the resulting nanocomposites are observed in the form of droplets [31], which are spherical and sparsely distributed particles with enhanced stability and prolonged drug release. The method is no longer employed as glutaraldehyde causes toxicity.

3.3.2 Microemulsion Method

This is another method employing covalent crosslinking and is also called the reversed micelles method. Water-in-oil reverse micelle structures aid in synthesis of chitosan-based nanocomposites. Aqueous phase of chitosan and glutaraldehyde and organic phase of lipophilic surfactant and organic solvent are mixed. Typically, cetyltrimethylammonium (CTAB) bromide or sodium bis(2-ethylhexyl) sulfosuccinate (AOT) are the surfactants, and the organic solvent employed is *n*-hexane [32, 33]. The micelle has a core composed of chitosan, where nanocomposites are formed via covalent crosslinking. The nanocomposites have a size less than 100 nm, which play important roles in efficient drug loading and precise release. Recently, the methodology has been improvised to incorporate non-toxic solvents and crosslinking agents. This overcomes limitations associated with the conventional method based on glutaraldehyde [34].

3.3.3 Ionic Gelation

Ionic crosslinking is the principle behind this procedure, which happens between opposite charges; amino groups in chitosan react with negative charge on sodium tripolyphosphate (TPP) [35]. For gelation, chitosan in acetic acid is added to a solution of TPP. After centrifugation and washing, chitosan nanoparticles are separated by freeze drying or oven drying. The process is uncomplicated, can be achieved at room temperature and nanocomposites are produced without involvement of toxic crosslinking agents or solvents. The particle size can be modified through variation in the chitosan-TPP ratio, which is significant for efficient drug loading and precise delivery to target tissues [36, 37].

3.3.4 Spray Drying

This method of producing chitosan nanocomposites involves dissolving chitosan in acetic acid and nanocomposites formed by passing this solution through a nozzle at high temperature [38]. Magnetic chitosan-based nanocomposites can also be synthesized by spray drying [39]. Supercritical-CO_2-assisted solubilization and atomization (SCASA) is a method that utilizes pressurized CO_2 and water for dissolving chitosan. After a long dissolving period of 48 h, chitosan is passed into a fluidized bed through a nozzle that results in atomization. Nanocomposites are formed upon spray drying and can be collected through a filter placed at the top of fluidized bed [40]. Major disadvantages are long processing time and the particle size. Ethyl cellulose and chitosan-based nanocomposites, which are used as carriers of neurotrophic factors, are also produced by spray drying. The nanoparticles obtained from spray drying are uniform and spherical in shape. They have a sustained release and are significant in treating tuberculosis and neurodegenerative disorders.

3.3.5 Precipitation/Flocculation

Chitosan-based nanocomposites can be created by procedures involving precipitation. Phase inversion precipitation employs emulsification in combination with precipitation. An oil-in-water emulsion consists of dichloromethane and acetone as organic phase and chitosan in an aqueous phase, along with a stabilizer (poloxamer). Homogenization under high pressure releases nanosized well-distributed emulsion drops. Methylene chloride is separated from emulsion at room temperature through volatilization at low pressure, which removes acetone from emulsion drops and simultaneously precipitates nanocomposites [41]. Emulsion–droplet coalescence is another method involving precipitation and is also called desolvation. It requires two water-in-oil emulsions, which coalesce leading to precipitation of nanocomposites.

Precipitation is brought about by NaOH, which is present in one of the two emulsions. Liquid paraffin and sorbitan sesquioleate are blended as a phase for the two emulsions, the first contains chitosan and the other has NaOH. Chitosan-containing emulsion is prepared by homogenization at high velocity. When the two emulsions are blended, NaOH molecules diffuse out into the nanodroplets, decreasing chitosan solubility and causing precipitation of the nanocomposite. Nanocomposites are disassociated upon centrifugation and washed by ethanol, toluene, water and freeze drying [42, 43]. Typically, precipitated nanocomposites are greater than 600–800 nm in size. Precipitation methods are not favoured for use because of requirement of solvents (organic) and high energy during homogenization, hence very few studies have been conducted. In one of the studies on phase inversion precipitation, chitosan-based nanocomposites had high effectiveness for encapsulating hydrophobic drugs like Cyclosporin A [34, 41].

3.3.6 Solvent Evaporation

This approach makes use of the variation in soluble phase along with sonication. The chitosan derivatives prepared by this technique exhibit amphipathic properties. The nanocomposite is obtained by mixing the oil phase with drug and finally distilling off the organic solvent by sonication. The solvent evaporation methodology has several advantages; it is preferred over spray drying, homogenization, etc., as it involves mild conditions like room temperature and stirring, without modifying the activity of drug. This style has been in vogue since the end of the 1970s and has been successfully used for preparing microspheres derived from many biocompatible polymers, for example, poly(D, L-lactide-co-glycolide) [44] and Eudragit [45]. These nanocomposites have a spherical shape and smooth surface.

3.3.7 Chitosan Solution Coating

In this technique, existing nanoparticles are added to a chitosan solution concentration. Chitosan encapsulates the nanoparticle owing to the presence of lone electrons in its structure and adhesiveness of chitosan. The resulting nanocomposites have a good shape, with size in between 75 and 85 nm. The nanoparticles are efficient in precision targeting and drug release.

3.3.8 Self-Assembly

Chitosan-based nanocomposites can be prepared by self-assembly, which is extensively used and is based on numerous electrostatic, hydrophobic interactions, hydrogen bonding or van der Waals interactions between chitosan and various other ligands [46, 47]. Chitosan can form complexes along with natural anionic molecules like hyaluronic acid or alginic acid. Nanocomposites prepared by self-assembly are conducive for carrying both hydrophilic and lipophilic drugs [47] (Table 3.2).

Table 3.2 Techniques involved in preparation of chitosan-based bionanocomposites [48]

Technique	Principle	Advantages	Disadvantages
Emulsion crosslinking	Covalent crosslinking	Uncomplicated procedure	Toxicity of crosslinking agents
Microemulsion method	Covalent crosslinking	Resulting nanocomposites are ultrafine, size below 100 nm	Complex procedure, toxic crosslinking agents, time-consuming process
Ionic gelation	Ionic crosslinking	No toxic chemicals used, simple procedure, flexibility in adjustment of nanocomposite size	
Spray drying	Atomization	Rapid and simple procedure, no additional separation/drying step	Particle size large, inappropriate for use with temperature labile substances
Phase inversion precipitation	Precipitation	Efficient encapsulation ability with specific compounds	Needs shear strength and toxic chemicals
Emulsion–droplet coalescence	Precipitation		Needs shear strength and toxic chemicals
Solvent evaporation	Vaporization	Requires mild conditions like room temperature and continuous stirring, without modifying the activity of drug	
Chitosan solution coating	Adhesive property of chitosan through lone pair of electrons	Nanocomposites have a smooth surface and shape, size appropriately between 75 and 85 nm, efficient in precision targeting and drug release	
Self-assembly	Electrostatic/hydrophobic interactions; hydrogen bonds/van der Waal's forces	Nanocomposites are highly stable, mild chemicals are used, flexible procedure	Difficult to regulate when executed at large scale

3.4 Biomedical Applications

In chitosan, reactive amino and hydroxyl functional groups provide considerable opportunities in several biomedical applications. Wide research has been conducted on chitosan-based nanocomposites and their applications in pharmaceutics and medicine. Use of chitosan-based nanocomposites is continuously growing in fields like medicine, tissue engineering, pharmaceutical research, biotechnology, wound dressing, as biosensors, in medical implants and for drug and vaccine delivery (Fig. 3.3).

3.4.1 *Drug Delivery*

Chitosan is biocompatible and readily encapsulates and grafts drugs and many active ligands. It prevents enzymatic lysis of drugs [49], and minimizes damage to normal tissues [50], making it a great asset in delivery of drugs and imaging in cancer theranostics [51]. Another advantage is the gradual degradation of chitosan nanocomposites, which ensures regulated and sustained release of drugs [52]. The high positive charge on surface gives stability to nanocomposites, which are capable of carrying many substances [53].

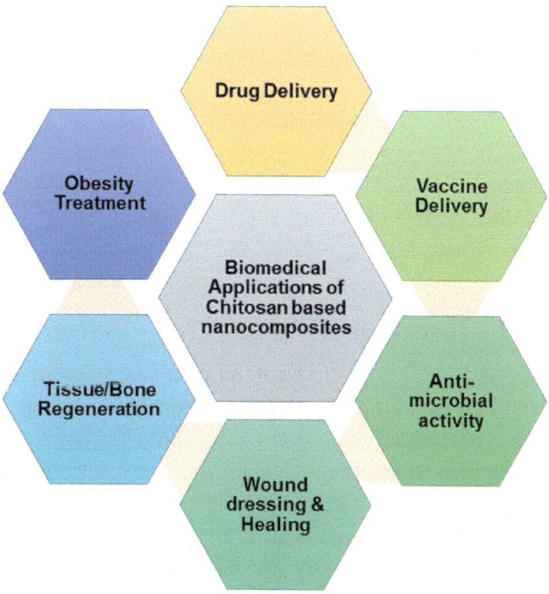

Fig. 3.3 Various biomedical applications of chitosan nanocomposites

3.4.2 Anti-Neoplastic Drug Delivery

Doxorubicin is frequently used in cancer therapy but is responsible for many adverse effects like cardiovascular toxicity. This can be prevented by encapsulation of the drug within chitosan nanocomposites. The nanocomposites are capable of reducing doxorubicin absorption in the entire small intestines [54]. This nanocomposite enhances survival of free drugs or their conjugates and also reduces adverse effects of drugs [54]. Chitosan tripolyphosphate (TPP) nanocomposites exhibit mucoadherence, aiding in retention of drugs on mucosa. An investigation exhibits that 46% chitosan/TPP nanocomposites are maintained in colon of rats after an incubation of 2 h at 37 °C, increasing the mucoadhesion of doxorubicin from 1.88% to 38.74% [55]. In another study on 5-fluorouracil carrying chitosan nanocomposites, a reduction in HT29 (human colorectal adenocarcinoma) and PC-3 (human prostate-3) cell proliferation was observed, restrained adhesivity to human umbilical vein endothelial cells was also confirmed [56]. Paclitaxel, a chemotherapeutic drug, is referred for non-small cell lung cancer at later stages of malignancy. Nanocomposites aggregate within the cell, enhance blood flow, resulting in enhanced drug trapping by pulmonary capillaries. Additionally, the authors observed that in acidic tumour environment, nanocomposites encapsulating paclitaxel interact strongly with negative charge on tumour cells [57]. Drug resistance in cancer therapy has resulted in failure of the existing chemotherapeutic methods [58]. A group of researchers developed folate-modified chitosan nanocomposites loaded with curcumin for breast cancer treatment [59]. In another study, chitosan ascorbate nanocomposites were used for inhibiting cervical cancer. The authors concluded that these nanocomposites decrease potentiality of malignant cells with no effect on normal cells [60]. Glycol-chitosan-based nanocomposites deliver drugs with great efficacy via permeability and retention effect [61]. Cisplatin was delivered by epidermal growth factor receptor-targeted chitosan nanocomposites for treatment of cisplatin-resistant lung cancer models [62]. Interferon-alpha-2b-loaded chitosan nanocomposites were employed for oral drug delivery in viral infections and few cancer types [63].

3.4.3 Protein and Peptide Drug Delivery

Proteolytic enzymes of the gastrointestinal tract can easily hydrolyse protein-based drugs, but when encapsulated by chitosan nanocomposites, these drugs are not hydrolysed, enhancing their stability. These nanocomposites enhance biodegradation of proteins, regulate release of drugs and improve diffusion of hydrophilic ligands through the epithelium. Currently, chitosan nanocomposites are being studied for their action with drugs in the stomach [64]. Chitosan nanocomposites loaded with insulin are prepared by crosslinking and membrane emulsification. The resulting nanocomposites are highly efficient in trapping drugs, steady insulin release and stability [65]. Chitosan and poly (ethylene glycol) dialdehyde were crosslinked forming a hydrogel, which facilitates release of protein. Hence, these nanocomposites are promising carriers for release of proteins and many large molecules are employed for use as oral drugs [66].

3.4.4 *Gene Delivery*

Chitosan forms a polyelectrolyte complex with negatively charged DNA, hence preventing degradation by nucleases. This enhances retention time of DNA in gastrointestinal tract and improves the process of transfection [67, 68]. Transfection is introduction of a foreign plasmid DNA into a target cell. Chitosan can enhance endocytosis and increase the immune response [69]. Chitosan nanocomposites enclosing plasmid DNA are synthesized by a complex coagulation process. Effective encapsulation and expression of the plasmid DNA was reported by a few in vivo studies [70]. This encapsulation of plasmids enhances regulation and expression of genes encoding interleukine-2 (IL-2), which have been researched for immune therapy based on genes. Plasmid remains unaltered after entrapment within the chitosan nanocomposite. In vitro studies confirm that IL-2 secreting cells are affected by molecular weight and mass of chitosan nanocomposites [71]. In another study, two DNA plasmids, pGL2 and pMK3, when entrapped within chitosan nanocomposites, exhibited no change in their function and structure [72]. In the process of gene delivery, plasmid DNA enters the target cells, followed by transcription and translation of genetic information into compatible protein. There are many hurdles in the process. Viruses can aid in delivery of genes but due to safety considerations, vectors other than viruses are preferred [73]. In a study by Li et al., quaternary chitosan was prepared by modifying quaternary ammonium group to enhance the solubility and DNA-bearing ability of chitosan [74]. In another study, miR-34a (endogenous tumour-suppressive molecule) and doxorubicin packed hyaluronic acid-chitosan nanocomposites were prepared by the ionotropic gelation technique, for the treatment of breast cancer, which exhibited promising results [75]. Chitosan and polyethylene glycol-grafted nanocomposites packed with anti-β-catenin siRNA were developed for transfection in malignant colon cells. The researchers confirmed successful entry of these nanocomposites into colon cells and reduced protein concentrations, were observed through western blot analysis, which restricts tumour progression [76].

3.4.5 *Antibiotic Delivery*

Antibiotic gentamicin encapsulated within a chitosan nanocomposite has antimicrobial and antioxidant characteristics and is employed for pulmonary delivery of the drug [77]. Fucoidan is a sulphated polysaccharide extracted from a seaweed *Ascophyllum nodosum*, which has antioxidant properties and removes the active oxygen species produced by gentamicin [78]. In this study, gentamicin and fucoidan were encapsulated within chitosan and delivery and release of drugs was observed. These chitosan-based nanocomposites increased antimicrobial activities.

3.4.6 Polyphenol Delivery

Polyphenols in diet exhibit wide therapeutic applications like antioxidant activities, anti-inflammatory properties, prevention of cardiovascular diseases, cancer and neurodegenerative diseases like Alzheimer's, though gradual assimilation and imperfect bioavailability restricts their oral administration [79]. To overcome this limitation, polyphenols derived from tea were encapsulated into chitosan nanocomposites as oral therapeutic agents [80]. This enhanced the stability of polyphenols and acted as a deterrent against degradation or oxidation by enzymes in the gastrointestinal tract. Encapsulation facilitated uptake of polyphenols by epithelial cells through the process of endocytosis. Rosmarinic acid-packed chitosan nanocomposites are prepared by ionic gelation for ocular use. In a study on retinal pigment epithelium (ARPE-19) and human corneal (HCE-T) cell lines, no cytotoxicity was induced by chitosan-rosmarinic acid composites. Enhanced penetration of rosmarinic acid was seen in combination with nanocomposites as compared to free solution [81].

3.4.7 Brain Drug Delivery

Efficient drug delivery to the brain is obstructed by the blood–brain barrier (BBB). Therapeutics can be delivered to the brain by the following three strategies [82]: (i) the BBB is bypassed through drug-encased ligands positioned in the tumour cavity during the course of surgical resection, administration of drugs is carried out through the nasal route and intradermal facial injection; (ii) surfactants and hyperosmotic inducers, polypeptides permeating the cell membrane and magnetic nanoparticles which cause hyperthermia cross the BBB with ease and (iii) increased non-invasive neural entry of drugs by employing internal transporters and receptors. The olfactory route to the brain facilitates drug targeting and decreases systemic adverse effects [83]. Chitosan nanocomposites packed with rivastigmine are used for brain targeting [83]. Alzheimer's disease requires rivastigmine therapy. The hydrophilic character decreases bioavailability and restricts its entry into the brain, necessitating persistent dosing, which causes adverse effects like nausea, dyspepsia, anorexia and bradycardia [84]. The maximum tolerated dose of drugs loaded on nanocomposites increases by 10%. Amnesia induced by scopolamine can be reversed by drug encapsulated nanoparticles, making them a potent therapeutic agent for precision drug delivery to the brain [83]. The antibodies-encased chitosan nanocomposites can permeate the BBB and impart neuroprotection. Receptors of transferrin recognize chitosan nanocomposites conjugated with antibodies. In a study on hCMEC/D3 cells (an in vitro BBB model), TfRmAb-chitosan nanocomposites conjugated with increased affinity to the hCMEC/D3 cells, exhibiting enhanced endocytosis and cytotoxicity. The encapsulation of nanocomposites by antibodies effectively crossed the BBB through macropinocytic and receptor-mediated endocytosis [85]. A few chitosan-based nanocomposites have the efficiency to enhance adequate neural intake of drugs, arresting apoptosis in the process. Chitosan-based nanospheres with OX26 monoclonal antibody and poly(ethylene glycol) recognize transferrin receptor (TfR) and induce receptor-mediated transfer through the BBB. In a study on rodents, the nanocomposites were fluorescently labelled and intravenously administered. Nanoparticles were accepted by the neural tissue, external to the intravascular region [86]. Chitosan nanocomposites can also be packed with thymoquinone [87], intranasal route of administration was followed by enhanced accumulation of drugs within the brain in comparison to the drug given intravenously. Olanzapine conjugated with chitosan nanocomposites is highly efficient for treating depression and schizophrenia [88, 89]. Donepezil, which is used to treat Alzheimer, is more effective in combination with chitosan nanocomposites [90]. In epilepsy, the BBB permeation of the drug is hindered by multiple drug resistance transporter and P-glycoprotein, thereby causing obstruction in therapy. Hence to increase therapeutic efficacy of drugs like carbamazepine, it was combined with carboxymethyl chitosan nanocomposites. Some properties of nanoparticles like the size, charge and surface characteristics play a significant role in the neural uptake and assimilation of the drug. Nanocomposites enhance permeability of drug across the BBB by relaxing junctions linking endothelial cells. Nanocomposites enveloped by surfactants have a higher efficiency to deliver drugs with precision, as they reduce uptake by reticuloendothelial system [91]. A variety of receptors on the BBB like transferrin and lipoprotein receptors are overexpressed. These receptors bind with nanoparticles to facilitate absorption into the BBB [92].

3.4.8 Vaccine Delivery

Chitosan has unique characteristics of mucoadhesion and osmosis, which facilitate adsorption and transfer of polypeptides across the nasal epithelial lining [93]. Many investigations reveal that chitosan helps macromolecules to cross mucosal barrier to connect with the nasal tissue [77]. Chitosan-based nanocomposites significantly enhance systemic and local immune responses to nasal diphtheria vaccine, as observed in mice [94]. Oral delivery of chitosan nanocomposites modified with glucomannan was recently studied in mice via both in vivo and in vitro tests. In this study, the lyophilized nanoparticles sustained biological activity exhibited by mediators and obstructed antigens. These nanocomposites significantly enhanced the immune responses by mucosal (secretory IgA), systemic (serum IgG), and cell-mediated (IL-2 and IFN-) pathways [95]. In another research, IgA-based chitosan and dextran sulphate nanocomposites were loaded with pertussis toxin. Both in vitro and in vivo investigations revealed that absorption of these nanocomposites occurs across the nasal epithelium followed by intranasal immunization [96]. In another investigation, N-2-hydroxypropyl trimethyl ammonium chloride chitosan (N-2-HACC) and N,O-carboxymethyl chitosan (CMC) nanocomposites were prepared to be used as a carrier and adjuvant for vaccine antigens against Newcastle disease and infectious bronchitis in chicken. The above chitosan-based nanocomposites induce a significantly improved immune response through better intranasal induction of IgG and IgA antibodies, along with enhanced generation of lymphocytes [96]. Chitosan nanocomposites with plasmid DNA increase antigen-specific immune responses [97]. A few investigations have studied action of intranasal DNA vaccines. Furthermore, T-cells which release IFN were observed in lungs, and CD8+ and CD4+ T-cells induced specific immune responses to facilitate synthesis of a DNA carrier in combination with polyethyleneimine, resulting in an improved metastatic rate of genes by 1000-fold [98].

3.4.9 Antimicrobial Activities

The antimicrobial features of chitosan and its nanocomposites depend on the molecular weight, pH, degree of deacetylation and the cellular functions. Various warm-blooded species are prone to infection by parasites (ticks and mites). Chitosan and nanocomposites are very effective against these, and also against many bacteria and fungi. Chitosan-based nanocomposites form a water-insoluble film on surfaces, which prolongs the presence of chitosan and enhances the antimicrobial action. Chitosan nanocomposites make use of their antimicrobial activity and prevent infection in cattle undergoing transition from lactation to drying phase. This not only prevents pathogenic intrusion into milk, but also enhances degradation of pathogen in breast. The chitosan hydrogel induces immune response, which controls the spread of active infection [99]. The phosphoryl group on chitosan interacts with lipopolysaccharide on the bacterial surface to induce an antimicrobial effect. This facilitates prevention of pulmonary infections. Rifampicin-loaded chitosan nanocomposites safely release the drug without inducing toxicity to healthy organs/cells. In vivo research confirms that chitosan nanocomposites enhance the retention time of drug and result in its gradual clearance [100].

3.4.10 Wound Dressings

Chitosan-based nanocomposites are antimicrobial, have low immunogenicity and have biodegradable and biocompatible properties, hence they facilitate wound healing [101]. They aid growth of a three-dimensional tissue matrix, trigger activity of macrophages and activate cellular proliferation [102]. Figure 3.4 represents the role of chitosan in wound healing mechanism. Chitosan stimulates polymorphonuclear leukocytes, fibroblasts and macrophages that promote granulation and organization of repaired tissues [103], which will gradually degrade into *N*-acetyl-β-D-glucosamine, stimulating fibroblast growth. Chitosan hastens the process of healing and prevents scar formation [104]. A recent innovation in use of chitosan-based bionanocomposites for wound healing has been in development of nanofibers and evolution of adhesive-based chitosan. The electrospun chitosan nanofibrous mats are porous with a high tensile strength and surface area, along with perfect rate of vapour and oxygen transmission. They are also compatible with stem cells derived from adipose, which aid in wound healing [105]. The adhesive-based dressing is applicable in post-surgical wound healing. Many clinical studies have exhibited positive results in patients of plastic surgery after using chitosan-based nanocomposites for healing [106]. Favourable effects were also observed in patients after skin grafting [107] and sinus surgery through endoscopy [108]. Currently, many types of chitosan-based wound dressings are available as nanofibers, non-woven mats, composites, sponges and films.

Chitosan-based carboxyl-modified poly(vinyl alcohol) hydrogels, prepared by amide crosslinking, facilitate presence of moisture over the wound. Another crosslinked hydrogel film combined with gentamicin sulphate exhibited precise and uninterrupted release of drugs, along with efficient control of bacterial growth to block infection at the site of trauma [109]. Chitosan-based nanocomposite enveloped by curcumin is employed for wound healing, as it removes 1,1-diphenyl-2-picrylhydrazyl and superoxide free radical. Curcumin-encapsulated composite exhibits significantly improved antibacterial action against *Staphylococcus aureus* [110].

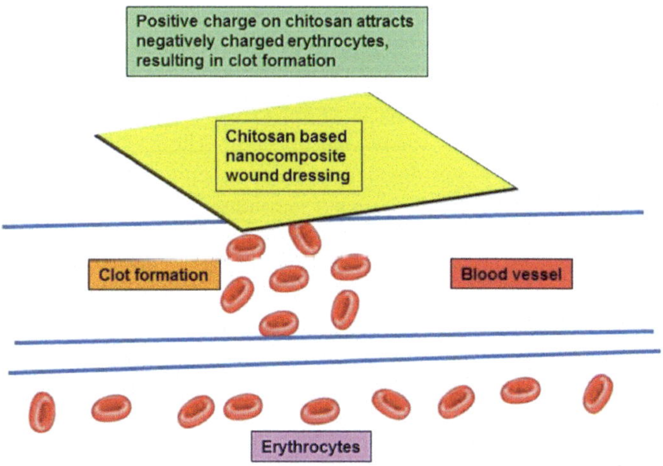

Fig. 3.4 Role of chitosan-based nanocomposites in wound healing

3.4.11 *Tissue/Bone Regeneration*

Tissue engineering is amalgamation of cells, materials and biochemical factors for amelioration or restoration of biological activities. It is extensively applied in repair or replacement of tissues like bone, cartilage, skin, muscle, blood vessels and bladder. Recently chitosan-based nanocomposites are leveraged for developing and progressing tissue engineering, which provides mechanical and structural strength for improvement of tissue functioning [111]. Chitosan nanocomposites expedite activation of pronuclear leukocytes and macrophage fibroblasts to augment granulation and tissue repair [103]. Gradual breakdown of N-acetyl-β-D-glucosamine enhances generation of fibroblasts, which causes collagen to precipitate synthesis of hyaluronic acid at the site of trauma, enhancing wound healing and checking formation of scars [104]. Glutaraldehyde lyophilized and crosslinked with chitosan sponge was employed for blood haemostasis [112]. Another chitosan-based nanocomposite with tricalcium phosphate is employed as a substitute for bone and as a scaffold in tissue engineering to enhance efficiency of bone formation. These nanocomposites have the ability to function as a bone substitute, fill up defective sites, enhance release of drugs and act as a scaffold for osteoblast cell culture [113].

3.5 Transpiring Trends and Conclusion

Chitosan and derivatives can be employed as carriers of nanomaterials for therapeutic delivery and have multiple biomedical applications, like precision delivery of drugs, vaccine delivery, antitumor and antimicrobial activity, wound healing, tissue engineering, etc. They also enhance the therapeutic efficiency of drugs and are degraded at a reasonable pace and prevent inflammation or formation of toxic products during formation of new tissue. Characteristics of chitosan-based nanocomposites are used as scaffolds during tissue engineering. Chitosan nanocomposites are employed as haemolytic agents for wound dressing. Their antibacterial properties enhance healing of wounds via regeneration and re-epithelialization of the granular layer in wounds. Their porous character facilitates diffusion of gases, metabolic waste and nutrients, which increases surface area for cellular migration, differentiation and attachment. Chitosan nanocomposites are efficient in ocular delivery of drugs owing to their tendency to form gels and mucoadhesive properties. They carry positive charge, which facilitates pulmonary delivery of drugs. Chitosan enhances drug permeability, thereby increasing prospects of nasal delivery. Mucosal drug delivery is facilitated by absorption of hydrophilic moieties. Chitosan nanocomposites have a significant application in vaccine delivery. They are successfully employed as a chemotherapeutic agent in malignancy. The antineoplastic activity of chitosan derivatives is attributed to generation of cytokines via increased proliferation of T-cells. Other investigations showed that chitosan nanocomposites inhibit MMP-9 and induce pro-apoptotic factors against malignant tissue. Physiochemical properties of chitosan nanocomposites are biodegradability, biocompatibility, mucoadhesivity, absorption enhancing property and in situ gelling characteristics. Chitosan and derivatives are considered safe for treatment of obesity; however, further research needs to be conducted to test the efficacy of chitosan treatment in obesity. Quaternized chitosan introduces a permanent positive charge on the hydroxyl or amino groups of polymers and intensifies antimicrobial activity across a wide pH scale. Furthermore, the quaternized chitosan is employed in orthopaedic and dental implants by forming an antimicrobial coating and is also used in surgeries as a material for antimicrobial wound dressing. Chitosan and nanocomposites are widely accepted as promising therapeutic agents in various biomedical applications. However, an elaborate scrutiny needs to be conducted, based on the toxic effects induced and the techniques employed for their preparation. Likewise, extensive investigations should be conducted on innovative use of chitosan and derivatives. The relationship between structure and activity of chitosan and derivatives requires further investigation and more clinical trials need to be conducted.

References

1. I. Aranaz, R. Harris, A. Heras, Chitosan amphiphilic derivatives. Chem. Appl. COC **14**, 308–330 (2010). https://doi.org/10.2174/138527210790231919
2. K. Ding, Y. Wang, H. Wang, L. Yuan, M. Tan, X. Shi, Z. Lyu, Y. Liu, H. Chen, 6-*O* -sulfated chitosan promoting the neural differentiation of mouse embryonic stem cells. ACS Appl. Mater. Interfaces **6**, 20043–20050 (2014). https://doi.org/10.1021/am505628g
3. C. Coquery, C. Negrell, N. Caussé, N. Pébère, G. David, Synthesis of new high molecular weight phosphorylated chitosans for improving corrosion protection. Pure Appl. Chem. **91**, 509–521 (2019). https://doi.org/10.1515/pac-2018-0509
4. J. Varshosaz, The promise of chitosan microspheres in drug delivery systems. Expert Opin. Drug Deliv. **4**, 263–273 (2007). https://doi.org/10.1517/17425247.4.3.263
5. G. Galed, B. Miralles, I. Panos, A. Santiago, A. Heras, Deacetylation and depolymerization reactions of chitin/chitosan: Influence of the source of chitin. Carbohyd. Polym. **62**, 316–320 (2005). https://doi.org/10.1016/j.carbpol.2005.03.019
6. M. Huang, E. Khor, L.-Y. Lim, Uptake and cytotoxicity of chitosan molecules and nanoparticles: Effects of molecular weight and degree of deacetylation. Pharm. Res. **21**, 344–353 (2004). https://doi.org/10.1023/B:PHAM.0000016249.52831.a5
7. C. Yan, J. Gu, D. Hou, H. Jing, J. Wang, Y. Guo, H. Katsumi, T. Sakane, A. Yamamoto, Synthesis of Tat tagged and folate modified N-succinyl-chitosan self-assembly nanoparticles as a novel gene vector. Int. J. Biol. Macromol. **72**, 751–756 (2015). https://doi.org/10.1016/j.ijbiomac.2014.09.031
8. N.R. Sudarshan, D.G. Hoover, D. Knorr, Antibacterial action of chitosan. Food Biotechnol. **6**, 257–272 (1992). https://doi.org/10.1080/08905439209549838
9. X. Fei Liu, Y. Lin Guan, D. Zhi Yang, Z. Li, K. De Yao, Antibacterial action of chitosan and carboxymethylated chitosan.J. Appl. Polym. Sci. **79**, 1324–1335 (2001). https://doi.org/10.1002/1097-4628(20010214)79:7<1324::AID-APP210>3.0.CO;2-L
10. A.J. Varma, S.V. Deshpande, J.F. Kennedy, Metal complexation by chitosan and its derivatives: a review. Carbohydr. Polym. **55**, 77–93 (2004). https://doi.org/10.1016/j.carbpol.2003.08.005

11. I. Younes, S. Sellimi, M. Rinaudo, K. Jellouli, M. Nasri, Influence of acetylation degree and molecular weight of homogeneous chitosans on antibacterial and antifungal activities. Int. J. Food Microbiol. **185**, 57–63 (2014). https://doi.org/10.1016/j.ijfoodmicro.2014.04.029
12. M. Kong, X.G. Chen, K. Xing, H.J. Park, Antimicrobial properties of chitosan and mode of action: A state of the art review. Int. J. Food Microbiol. **144**, 51–63 (2010). https://doi.org/10.1016/j.ijfoodmicro.2010.09.012
13. H.K. No, S.H. Kim, S.H. Lee, N.Y. Park, W. Prinyawiwatkul, Stability and antibacterial activity of chitosan solutions affected by storage temperature and time. Carbohydr. Polym. **65**, 174–178 (2006). https://doi.org/10.1016/j.carbpol.2005.12.036
14. I. Aranaz, A.R. Alcántara, M.C. Civera, C. Arias, B. Elorza, A. Heras Caballero, N. Acosta, Chitosan: An overview of its properties and applications. Polymers **13**, 3256 (2021). https://doi.org/10.3390/polym13193256
15. Y. Okamoto, R. Yano, K. Miyatake, I. Tomohiro, Y. Shigemasa, S. Minami, Effects of chitin and chitosan on blood coagulation. Carbohyd. Polym. **53**, 337–342 (2003). https://doi.org/10.1016/S0144-8617(03)00076-6
16. A. Busilacchi, A. Gigante, M. Mattioli-Belmonte, S. Manzotti, R.A.A. Muzzarelli, Chitosan stabilizes platelet growth factors and modulates stem cell differentiation toward tissue regeneration. Carbohyd. Polym. **98**, 665–676 (2013). https://doi.org/10.1016/j.carbpol.2013.06.044
17. A. Tokoro, N. Takewaki, K. Suzuki, T. Mikami, S. Suzuki, M. Suzuki, Growth-inhibitory effect of hexa-N-acetylchitohexaose and chitohexaose against Meth-A solid tumor. Chem. Pharm. Bull. **36**, 784–790 (1988). https://doi.org/10.1248/cpb.36.784
18. L. Bugnicourt, C. Ladavière, Interests of chitosan nanoparticles ionically cross-linked with tripolyphosphate for biomedical applications. Prog. Polym. Sci. **60**, 1–17 (2016). https://doi.org/10.1016/j.progpolymsci.2016.06.002
19. H. Yamamoto, Y. Kuno, S. Sugimoto, H. Takeuchi, Y. Kawashima, Surface-modified PLGA nanosphere with chitosan improved pulmonary delivery of calcitonin by mucoadhesion and opening of the intercellular tight junctions. J. Control. Release **102**, 373–381 (2005). https://doi.org/10.1016/j.jconrel.2004.10.010
20. D.-H. Ngo, S.-K. Kim, Antioxidant effects of chitin, chitosan, and their derivatives, in *Advances in Food and Nutrition Research* (Elsevier, 2014), pp. 15–31. https://doi.org/10.1016/B978-0-12-800268-1.00002-0
21. P.-J. Park, J.-Y. Je, S.-K. Kim, Free radical scavenging activity of chitooligosaccharides by electron spin resonance spectrometry. J. Agric. Food Chem. **51**, 4624–4627 (2003). https://doi.org/10.1021/jf034039+
22. I. Younes, M. Rinaudo, Chitin and chitosan preparation from marine sources. Struct. Prop. Appl. Mar. Drugs **13**, 1133–1174 (2015). https://doi.org/10.3390/md13031133
23. D. Fong, C.D. Hoemann, Chitosan immunomodulatory properties: perspectives on the impact of structural properties and dosage. Fut. Sci. OA **4**, FSO225 (2018). https://doi.org/10.4155/fsoa-2017-0064
24. D.A. Zaharoff, C.J. Rogers, K.W. Hance, J. Schlom, J.W. Greiner, Chitosan solution enhances both humoral and cell-mediated immune responses to subcutaneous vaccination. Vaccine **25**, 2085–2094 (2007). https://doi.org/10.1016/j.vaccine.2006.11.034
25. S. Mushtaq, J.A. Khan, F. Rabbani, U. Latif, M. Arfan, M.A. Yameen, Biocompatible biodegradable polymeric antibacterial nanoparticles for enhancing the effects of a third-generation cephalosporin against resistant bacteria. J. Med. Microbiol. **66**, 318–327 (2017). https://doi.org/10.1099/jmm.0.000445
26. Y. Ohya, T. Takei, H. Kobayashi, T. Ouchi, Release behaviour of 5-fluorouracil from chitosan-gel microspheres immobilizing 5-fluorouracil derivative coated with polysaccharides and their cell specific recognition. J. Microencapsul. **10**, 1–9 (1993). https://doi.org/10.3109/02652049309015307
27. M.S. Verma, S. Liu, Y.Y. Chen, A. Meerasa, F.X. Gu, Size-tunable nanoparticles composed of dextran-b-poly(D, L-lactide) for drug delivery applications. Nano Res. **5**, 49–61 (2012). https://doi.org/10.1007/s12274-011-0184-z

28. A. Anitha, K.P. Chennazhi, S.V. Nair, R. Jayakumar, 5-Flourouracil loaded <I>N O</I>-carboxymethyl chitosan nanoparticles as an anticancer nanomedicine for breast cancer. J. Biomed. Nanotechnol. **8**, 29–42 (2012). https://doi.org/10.1166/jbn.2012.1365
29. P.-R. Wei, S.-H. Cheng, W.-N. Liao, K.-C. Kao, C.-F. Weng, C.-H. Lee, Synthesis of chitosan-coated near-infrared layered double hydroxide nanoparticles for in vivo optical imaging. J. Mater. Chem. **22**, 5503 (2012). https://doi.org/10.1039/c2jm16447g
30. F.S. Majedi, M.M. Hasani-Sadrabadi, J.J. VanDersarl, N. Mokarram, S. Hojjati-Emami, E. Dashtimoghadam, S. Bonakdar, M.A. Shokrgozar, A. Bertsch, P. Renaud, On-chip fabrication of paclitaxel-loaded chitosan nanoparticles for cancer therapeutics. Adv. Funct. Mater. **24**, 432–441 (2014). https://doi.org/10.1002/adfm.201301628
31. S.R. Jameela, T.V. Kumary, A.V. Lal, A. Jayakrishnan, Progesterone-loaded chitosan microspheres: A long acting biodegradable controlled delivery system. J. Control. Release **52**, 17–24 (1998). https://doi.org/10.1016/S0168-3659(97)00187-9
32. T. Banerjee, S. Mitra, A. Kumar Singh, R. Kumar Sharma, A. Maitra, Preparation, characterization and biodistribution of ultrafine chitosan nanoparticles. Int. J. Pharmaceut. **243**, 93–105 (2002). https://doi.org/10.1016/S0378-5173(02)00267-3
33. M.H. Kafshgari, M. Khorram, M. Mansouri, A. Samimi, S. Osfouri, Preparation of alginate and chitosan nanoparticles using a new reverse micellar system, Iran. Polym. J. **21**, 99–107 (2012). https://doi.org/10.1007/s13726-011-0010-1
34. A. Grenha, Chitosan nanoparticles: A survey of preparation methods. J. Drug Target. **20**, 291–300 (2012). https://doi.org/10.3109/1061186X.2011.654121
35. W. Fan, W. Yan, Z. Xu, H. Ni, Formation mechanism of monodisperse, low molecular weight chitosan nanoparticles by ionic gelation technique. Colloids Surf. B **90**, 21–27 (2012). https://doi.org/10.1016/j.colsurfb.2011.09.042
36. H. Liu, C. Gao, Preparation and properties of ionically cross-linked chitosan nanoparticles: Ionically cross-linked chitosan nanoparticles. Polym. Adv. Technol. **20**, 613–619 (2009). https://doi.org/10.1002/pat.1306
37. E.M.A. Hejjaji, A.M. Smith, G.A. Morris, Evaluation of the mucoadhesive properties of chitosan nanoparticles prepared using different chitosan to tripolyphosphate (CS:TPP) ratios. Int. J. Biol. Macromol. **120**, 1610–1617 (2018). https://doi.org/10.1016/j.ijbiomac.2018.09.185
38. L.T.K. Ngan, S.-L. Wang, ĐM. Hiep, P.M. Luong, N.T. Vui, T.M. Đinh, N.A. Dzung, Preparation of chitosan nanoparticles by spray drying, and their antibacterial activity. Res. Chem. Intermed. **40**, 2165–2175 (2014). https://doi.org/10.1007/s11164-014-1594-9
39. H.-Y. Huang, Y.-T. Shieh, C.-M. Shih, Y.-K. Twu, Magnetic chitosan/iron (II, III) oxide nanoparticles prepared by spray-drying. Carbohyd. Polym. **81**, 906–910 (2010). https://doi.org/10.1016/j.carbpol.2010.04.003
40. N. Hijazi, E. Rodier, J.-J. Letourneau, H. Louati, M. Sauceau, N. Le Moigne, J.-C. Benezet, J. Fages, Chitosan nanoparticles generation using CO_2 assisted processes. J. Supercrit. Fluids **95**, 118–128 (2014). https://doi.org/10.1016/j.supflu.2014.08.003
41. M.H. El-Shabouri, Positively charged nanoparticles for improving the oral bioavailability of cyclosporin-A. Int. J. Pharm. **249**, 101–108 (2002). https://doi.org/10.1016/S0378-5173(02)00461-1
42. H. Tokumitsu, H. Ichikawa, Y. Fukumori, Chitosan-gadopentetic acid complex nanoparticles for gadolinium neutron-capture therapy of cancer: Preparation by novel emulsion-droplet coalescence technique and characterization. Pharm. Res. **16**, 1830–1835 (1999). https://doi.org/10.1023/A:1018995124527
43. H. Ichikawa, H. Tokumitsu, M. Miyamoto, Y. Fukumori, in Nanoparticles for neutron capture therapy of cancer, ed. by C.S.S.R. Kumar. *Nanotechnologies for the Life Sciences* (Wiley-VCH Verlag GmbH & Co. KGaA, Weinheim, Germany, 2007), p. ntls0062. https://doi.org/10.1002/9783527610419.ntls0062
44. T. Jung, Biodegradable nanoparticles for oral delivery of peptides: is there a role for polymers to affect mucosal uptake? Eur. J. Pharm. Biopharm. **50**, 147–160 (2000). https://doi.org/10.1016/S0939-6411(00)00084-9

45. K. Bouchemal, S. Briançon, E. Perrier, H. Fessi, I. Bonnet, N. Zydowicz, Synthesis and characterization of polyurethane and poly(ether urethane) nanocapsules using a new technique of interfacial polycondensation combined to spontaneous emulsification. Int. J. Pharm. **269**, 89–100 (2004). https://doi.org/10.1016/j.ijpharm.2003.09.025
46. Y. Yang, S. Wang, Y. Wang, X. Wang, Q. Wang, M. Chen, Advances in self-assembled chitosan nanomaterials for drug delivery. Biotechnol. Adv. **32**, 1301–1316 (2014). https://doi.org/10.1016/j.biotechadv.2014.07.007
47. J.P. Quiñones, H. Peniche, C. Peniche, Chitosan based self-assembled nanoparticles in drug delivery. Polymers **10**, 235 (2018). https://doi.org/10.3390/polym10030235
48. M. Yanat, K. Schroën, Preparation methods and applications of chitosan nanoparticles; with an outlook toward reinforcement of biodegradable packaging. React. Funct. Polym. **161**, 104849 (2021). https://doi.org/10.1016/j.reactfunctpolym.2021.104849
49. X. Huang, Y.-Z. Du, H. Yuan, F.-Q. Hu, Preparation and pharmacodynamics of low-molecular-weight chitosan nanoparticles containing insulin. Carbohyd. Polym. **76**, 368–373 (2009). https://doi.org/10.1016/j.carbpol.2008.10.025
50. A. des Rieux, V. Fievez, M. Garinot, Y.-J. Schneider, V. Préat, Nanoparticles as potential oral delivery systems of proteins and vaccines: A mechanistic approach. J. Control. Release **116**, 1–27 (2006). https://doi.org/10.1016/j.jconrel.2006.08.013
51. Y. Yin, Q. Dang, C. Liu, J. Yan, D. Cha, Z. Yu, Y. Cao, Y. Wang, B. Fan, Itaconic acid grafted carboxymethyl chitosan and its nanoparticles: Preparation, characterization and evaluation. Int. J. Biol. Macromol. **102**, 10–18 (2017). https://doi.org/10.1016/j.ijbiomac.2017.04.005
52. P. Rajitha, D. Gopinath, R. Biswas, M. Sabitha, R. Jayakumar, Chitosan nanoparticles in drug therapy of infectious and inflammatory diseases. Expert Opin. Drug Deliv. **13**, 1177–1194 (2016). https://doi.org/10.1080/17425247.2016.1178232
53. L. Mu, S.S. Feng, Fabrication, characterization and in vitro release of paclitaxel (Taxol®) loaded poly (lactic-co-glycolic acid) microspheres prepared by spray drying technique with lipid/cholesterol emulsifiers. J. Control. Release **76**, 239–254 (2001). https://doi.org/10.1016/S0168-3659(01)00440-0
54. S. Mitra, U. Gaur, P.C. Ghosh, A.N. Maitra, Tumour targeted delivery of encapsulated dextran–doxorubicin conjugate using chitosan nanoparticles as carrier. J. Control. Release **74**, 317–323 (2001). https://doi.org/10.1016/S0168-3659(01)00342-X
55. C. Feng, J. Li, M. Kong, Y. Liu, X.J. Cheng, Y. Li, H.J. Park, X.G. Chen, Surface charge effect on mucoadhesion of chitosan based nanogels for local anti-colorectal cancer drug delivery. Colloids Surf. B **128**, 439–447 (2015). https://doi.org/10.1016/j.colsurfb.2015.02.042
56. R. Cavalli, F. Leone, R. Minelli, R. Fantozzi, C. Dianzani, New chitosan nanospheres for the delivery of 5-fluorouracil: Preparation characterization and in vitro studies. CDD **11**, 270–278 (2014). https://doi.org/10.2174/1567201811666140206103609
57. R. Yang, W.-S. Shim, F.-D. Cui, G. Cheng, X. Han, Q.-R. Jin, D.-D. Kim, S.-J. Chung, C.-K. Shim, Enhanced electrostatic interaction between chitosan-modified PLGA nanoparticle and tumor. Int. J. Pharm. **371**, 142–147 (2009). https://doi.org/10.1016/j.ijpharm.2008.12.007
58. I. Ferrera, O. Sánchez, Insights into microbial diversity in wastewater treatment systems: How far have we come? Biotechnol. Adv. **34**, 790–802 (2016). https://doi.org/10.1016/j.biotechadv.2016.04.003
59. S. Esfandiarpour-Boroujeni, S. Bagheri-Khoulenjani, H. Mirzadeh, S. Amanpour, Fabrication and study of curcumin loaded nanoparticles based on folate-chitosan for breast cancer therapy application. Carbohyd. Polym. **168**, 14–21 (2017). https://doi.org/10.1016/j.carbpol.2017.03.031
60. V. Sekar, K. Rajendran, S. Vallinayagam, V. Deepak, S. Mahadevan, Synthesis and characterization of chitosan ascorbate nanoparticles for therapeutic inhibition for cervical cancer and their in silico modeling. J. Ind. Eng. Chem. **62**, 239–249 (2018). https://doi.org/10.1016/j.jiec.2018.01.001
61. S.J. Lee, H.S. Min, S.H. Ku, S. Son, I.C. Kwon, S.H. Kim, K. Kim, Tumor-targeting glycol chitosan nanoparticles as a platform delivery carrier in cancer diagnosis and therapy. Nanomedicine **9**, 1697–1713 (2014). https://doi.org/10.2217/nnm.14.99

62. A.V. Nascimento, A. Singh, H. Bousbaa, D. Ferreira, B. Sarmento, M.M. Amiji, Overcoming cisplatin resistance in non-small cell lung cancer with Mad2 silencing siRNA delivered systemically using EGFR-targeted chitosan nanoparticles. Acta Biomater. **47**, 71–80 (2017). https://doi.org/10.1016/j.actbio.2016.09.045
63. K.H. Min, K. Park, Y.-S. Kim, S.M. Bae, S. Lee, H.G. Jo, R.-W. Park, I.-S. Kim, S.Y. Jeong, K. Kim, I.C. Kwon, Hydrophobically modified glycol chitosan nanoparticles-encapsulated camptothecin enhance the drug stability and tumor targeting in cancer therapy. J. Control. Release **127**, 208–218 (2008). https://doi.org/10.1016/j.jconrel.2008.01.013
64. V.R. Sinha, A.K. Singla, S. Wadhawan, R. Kaushik, R. Kumria, K. Bansal, S. Dhawan, Chitosan microspheres as a potential carrier for drugs. Int. J. Pharm. **274**, 1–33 (2004). https://doi.org/10.1016/j.ijpharm.2003.12.026
65. L.-Y. Wang, Y.-H. Gu, Q.-Z. Zhou, G.-H. Ma, Y.-H. Wan, Z.-G. Su, Preparation and characterization of uniform-sized chitosan microspheres containing insulin by membrane emulsification and a two-step solidification process. Colloids Surf. B **50**, 126–135 (2006). https://doi.org/10.1016/j.colsurfb.2006.05.006
66. Z.-W. Jing, Z.-W. Ma, C. Li, Y.-Y. Jia, M. Luo, X.-X. Ma, S.-Y. Zhou, B.-L. Zhang, Chitosan cross-linked with poly(ethylene glycol)dialdehyde via reductive amination as effective controlled release carriers for oral protein drug delivery. Bioorg. Med. Chem. Lett. **27**, 1003–1006 (2017). https://doi.org/10.1016/j.bmcl.2016.12.072
67. S.A. Agnihotri, N.N. Mallikarjuna, T.M. Aminabhavi, Recent advances on chitosan-based micro- and nanoparticles in drug delivery. J. Control. Release **100**, 5–28 (2004). https://doi.org/10.1016/j.jconrel.2004.08.010
68. A. Masotti, G. Ortaggi, Chitosan micro- and nanospheres: Fabrication and applications for drug and DNA delivery. MRMC. **9**, 463–469 (2009). https://doi.org/10.2174/138955709787847976
69. G. Feng, Q. Jiang, M. Xia, Y. Lu, W. Qiu, D. Zhao, L. Lu, G. Peng, Y. Wang, Enhanced immune response and protective effects of nano-chitosan-based DNA vaccine encoding T cell epitopes of Esat-6 and FL against mycobacterium tuberculosis infection. PLoS ONE **8**, e61135 (2013). https://doi.org/10.1371/journal.pone.0061135
70. U. Guliyeva, F. Oner, S. Ozsoy, R. Haziroglu, Chitosan microparticles containing plasmid DNA as potential oral gene delivery system. Eur. J. Pharm. Biopharm. **62**, 17–25 (2006). https://doi.org/10.1016/j.ejpb.2005.08.006
71. J. Akbuğa, S. Özbaş-Turan, N. Erdoğan, Plasmid-DNA loaded chitosan microspheres for in vitro IL-2 expression. Eur. J. Pharm. Biopharm. **58**, 501–507 (2004). https://doi.org/10.1016/j.ejpb.2004.04.015
72. S. Ozbas-Turan, C. Aral, L. Kabasakal, M. Keyer-Uysal, J. Akbuga, Co-encapsulation of two plasmids in chitosan microspheres as a non-viral gene delivery vehicle. J. Pharm. Pharm. Sci. **6**, 27–32 (2003)
73. L. De Laporte, J. Cruz Rea, L.D. Shea, Design of modular non-viral gene therapy vectors. Biomaterials **27**, 947–954 (2006). https://doi.org/10.1016/j.biomaterials.2005.09.036
74. G.-F. Li, J.-C. Wang, X.-M. Feng, Z.-D. Liu, C.-Y. Jiang, J.-D. Yang, Preparation and testing of quaternized chitosan nanoparticles as gene delivery vehicles. Appl. Biochem. Biotechnol. **175**, 3244–3257 (2015). https://doi.org/10.1007/s12010-015-1483-8
75. X. Deng, M. Cao, J. Zhang, K. Hu, Z. Yin, Z. Zhou, X. Xiao, Y. Yang, W. Sheng, Y. Wu, Y. Zeng, Hyaluronic acid-chitosan nanoparticles for co-delivery of MiR-34a and doxorubicin in therapy against triple negative breast cancer. Biomaterials **35**, 4333–4344 (2014). https://doi.org/10.1016/j.biomaterials.2014.02.006
76. W.E. Rudzinski, A. Palacios, A. Ahmed, M.A. Lane, T.M. Aminabhavi, Targeted delivery of small interfering RNA to colon cancer cells using chitosan and PEGylated chitosan nanoparticles. Carbohyd. Polym. **147**, 323–332 (2016). https://doi.org/10.1016/j.carbpol.2016.04.041
77. N. Islam, V. Ferro, Recent advances in chitosan-based nanoparticulate pulmonary drug delivery. Nanoscale **8**, 14341–14358 (2016). https://doi.org/10.1039/C6NR03256G

78. Y. Yuan, D. Macquarrie, Microwave assisted extraction of sulfated polysaccharides (fucoidan) from Ascophyllum nodosum and its antioxidant activity. Carbohyd. Polym. **129**, 101–107 (2015). https://doi.org/10.1016/j.carbpol.2015.04.057
79. H. Rastogi, S. Jana, Evaluation of physicochemical properties and intestinal permeability of six dietary polyphenols in human intestinal colon adenocarcinoma Caco-2 cells. Eur. J. Drug Metab. Pharmacokinet. **41**, 33–43 (2016). https://doi.org/10.1007/s13318-014-0234-5
80. J. Liang, H. Yan, P. Puligundla, X. Gao, Y. Zhou, X. Wan, Applications of chitosan nanoparticles to enhance absorption and bioavailability of tea polyphenols: A review. Food Hydrocolloids **69**, 286–292 (2017). https://doi.org/10.1016/j.foodhyd.2017.01.041
81. S.B. da Silva, D. Ferreira, M. Pintado, B. Sarmento, Chitosan-based nanoparticles for rosmarinic acid ocular delivery—In vitro tests. Int. J. Biol. Macromol. **84**, 112–120 (2016). https://doi.org/10.1016/j.ijbiomac.2015.11.070
82. C. Sun, Y. Ding, L. Zhou, D. Shi, L. Sun, T.J. Webster, Y. Shen, Noninvasive nanoparticle strategies for brain tumor targeting, Nanomedicine: Nanotechnology. Biol. Med. **13**, 2605–2621 (2017). https://doi.org/10.1016/j.nano.2017.07.009
83. K. Nagpal, S.K. Singh, D.N. Mishra, Optimization of brain targeted chitosan nanoparticles of Rivastigmine for improved efficacy and safety. Int. J. Biol. Macromol. **59**, 72–83 (2013). https://doi.org/10.1016/j.ijbiomac.2013.04.024
84. R. Khoury, J. Rajamanickam, G.T. Grossberg, An update on the safety of current therapies for Alzheimer's disease: focus on rivastigmine. Therapeut. Adv. Drug Saf. **9**, 171–178 (2018). https://doi.org/10.1177/2042098617750555
85. A. Sahin, D. Yoyen-Ermis, S. Caban-Toktas, U. Horzum, Y. Aktas, P. Couvreur, G. Esendagli, Y. Capan, Evaluation of brain-targeted chitosan nanoparticles through blood–brain barrier cerebral microvessel endothelial cells. J. Microencapsul. **34**, 659–666 (2017). https://doi.org/10.1080/02652048.2017.1375039
86. Y. Aktaş, M. Yemisci, K. Andrieux, R.N. Gürsoy, M.J. Alonso, E. Fernandez-Megia, R. Novoa-Carballal, E. Quiñoá, R. Riguera, M.F. Sargon, H.H. Çelik, A.S. Demir, A.A. Hıncal, T. Dalkara, Y. Çapan, P. Couvreur, Development and brain delivery of chitosan−PEG nanoparticles functionalized with the monoclonal antibody OX26. Bioconjugate Chem. **16**, 1503–1511 (2005). https://doi.org/10.1021/bc050217o
87. S. Alam, G. Mustafa, Z.I. Khan, F. Islam, A. Bhatnagar, F. Ahmad, Kumar, Development and evaluation of thymoquinone-encapsulated chitosan nanoparticles for nose-to-brain targeting: A pharmacoscintigraphic study. IJN 5705 (2012). https://doi.org/10.2147/IJN.S35329
88. S. Heres, S. Kraemer, R.F. Bergstrom, H.C. Detke, Pharmacokinetics of olanzapine long-acting injection: The clinical perspective. Int. Clin. Psychopharmacol. **29**, 299–312 (2014). https://doi.org/10.1097/YIC.0000000000000040
89. J. Ruby, V. Pandey, Formulation and evaluation of olanzapine loaded chitosan nanoparticles for nose to brain targeting an in vitro and ex vivo toxicity study. J. App. Pharm. Sci. 034–040 (2016). https://doi.org/10.7324/JAPS.2016.60905
90. D.A. Casey, D. Antimisiaris, J. O'Brien, Drugs for Alzheimer's disease: Are they effective? PT **35**, 208–211 (2010)
91. B.J. Chacko, S. Palanisamy, N.L. Gowrishankar, J. Honeypriya, A. Sumathy, Effect of surfactant coating on brain targeting polymeric nanoparticles; a review. Pharmaceut. Sci. **80** (2018). https://doi.org/10.4172/pharmaceutical-sciences.1000348
92. W.J. Geldenhuys, D.D. Allen, The blood-brain barrier choline transporter. CNSAMC **12**, 95–99 (2012). https://doi.org/10.2174/187152412800792670
93. V. Dodane, Effect of chitosan on epithelial permeability and structure. Int. J. Pharm. **182**, 21–32 (1999). https://doi.org/10.1016/S0378-5173(99)00030-7
94. I.M. van der Lubben, G. Kersten, M.M. Fretz, C. Beuvery, J. Coos Verhoef, H.E. Junginger, Chitosan microparticles for mucosal vaccination against diphtheria: Oral and nasal efficacy studies in mice. Vaccine **21**, 1400–1408 (2003). https://doi.org/10.1016/S0264-410X(02)00686-2
95. H. Harde, A.K. Agrawal, S. Jain, Development of stabilized glucomannosylated chitosan nanoparticles using tandem crosslinking method for oral vaccine delivery. Nanomedicine **9**, 2511–2529 (2014). https://doi.org/10.2217/nnm.13.225

96. K. Zhao, S. Li, W. Li, L. Yu, X. Duan, J. Han, X. Wang, Z. Jin, Quaternized chitosan nanoparticles loaded with the combined attenuated live vaccine against Newcastle disease and infectious bronchitis elicit immune response in chicken after intranasal administration. Drug Deliv. **24**, 1574–1586 (2017). https://doi.org/10.1080/10717544.2017.1388450
97. W. Tao, K.S. Ziemer, H.S. Gill, Gold nanoparticle–M2e conjugate coformulated with CpG induces protective immunity against influenza A virus. Nanomedicine **9**, 237–251 (2014). https://doi.org/10.2217/nnm.13.58
98. L. Torrieri-Dramard, B. Lambrecht, H.L. Ferreira, T. Van den Berg, D. Klatzmann, B. Bellier, Intranasal DNA vaccination induces potent mucosal and systemic immune responses and cross-protective immunity against influenza viruses. Mol. Ther. **19**, 602–611 (2011). https://doi.org/10.1038/mt.2010.222
99. S. Lanctôt, P. Fustier, A.R. Taherian, B. Bisakowski, X. Zhao, P. Lacasse, Effect of intramammary infusion of chitosan hydrogels at drying-off on bovine mammary gland involution. J. Dairy Sci. **100**, 2269–2281 (2017). https://doi.org/10.3168/jds.2016-12087
100. T. Rawal, R. Parmar, R.K. Tyagi, S. Butani, Rifampicin loaded chitosan nanoparticle dry powder presents an improved therapeutic approach for alveolar tuberculosis. Colloids Surf. B **154**, 321–330 (2017). https://doi.org/10.1016/j.colsurfb.2017.03.044
101. J. Rhoades, S. Roller, Antimicrobial actions of degraded and native chitosan against spoilage organisms in laboratory media and foods. Appl. Environ. Microbiol. **66**, 80–86 (2000). https://doi.org/10.1128/AEM.66.1.80-86.2000
102. R.S. Jayasree, K. Rathinam, C.P. Sharma, Development of artificial skin (template) and influence of different types of sterilization procedures on wound healing pattern in rabbits and guinea pigs. J. Biomater. Appl. **10**, 144–162 (1995). https://doi.org/10.1177/088532829501000205
103. H. Ueno, T. Mori, T. Fujinaga, Topical formulations and wound healing applications of chitosan. Adv. Drug Deliv. Rev. **52**, 105–115 (2001). https://doi.org/10.1016/S0169-409X(01)00189-2
104. R.A.A. Muzzarelli, M. Mattioli-Belmonte, A. Pugnaloni, G. Biagini, in Biochemistry, histology and clinical uses of chitins and chitosans in wound healing, ed. by P. Jollès, R.A.A. Muzzarelli. *Chitin and Chitinases, Birkhäuser Basel, Basel* (1999), pp. 251–264. https://doi.org/10.1007/978-3-0348-8757-1_18
105. N. Naseri, C. Algan, V. Jacobs, M. John, K. Oksman, A.P. Mathew, Electrospun chitosan-based nanocomposite mats reinforced with chitin nanocrystals for wound dressing. Carbohyd. Polym. **109**, 7–15 (2014). https://doi.org/10.1016/j.carbpol.2014.03.031
106. G. Biagini, A. Bertani, R. Muzzarelli, A. Damadei, G. DiBenedetto, A. Belligolli, G. Riccotti, C. Zucchini, C. Rizzoli, Wound management with N-carboxybutyl chitosan. Biomaterials **12**, 281–286 (1991). https://doi.org/10.1016/0142-9612(91)90035-9
107. C.A. Stone, H. Wright, V.S. Devaraj, T. Clarke, R. Powell, Healing at skin graft donor sites dressed with chitosan. Br. J. Plast. Surg. **53**, 601–606 (2000). https://doi.org/10.1054/bjps.2000.3412
108. R. Valentine, T. Athanasiadis, S. Moratti, L. Hanton, S. Robinson, P.-J. Wormald, The efficacy of a novel chitosan gel on hemostasis and wound healing after endoscopic sinus surgery. Am. J. Rhinol. Allergy **24**, 70–75 (2010). https://doi.org/10.2500/ajra.2010.24.3422
109. D. Zhang, W. Zhou, B. Wei, X. Wang, R. Tang, J. Nie, J. Wang, Carboxyl-modified poly(vinyl alcohol)-crosslinked chitosan hydrogel films for potential wound dressing. Carbohyd. Polym. **125**, 189–199 (2015). https://doi.org/10.1016/j.carbpol.2015.02.034
110. S. Jebahi, M. Saoudi, L. Farhat, H. Oudadesse, T. Rebai, A. Kabir, A. El Feki, H. Keskes, Effect of novel curcumin-encapsulated chitosan-bioglass drug on bone and skin repair after gamma radiation: experimental study on a Wistar rat model: Biomaterials and tissue engineering. Cell Biochem. Funct. **33**, 150–159 (2015). https://doi.org/10.1002/cbf.3098
111. I.-Y. Kim, S.-J. Seo, H.-S. Moon, M.-K. Yoo, I.-Y. Park, B.-C. Kim, C.-S. Cho, Chitosan and its derivatives for tissue engineering applications. Biotechnol. Adv. **26**, 1–21 (2008). https://doi.org/10.1016/j.biotechadv.2007.07.009

112. P.C. Kavitha Sankar, G. Rajmohan, M.J. Rosemary, Physico-chemical characterisation and biological evaluation of freeze dried chitosan sponge for wound care. Mater. Lett. **208**, 130–132 (2017). https://doi.org/10.1016/j.matlet.2017.05.010
113. J.-Y. Lee, Y.-J. Seol, K.-H. Kim, Y.-M. Lee, Y.-J. Park, I.-C. Rhyu, C.-P. Chung, S.-J. Lee, Transforming growth factor (TGF)—1 releasing tricalcium phosphate/chitosan microgranules as bone substitutes. Pharm. Res. **21**, 1790–1796 (2004). https://doi.org/10.1023/B:PHAM.0000045230.14287.73

Chapter 4
Biomedical Applications of Chitosan-Based Nanostructured Composite Materials

Chirâa El Idrissi El Hassani, Hamza Daoudi, Mounir El Achaby, and Zineb Kassab

Abstract Chitosan is a natural polysaccharide that has demonstrated distinct characteristics for many applications, especially in the biomedical field. This natural polymer offers a variety of properties such as biocompatibility, biodegradability, and non-toxicity, as well as its antibacterial activity and low immunogenicity. Having various active groups, the structure of chitosan may be easily modified by its combination with various nanostructured materials imbibing it with new characteristics that can meet the growing demands of regenerative medicine, to overcome the problems requiring the development of biodegradable and biocompatible nanocomposites. Moreover, chitosan-based nanostructured composites, with improved physical and chemical characteristics, have attracted great attention from researchers in the different biomedical fields such as bone tissue engineering, drug delivery, wound healing, and biosensors, among others. This chapter sheds light on recent research scenario on the different aspects of chitosan-based nanostructured materials, their preparation, their characterization, and their effective utilization in the biomedical area.

4.1 Introduction

Several natural polymers are gaining interest from researchers in the biomedical area due to their exceptional qualities, high durability, and low cost, in order to address the expanding therapeutic demands [1, 2]. They have wide applications in pharmaceuticals, medicine, drug delivery, wound healing, gene delivery, and bone

C. El Idrissi El Hassani
Laboratory of Materials, Catalysis, and Valorization of Natural Resources, URAC 24, Faculty of Science and Techniques of Mohammedia, University Hassan II, B.P. 146, Hassan, Casablanca, Morocco

H. Daoudi · M. El Achaby · Z. Kassab (✉)
Materials Science, Energy and Nano-Engineering (MSN) Department, Mohammed VI Polytechnic University (UM6P), Lot 660 – Hay Moulay Rachid, 43150 Benguerir, Morocco
e-mail: zineb.kassab@um6p.ma

S. K. Swain and A. Biswal (eds.), *Chitosan Nanocomposites*,
Biological and Medical Physics, Biomedical Engineering,
https://doi.org/10.1007/978-981-19-9646-7_4

tissue engineering as well as others [3]. Chitosan, a natural polysaccharide, is the deacetylated form of chitin, consisting of β-(1 → 4)-2acetamido-D-glucose and β-(1 → 4)-2-amino-D-glucose units [4]. Chitosan is one of the most versatile polysaccharides, which holds importance in pharmaceuticals and therapeutics due to the presence of outstanding properties such as biodegradability, low toxicity, biocompatibility, antibacterial and antifungal activities [5, 6]. Moreover, human tissues can absorb a harmless amino sugar produced after the degradation of chitosan. Accordingly, most of these properties make it a promising candidate for various biomedical practices, such as drug delivery carrier and tissue engineering, etc. [7, 8]. According to the literature, chitosan could help improve the body's immunity, lower cholesterol levels, and inhibit fat absorption [9]. The reactive functional groups (-OH) and ($-NH_2$) present in the backbone of chitosan allow it to be adapted physically or chemically, to create scaffolds with characteristics suitable for medicinal applications [7, 10]. Nevertheless, because of the accumulation of degradation products, the risk of infection, and low mechanical properties, these materials do not match the standards of biomedical engineering. Chitosan has challenges due to its poor functional efficacy and limited water solubility while possessing several beneficial qualities [1, 9]. Researchers have developed composites to overcome the problems mentioned above by combining biopolymers with appropriate nanofillers or inorganic materials with desirable properties [7]. By modifying the structure of chitosan, a wide variety of unique bio-scaffold materials could be created. With the advance in nanotechnology, chitosan can be combined with various nanostructures, added into the bulk material, or deposited on the surface, to develop chitosan-based nanocomposites with improved thermal, mechanical and barrier properties [6, 11]. Chitosan-based nanocomposites continued to generate considerable interest especially due to their distinctive physicochemical and biological properties. These nanoscale biocomposites of chitosan present stronger and more stable nanocomposite materials [12]. Chitosan-based nanocomposites are non-toxic and have a good affinity to proteins. Due to their remarkable non-toxicity, biocompatibility and biological activity, these nanostructured composites have attracted immense interest in the biological sciences. They are widely used as therapeutic care, and they are also being used in biomedical fields as biomaterials for drug delivery systems, tissue engineering, wound healing among others [13].

This chapter aims to emphasize the different functional aspects of the chitosan biopolymer combined with nanostructured materials in various functional forms to have chitosan-based nanostructured composites with new functional characteristics and properties, as well as to spot the different interesting practices in biomedical applications in which they have been used over the last decade. Moreover, this book chapter's first section contains comprehensive information about chitosan chemistry, describing its structure, physical, and antibacterial properties, the second part discusses the chitosan-based nanostructured composites, their preparation, modification, and characterization. The final part focuses on the potential applications of these nanostructured composites in the biomedical field, the conclusion, and future perspectives.

4.2 Chemical Structure, Physical and Antibacterial Properties of Chitosan

Chitin is the most prevalent biosourced polymer (β-(1–4)-poly-N-acetyl-D-glucosamine); it is present in the exoskeletons of shrimp, crabs, insect cuticles, and fungal cell walls [14]. Chitin is hydrolyzed chemically or enzymatically to produce its deacetylated derivative, chitosan, which is used in biomedical applications [15]. A copolymer with less crystallinity and has a molecular weight (M_W) of essentially 100 Kilodalton (kDa) than chitin, chitosan is made of -1,4-linked 2-amino-2-deoxy-D-glucose and N-acetyl-D-glucosamine groups. Its structure is like that of cellulose, with the exception that chitosan consists of amino units (-NH_2) in place of cellulose's hydroxyl group (-OH) at position C_2 [16, 17]. Scientists have thoroughly investigated the chitosan system's origin, composition, structure, and physicochemical properties. Chitosan and its derivatives have found usage in the food sector among other industries because of their unique properties [18, 19], cosmetics [20], agriculture [21], and biomedicine [22]. Figure 4.1 depicts the chemical structure of chitosan and its biological characteristics.

4.2.1 *Physicochemical Properties of Chitosan*

Chitin and chitosan are heteropolymers, unlike cellulose, which is a homopolymer for comparison. For chitin and chitosan, neither their orientation is intended to be inferred. The principal physicochemical properties of the chitosan material are on the basis of the substance, the degree of N-deacetylation, the molecular weight, and the solvent and solution properties. Therefore, in the subsequent sections, we will discuss regarding the four characteristics of the chitosan.

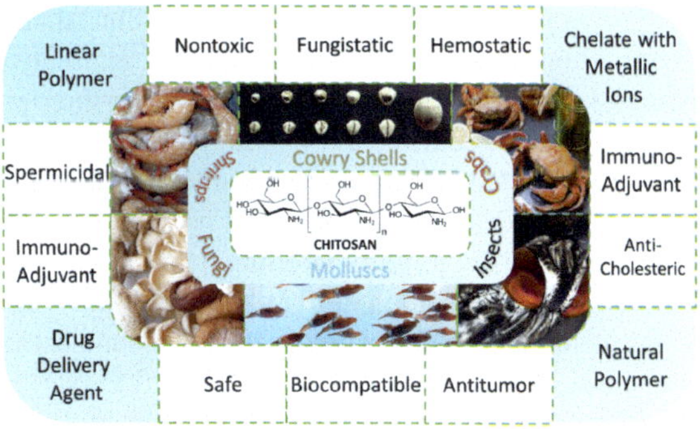

Fig. 4.1 Chitosan's chemical composition and its different sources and properties

4.2.1.1 Degree of N-Acetylation

The degree of N-acetylation in chitosan, for instance, the ratio of 2-acetamido-2-deoxy-d-glucopyranose to 2-amino-2-deoxy-d-glucopyranose structural parts, is a crucial characteristic to pay particular attention for characterization of chitosan [23]. This ratio significantly impacts the solubility and solution characteristics of chitosan. Meanwhile, N-deacetylated offshoot of chitin that is commonly recognized as being non-toxic and soluble in diluted aqueous formic and acetic acids is called chitosan. The majority of the units in chitin are acetylated, with an average acetylation level of 0.90 [24, 25]. The N-deacetylated from chitin, chitosan, generally has a deacetylating grade significantly lower to 0.35 and can be fully or partially deacetylated [26, 27].

4.2.1.2 Molecular Weight

Whereas conventionally, the HPLC is employed to define the M_w distributions of chitosan, light scattering has allowed for the identification of the average molecular weight, M_w of chitosan [28]. The Mark-Houwink equation's constants and K have been calculated in 1/10 M CH_3COOH and 1/5 M NaCl solution by the quick and easy technique of viscometry [29]. The inherent viscosity is defined as follows: $[\eta] = KM^{\alpha} = 1.81 \times 10^{-3}\ M^{0.93}$

Due to chitosan's charged nature in acidic solvents, in addition to its propensity to form clustering multiplexes, using these constants requires care.

4.2.1.3 Crystallinity

Chitosan's crystallinity is determined by how much of a biopolymer is crystalline and how much is amorphous. This feature is measured as a crystallinity index (*CrI*) based on the relationship between the distinctive X-ray diffraction peaks (XRD). Chitosan is a polymorphic biopolymer, semi-crystalline that is solid at room temperature. Its orthorhombic crystal system cell has two alternatively spliced chains without any hydroxyl groups. In addition, chitosan exists in three polymorphic phases in conjunction with variable degrees of crystallinity, centered on the variety and configuration of the chains [30]. Chitosan used for commercial purposes is a semi-crystalline polymer, and *CrI* depends on how much of it has been deacetylated. Additionally, because these characteristic impacts swelling, porosity, hydration, and absorption qualities, the quantitative evaluation of *CrI* are crucial. According to Thakhiew et al., different drying techniques impact chitosan films' mechanical and physical characteristic [31]. They employed low-pressure steam drying (LPSSD) and hot air drying (HAD). They discovered that *CrI* values for LPSSD films are lower (3–5%) than those for HAD films (10–16 percent). The swelling degree was increased as a result to 48–100 percent (LSSPD) and 75–110 percent of HAD [31, 32].

4.2.1.4 Solubility

Chitin like cellulose, there are only a few solvents that can be used as reaction solvents, and materials that are very crystalline and difficult to work with. Chitosan breaks down before melting, which is characteristic of polysaccharides with a lot of hydrogen bonds [33]. In order to impart functionality, it is important to dissolve chitin and chitosan in a suitable solvent. The effects of polymer content, counterion concentration, pH, and temperature on the solvent viscosity must be recognized for each solvent system. There are no data that compare one solvent to another [34]. The most polymer possible is typically dissolved in each solvent solution to create a homogenous liquid. The polymer is then recreated in the necessary form. For polymer regeneration or solidification, a coagulant is necessary. The parameters of the solvent, solution, and polymer utilized have a significant impact on the coagulant's nature.

4.2.2 Antibacterial Properties

Chitosan has many biological effects that make it a potential therapeutic polymer. These effects include antitumoral, antibacterial, antioxidant, and anti-inflammatory effects. It is amazing that regulatory organizations only approve chitosan and chitosan hydrochloride as excipients, not as drugs for treating illnesses. Finding antibiotic alternatives is urgent because bacterial antibiotic resistance remains a serious public health issue. Antimicrobial action is produced by chitosan, chitosan derivatives, and chitooligosaccharides against various types of yeast, filamentous fungus, and bacteria, or generally talking microorganisms [35, 36]. Bacteria can grow after the polymer has been removed from the medium, suggesting that chitosan has a growth-inhibitory effect. This is crucial because resistant populations might develop if the cells acclimate to the chitosan.

The method by which chitosan employs its antibacterial action is still up for debate. This phenomenon may be explained by considering the dearth of suitable polymer characterization, problems with purity, the employment of various microorganisms, and the absence of standard methodology [37]. According to several research, the polymer coating on the cell surface reduces the permeability of their cell membranes, preventing the cells from accessing nutrients [38]. This reaction occurs due to the contact between -NH_2 groups from chitosan chains and -COO- groups on microorganisms' outer cell membranes. As a result, the level of acetylation determines the antibacterial activity. According to another theory, Chitosan may enter cells and prevent RNA transcription because of adsorption with bacterial deoxyribonucleic acid (DNA) [36]. These processes are probably not reciprocally limited, and multiple actions are connected to the suppression of cell development. The following Table 4.1 displays some instances of the bacterial strains that are sensitive to chitosan:

Table 4.1 Chitosan's capacity to inhibit microorganisms

Chitosan Precursor	Affected bacterial System	References
Chitosan	*Hydrophilic Aeromonas* *Aspergillus ictalurid* *Columnar Flavobacterium* *Albicans Candida* *Gram-positive microorganisms* *Gram-negative microorganisms*	[39, 40]
Chitosan Sponges	*S. aureus* *E. coli*	[41, 42]
Chitosan wound dressing	*P. aeruginosa* *B. cereus* *L. monocytogenes*	[43, 44]
Nano- and microparticles of chitosan	*Vibrio cholerae, E. coli, and S. enterica* *Enteric Streptococcus uberis S. uberis* *B. pneumoniae* *S. cholerae V. aureus* *Salmonella typhimurium and choleraesuis*	[37, 45, 46]
Chlorhydrate of chitosan Chitosan carboxymethyl Oligosaccharide of chitosan N-acetyl-D-glucosamine	*Candida glabrata, C. albicans, Candida krusei*	[47]

4.3 Chitosan-Based Nanostructured Composite Preparation

Nanofillers with dimensions under 100 nm can be added to the biopolymers, to form "nano-biocomposites", the same as chitosan, to produce materials with enhanced characteristics that are primarily employed in biomedical applications. A powerful method of overcoming the typical limitations of biopolymers has been the inclusion of nanomaterials integrated into the chitosan matrix. The nanocomposites made of chitosan have the potential to be used in a variety of industries, notably the biomedical ones. Measure up to traditional composites containing 40–50% conventional fillers, chitosan-based nanocomposites exhibit or reveal a substantial increase in properties even at relatively low nanofillers weight amounts like 5–10 wt.% [48]. In composites with high chitosan content, the resulting products exhibit strong bioactivity and biocompatibility. Some additional characteristics are better thermal stability, transparency, enhanced mechanical, and barrier quality [48].

According to their shape and aspect ratio, there are typically three different types of nanofillers. The aspect ratio refers to the ratio of particle length to thickness, which enhances the characteristics of composite materials. Layered or platelet particles (nanoplate), spherical particles (nanoparticle), and fibrous ones (nanofiber) are the existing types of these nanofillers; Fig. 4.2 shows the creation of chitosan-based

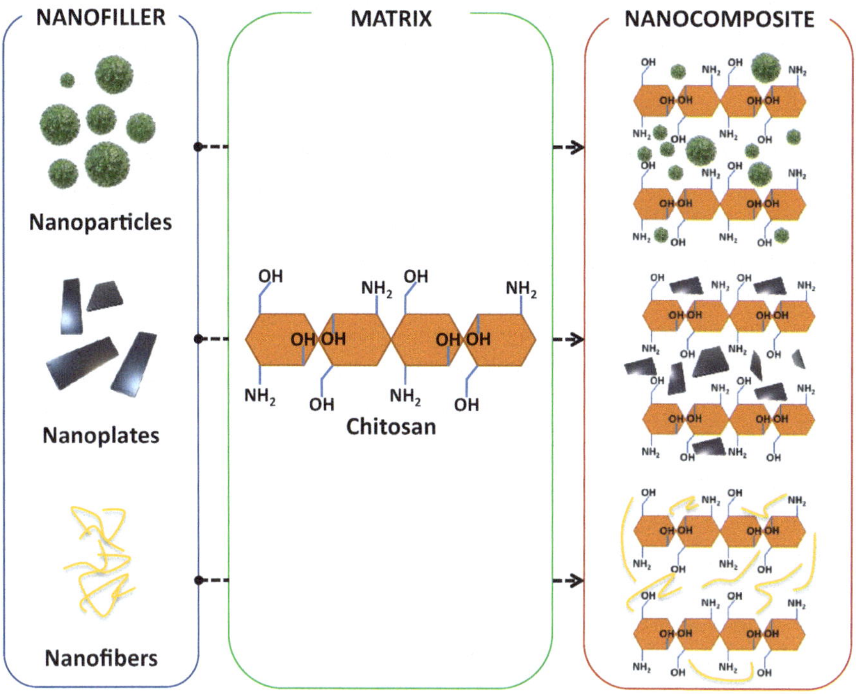

Fig. 4.2 Chitosan based bio-nanocomposites employing different types of nanofillers

bio-nanocomposites using different types of nanofillers. Because they provided a non-toxic, biocompatible, stable, target-specific, and biodegradable delivery system, chitosan nanocomposites were crucial in various industries [36]. Similar to nanocomposites, target-specific diagnostic applications are receiving a lot of interest due to their ease of formulation through the introduction of medicinal agents and imaging agents, which are recognized and designate a significant community concern worldwide. Owing to their unique size-dependent electrical, magnetic, and optical capabilities, some nanoparticles have drawn considerable interest, especially in therapeutic diagnostics, cancer imaging and treatment. The targeted cure becomes a crucial step, particularly in cancer treatment, as it helps in advance imaging- and therapy-based anti-cancer therapies.

4.3.1 Techniques and Methods of Chitosan Nanocomposites Preparation

The preparation procedures for the chitosan-based nanocomposites are given in the following typical approaches.

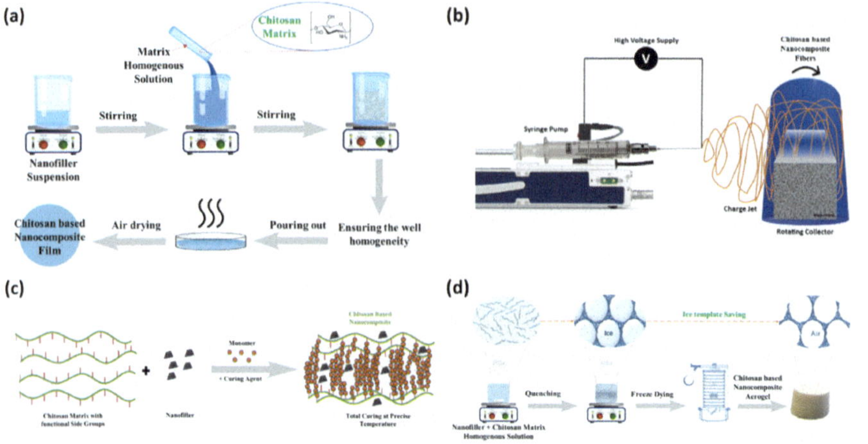

Fig. 4.3 Chitosan nanostructures and their preparation methods

4.3.1.1 Solution Casting Technique

The most straightforward approach for creating chitosan nanocomposites is solution casting, as depicted in Fig. 4.3a. There are three steps in this approach for creating nanocomposites. Initially, mechanical agitation or sonication dissolves or distribute the fillers in a suitable solvent. Later, the polymer is melted in a related solvent. Finally, the polymer and filler solutions are combined at the proper temperature. Nanocomposites are created when this combination is cast or precipitated. Using a solution-casting method, Regiel-Futyra and his research group created chitosan-gold nanocomposites in 2015 [49].

4.3.1.2 Electrospinning Technique

Superior and fragile fibers may be created using electrospinning, with diameters reaching down to tens of nanometers. Figure 4.3b illustrates that electrospinning instruments consist of three basic parts: a grounded conductor that serves as a collector, a multi-needle holding spinneret, and a high-voltage electricity source. The electrospinning liquid is fed into a syringe at a predetermined pace controlled by the syringe pump. Immobilizing charge contents on the shell of a liquid droplet results in a continuous jet. Nanofibers are created when the fluid filament is whipped and stretched quickly. By strengthening chitin nanocrystals with this method, Naseri's research group created fiber mats made of chitosan and polyethylene oxide [50].

4.3.1.3 In-Situ Method

As seen in Fig. 4.3c, the in-situ process is an efficient way to create nanocomposites with evenly distributed filler components. Taking part in this method, filler chemicals are spread across monomers whether a solvent is present. The curing ingredient is then added at the precise temperature needed for optimal polymerization. Hebeish's research team used this method in 2014 to create nanocomposites based on chitosan. Silver ions were chemically reduced in implant copolymerization of acrylonitrile onto chitosan films to create poly acrylonitrile silver nanocomposites [51].

4.3.1.4 Freeze-Drying Technique

A homogenous combination of polymer/filler solution is flowed into a copper mold, which is then quenched in nitrogen liquid to chill it to extremely low temperatures. The solvents are subsequently removed from the polymer scaffold via freeze-drying, as shown in Fig. 4.3d. The porosity of the constructed scaffold is 90%, with pore sizes ranging from 15 to 35 m. The use of these scaffolds in tissue engineering has skyrocketed. In 2013, Liu's research team created a chitosan-based nanocomposite containing halloysite nanotubes by combining the freeze-drying and solution-casting techniques [52].

4.3.1.5 Additional Methods

Since chitosan is still a polymer material derived from natural sources, it may be subjected to conventional polymer shaping techniques, including three dimensions (3D) printing, extrusion, injection molding, rotational molding, and thermoforming. Choo et al. and Amouzgar et al. effectively created extruded chitosan nanocomposites [53, 54]. They investigated the utilization of chitosan within nano-activated carbon in the first study, while halloysite nanotubes were the chosen nanofillers in the second. An extra ultrasonic dispersion was necessary for the extrusion procedure to be effective in both works.

4.3.2 Nanofillers of the Chitosan-Based Nanostructured Composites

As indicated in Table 4.2, various nanostructures, including nanoparticles, nanosheets, nanorods, nano-capsules, nanowires, and nanofibers, can be used and disseminated as additives in chitosan. These include inorganic, organic, metallic, and semiconducting nanostructures. In addition to being used as a pure biomaterial matrix, chitosan may now be combined with various nanostructures, either

Table 4.2 Different nanoparticles used as fillers and reinforcements in chitosan nanocomposites

Filler category	Nanostructure system	Size (nm)	References
Cellulose	Nanocrystals	100	[55]
Chitin	Nanofibers	50–500	[56]
Graphene	Nanosheets	<400	[57]
Nanoclay	Nanoparticles	<100	[58]
PVA	Nanocapsules	113	[59]
Silica	Nanocapsules	6–50	[60]
Metals and Metals oxide	Nanoparticles	10–125	[61] [62] [63]

incorporated in the bulk material or coated on the surface, thanks to advancements in nanotechnology. As pure single materials, biopolymers like chitosan may have significant limitations, weak mechanical strength, low barrier properties, and limited thermal stability. Inorganic and organic substances of nanoscale sizes have been added to chitosan to solve several of these structural faults, either as a filler distributed throughout the entire matrix or as a layer at the material surface.

4.4 Chitosan-Based Nanostructured Composite Characterization

Recent and significant efforts in the biomedical sphere are receiving growing attention due to all the application possibilities for chitosan nanocomposites attributable to their versatile properties and assets. The literature is concentrated chiefly on chitosan nanocomposites in tissue engineering, biomedical applications, and other industrial applications due to this material's quality of adequation. The overall properties of the chitosan nanocomposites on the structural, morphological, chemical, and antibacterial attributes will therefore be the focus of the following sections.

4.4.1 Microscopy-Based Analysis

Because they directly affect the material's ability to absorb fluids, the surface morphology of chitosan nanocomposite materials is crucial, particularly for wound dressings. For instance, Shao et al. produced freeze-dried chitosan sponges with embedded silver sulfadiazine nanoparticles as possible materials for wound dressings [64]. Figure 4.4 depicts the surface morphology of various sponge preparations employing the Scanning Electron Microscope (SEM). Sponge's open porous three-dimensional network architecture and the homogeneous pore diameters are readily

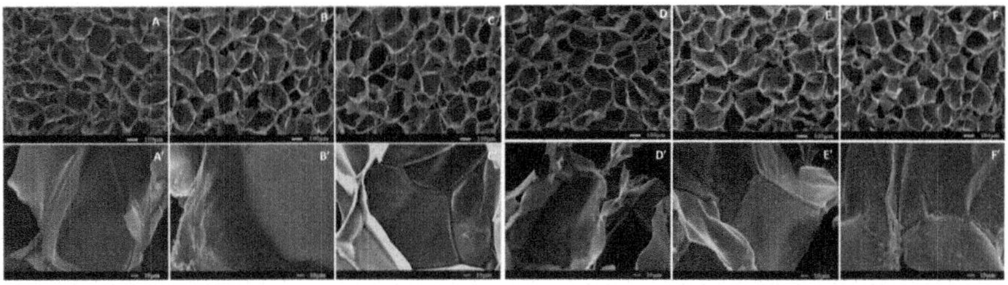

Fig. 4.4 The cross-section shows chitosan nanocomposite sponges reinforced with varying silver sulfadiazine nanoparticles. The witness chitosan is presented by (A), (A'), and the chitosan nanocomposites with filler quantities of 0.01 wt.%, 0.02 wt.%, 0.04 wt.%, 0.08 wt.%, and 0.12 wt.% are presented by (B), (B'), (C), (C'), (D), (D'), (E), (E'), and (F), (F'), respectively. Reproduced with permission from Elsevier [64]

seen in the cross-sectional image. The nanocomposites demonstrated better cell survival of human embryonic kidney 293 cells coupled with higher swelling ratios, porosity, and antibacterial properties. The cell shape, however, abruptly changed when the nano-silver particle concentration was raised over 5%. This may be because of harmful repercussions, according to speculation.

4.4.2 *Biocompatibility Analysis*

Chitosan nanocomposites must be biocompatible to stimulate the resilience or stability of the material within the human body. Jiang et al. created spiral-cylindrical nanocomposite scaffolds for bone regeneration applications using nano-hydroxyapatite mixed with chitosan and cellulose [65]. The chitosan nanocomposite's biocompatibility with medium was demonstrated by an in vitro tetrazolium-based colorimetric assay that revealed there was no difference in cell proliferation between the control sample and nanocomposites after 5 days of culture. However, when 60 weight percent of nano-hydroxyapatite filler was added to the blends, there was an increase in cellular absorption and osteocalcin expression. Including the nanocomposite scaffolds significantly increased osseointegration, total intrusion of bone tissues, and NB formation compared to the control group in vivo was studied using New Zealand white rabbits. These characteristics could make it easier to repair and rebuild bone abnormalities quickly. Figure 4.5 displays the main conclusions drawn from X-ray images, 3D Micro-computed tomography (micro-CT) scans of the control groups, and nanocomposite scaffolds. The manufactured substrate is a good option for bone tissue regeneration, according to these combined data.

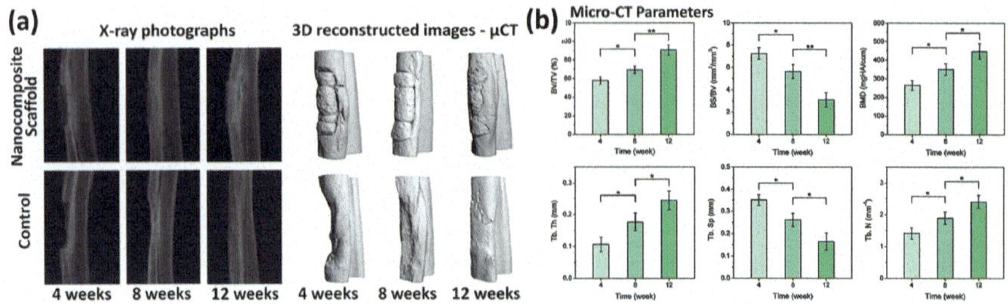

Fig. 4.5 **a** Images of the rabbit forelimb's bone taken using X-rays and 3D reconstruction, as well as, **b** parameters measured using micro-CT scans of the spiral-cylindrical chitosan nanocomposite scaffold during the production of new bone. Reproduced with permission from American Chemical Society [65]

4.4.3 Mechanical Characterization

In addition to addressing the tendency of nanoparticles to aggregate, which reduces the interfacial area and leads to poor mechanical characteristics, a good nanofiller dispersion within the matrix must be attained to obtain the required increase in mechanical capabilities. The interfacial strength significantly influences the final mechanical qualities of the chitosan nanocomposites, for instance, toughness, tensile strength, and elastic modulus. Graphene oxide (GO) skeletons blending chitosan with sodium alginate (SA) and collagen (Col) for bone tissue engineering pose as a good potential material, according to studies conducted by Kolanthai et al. [66]. In the same context, Ca^{2+} provided a second crosslink among the alginate chains of the crosslinked Sodium Alginate-Chitosan-Collagen-Graphene Oxide (SA-CS-Col-GO) scaffold, in which the primary contact between SA, chitosan, Col molecules, and GO particles in the SA-CS-Col-GO scaffold was a hydrogen bond. Photographs of the SA-CS-Col-GO scaffolds are shown in Fig. 4.6a. Also, according to Fig. 4.6b, compressive stress–strain curves demonstrated that adding GO and Ca^{2+} to the Sodium Alginate-Chitosan-Collagen (SA-CS-Col) matrix enhanced mechanical strength.

4.4.4 Chemical Structure

By using Fourier transform infrared spectroscopy (FT-IR), Hydrogen-nuclear magnetic resonance (^{1}H-NMR), and Carbon-nuclear magnetic resonance (^{13}C-NMR), the structural alterations of chitosan-based nanocomposites are primarily identified. Alterations are often applied to nanofillers to enable different physicochemical features to the resultant chitosan nanocomposite, identify interactions with active ingredients, and investigate how chitosan binds to specific biological receptors. Following the synthesis of 4-pyridinecarboxaldehyde, Omidi et al. grafted

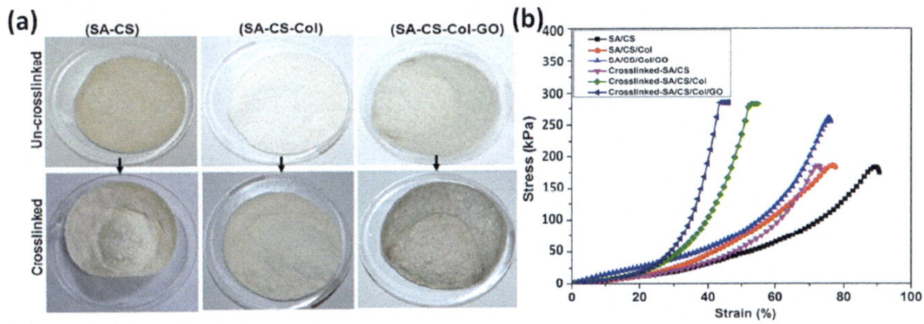

Fig. 4.6 **a** Photographic images of non-crosslinked and chemically cross-linked scaffolds prepared by freeze-drying technique and **b** their compressive stress–strain curves. Reproduced with permission from American Chemical Society [66]

chitosan and nanoparticles onto the series of quaternary pyridinium derivatives [67]. Quaternary ammonium compounds-chitosan crossbreeds were tested for antibacterial activity towards two Gram-positive and two Gram-negative bacteria. Consequently, they have demonstrated a substantial impact on the Gram-positive bacteria, as shown in Fig. 4.7. The development of chitosan nanocomposite was confirmed by FT-IR, ^{1}H, and ^{13}C NMR. New chitosan derivatives exhibited marginally improved antioxidant activity, according to the 2,2-diphenyl-1-picryl-hydrazyl-hydrate radical essay (DPPH-radical) scavenging experiment.

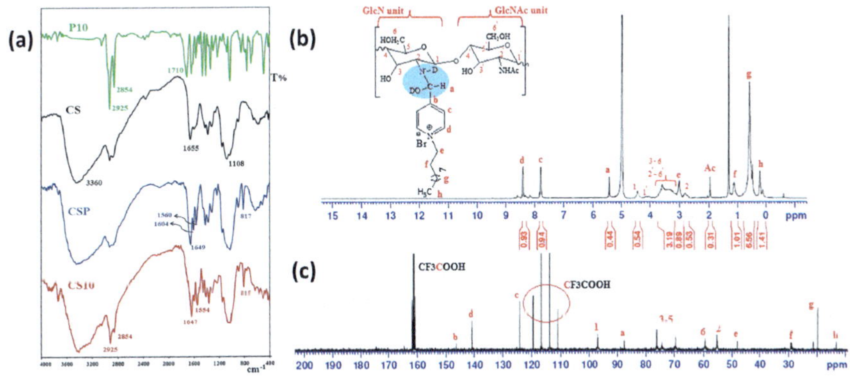

Fig. 4.7 The **a** FT-IR, **b** 1H-NMR, and **c** 13C-NMR spectra of chitosan derivatives with nanofillers. Reproduced with permission from Elsevier [67]

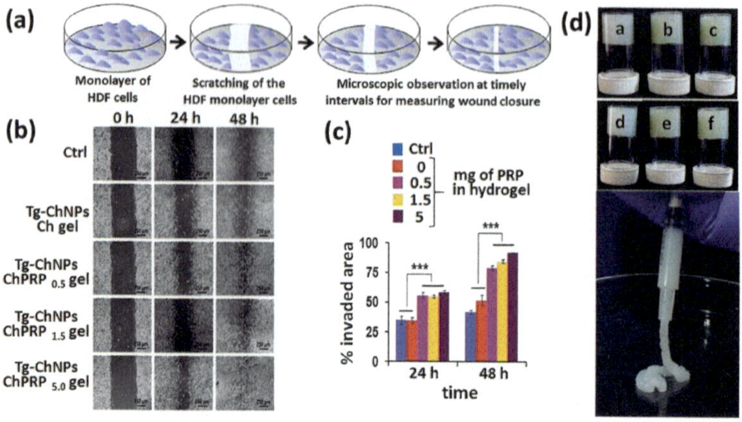

Fig. 4.8 **a** A graphic illustration of the culture dish-based in vitro wound closure experiment. **b** Photographs showing the migration of L929 cells 48 h after they were cultured with tigecycline nanoparticles reinforced chitosan-platelet-rich plasma (PRP) nanocomposite hydrogels at various PRP concentrations in comparison to a control chitosan gel. **c** Area of the nanocomposite hydrogel's scratch wound closure percentage in comparison to a chitosan gel control. **d** An inversion test for the integration of tigecycline nanoparticles in chitosan-platelet-rich plasma (PRP) nanocomposite hydrogels to determine gel formation and enhanced rheological properties. Reproduced with permission from American Chemical Society [68]

4.4.5 *Antibacterial Properties*

Chitosan's cationic kind, which enables the communication between negatively charged lipids and proteins found in the bacterial cell wall, has been shown in several trials to have a remarkable antibacterial impact on various disease-causing bacteria. Nimal et al. created injectable nanostructured chitosan hydrogels containing tigecycline and platelet-rich plasma to treat infected wounds [68]. The best outcomes in antibacterial behavior were achieved in the presence of the hybrid nanogels (Tigecycline Loaded Chitosan Nanoparticles (tg-ChNPs-ChPRP) gels) in an in vitro wound closure research utilizing L929 cell migration. Figure 4.8 illustrates in vitro scratch wound closure tests that help us better grasp the chitosan nanocomposites' wound regeneration properties.

4.5 Biomedical Applications of Chitosan-Based Nanostructured Composite

Chitosan is a biopolymer with interesting properties that make it a revolutionary material in the pharmaceutical and medical industry. According to investigations, the structure of chitosan helps with better collagen deposition and three-dimensional cell growth which accelerates wound healing [69]. Usually, sponges' fiber and

Table 4.3 Nanostructured additives applied in chitosan-based nanocomposite materials for biomedical applications

Nanostructured additives	Nanostructure	Biomedical application	References
Ag	Nanoparticles	Anticancer	[71]
Nanocellulose	Nanocrystals/ nanofibrils	Wound dressings	[72]
Nano-hydroxyapatite (nHAP)	Nanoparticles	Bone tissue engineering	[73]
Fe_3O_4	Nanoparticles	Magnetic hyperthermia for cancer therapy	[74]
(Pt)	Nanoparticles	Drug delivery	[75]
Graphene oxide	Nanosheets	Biosensor	[76]
ZnO	Nanoparticles	Cancer therapy	[77]
Gold	Nanoparticles	Biosensor	[78]
Gold	Nanorods	Drug delivery	[79]
Carbon nanotube	Nanofibers	Cartilage tissue engineering	[80]
Nanocellulose	Nanofiber	Wound dressing	[81]
ZrO_2	Nanoparticles	Biosensor	[82]
Ag	Nanoparticles	Drug delivery	[83]
TiO_2	Nanoparticles	Wound healing	[84]

film/membranes forms of chitosan have been utilized for wound healing and tissue engineering applications. Chitosan has great capacity as a delivery agent for biologically active molecules, as well as in the organization of tissues and cells [7]. As pure material, chitosan can have some drawbacks such as weak mechanical, thermal, and barrier properties [14]. Several nanometric compounds of inorganic, organic, or metallic nature may be used and dispersed as nanoadditives in chitosan, such as nanoparticles, nanosheets, nanofibers, nanorods, nanocrystals among others, to solve most of the disadvantages related to chitosan in order to be applied in various fields of biomedical research [70], as presented in Table 4.3. Thanks to their non-toxicity, biodegradability, and antibacterial properties, chitosan-based nanostructured composites have interested many researchers for several biomedical applications such as tissue engineering, drug delivery systems, dressing for wounds and burns [13].

4.5.1 Drug and Bio-Macromolecules Delivery

The use of nanomaterials in the pharmaceutical field is increasing more and more. Using materials or biopolymers of size $\leq$ 100 nm to formulate chitosan-based nanostructured composites improves the hemostatic potential. These materials have the ability to cross biological barriers in order to transport drugs to the specific site efficiently [7]. Various nanomaterials/nanoparticles have been applied to treat infections and various eye diseases, diabetes, and cancer [7]. The use of nanoparticles/nanomaterials can provide better drug stability with a slower, continuous release approach [10]. Hua et al. studied the controlled release of ofloxacin using nanostructured chitosan/MMT-based hydrogels. The results found showed that the nanomaterial enhances drug entrapment with high potential as drug carriers for sustained release [85]. George et al. [86] investigated the design of a nanomaterial in a hydrogel form based on ZnO nanoparticles in which an extract of Cucumis melo seeds was added in the matrix followed by cross-linking with dialdehyde cellulose for curcumin delivery. New methods of treating cancer are currently possible through the use of nanostructured chitosan composites, specifically through drug delivery. Kankala et al. [75] developed an interesting drug delivery platform by coating a chitosan polymer composite layer embedded with platinum (Pt) nanoparticles onto zinc (Zn) doped siliceous matrices loaded with doxorubicin, generally used against multidrug resistance for treating cancer. Zhao et al. [87] developed a new graphene oxide nanoparticles (GON)-based system with GON and chitosan/dimethylmaleic anhydride-modified chitosan for drug and biomacromolecules delivery. According to the results, chitosan-based nanomaterials show hope for gene therapy and biomacromolecules delivery for treating genetic diseases. Long et al. [88] synthesized magnetite chitosan/carrageenan nanomaterials via in-situ preparation of iron ions in carrageenan and subsequently incorporated with chitosan. The results showed that the prepared nanocomposites could effectively be applied for targeted delivery of therapeutic macromolecules.

4.5.2 Wound Healing

Antibacterial resistance of microorganisms is considered one of the major obstacles encountered in wound care, leading to complications such as infection and delayed healing [89]. Several studies are currently devoted to develop newer antibacterials for treating wounds infected with antibacterial resistant microorganisms. Thanks to its antibacterial capacity, chitosan could solve wound healing problems [90]. Chitosan-based nanostructured materials provide anti-microbial agents for the treatment of wound or burn [13]. Figure 4.9 presents in detail the stages of the skin wound healing process using chitosan-based nanocomposite materials. Ding et al. [91] studied a nanocomposite system with silver nanoparticles and chitosan for external skin wound healing. The film required higher mechanical strength to attach to the skin and was crosslinked with genipin, which led to a decrease in the active sites of chitosan as well as its antibacterial capacity, which was replaced by the incorporation of silver nanoparticles. These results showed that chitosan nanocomposites could also be used for the treatment of internal wounds. The study by Sundaram et al. [92] has developed an injectable composite with chitosan and nano-bio-glass. Their study provides safe blood clotting, thanks to the synergistic interactions existing between the chitosan biopolymer and the nano-glass in contact with blood. Riccio et al. [72] designed

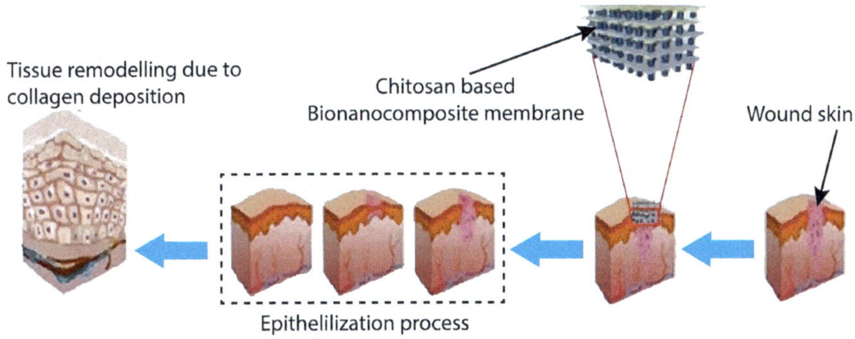

Fig. 4.9 Wound skin healing process using chitosan-based nanocomposite materials. Reproduced with permission from Elsevier [13]

novel chitosan/nanocellulose-based nanocomposite films filled by betamethasone or silver sulfadiazine as functional dressings. This work showed that the designed films have a great potential as active dressings for controlled drug delivery in the wound site to treat inflamed and purulent wounds, noninfectious dry, and infectious wounds.

4.5.3 Bone Tissue Engineering

Tissue engineering is a reparative medicinal area that has undergone a great evolution in recent years. It emerged from research on the development of biomaterials. This practice involves using cells, 3D scaffolds, and biologically active molecules to maintain, repair and improve the function of a specific injured tissue or organ [89]. The use of biopolymer-based nanocomposites has got huge attention in the discipline of tissue engineering, thus enforcing researchers to enhance the techniques of biopolymer design development to have materials with a high potential and ablilty to repair damaged tissue. Thanks to their anti-microbial, hemostatic, and biodegradable abilities, chitosan-based nanomaterials are the subject of major applications in the field of tissue engineering [7, 93]. In bone tissue engineering, chitosan-based nanostructured composites may replace synthetic polymers. The chitosan molecules have a hydrophilic surface as contrary to several synthetic polymers that promote cell adhesion and proliferation; they are designed as films and scaffolds, with multiple applications in tissue regeneration and cell transplantation [94]. Owing to the crosslinking capacity of the chitosan biopolymer, the combination of chitosan-based composites with other synthetic biopolymers, the use of nanoparticles, expanded the chitosan shaping methods [16, 95], to create new scaffolds for bone tissue engineering with controlled biocompatibility and mechanical strength of the produced material [96]. Moreira et al. [97] developed new in-situ forming hydrogels from chitosan and gelatine combined with bioactive glass nanoparticles. The nanomaterials cytocompatibility was studied based on the cell viability of the human osteosarcoma cell line.

Carbon-based nanomaterials such as carbon nanotubes [98], graphene and graphene oxide [7] have been successfully combined with chitosan for bone tissue engineering. Carbon-based nanocomposites have shown remarkable potential for the reinforcement of hybrid scaffolds due to their good mechanical stability [99].

4.5.4 Biosensor

Biosensors are chemical sensors with the ability to recognize the properties of biological components [100]. Recently, the pharmaceutical field has increasingly seen the rise in use of biosensors. A better quality of immobilization of biological recognition elements is very important to develop reliable biosensors. Chitosan-based nanocomposites are considered an excellent immobilization matrix. Chitosan is a natural polymer with many functional properties, such as crosslinking ability with multiple nanomaterials, and high permeability [101]. The biosensors based on chitosan nanocomposites have served as new instruments for the early identification of diseases and detecting of various biomarkers [13]. Shan's research group developed nanocomposite films of graphene/gold nanoparticle/chitosan (GAuCS) for glucose biosensing. The biosensors aim to fix glucose oxidase on the gold electrode in the films of the GAuCS nanomaterial [102]. Ahmet Güner et al. [103] studied a hybrid detection platform using gold nanoparticles and chitosan as well as multi-walled carbon nanotubes (MWCNTs) for the detection of O157:H7 Escherichia coli (E. coli). The nanomaterial is modified with polypyrrole in order to have a branched pyrrole chitosan structure for good antibody binding. Nanomaterials such as graphene nanosheets and carbon nanotubes are associated with chitosan as materials for improving the properties of the composite [101]. Graphene nanosheets can improve the homogeneity of chitosan film and promote DNA hybridization reaction [101]. AL-Mokaram et al. [104] investigated biosensors consisting of chitosan polypyrrole-iron oxide composite for the detection of glucose. The incorporation of iron oxide nanoparticles aims to achieve the objective of non-enzymatic detection.

4.5.5 Other Biomedical Applications

Besides the use of chitosan-based nanocomposites for drug delivery, tissue engineering, wound healing and biosensors, chitosan as a composite nanostructured material has been used in other biomedical fields such as bioimaging [70]. This practice has been used extensively in research and clinical trials. Bioimaging makes it possible to highlight biological problems due to its ability to quantify and detect phenotypic changes [7]. Currently, chitosan nanocomposite materials have gradually emerged for bioimaging as excellent candidates [13]. Salehizadeh's research group [105] has designed a magnetic core–shell nanocomposite for bioimaging, combining Fe_3O_4-gold nanoparticles with stabilizing agent chitosan. Fe_3O_4 showed excellent magnetic properties providing magnetic resonance for imaging. Kumar et al. [106] have devised carboxymethyl CS-wrapped Fe_3O_4 nanoparticles (NPs) decorated with carbon dots to have the fluorescent magnetic nanoparticles for cancer cells imaging. This demonstrates that chitosan-based fluorescent magnetic nanoparticles have an outstanding capacity for diagnostic aims. Banerjee et al. [107] have designed chitosan-hydroxyapatite luminescent nanocomposite films that can be used as a fluorescent cell regeneration composite material or as an antibacterial wound dressing.

Among the various applications of chitosan-based nanocomposites in the biomedical field, mention may be made of hemoperfusion, which consists of eliminating toxins from the blood using adsorbent in the purification of blood for treating diseases. The excess bilirubin, which is a pathogenic toxin, can damage various tissues which can lead to death [108]. In the last decades, several types of adsorbents have been widely investigated to eliminate the excess of bilirubin, such us carbon nanotubes, and cross-linked chitosan [109]. Chen et al. [110] synthesized porous nanomaterials with improved adsorption properties. Cross-linked chitosan/SiO_2-loaded graphene (CS/graphene-SiO_2) nanocomposite beads were synthesized for the adsorption of bilirubin. According to results, the developed CS/graphene-SiO_2 nanocomposites as an adsorbent for bilirubin could be promising materials in blood purification applications.

4.6 Conclusion and Perspectives

The growing demand for biomaterials over the recent decades in the medical area has driven the use of various natural polymers. Being biodegradable, non-toxic, and biocompatible, chitosan is a promising biopolymer with many functionalities. As discussed in this chapter, chitosan has attracted great interest in the biomedical field, including tissue engineering, bioimaging, biosensors, and drug delivery among others. The reactive functional groups present in the skeleton of chitosan facilitate its physical or chemical modification, in order to give it interesting new properties. Chitosan can be combined with various inorganic and organic nanostructures to provide other properties, useful in the biomedical area, especially to meet new challenges in applications. This chapter includes a comprehensive list of the development, preparation, microscopic, mechanical, and structural characterization of nanocomposite materials based on chitosan. Chitosan-based nanostructured composites have become a focus of research interest in several medical practices such as drug delivery, bone tissue engineering, and biosensors, due to their distinct advantages such as biodegradability, biocompatibility and non-toxicity and their enhanced structural and functional characteristics. In addition, chitosan is considered as an interesting matrix for the design of wound healing fabrics thanks to its low cost, high biocompatibility, and antibacterial capacity. All these features combine to make chitosan based nanostructured materials potential candidates for several biomedical areas, but few limitations still need to be considered to meet the clinical criteria. Additionally, drawbacks related to biomedical applications still remain as challenges, such as drug delivery efficiency, delivery duration, and drug loading ability. This chapter also highlights some biomedical areas and motivates researchers to improve the physicochemical attributes of chitosan-based nanocomposites. Further research is needed to forward chitosan-based nanocomposites from the laboratory to human clinical trials. Future studies are anticipated in fields such as bioimaging, biosensors, and anti-cancer drug applications. The points investigated in this chapter underline the nanostructure-activity-application trade-off in the preparation and modification of chitosan-based nanostructured composites. This perspective may guide researchers toward the most suitable preparation and modification strategies to obtain optimal performance for their targeted applications. In addition, the collaboration of bioengineers and clinical researchers can overcome the problems related to unknown regenerative therapies by studying the morphologies, sizes, and diameters of the materials developed according to the desired clinical application.

References

1. F.S. Rezaei, F. Sharifianjazi, A. Esmaeilkhanian, E. Salehi, Chitosan films and scaffolds for regenerative medicine applications: a review, in *Carbohydrate Polymers*, vol. 273. (Elsevier Ltd., 2021). https://doi.org/10.1016/j.carbpol.2021.118631
2. M. Xie, K. Huang, F. Yang, R. Wang, L. Han, H. Yu, Z. Ye, F. Wu, Chitosan nanocomposite films based on halloysite nanotubes modification for potential biomedical applications. Int. J. Biol. Macromol. **151**, 1116–1125 (2020). https://doi.org/10.1016/j.ijbiomac.2019.10.154
3. E. Khademian, E. Salehi, H. Sanaeepur, F. Galiano, A. Figoli, A systematic review on carbohydrate biopolymers for adsorptive remediation of copper ions from aqueous environments-part A: classification and modification strategies, in *Science of the Total Environment*, vol. 738 (Elsevier B.V., 2020). https://doi.org/10.1016/j.scitotenv.2020.139829
4. M. Jouyandeh, S.M.R. Paran, M. Shabanian, S. Ghiyasi, H. Vahabi, M. Badawi, K. Formela, D. Puglia, M.R. Saeb, Curing behavior of epoxy/Fe3O4 nanocomposites: a comparison between the effects of bare Fe3O4, Fe3O4/SiO2/chitosan and Fe3O4/SiO2/chitosan/imide/phenylalanine-modified nanofillers. Prog. Org. Coat. **123**, 10–19 (2018). https://doi.org/10.1016/j.porgcoat.2018.06.006
5. W. Deng, Y. Tan, M.S. Riaz Rajoka, Q. Xue, L. Zhao, Y. Wu, A new type of bilayer dural substitute candidate made up of modified chitin and bacterial cellulose. Carbohydr. Polym. 256 (2021). https://doi.org/10.1016/j.carbpol.2020.117577
6. A.O. Silva, R.S. Cunha, D. Hotza, R.A.F. Machado, Chitosan as a matrix of nanocomposites: a review on nanostructures, processes, properties, and applications, in *Carbohydrate Polymers*, vol. 272 (Elsevier Ltd., 2021). https://doi.org/10.1016/j.carbpol.2021.118472
7. A. Khan, K.A. Alamry, Recent advances of emerging green chitosan-based biomaterials with potential biomedical applications: a review, in *Carbohydrate Research*, vol. 506 (Elsevier Ltd., 2021). https://doi.org/10.1016/j.carres.2021.108368
8. M.M. Rahman, M. Shahruzzaman, M.S. Islam, M.N. Khan, P. Haque, Preparation and properties of biodegradable polymer/nano-hydroxyapatite bioceramic scaffold for spongy bone regeneration. J. Polym. Eng. **39**(2), 134–142 (2019). https://doi.org/10.1515/polyeng-2018-0103

9. G. Yang, H. Sun, R. Cao, Q. Liu, X. Mao, Characterization of a novel glycoside hydrolase family 46 chitosanase, Csn-BAC, from Bacillus sp. MD-5. Int. J. Biol. Macromol. **146**, 518–523 (2020). https://doi.org/10.1016/j.ijbiomac.2020.01.031
10. Annu, K. Manzoor, S. Ahmad, A. Soundarajan, S. Ikram, S. Ahmed, Chitosan based nanomaterials for biomedical applications, in *Handbook of Nanomaterials for Industrial Applications* (Elsevier, 2018), pp. 543–562. https://doi.org/10.1016/B978-0-12-813351-4.00031-6
11. C. Rodrigues, J.M.M. de Mello, F. Dalcanton, D.L.P. Macuvele, N. Padoin, M.A. Fiori, C. Soares, H.G. Riella, Mechanical, thermal and antimicrobial properties of chitosan-based-nanocomposite with potential applications for food packaging. J. Polym. Environ. **28**(4), 1216–1236 (2020). https://doi.org/10.1007/s10924-020-01678-y
12. S.K. Pillai, S.S. Ray, *Chitosan-based Nanocomposites* (2012), pp. 33–68. https://doi.org/10.1039/9781849735315-00033
13. M. Azmana, S. Mahmood, A.R. Hilles, A. Rahman, M.A. Arifin, Bin, S. Ahmed, A review on chitosan and chitosan-based bionanocomposites: promising material for combatting global issues and its applications. Int. J. Biol. Macromol. **185**, 832–848 (2021); Elsevier B.V. https://doi.org/10.1016/j.ijbiomac.2021.07.023
14. P.S. Bakshi, D. Selvakumar, K. Kadirvelu, N.S. Kumar, Chitosan as an environment friendly biomaterial – a review on recent modifications and applications. Int. J. Biol. Macromol. **150**, 1072–1083 (2020); Elsevier B.V. https://doi.org/10.1016/j.ijbiomac.2019.10.113
15. H. El Knidri, A. Laajeb, A. Lahsini, Chitin and chitosan: chemistry, solubility, fiber formation, and their potential applications, in *Handbook of Chitin and Chitosan: Volume 1: Preparation and Properties. INC* (2020). https://doi.org/10.1016/B978-0-12-817970-3.00002-X
16. M. El Achaby, Z. Kassab, A. Barakat, A. Aboulkas, Alfa fibers as viable sustainable source for cellulose nanocrystals extraction: application for improving the tensile properties of biopolymer nanocomposite films. Ind. Crop. Prod. 112(December 2017), 499–510 (2018). https://doi.org/10.1016/j.indcrop.2017.12.049
17. J.A.S. Moreno, A.C. Mendes, K. Stephansen, C. Engwer, F.M. Goycoolea, A. Boisen, L.H. Nielsen, I.S. Chronakis, Development of electrosprayed mucoadhesive chitosan microparticles. Carbohyd. Polym. **190**, 240–247 (2018). https://doi.org/10.1016/j.carbpol.2018.02.062
18. F. Devlieghere, A. Vermeulen, J. Debevere, Chitosan: antimicrobial activity, interactions with food components and applicability as a coating on fruit and vegetables. Food Microbiol. **21**(6), 703–714 (2004). https://doi.org/10.1016/j.fm.2004.02.008
19. F. HamPichavant, G. Sèbe, P. Pardon, V. Coma, Fat resistance properties of chitosan-based paper packaging for food applications. Carbohyd. Polym. **61**(3), 259–265 (2005). https://doi.org/10.1016/j.carbpol.2005.01.020
20. X.Y. Sun, J.M. Wang, J.M. Ouyang, L. Kuang, Antioxidant Activities and Repair Effects on Oxidatively Damaged HK-2 Cells of Tea Polysaccharides with Different Molecular Weights. *Oxidative Medicine and Cellular Longevity* (2018). doi.org/https://doi.org/10.1155/2018/5297539
21. M. Rinaudo, Chitin and chitosan: properties and applications. Prog. Polym. Sci. (Oxford) **31**(7), 603–632 (2006). https://doi.org/10.1016/j.progpolymsci.2006.06.001
22. Y.C. Chung, H.L. Wang, Y.M. Chen, S.L. Li, Effect of abiotic factors on the antibacterial activity of chitosan against waterborne pathogens. Biores. Technol. **88**(3), 179–184 (2003). https://doi.org/10.1016/S0960-8524(03)00002-6
23. G. Maghami, G. Roberts, Studies on the adsorption of anionic dyes on chitosan. Die Makromolekulare Chemie **189**(10), 2239–2243 (1988)
24. A. Baxter, M. Dillon, K.D. Anthony Taylor, G.A.F. Roberts, Improved method for i.r. determination of the degree of N-acetylation of chitosan. Int. J. Biol. Macromol. **14**(3), 166–169 (1992). https://doi.org/10.1016/S0141-8130(05)80007-8
25. A. Domard, Circular dichroism study on N-acetylglucosamine oligomers. Int. J. Biol. Macromol. **8**(4), 243–246 (1986). https://doi.org/10.1016/0141-8130(86)90034-6
26. L. Raymond, F.G. Morin, R.H. Marchessault, Degree of deacetylation of chitosan using conductometric titration and solid-state NMR. Carbohyd. Res. **246**(1), 331–336 (1993). https://doi.org/10.1016/0008-6215(93)84044-7

27. H. Sashiwa, H. Saimoto, Y. Shigemasa, R. Ogawa, S. Tokura, Distribution of the acetamide group in partially deacetylated chitins. Carbohyd. Polym. **16**(3), 291–296 (1991). https://doi.org/10.1016/0144-8617(91)90114-R
28. A.C.M. Wu, Determination of molecular-weight distribution of chitosan by high-performance liquid chromatography. Methods Enzym. **161**(C), 447–452 (1988). https://doi.org/10.1016/0076-6879(88)61055-X
29. R.A.A. Muzzarelli, C. Lough, M. Emanuelli, The molecular weight of chitosans studied by laser light-scattering. Carbohydr. Res. **164**(C), 433–442 (1987). https://doi.org/10.1016/0008-6215(87)80146-5
30. Y. Zhong, C. Zhuang, W. Gu, Y. Zhao, Effect of molecular weight on the properties of chitosan films prepared using electrostatic spraying technique. Carbohyd. Polym. **212**(February), 197–205 (2019). https://doi.org/10.1016/j.carbpol.2019.02.048
31. W. Thakhiew, S. Devahastin, S. Soponronnarit, Physical and mechanical properties of chitosan films as affected by drying methods and addition of antimicrobial agent. J. Food Eng. **119**(1), 140–149 (2013). https://doi.org/10.1016/j.jfoodeng.2013.05.020
32. A. Dodero, E. Brunengo, M. Alloisio, A. Sionkowska, S. Vicini, M. Castellano, Chitosan-based electrospun membranes: Effects of solution viscosity, coagulant and crosslinker. Carbohyd. Polym. **235**(February), 115976 (2020). https://doi.org/10.1016/j.carbpol.2020.115976
33. S. Wei, Y.C. Ching, C.H. Chuah, Synthesis of chitosan aerogels as promising carriers for drug delivery: a review. Carbohyd. Polym. **231**, 115744 (2020). https://doi.org/10.1016/j.carbpol.2019.115744
34. N.D. Aljbour, M.D.H. Beg, J. Gimbun, Acid hydrolysis of chitosan to oligomers using hydrochloric acid. Chem. Eng. Technol. **9**, 1741–1746 (2019). https://doi.org/10.1002/ceat.201800527
35. A.A. El-Bindary, E.A.Toson, K.R. Shoueir, H.A. Aljohani, M.M. Abo-Ser, Metal–organic frameworks as efficient materials for drug delivery: Synthesis, characterization, antioxidant, anticancer, antibacterial and molecular docking investigation. Appl. Organomet. Chem. **34**(11), 1–15 (2020). https://doi.org/10.1002/aoc.5905
36. M. Shahini Shams Abadi, E. Mirzaei, A. Bazargani, A. Gholipour, H. Heidari, N. Hadi, Antibacterial activity and mechanism of action of chitosan nanofibers against toxigenic Clostridioides (Clostridium) difficile Isolates. Annali Di Igiene : Medicina Preventiva e Di Comunita **32**, 72–80 (2020). https://doi.org/10.7416/AI.2020.2332
37. B.C. Lee, M.S. Kim, S.H. Choi, K.Y. Kim, T.S. Kim, In vitro and in vivo antimicrobial activity of water-soluble chitosan oligosaccharides against Vibrio vulnificus. Int. J. Mol. Med. **24**, 327–333 (2009). https://doi.org/10.3892/IJMM_00000236/HTML
38. C. Duan, X. Meng, J. Meng, M.I.H. Khan, L. Dai, A. Khan, X. An, J. Zhang, T. Huq, Y. Ni, Chitosan as a preservative for fruits and vegetables: a review on chemistry and antimicrobial properties. J. Bioresour. Bioprod. **4**(1), 11–21 (2019). https://doi.org/10.21967/jbb.v4i1.189
39. M. Hosseinnejad, S.M. Jafari, Evaluation of different factors affecting antimicrobial properties of chitosan. Int. J. Biol. Macromol. **85**, 467–475 (2016). https://doi.org/10.1016/j.ijbiomac.2016.01.022
40. M. Yildirim-Aksoy, B.H. Beck, Antimicrobial activity of chitosan and a chitosan oligomer against bacterial pathogens of warmwater fish. J. Appl. Microbiol. **122**, 1570–1578 (2017). https://doi.org/10.1111/JAM.13460
41. F. Han, Y. Dong, A. Song, R. Yin, S. Li, Alginate/chitosan based bi-layer composite membrane as potential sustained-release wound dressing containing ciprofloxacin hydrochloride. Appl. Surf. Sci. **311**, 626–634 (2014). https://doi.org/10.1016/j.apsusc.2014.05.125
42. W.A. Sarhan, H.M.E. Azzazy, I.M. El-Sherbiny, Honey/chitosan nanofiber wound dressing enriched with *Allium sativum* and *Cleome droserifolia*: enhanced antimicrobial and wound healing activity. ACS Appl. Mater. Interfaces **8**(10), 6379–6390 (2016). https://doi.org/10.1021/acsami.6b00739
43. R.C. Chien, M.T. Yen, J.L. Mau, Antimicrobial and antitumor activities of chitosan from shiitake stipes, compared to commercial chitosan from crab shells. Carbohyd. Polym. **138**, 259–264 (2016). https://doi.org/10.1016/j.carbpol.2015.11.061

44. M.E. Abd El-Hack, M.T. El-Saadony, M.E. Shafi, N.M. Zabermawi, M. Arif, G.E. Batiha, A.F. Khafaga, Y.M. Abd El-Hakim, A.A. Al-Sagheer, Antimicrobial and antioxidant properties of chitosan and its derivatives and their applications: a review. Int. J. Biol. Macromol. **164**, 2726–2744 (2020). https://doi.org/10.1016/j.ijbiomac.2020.08.153
45. H.K. No, S.H. Kim, S.H. Lee, N.Y. Park, W. Prinyawiwatkul, Stability and antibacterial activity of chitosan solutions affected by storage temperature and time. Carbohyd. Polym. **65**(2), 174–178 (2006). https://doi.org/10.1016/j.carbpol.2005.12.036
46. K.B. Mukhtar Ahmed, M.M.A. Khan, H. Siddiqui, A. Jahan, Chitosan and its oligosaccharides, a promising option for sustainable crop production- a review. Carbohyd. Polym. **227**, 115331 (2020). https://doi.org/10.1016/j.carbpol.2019.115331
47. F. Seyfarth, S. Schliemann, P. Elsner, U.C. Hipler, Antifungal effect of high- and low-molecular-weight chitosan hydrochloride, carboxymethyl chitosan, chitosan oligosaccharide and N-acetyl-d-glucosamine against Candida albicans, Candida krusei and Candida glabrata. Int. J. Pharm. **353**(1–2), 139–148 (2008). https://doi.org/10.1016/j.ijpharm.2007.11.029
48. E.K. Lim, B.H. Chung, Preparation of pyrenyl-based multifunctional nanocomposites for biomedical applications. Nat. Protoc. **11**(2), 236–251 (2016). https://doi.org/10.1038/nprot.2015.135
49. A. Regiel-Futyra, M. Kus-Liśkiewicz, V. Sebastian, S. Irusta, M. Arruebo, G. Stochel, A. Kyzioł, Development of noncytotoxic chitosan-gold nanocomposites as efficient antibacterial materials. ACS Appl. Mater. Interfaces **7**(2), 1087–1099 (2015). https://doi.org/10.1021/am508094e
50. N. Naseri, C. Algan, V. Jacobs, M. John, K. Oksman, A.P. Mathew, Electrospun chitosan-based nanocomposite mats reinforced with chitin nanocrystals for wound dressing. Carbohyd. Polym. **109**, 7–15 (2014). https://doi.org/10.1016/j.carbpol.2014.03.031
51. A.A. Hebeish, M.A. Ramadan, A.S. Montaser, A.M. Farag, Preparation, characterization and antibacterial activity of chitosan-g-poly acrylonitrile/silver nanocomposite. Int. J. Biol. Macromol. **68**, 178–184 (2014). https://doi.org/10.1016/j.ijbiomac.2014.04.028
52. M. Liu, C. Wu, Y. Jiao, S. Xiong, C. Zhou, Chitosan-halloysite nanotubes nanocomposite scaffolds for tissue engineering. J. Mater. Chem. B **1**(15), 2078–2089 (2013). https://doi.org/10.1039/c3tb20084a
53. C.K. Choo, X.Y. Kong, T.L. Goh, G.C. Ngoh, B.A. Horri, B. Salamatinia, Chitosan/halloysite beads fabricated by ultrasonic-assisted extrusion-dripping and a case study application for copper ion removal. Carbohyd. Polym. **138**, 16–26 (2016). https://doi.org/10.1016/j.carbpol.2015.11.060
54. P. Amouzgar, E.S. Chan, B. Salamatinia, Effects of ultrasound on development of Cs/NAC nano composite beads through extrusion dripping for acetaminophen removal from aqueous solution. J. Clean. Prod. **165**, 537–551 (2017). https://doi.org/10.1016/j.jclepro.2017.07.148
55. D.A. Marín-Silva, S. Rivero, A. Pinotti, Chitosan-based nanocomposite matrices: development and characterization. Int. J. Biol. Macromol. **123**, 189–200 (2019). https://doi.org/10.1016/j.ijbiomac.2018.11.035
56. H. Jafari, M.K. Pirouzifard, M.A. Khaledabad, H. Almasi, Effect of chitin nanofiber on the morphological and physical properties of chitosan/silver nanoparticle bionanocomposite films. Int. J. Biol. Macromol. **92**, 461–466 (2016). https://doi.org/10.1016/j.ijbiomac.2016.07.051
57. J. Jia, Y. Gai, W. Wang, Y. Zhao, Green synthesis of biocompatiable chitosan-graphene oxide hybrid nanosheet by ultrasonication method. Ultrason. Sonochem. **32**, 300–306 (2016). https://doi.org/10.1016/j.ultsonch.2016.03.027
58. Y. Huang, N. Dan, W. Dan, W. Zhao, Reinforcement of polycaprolactone/chitosan with nanoclay and controlled release of curcumin for wound dressing. ACS Omega **4**(27), 22292–22301 (2019). https://doi.org/10.1021/acsomega.9b02217
59. C.D.G. Tovar, J.I. Castro, C.H. Valencia, D.P.N. Porras, J.H.M. Hernandez, M.E. Valencia, J.D. Velásquez, M.N. Chaur, Preparation of chitosan/poly(Vinyl alcohol) nanocomposite films incorporated with oxidized carbon nano-onions (multi-layer fullerenes) for tissue-engineering applications. Biomolecules **9**(11) (2019). https://doi.org/10.3390/biom9110684

60. S.A. Gaware, K.A. Rokade, S.N. Kale, Silica-chitosan nanocomposite mediated pH-sensitive drug delivery. J. Drug Deliv. Sci. Technol. **49**, 345–351 (2019). https://doi.org/10.1016/j.jddst.2018.11.022
61. E. Cheraghipour, M. Pakshir, Process optimization and modeling of Pb (II) ions adsorption on chitosan-conjugated magnetite nano-biocomposite using response surface methodology. Chemosphere **260**, 127560 (2020). https://doi.org/10.1016/j.chemosphere.2020.127560
62. H. V. Tran, L. D. Tran, T. N. Nguyen, Preparation of chitosan/magnetite composite beads and their application for removal of Pb (II) and Ni (II) from aqueous solution. Mater. Sci. Eng. C. **30**(2), 304–310 (2010). https://doi.org/10.1016/J.MSEC.2009.11.008
63. M. Keshvardoostchokami, L. Babaei, A. A. Zamani, A. H. Parizanganeh, F. Piri, Synthesized chitosan/ iron oxide nanocomposite and shrimp shell in removal of nickel, cadmium and lead from aqueous solution. Glob. J. Env. Sci. Manage. **3**(3), 267–278 (2017). https://doi.org/10.22034/GJESM.2017.03.03.004
64. W. Shao, J. Wu, S. Wang, M. Huang, X. Liu, R. Zhang, Construction of silver sulfadiazine loaded chitosan composite sponges as potential wound dressings. Carbohyd. Polym. **157**, 1963–1970 (2017). https://doi.org/10.1016/j.carbpol.2016.11.087
65. H. Jiang, Y. Zuo, Q. Zou, H. Wang, J. Du, Y. Li, X. Yang, Biomimetic spiral-cylindrical scaffold based on hybrid chitosan/cellulose/nano-hydroxyapatite membrane for bone regeneration. ACS Appl. Mater. Interfaces **5**(22), 12036–12044 (2013). https://doi.org/10.1021/am4038432
66. E. Kolanthai, P.A. Sindu, D.K. Khajuria, S.C. Veerla, D. Kuppuswamy, L.H. Catalani, D.R. Mahapatra, Graphene oxide - a tool for the preparation of chemically crosslinking free alginate-chitosan-collagen scaffolds for bone tissue engineering. ACS Appl. Mater. Interfaces **10**(15), 12441–12452 (2018). https://doi.org/10.1021/acsami.8b00699
67. S. Omidi, A. Kakanejadifard, Modification of chitosan and chitosan nanoparticle by long chain pyridinium compounds: synthesis, characterization, antibacterial, and antioxidant activities. Carbohydr. Polym. **208**(December 2018), 477–485 (2019). https://doi.org/10.1016/j.carbpol.2018.12.097
68. T.R. Nimal, G. Baranwal, M.C. Bavya, R. Biswas, R. Jayakumar, Anti-staphylococcal activity of injectable nano tigecycline/chitosan-PRP composite hydrogel using drosophila melanogaster model for infectious wounds. ACS Appl. Mater. Interfaces **8**(34), 22074–22083 (2016). https://doi.org/10.1021/acsami.6b07463
69. S. Sharma, S. Batra, Recent advances of chitosan composites in artificial skin: the next era for potential biomedical application, in *Materials for Biomedical Engineering: Nanobiomaterials in Tissue Engineering* (Elsevier, 2019), pp. 97–119. https://doi.org/10.1016/B978-0-12-816909-4.00005-1
70. S. Dey, M.S. Hasnain, A.K. Nayak, Chitosan nanocomposites for biomedical applications, in *Chitosan in Biomedical Applications* (2021), pp. 111–138. https://doi.org/10.1016/B978-0-12-821058-1.00011-3
71. Annu, S. Ahmed, R.K. Nirala, R. Kumar, S. Ikram, Green synthesis of chitosan/nanosilver hybrid bionanocomposites with promising antimicrobial, antioxidant and anticervical cancer activity. Polym. Polym. Compos. **29**(9_suppl), S199–S210 (2021). https://doi.org/10.1177/0967391121993977
72. B.V.F. Riccio, A.B. Klosowski, E. Prestes, T.B. de Sousa, L.C. de Assunção Morais, B.M. Lemes, F.L. Beltrame, P.M. Campos, P.C. Ferrari, Chitosan/nanocellulose-based bionanocomposite films for controlled betamethasone and silver sulfadiazine delivery. J. Appl. Polym. Sci. **138**(21) (2021). https://doi.org/10.1002/app.50468
73. R. Ying, H. Wang, R. Sun, K. Chen, Preparation and properties of a highly dispersed nano-hydroxyapatite colloid used as a reinforcing filler for chitosan. Mater. Sci. Eng. C **110**, 110689 (2020). https://doi.org/10.1016/j.msec.2020.110689
74. A. Barra, Z. Alves, N.M. Ferreira, M.A. Martins, H. Oliveira, L.P. Ferreira, M.M. Cruz, M.D.D. Carvalho, S.M. Neumayer, B.J. Rodriguez, C. Nunes, P. Ferreira, Biocompatible chitosan-based composites with properties suitable for hyperthermia therapy. J. Mater. Chem. B **8**(6), 1256–1265 (2020). https://doi.org/10.1039/c9tb02067e

75. R.K. Kankala, C.-G. Liu, D.-Y. Yang, S.-B. Wang, A.-Z. Chen, Ultrasmall platinum nanoparticles enable deep tumor penetration and synergistic therapeutic abilities through free radical species-assisted catalysis to combat cancer multidrug resistance. Chem. Eng. J. **383**, 123138 (2020). https://doi.org/10.1016/j.cej.2019.123138
76. A. Singh, G. Sinsinbar, M. Choudhary, V. Kumar, R. Pasricha, H.N. Verma, S.P. Singh, K. Arora, Graphene oxide-chitosan nanocomposite based electrochemical DNA biosensor for detection of typhoid. Sens. Actuators B Chem. **185**, 675–684 (2013). https://doi.org/10.1016/j.snb.2013.05.014
77. N. Ahmad, S. Sultana, S.M. Faisal, A. Ahmed, S. Sabir, M.Z. Khan, Zinc oxide-decorated polypyrrole/chitosan bionanocomposites with enhanced photocatalytic, antibacterial and anticancer performance. RSC Adv. **9**(70), 41135–41150 (2019). https://doi.org/10.1039/c9ra06493a
78. P. Gomathi, D. Ragupathy, J.H. Choi, J.H. Yeum, S.C. Lee, J.C. Kim, S.H. Lee, H.D. Ghim, Fabrication of novel chitosan nanofiber/gold nanoparticles composite towards improved performance for a cholesterol sensor. Sens. Actuators B Chem. **153**(1), 44–49 (2011). https://doi.org/10.1016/j.snb.2010.10.005
79. A. Pourjavadi, M. Bagherifard, M. Doroudian, Synthesis of micelles based on chitosan functionalized with gold nanorods as a light sensitive drug delivery vehicle. Int. J. Biol. Macromol. **149**, 809–818 (2020). https://doi.org/10.1016/j.ijbiomac.2020.01.162
80. M.H. Mirmusavi, M. Ahmadian, S. Karbasi, Polycaprolactone-chitosan/multi-walled carbon nanotube: a highly strengthened electrospun nanocomposite scaffold for cartilage tissue engineering. Int. J. Biol. Macromol. **209**, 1801–1814 (2022). https://doi.org/10.1016/j.ijbiomac.2022.04.152
81. N. Ardila, N. Medina, M. Arkoun, M.C. Heuzey, A. Ajji, C.J. Panchal, Chitosan–bacterial nanocellulose nanofibrous structures for potential wound dressing applications. Cellulose **23**(5), 3089–3104 (2016). https://doi.org/10.1007/s10570-016-1022-y
82. M.T. Beleno Cabarcas, M. Stoytcheva, R. Zlatev, G. Montero, Z. Velkova, V. Gochev, Chitosan nanocomposite modified OPH-based amperometric sensor for organophosphorus pesticides determination. Curr. Anal. Chem. **14**(1) (2018). https://doi.org/10.2174/1573411013666170307094412
83. M.W. El-Maadawy, R.R. Mohamed, D.H. Hanna, Preparation of carrageenan/chitosan-based (N,N,N-trimeth(yl chitosan chloride) silver nanocomposites as pH sensitive carrier for effective controlled curcumin delivery in cancer cells. OpenNano**7**, 100050 (2022). https://doi.org/10.1016/j.onano.2022.100050
84. A. Rahmanpour, M.R. Farahpour, R. Shapouri, S. Jafarirad, P. Rahimi, Synthesis and characterization of alumina-based nanocomposites of TiO2/Al2O3/Chitosan with antibacterial properties accelarate healing of infected excision wounds. Colloids Surf. A **644**, 128839 (2022). https://doi.org/10.1016/j.colsurfa.2022.128839
85. S. Hua, H. Yang, W. Wang, A. Wang, Controlled release of ofloxacin from chitosan–montmorillonite hydrogel. Appl. Clay Sci. **50**(1), 112–117 (2010). https://doi.org/10.1016/j.clay.2010.07.012
86. D. George, P.U. Maheswari, K.M.M. SheriffaBegum, G. Arthanareeswaran, Biomass-derived dialdehyde cellulose cross-linked chitosan-based nanocomposite hydrogel with phytosynthesized zinc oxide nanoparticles for enhanced curcumin delivery and bioactivity. J. Agric. Food Chem. **67**(39), 10880–10890 (2019). https://doi.org/10.1021/acs.jafc.9b01933
87. X. Zhao, Z. Wei, Z. Zhao, Y. Miao, Y. Qiu, W. Yang, X. Jia, Z. Liu, H. Hou, Design and development of graphene oxide nanoparticle/chitosan hybrids showing pH-sensitive surface charge-reversible ability for efficient intracellular doxorubicin delivery. ACS Appl. Mater. Interfaces **10**(7), 6608–6617 (2018). https://doi.org/10.1021/acsami.7b16910
88. J. Long, X. Yu, E. Xu, Z. Wu, X. Xu, Z. Jin, A. Jiao, In situ synthesis of new magnetite chitosan/carrageenan nanocomposites by electrostatic interactions for protein delivery applications. Carbohyd. Polym. **131**, 98–107 (2015). https://doi.org/10.1016/j.carbpol.2015.05.058
89. R. Jayakumar, D. Menon, K. Manzoor, S.V. Nair, H. Tamura, Biomedical applications of chitin and chitosan based nanomaterials—a short review. Carbohyd. Polym. **82**(2), 227–232 (2010). https://doi.org/10.1016/j.carbpol.2010.04.074

90. M.A. Matica, F.L. Aachmann, A. Tøndervik, H. Sletta, V. Ostafe, Chitosan as a wound dressing starting material: Antimicrobial properties and mode of action. Int. J. Mol. Sci. **20**(23). MDPI AG (2019). https://doi.org/10.3390/ijms20235889
91. L. Ding, X. Shan, X. Zhao, H. Zha, X. Chen, J. Wang, C. Cai, X. Wang, G. Li, J. Hao, G. Yu, Spongy bilayer dressing composed of chitosan–Ag nanoparticles and chitosan–Bletilla striata polysaccharide for wound healing applications. Carbohyd. Polym. **157**, 1538–1547 (2017). https://doi.org/10.1016/j.carbpol.2016.11.040
92. M.N. Sundaram, S. Amirthalingam, U. Mony, P.K. Varma, R. Jayakumar, Injectable chitosan-nano bioglass composite hemostatic hydrogel for effective bleeding control. Int. J. Biol. Macromol. **129**, 936–943 (2019). https://doi.org/10.1016/j.ijbiomac.2019.01.220
93. M.L. Pita-López, G. Fletes-Vargas, H. Espinosa-Andrews, R. Rodríguez-Rodríguez, Physically cross-linked chitosan-based hydrogels for tissue engineering applications: a state-of-the-art review. Eur. Polym. J. **145** (2021). Elsevier Ltd. https://doi.org/10.1016/j.eurpolymj.2020.110176
94. I. Younes, M. Rinaudo, D. Harding, S. H., Chitin and chitosan preparation from marine sources. Structure, properties and applications. Mar. Drugs **13**, 1133–1174 (2015). https://doi.org/10.3390/MD13031133
95. E. Bulanov, N. Silina, M. Lelet, A. Knyazev, L. Smirnova, D. Aleynik, I. Charykova, Study of physicochemical properties of nanohydroxyapatite–chitosan composites. Bull. Mater. Sci. **43**(1), 1–6 (2020). https://doi.org/10.1007/s12034-020-2065-0
96. W. Cui, Y. Zhou, J. Chang, Electrospun nanofibrous materials for tissue engineering and drug delivery. Sci. Technol. Adv. Mater. **11**(1), 014108 (2010). https://doi.org/10.1088/1468-6996/11/1/014108
97. C.D.F. Moreira, S.M. Carvalho, R.M. Florentino, A. França, B.S. Okano, C.M.F. Rezende, H.S. Mansur, M.M. Pereira, Injectable chitosan/gelatin/bioactive glass nanocomposite hydrogels for potential bone regeneration: In vitro and in vivo analyses. Int. J. Biol. Macromol. **132**, 811–821 (2019). https://doi.org/10.1016/j.ijbiomac.2019.03.237
98. S. Türk, I. Altınsoy, G. Çelebi Efe, M. Ipek, M. Özacar, C. Bindal, 3D porous collagen/functionalized multiwalled carbon nanotube/chitosan/hydroxyapatite composite scaffolds for bone tissue engineering. Mater. Sci. Eng. C **92**, 757–768 (2018). https://doi.org/10.1016/j.msec.2018.07.020
99. P.R. Sivashankari, M. Prabaharan, Chitosan/carbon-based nanomaterials as scaffolds for tissue engineering, in *Biopolymer-Based Composites: Drug Delivery and Biomedical Applications* (2017), pp. 381–397. Elsevier Inc. https://doi.org/10.1016/B978-0-08-101914-6.00012-0
100. Z. Wang, J. Yu, R. Gui, H. Jin, Y. Xia, Carbon nanomaterials-based electrochemical aptasensors, in *Biosensors and Bioelectronics*, vol. 79 (2016), pp. 136–149. Elsevier Ltd. https://doi.org/10.1016/j.bios.2015.11.093
101. Y. Jiang, J. Wu, Recent development in chitosan nanocomposites for surface-based biosensor applications. Electrophoresis **40**(16–17), 2084–2097 (2019). https://doi.org/10.1002/elps.201900066
102. C. Shan, H. Yang, D. Han, Q. Zhang, A. Ivaska, L. Niu, Graphene/AuNPs/chitosan nanocomposites film for glucose biosensing. Biosens. Bioelectron. **25**(5), 1070–1074 (2010). https://doi.org/10.1016/j.bios.2009.09.024
103. A. Güner, E. Çevik, M. Şenel, L. Alpsoy, An electrochemical immunosensor for sensitive detection of Escherichia coli O157:H7 by using chitosan, MWCNT, polypyrrole with gold nanoparticles hybrid sensing platform. Food Chem. **229**, 358–365 (2017). https://doi.org/10.1016/j.foodchem.2017.02.083
104. A.M.A.A. AL-Mokaram, R. Yahya, M.M. Abdi, H.N.M.E. Mahmud, One-step electrochemical deposition of Polypyrrole–Chitosan–Iron oxide nanocomposite films for non-enzymatic glucose biosensor. Mater. Lett. **183**, 90–93 (2016). https://doi.org/10.1016/j.matlet.2016.07.049
105. H. Salehizadeh, E. Hekmatian, M. Sadeghi, K. Kennedy, Synthesis and characterization of core-shell Fe3O4-gold-chitosan nanostructure. J. Nanobiotechnology **10**(1), 3 (2012). https://doi.org/10.1186/1477-3155-10-3

106. A. Kumar, A.R. Chowdhuri, D. Laha, S. Chandra, P. Karmakar, S.K. Sahu, One-pot synthesis of carbon dot-entrenched chitosan-modified magnetic nanoparticles for fluorescence-based Cu^{2+} ion sensing and cell imaging. RSC Adv. **6**(64), 58979–58987 (2016). https://doi.org/10.1039/C6RA10382K
107. S. Banerjee, B. Bagchi, S. Bhandary, A. Kool, N.A. Hoque, P. Biswas, K. Pal, P. Thakur, K. Das, P. Karmakar, S. Das, Antimicrobial and biocompatible fluorescent hydroxyapatite-chitosan nanocomposite films for biomedical applications. Colloids Surf. B **171**, 300–307 (2018). https://doi.org/10.1016/j.colsurfb.2018.07.028
108. G. Tao, L. Zhang, Z. Hua, Y. Chen, L. Guo, J. Zhang, Z. Shu, J. Gao, H. Chen, W. Wu, Z. Liu, J. Shi, Highly efficient adsorbents based on hierarchically macro/mesoporous carbon monoliths with strong hydrophobicity. Carbon **66**, 547–559 (2014). https://doi.org/10.1016/j.carbon.2013.09.037
109. Y. Yu, Adsorbents in blood purification: from lab search to clinical therapy. Chin. Sci. Bull. **58**(35), 4357–4361 (2013). https://doi.org/10.1007/s11434-013-6071-0
110. J. Chen, Y. Ma, L. Wang, W. Han, Y. Chai, T. Wang, J. Li, L. Ou, Preparation of chitosan/SiO2-loaded graphene composite beads for efficient removal of bilirubin. Carbon **143**, 352–361 (2019). https://doi.org/10.1016/j.carbon.2018.11.045
111. S. Chahal, A. Kumar, F.S.J. Hussian, Development of biomimetic electrospun polymeric biomaterials for bone tissue engineering. A review. J. Biomater. Sci., Polym. Ed. **30**(14), 1308–1355 (2019). https://doi.org/10.1080/09205063.2019.1630699

Chapter 5
Functionalized Chitosan and Biomedical Devices

Damaris-Cristina Gheorghe, Ruxandra -Maria Ilie-Mihai, Catalina Cioates Negut, and Raluca-Ioana Stefan-van Staden

Abstract Chitosan is the polymer that results from the deacetylation of chitin taken from the exoskeleton of crustaceans. This chapter aims to introduce some of the most well-known chitosan derivatives, including carboxymethyl chitosan, N-trimethyl chitosan, thiolated chitosan, quaternary ammonium chitosan, and sulfated-chitosan oligomers, as well as their unique properties: anti-allergic, anti-bacterial, anti-coagulant, anti-obesity, anti-cancer, anti-oxidant, analgesic, anti-cholesterolemic, anti-human immunodeficiency virus, anti-diabetic, anti-hypersensitive, anti-Alzheimer's and matrix metalloproteinase-inhibiting properties. Their utilization for the design of biomedical devices used for analysis of nutrition, and pharmacology products, and in medicine, biotechnology, materials science, and other fields was proven.

5.1 Introduction

Biomedical devices are quickly replacing traditional medical instruments. The goal is to use biomaterials to create secure, excellent, and affordable composite materials in the medical field. Biomaterials are multi-component systems that are safe to interact with biological systems and can be used to supplement or substitute a normal component or function in a living body. Biomaterials are composed of an extensive series of chemical elements, including ceramics, polymers, and metals. The biomaterial-based composites are capable of being transformed into intelligent, multifunctional materials that are appropriate for biomedical devices. Natural polymers like polysaccharides, polypeptides, and polynucleotides [1] received by the United States Food and Drug Administration approval are thought to be harmless

D.-C. Gheorghe · R. .-M. Ilie-Mihai · C. Cioates Negut · R.-I. Stefan-van Staden (✉)
Laboratory of Electrochemistry and PATLAB, National Institute of Research for Electrochemistry and Condensed Matter, 202 Splaiul Independentei Str, 060021 Bucharest, Romania
e-mail: ralucavanstaden@gmail.com

S. K. Swain and A. Biswal (eds.), *Chitosan Nanocomposites*,
Biological and Medical Physics, Biomedical Engineering,
https://doi.org/10.1007/978-981-19-9646-7_5

for use in tissue engineering, medication delivery, orthopaedic and dental care, and wound healing [2]. The widespread use of polysaccharides such as alginate, chitosan (C), starch, cellulose, agar, gums, hyaluronic acid, dextran, and heparin has been limited by high-temperature oxidation above the melting point, microbial contamination, high hydrophilicity, pH-dependent solubility, and poor mechanical strength [3]. Chitosan has been chosen by researchers as the right polymer for prospective applications in nutrition, pharmaceuticals, biotechnology, and material sciences [4]. Chitosan is a recyclable, biocompatible, and harmless polymer obtained after chitin's deacetylation [5]. Sources for extracting C from chitin are crabs, oysters, lobsters, crayfish, and prawns [6]. In recent years, interest in C materials and nanocomposites has increased [7]. The subjects that are going to be covered in this chapter are shown in Scheme 5.1.

Chitosan contains a number of functional groups, including hydroxyl (-OH) and amine ($-NH_2$), which enables it to react with crosslinking agents to produce chemical crosslinking at the site of application [5]. Chitosan also contains a number of other functional groups. Chitosan is appropriate for use in clinical settings because it does

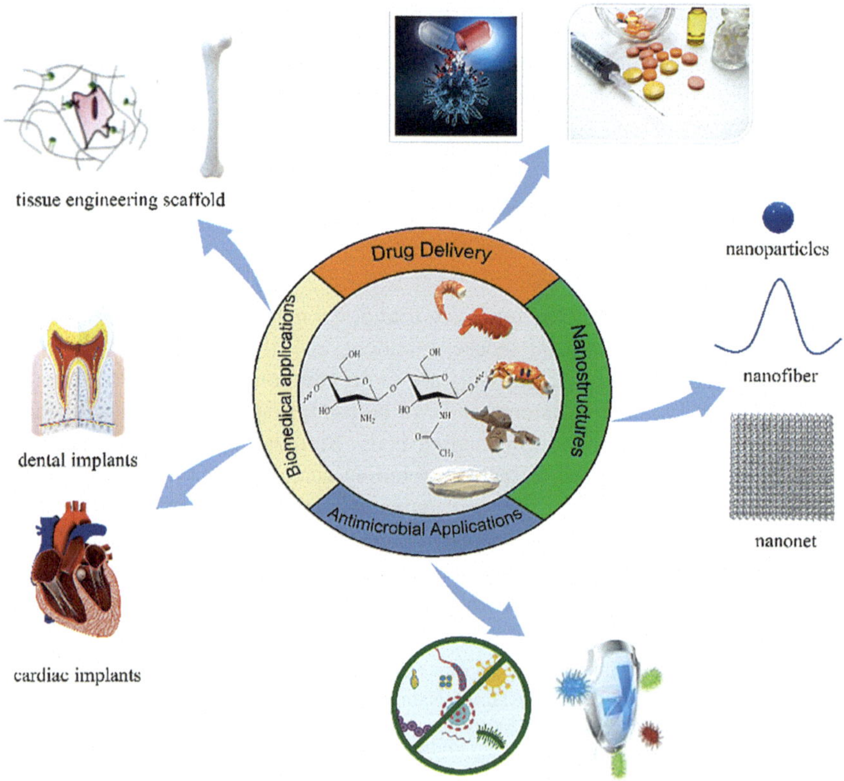

Scheme 5.1 Schematic diagram of chitosan-based biomedical applications

not cause toxicity, is compatible with living organisms, and can be broken down by specific enzymes into oligosaccharides that do not cause toxicity [6]. Chitosan can be utilized in the conduct of scientific research due to the aforementioned qualities. In addition, chitosan nanoparticles (CNPs) and their derivatives demonstrated extraordinary performance in the fields of ophthalmology, dentistry, bioimaging, biosensing, and diagnostics [6]. Chitosan, its derivatives, and CNPs are among the families of natural biopolymer materials that have been studied the most for possible applications in biology. This is because C is one of the most abundant naturally occurring polysaccharides.

5.2 Chitosan and Its Derivatives

The areas of medicine and other academic subjects associated with health have also developed over the course of time as a direct result of the ongoing development of civilization all throughout the course of history. Recently, over the past few years, polymeric compounds that are originated from the liver, as well as inulin, alginates, and chitin, had also emerged as topics of study in a variety of fields, including medicine, health care, and biology, among others [8–10]. Among them, chitin, in conjunction with C and its derivatives, as a result of the following distinctive properties: anti-allergic, anti-bacterial, anti-coagulant, anti-cancer, anti-oxidant, anti-human immunodeficiency virus, anti-diabetic, anti-hypersensitive, anti-Alzheimer's, and matrix metalloproteinase-inhibiting properties, has attracted a significant amount of attention [4, 11]. Due to the insolubility of C in water and in most organic solvents, its applicability in diverse disciplines is limited. Derivatives of C have the potential to fill a wide variety of roles in the drug industry, including those of targeted drug delivery vehicles, vaccine adjuvants, and vaccine delivery carriers [12, 13]. Chitosan has been subjected to a number of modifications, both chemical and physiological, such as carboxylation, quaternization, acylation, and alkylation, in attempt to enhance the solubilization of the compound and develop derivative products that can be used in a wide variety of contexts. These modifications have been carried out in order to improve the solubility of C. According to Wang et al., N-acylated chitosan's biocompatibility and anti-coagulant properties have seen tremendous improvements during the course of time, and the substance can now be employed as a sustained-release medication in clinical settings [14]. At a pH of 7.00, the research that was conducted in 2005 by Yang and colleagues demonstrated that water-soluble N-alkylated disaccharide chitosan derivatives seem to remove pathogens such as *Escherichia coli* and *Staphylococcus aureus* much more effectively than native C [15].

In 2018, Shariatini stated that by altering the level of carboxymethylation, CMC has the potential to change the degree to which it is miscible in water across a broad range of pH levels. As a result of this, The reaction time of the drug administration system would consequently lengthen [16]. According to Freitas et al., modifying C through the process of quaternization has the potential to significantly improve the substance's hydrophobic character, antimicrobial activities, mucosal adhesion, and permeability, all of which are desirable qualities for the development of medical dressings and drug carriers [17].

As therapeutic carriers, C and its derivatives display anti-bacterial, anti-oxidant, and anti-cancer properties in vivo, indicating its potential application in the treatment of clinical disorders. Due to the fact that C has the potential to be utilized in a wide variety of biomedical applications, it has been the subject of extensive research. Chitosan is a naturally occurring amino polysaccharide that possesses an innovative structure and highly complex characteristics [18]. Chitosan's exceptional biocompatibility and biodegradability under physiological settings [19] have led to its widespread usage in medication delivery and gene therapy, in addition to its use in bone tissue regeneration. Because of its flexible structure, C can be modified chemically or mechanically to create new qualities, functions, and uses [18].

Chitosan has a pKa with value equivalent to 6.3, and is a strong base due to its primary groups [20]. A C solution can be created under aqueous conditions with a pH of less than 6. This will protonate the amino groups of C, resulting in a positively charged polymer and allowing it to overcome the associative forces between chains [21]. Increasing the value of the pH of a C solution higher than 6, the amino groups of the C become deprotonated and the charge on the polymer chain is lost; this leads to C insolubility. The degree and type of deacetylation as well as the molecular weight have a significant impact on solubility. It is possible to increase the solubility of C by using chemical modifications that result in the presence of two hydroxyl functional groups in the polymer chain. A comprehensive study described by Upadhyaya et al. [22] provides a summary of a promising alternative to classic C, namely, carboxymethyl chitosan, a material that has better biological and physicochemical features and is also water soluble. When introducing carboxymethyl groups to the configuration of C, at pH levels that range from neutral to alkaline, chitosan's solubility improves in an exponential manner, and by doing so, the chitosan's distinctive features remain intact. Anti-bacterial qualities that do not harm healthy cells are at the top of the wish list for biomaterials that are specifically developed for human use. The polyacetylated chain composition of C is important for anti-bacterial activity, without it, there would be no anti-bacterial effect. Most likely, the mechanism by which C kills bacteria involves binding the compound to the surface of the cell wall of bacteria, which is negatively charged, which causes the membrane of the cell to rupture. This is then followed by the attachment of the compound to deoxyribonucleic acid (DNA), which prevents DNA replication and ultimately leads to cell death. Anti-bacterial activity depends largely on the electrostatic interaction that occurs among the anionic elements that can be found on the surface of microorganisms and the polycationic structure [23]. Lipopolysaccharide that is gram negative and the proteins that are on top of cell surface are two examples of these types of components.

5.2.1 *Chitosan Derivatives*

In order to obtain C derivatives, its modification is required. The outcomes are intended to endow C with novel properties like mucoadhesiveness and solubility, thereby changing this biomaterial into substrates suitable for the transport of biomolecules. Some of the most known derivatives are carboxymethyl chitosan (CMC), N-trimethyl chitosan (TMC), thiolated chitosan (TC), quaternary ammonium chitosan (QNHC), and sulfated-chitosan oligomers (S-COS), which can also be found in Fig. 5.1.

5.2.1.1 Carboxymethyl Chitosan

The addition of carboxyalkyl groups into the structure of C, primarily in the form of carboxymethyl, is a method that has been developed primarily to enhance the miscibility of C. The reaction takes place at the C6 hydroxyl group or the amino moiety, producing either N-CMC, O-CMC, or N, O-CMC as a by-product of the process. These derivatives are amphoteric polymers whose solubility varies with pH, features that allow them to retain water, are biodegradable, biocompatible, and have anti-oxidant activity [16, 24]. CMC is capable of efficiently chelating to metal ions [25], which is necessary for the production of self-polishing antifouling paints like CMC copper ion [26]. Additionally, the anti-bacterial activity of CMC can be increased by producing its composite materials employing a variety of nanoparticles [27]. It has been demonstrated that O-CMC scaffolds can be used to advance regenerative medicine [28]. For the purpose of engineering adipose tissue, these were injectable hydrogels that had been made by loading fibrin nanoparticles with alginate-O-CMC [28]. These hydrogels might expand to enlarge their pores, facilitating the entry of cells into their three-dimensional networks. Hydrogel scaffolds cultivated using human adipose-derived stem cells promoted cell development, while the addition of fibrin nanoparticles enhanced cell adherence and proliferation on the scaffold [28]. The CMC derivatives were utilized to prevent post-surgical adhesion,

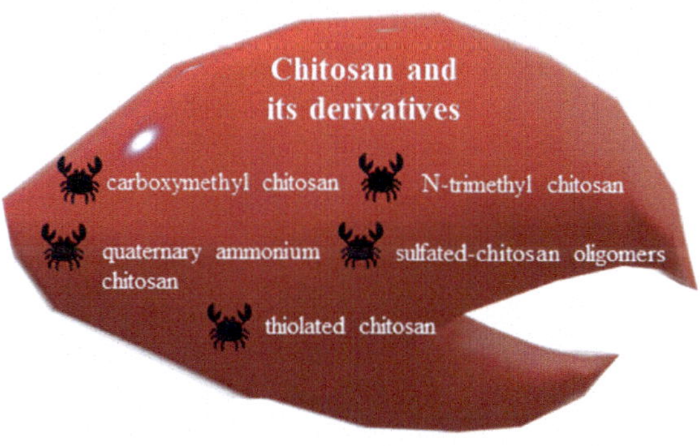

Fig. 5.1 Schematic illustration of the derivatives of chitosan that are to be discussed in this chapter

which might affect a patient's future health by causing organ dysfunction, intestinal blockage, diminished fertility, difficulty with reoperation, and possibly discomfort [29] or illness [30]. In operative models of abdominal aortic anastomosis and abdominal skin incision, N, O-CMC was tested to see how well it would perform [31]. Prior administration of a diluted dispersion to the surgical site did not result in a substantial reduction in the formation of adhesions. After surgery, administering the gel and solution post operatively proved to be the most effective treatment [32].

5.2.1.2 N-trimethyl Chitosan

The amino groups present in chitosan are modified by a process called methylation by the addition of a variety of distinct alkyl groups to the chitosan molecule. Due to the sheer cationic character of its structure, TMC is regarded as one of the polymers with the highest mucoadhesive strength [33]. Because it contains protonated units (–$N^+(CH_3)_3$), it is soluble in a substantially wider range of pH than unmodified chitosan is [34]. The solubility of TMC can be affected by the deacetylation degree, the deacetylation quantity, the percentage of O-methylation, and the molecular weight of the original chitosan. TMC is more soluble than chitosan because, following quaternization, it has a greater number of hydrophilic amino groups on the main chain. Chitosan does not have as many of these groups. In general, the degree of deacetylation is inversely proportional to the likelihood of the formation of quaternary ammonium salt during methylation. This likelihood, in turn, is directly correlated with the degree to which solubility is reduced. As a result of the quaternization process, the amount of positive charge that is carried by TMC is more than that of the parent chitosan. This contributes to the TMC's enhanced anti-bacterial activity. Rapid evacuation from the site of drug action is one of the most important aspects that can reduce the effectiveness of drugs that are administered to the mucosa of the ocular, nasal, or vaginal cavities. It is a widely held belief that increasing the viscosity of a drug formulation will result in a reduction in the clearance of the medication, which will, in turn, result in an increase in the medication's bioavailability. The manufacture of medicinal formulations with a viscosity that is appropriate using in situ gels is an approach that shows a great deal of promise. The production of a gel at the site of drug delivery combines the benefits of an easily administered solution with the viscoelastic characteristics of a gel, thereby extending the residence duration of the formulation. This is achieved by combining the advantages of an easily administered solution with the characteristics of a gel.

5.2.1.3 Thiolated Chitosan

Chitosan's primary amino groups undergo thiolation when they react with coupling chemicals that have thiol groups. This process is known as thiolation. This compound exhibits strong penetration and mucoadhesion, as well as enhanced solubilization at physiological pH, and it also demonstrates in situ gelling capabilities [35]. Due to the sheer development of covalent connections between the thiol groups of the polymer and the cysteine-rich subdomains of glycoproteins in the mucus layer, TC has increased mucoadhesive characteristics. Noncovalent bonds, such as the ionic interactions of C with the anionic substructures of the mucus layer, are thought to be weaker than their covalent counterparts, which are believed to be significantly stronger. Thiolated chitosan, in addition to exhibiting powerful mucoadhesive qualities and characteristics that enhance permeability, also demonstrates outstanding cohesive capabilities. As the backbone of C has reduced thiol functionality, TC is able to form disulfide bonds not only with mucus glycoproteins but also between molecules and within molecules. In other words, it is able to form disulfide bonds in all three configurations. Rapid evacuation from the site of drug action is one of the most important aspects that can reduce the effectiveness of drugs that are administered to the mucosa of the ocular, nasal, or vaginal cavities. It is a widely held belief that increasing the viscosity of a drug formulation will result in a reduction in the clearance of the medication, which will, in turn, result in an increase in the medication's bioavailability. The manufacture of medicinal formulations with a viscosity that is appropriate using in situ gels is an approach that shows a great deal of promise. The production of a gel at the site of drug delivery combines the benefits of an easily administered solution with the viscoelastic characteristics of a gel, thereby extending the residence duration of the formulation. This is achieved by combining the advantages of an easily administered solution with the characteristics of a gel. Because of these qualities, TC makes an excellent substrate for the transport of drugs to the brain, where it is most commonly utilized in the form of nanoparticles.

5.2.1.4 Quaternary Ammonium Chitosan

The quaternary ammonium group is an example of a hydrophilic group that also carries a positive charge. Not only does the addition of a quaternary ammonium salt group makes the substance more soluble in water, but it also makes it more chargeable. The C_2–NH_2 system is responsible for the quaternization reaction. The basic process of quaternization can be broken down into three distinct steps: the direct quaternary ammonium substitution approach, the N-alkylation method, and the epoxy derivative open-loop method [36]. 2,3-epoxypropyl trimethyl ammonium chloride and 3-chloro-2-hydroxypropyl trimethyl have been the quaternary ammonium salts that have seen the most widespread application over the past few years [37]. A higher charging strength is achieved with the use of quaternary ammonium salt, which also reduces the strength of hydrogen bonds. This increases water solubility. In addition to this, a higher degree of substitution results in an improved water solubility and an increased potential [38]. In addition to having superior anti-bacterial properties, QNHC salt also possesses superior biocompatibility, biodegradability, non-toxicity, and biological impacts, as well as intrinsic mucoadhesiveness and the ability to permeate mucus layers and adhere to epithelial surfaces. As a result of this, it has found widespread application in the medical industry [39]. Quaternary ammonium chitosan, which has anti-bacterial qualities, can be utilized in anti-inflammatory medications or as a filler fiber in materials for treating wounds [40]. This is possible because of the chitosan's high molecular weight.

5.2.1.5 Sulfated-Chitosan Oligomers

Chitosan oligomers (COS) have a wide variety of functional groups at the C2, C3, and C6 locations, respectively. Due to the existence of such functional groups, C has been exploited in a vast array of biological processes, which include the suppression of angiotensin-1-converting enzyme (ACE) [41]. In a recent study, it was shown that COS can operate as ACE inhibitors by attaching to chloride ions. Since activating the ACE helps to lower the severity of cardiovascular disorders, this is a highly useful property of COS.

In order to increase the ACE inhibitory action of the COS that were used in the preparation of S-COS, grafts of sulfate were added to the COS [42]. Even when employing the same reagents, it is common knowledge that the hydroxyl group of pyranose moieties possesses a unique chemical activity in various locations. This is the case even when the chemicals used are the same. Because the hydroxyl groups at the C6 position of COS have the highest reactivity for sulfates, it is possible to effectively and selectively replace the hydroxyl group of COS at the C6 position with a sulfate group in order to facilitate the synthesis of S-COS. This can be done while the structure of COS is preserved. For example, Qian et al. (2010) developed three different forms of S-COS with different molecular weights ranging from 3–1 to 5–3 and 10–5 kDa. All of these S-COS molecules show great potential for use as ACE inhibitors. Although the S-COS with a size range of 5–3 kDa was the one that was most successful in inhibiting ACE activity, none of the three S-COS had a negative effect on human lung fibroblast cells MRC-5 [43]. Recent studies have shown that the ACE inhibitory activity of COS and S-COS can be improved by adding a sulfate group in place of the hydrogen atom that is located at the C6 position. Because of this, C and its derivatives show promise as potential additions to pharmaceuticals and nutraceuticals that are intended to treat hypertension and the ailments that are linked to it.

5.3 Biomedical Tools Based on Chitosan

With advances in nanotechnology, C, which was used as a pure biomaterial, could be combined with some types of nanostructures, incorporated into the base material, or deposited on the surface. To eliminate the disadvantages of the structure of pure C [44], organic and inorganic nanosized compounds were included as dispersed fillers within the entire matrix and/or as a covering for the material's surface [45]. In addition, in C, a selection of nanostructures, for example, nanosheets, nanorods, nanocapsules, nanowires, and nanofibers, can be used. These nanostructures can be metallic, organic, inorganic, or semiconducting.

The demand for nanomaterials in the pharmaceutical industry is increasing daily. Because of their small nanoparticle size, they can permit through a variety of obstacles and transport medication to the destination [46]. Therefore, these are used for cancer and diabetes therapy, etc. Chitosan nanoparticles have been employed as DNA carriers, vaccination carriers, and low molecular weight medication carriers [47]. Due to nanoparticles exhibiting unique properties such as well-defined optics, surface properties, high biocompatibility [48], ease of fabrication, and customizable size [49], they have been employed as drug delivery systems [50]. Thus, CNPs are the most prevalent polysaccharides in nature and promising in cancer therapy [51, 52]. A synergistic effect against cancer cell growth has been observed by adding chemotherapeutic agents like ponatinib, doxorubicin, paclitaxel, and imatinib (I) to silver nanoparticles [53]. Due to the mucoadhesive properties generated by the existence of disulfide bonds between thiolated polymers or with mucus glycoproteins, thiolate derivatives of C have also been demonstrated to offer the potential for oral drug delivery [54]. Using thiolated chitosan-based nanoparticles (TCNPs) laden with insulin using the Schiff-base technique, Sudhakar et al. [55] developed an oral carrier for insulin administration. Based on what they found, giving insulin-TCNPs by mouth to a model of a diabetic rat could control blood sugar and insulin levels.

5.3.1 Electrochemical Sensors Based on Chitosan

Electrical conductivity is not possessed by C. Because of this, it is mixed with conductive polymers (CPs), polyaniline [56] and polypyrrole [57], nanoparticles like multiwalled carbon nanotubes [58], and graphene [59] to improve its electrical properties for sensing applications. Researchers have developed high-performance electrochemical biosensors for glucose detection by covalently immobilizing glucose-oxidase in Fe_3O_4/CG hybrids. The hybrids that were generated can be used not only for biosensors but also in magnetic resonance imaging.

5.3.1.1 Electrochemical Sensors Based on Chitosan and Conductive Polymers

Due to the non-conductive nature of C, the researchers mixed it with other CPs to improve charge transfer. Polypyrrole and polyaniline were used to design the sensors because they can form hybrid polymers with chitosan [60]. A novel sensitive conductive nanocomposite sensor for the recognition of isoxsuprine hydrochloride was designed by Hassanein et al. [61] using polypyrrole, ZnO NPs, and C. In addition to C having high conductivity and surface area, its hydrophobicity has been modified in the ternary nanocomposite to serve as a selective adsorbent in medications. The ternary nanocomposite-containing carbon paste electrode showed selective adsorptive properties in addition to the ZnO NPs electrochemical sensor [62]. In the literature, various conducting polymers have been combined with C. Therefore, Shen et al.

[63] prepared an electrochemical sensor for dopamine recognition, employing C to enable film development and poly(3,4-ethylene dioxythiophene) as the conducting polymer and graphene to improve electron transfer. The resulting sensor demonstrated good sensitivity by the low limit of detection of 0.29 μM for dopamine. Zad et al. [56] developed an efficient electrochemical sensor for the recognition of fluconazole from biological and tablet samples using a C-polyaniline-Fe_3O_4-Ni-Pd. Due to the coupling of nanoparticles with polyaniline, the electrochemical signal was improved.

5.3.1.2 Electrochemical Sensors Based on Chitosan and Nanomaterials

Chitosan and gold nanoparticles (AuNPs) are suitable for the preparation of electrochemical sensors due to their high absorbance and conductivity. Recently, Diouf et al. [64] obtained a new electrochemical sensor based on a screen-printing carbon electrode coated with AuNPs and C for aspirin recognition. The sensor showed good electrochemical performance toward aspirin in biological and tablet samples, compared to previously prepared sensors based on carbon paste electrodes modified by ionic liquid composite with ZnO nanoparticles [65] or glassy carbon electrodes modified with graphene [66]. The development of nanocomposite films has been made possible by easily modifying the polymer chain of C with NPs. Chitosan nanocomposites are extensively employed in sensor design because of their high charge transfer and specific surface of nanoparticles. The development of C nanocomposite-based platforms has led to high sensitivities for the analyte. Thus, Lou et al. [67] developed an electrochemical sensor using a glassy carbon electrode modified with porous nickel molybdate nanocomposite and chitosan nanosheets for the identification of amlodipine. The electrochemical performance of this device was better than that of sensors made before, like the gold electrode based on multi-wall carbon nanotubes (MWCNTs) [68] or the paste electrode based on graphite and MWCNTs [69]. Electrostatic connections among the positively charged amino groups in C and the negatively charged carboxyl groups in functionalized MWCNTs permit C to self-assemble. Based on these electrostatic interactions, Akhter et al. [70] designed an electrochemical sensor based on MWCNTs-C-Cobalt for the analysis of paracetamol from tablets and human serum samples. Screen-printed electrodes can also be used with C nanocomposites. Reddy et al. [71] designed an electrochemical sensor based on SPE modified with a conducting film of the C-gold nanocomposite to recognize ciprofloxacin in biological samples with a value of micromolar magnitude order as a limit of detection.

5.3.1.3 Electrochemical Biosensors Based on Chitosan

Real-time and non-invasive determination of glucose in tear fluids holds promise for the early detection and treatment of diabetes. Zou et al. [72] developed an electrochemically sensitive biosensor electrode suitable for the eye. Thus, nitrogen-doped graphene (N-G) was employed as an ophthalmic electrode in a high-performance intraocular biosensor. The carboxyl-chitosan-functionalized nitrogen-containing graphene was biocompatible with ophthalmic cells. When the sensor was mounted on animals for more than 24 h, it did not hurt them. This shows that it could be used in clinics for intraocular applications.

A single-step bead milling method [73] was presented by Zhang et al. [74] to produce N-G for electrochemical biosensors to determine glucose levels. For multifunctional applications, Fe_3O_4 nanoparticles were added to N-G. The three-dimensional (3D) hybrids were formed from nanoparticle-modified nanosheets. When the magnetic properties of the nanosheets and the catalytic activity of the nanoparticles are mixed, supplementary benefits are produced in the hybrid materials. The supplementary benefits include higher active surface areas and improved electron transport, which are useful for creating electrochemical sensing devices [75, 76].

5.3.2 *Stochastic Sensors Based on Chitosan*

In biomedical analysis, it is important to develop new highly selective and sensitive nanomaterial-based devices that can be obtained at low cost, such as electrochemical sensors [77]. Stochastic sensors are a new kind of electrochemical sensor that Bayley and Cremer [78] described in 2001. These are described as sensory systems that can distinguish between various chemicals using a range of membrane-bound receptors, with sensitive ion channels. According to the authors, these sensors have a varied series of possible uses, including medication and biomedical screening in addition to the detection of biological agents. The only type of sensor that can perform both qualitative and quantitative analyses is stochastic sensors, which can be utilized for single or panel molecule detection. The reduction of apparatus and sample size (from droplet to needle) has made these sensors practical and inexpensive for the rapid examination of biological samples. Screening tests based on stochastic sensors are becoming a reality and are increasingly used for the fast recognition of a variety of illnesses. For screening tests in biological samples such as whole blood, urine, saliva, and tissues, microsensors, disposable two-dimensional stochastic sensors, and needle stochastic sensors have been designed and proven to be very reliable when used for the early diagnosis of viruses [79], diabetes [80], as well as various cancers such as the brain [81], gastric [82–84], breast [85], and bladder cancer [86]. Electrochemical sensors are valuable biomedical analysis tools. Chitosan in various forms, such as C I (which has a low molecular weight), C II (which has a medium molecular weight), and C III (which has the highest molecular weight), has been successfully used in the development of stochastic sensors, whereby the introduction of a certain amount of it has led to improved properties of the modified sensor compared to the bare electrode. Researchers have shown that stochastic sensors are better than other types of electrochemical sensors in several ways. For example, stochastic sensors can recognize biomarkers at the molecular level even in very complex matrices and at concentrations in the order of fg mL^{-1} [81]. The shape of the matrix in which the biomarker is found doesn't affect how the stochastic sensor responds, and accurate quantification could be achieved over an extensive range of concentrations with high selectivities and sensitivities.

5.4 Biomedical Applications of Chitosan

One of the most effective medical techniques now in use, tissue engineering, offers hope to millions of people who experience various dysfunctions brought on by lesions, illnesses, or trauma. The primary goal of this application is to maintain or refine the biological activities of deteriorated tissue to the greatest extent possible [87]. This method utilizes the advantages of the scaffold, by providing the cells with the necessary support due to its chemical structure. To enhance their capabilities, scaffolds are routinely functionalized with bioactive substances like growth factors or antibiotics. Scaffolds for human use are made from diverse raw items, including natural or synthetic polymers (calcium carbonate, calcium phosphates, bioactive glasses), inducers (such as molecules that have the function of signaling), mechanical signals (like hydrogels), and CPs [87–89].

Various approaches are used to prepare some materials for bone regeneration that can be in the formed from three-dimensional structures that mirror native tissues. Porous materials can be impregnated with a variety of chemicals that induce osteoconduction. An optimum scaffold must satisfy some criteria to be able to be used for human employment: a proper durability, a long-term biocompatibility, high bioactivity, good biodegradability, and fitting structure [90, 91].

5.4.1 The Potential Uses of Chitosan in Bone Tissue Engineering

A bone tissue engineering (BTE) scaffold is meant to be a short-term mechanical aid for the wounded region, and a catalyst for tissue regrowth [92, 93]. For bone ingrowth and neovascularization to take place, the structure must be very porous. The scaffold's primary function is to create a surface on which osteoblasts can adhere and grow and an environment favorable to the production of extracellular matrix. These bone tissue scaffolds can be made out from different forms of metal alloys, nanofibers, β-tricalcium phosphate, hydrogels, bioactive glasses, some forms of hydroxyapatite like powders, granules [94, 95].

Systems for drug delivery and release as well as scaffold components utilize natural polymers undergoing enzymatic biodegradation, the pace of which can be modified [96]. Biodegradable polymers like poly-lactic-co-glycolic acid/polylactic acid (PLA) and polycaprolactone, and other substances such as alginate, gelatine, fibrin, collagen, and chitin, together with its derivatives, are among the most frequently utilized materials [97, 98]. Most importantly, they need to be used with other materials because of their limited durability and rapid degradability [99, 100].

Every year, nearly two million bone graft surgeries are performed around the world, making bone grafting among the most frequently used treatments for bone regrowth. Over the course of last decades, bone tissue engineering has used many kinds of bone grafts, but recently the focus has shifted to the biomimetic approach in scaffold design [101, 102]. Studies switched to biomimicry and biomaterials that activate distinctive biological reaction at the molecular biology stage when the fundamental limits of first- and second-generation biomaterials were revealed [102]. Natural polymers like glycosaminoglycans (GAGs), collagen, gelatine, C, and fibroin are frequently employed in the field of tissue engineering because these particular polymers work similar to the extracellular matrix in the body. Natural polymers, being essential structural components of living organisms, are highly osteogenic and osteoconductive [101] due to their wide spectrum of ligands and peptides that promote cell adhesion.

Chitosan has in common with GAGs a similar structure and chemical composition too—an essential part of the extracellular matrix that facilitates cell–cell adhesion through its interactions with collagen fibers. During the depolymerization process, C produces bioactive chitooligosaccharides that have higher antimicrobial characteristics. In addition, the monomeric components of C (glucosamine) are metabolized or eliminated from the body. Chitosan is a polymer made up of glucose units, it can be broken down by the body's natural processes and is highly biocompatible with almost all body tissues. Chitosan has shown high osteoconductivity. It causes neovascularization in vivo, as well as proliferation of osteoblast and mesenchymal cells [103]. Chitosan is used in different shapes, like films, fibers, sponges, and other complex structures, especially for orthopaedic applications. Therefore, C meets most of the requirements that are needed for it to be used in tissue engineering [104–106].

The basis of bone tissue engineering is the use of a three-dimensional biodegradable polymeric scaffold. The goal of this technique is to encourage the growth and remodeling of bone tissue. Because of its inherent qualities, C may be molded into a wide variety of structures, including membranes, sponges, fibers, and porous scaffolds, all of which are suitable for use in BTE. There are many ways to make scaffolds out of C, but only a few of them are usually used. When compared to scaffolds made of pure C, scaffolds made from a hybrid that contains C mixed with synthetic and/or natural polymers, and even ceramic fragments have been discovered to possess superior bioactivity as well as improved mechanical features.

Bone-tissue engineering applications benefit from the incorporation of additional polymers and inorganic components into C, which is then employed in the form of 3D lyophilized scaffolds, hydrogels/film, and electrospun mats to improve the material's characteristics. The addition of nano-hydroxyapatite (nHAp) to C hydrogel membranes resulted in a significant increase in the crystallinity of the composite and showed that MG-63 cells might be biocompatible with the material [107]. Osteoblast-like MG-63 cells' ability to adhere to and proliferate on collagen was greatly enhanced by using nano-bioactive glass ceramics and nHAp onto C/gelatin [108, 109]. The disintegration rate of unmodified C scaffolds was drastically slowed by the incorporation of zirconia particles and nanoscale silicon dioxide inside the C matrix, and the scaffolds' capacity to adsorb proteins was greatly improved [110]. The bioactivity of a material utilized in BTE is mostly determined by its ability to adsorb proteins and undergo biomineralization. Increased apatite deposition by including carbohydrate anionic moieties like chondroitin 4-sulfate into the C matrix aided in the migration of bone marrow stromal cells and greatly raised the compressive modulus [111]. In addition, placing some diopside fragments ($CaMgSi_2O_6$) on top of a C matrix increased the term of osteoblast development-specific genes such as alkaline phosphatase and type I collagen and showed biocompatibility in a rat model that has been used for in vivo studies [112]. When nHAp was combined with C, the compressive strength increased by 33.07%, and the proliferation of mouse preosteoclast cells (MC3T3-E1) was stimulated [113].

Radwan-Pragłowska et al. [114] developed hybrid three-dimensional scaffolds capable of stimulating cell responses and incorporating nHAp. After electrospinning PLA and crosslinking C with the assistance of a microwave, they were able to successfully create the matrices, which were then doped with distinct forms of metallic nanoparticles (gold, platinum, and titanium dioxide) (Fig. 5.2). The physicochemical properties of the substances were examined to characterize the products and semi-components. These properties include chemical structure, crystallinity, and the degree of swelling.

Techniques like scanning electron microscopy and the transmission electron microscopy were utilized in order to investigate the nanoparticles in addition to biomaterials. In the final stage of the process, the scaffolds were tested to determine both their effect on current-stimulated biomineralization and their bioactivity on the MG-63 cells. According to the findings, it is possible to create biomimicking matrices that are flexible and offer promise as a resource for bone tissue engineering.

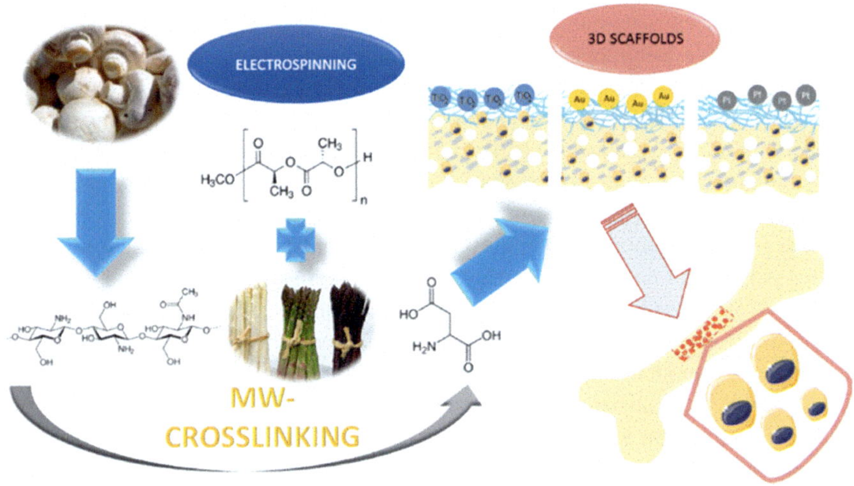

Fig. 5.2 The general 3D scaffolds' obtainment strategy and application. Reproduced with permission from MDPI [114]

5.4.2 Chitosan's Potential Uses in Vascular Tissue Engineering

A surgical procedure known as a vascular patch may be required to repair blood vessel damage caused by heart disease [115, 116]. Due to the rising incidence of cardiovascular problems, vascular surgery is required in a growing number of patients, but a severe lack of vascular patches poses a serious risk to their lives [117]. This has made the study of vascular patches an immediate concern [118, 119]. In order to prevent complications such as thrombosis, inflammation, calcification, or neointimal hyperplasia after surgery, vascular patches, which come into touch with blood, must be biocompatible on the outside [120, 121]. Researchers have typically relied on a single component when treating in order to create a surface of this kind, either the original material or a substrate must be used. This is because the procedures that are being used have limitations, such as the fact that they are laborious to fabricate and that numerous components are unstable [122, 123]. The biocompatibility and endothelialization of single-component-modified vascular patches have been improved; however, these patches have not been successful in resolving patient difficulties such as blood clotting before surgery or infection at the surgical site [116]. As a result, there is an immediate need for the creation of vascular patches that are capable of quickly including and stabilizing a large number of elements. Given these circumstances, the layer-by-layer (LbL) self-assembly method suggests fascinating new resources for investigation and exploration.

In the research on artificial vascular prostheses, the emphasis has been placed on non-thrombogenic biomaterials, reducing the risk of intimal hyperplasia, having physical qualities that are analogous to those of native vessels, and enabling the regeneration of endogenous artery tissue. The most recent findings from studies on biodegradable polymers have helped illuminate questions about the development of artificial vascular prostheses. In particular, when utilized as a vessel with a narrow diameter, the biodegradable polymers displayed superior biocompatibility when compared to the other synthetic materials. Also, biodegradable polymers can help stimulate the growth of new vascular tissue, which will eventually replace the artificial conduit [124, 125].

Biodegradable polymers [96] may be the optimal answer since they maintain luminal patency for an extended period. On the other hand, biodegradable polymers are capable of being absorbed during angiogenesis and vascular remodeling. The idea that bioabsorbable polymeric stents are employed as vehicles for drug administration for suppressing in-stent restenosis was developed further by Zamiri et al. [127]. The biodegradable materials are intended to decrease the reaction of inflammation at the place of implantation as an addition to enhancing adhesion between blood vessel cells (endothelial cells, vascular smooth muscle cells). Chitosan, its derivatives, and heparin, because of the highly plastic nature of their chemical composition, all of these materials make excellent prospects for the creation of artificial vascular prostheses. Because of this, they can work well together, possess desirable biophysical properties, and degrade at rates that can be managed, resulting in a consistent set of physical parameters across the complete remodeling process [128].

Kong et al. [129] created a synthetic vascular prosthesis composing of biodegradable C, heparin, and C derivatives, having a two-millimeter internal diameter and varying degrees of crosslinking. Degradation investigations, both in vivo and in vitro, an analysis of the reaction to an inflammation process, and scanning with an electron microscope were analyses that were carried out on this synthetic vascular prosthesis. Results showed that approximately 50% of the prosthesis disintegrated when in vivo analysis was performed, and normal tissues took their place. Altering the degree to which the synthetic prosthesis is composed of C-heparin, a crosslinking process enabled the researchers to exert control over the deterioration process. This type of artificial vascular prosthesis works well with the body and can be made so that it can be controlled, which you want in a vascular replacement application. Because of their high level of biocompatibility, heparin and C are frequently utilized in the process of surface modification for biomaterials. Both heparin and chitosan are examples of natural polysaccharides. However, C is a polycation while heparin is a polyanion [130]. Because it contains a high level of anti-coagulant activity, heparin is popularly used in a variety of biomedical utilities. These applications include the prevention of blood clots before surgery [131, 132], as well as the inhibition of deep venous thrombosis. Chitosan is administered during the preliminary stages of surgical procedures in order to reduce the risk of bacterial infection because it has a high anti-bacterial ability and is biodegradable in vivo [133, 134]. The combination of heparin and C, demonstrated significant aptitude in a range of domains [135–137], involving the transport of drugs, the repair of tooth damage, and the treatment of vascular disease.

Therefore, in their research, Zhang et al. [138] used LbL self-assembly technique in order to alternatively deposit heparin and C on decellularized scaffolds (DCS) that were previously coated with a layer of polyurethane (PU), the end result being the obtaining of some pieces of polyelectrolyte multilayer (PEM) for vascular diseases. In order to test how the compatibility of the patches with a human body would be, a pig model was used for in vivo testing using some different methods such as Doppler spectrum, computed tomography angiography, and B ultrasonography. By comparing the PEM patches with the single PU/DCS substrate, the researchers could observe that the biocompatibility was changed in a very positive way. The result was sustained by other secondary effects such as enhanced resistance to platelet adhesion, a lengthened in vitro coagulation time, and decreased haemolysis rate. Not only that, but they also showed an increase in endothelial progenitor cell adhesion and proliferation. Computed tomography angiography test revealed that the initial PU/DCS substrate obstructed the surgical arteries after only 2 weeks after the implantation was performed; the PEM patches proved instead that they can be used for a period up to 5 months, without damaging the arteries. As a consequence of this, the PEM vascular patches have demonstrated a considerable amount of promise for their potential application in vascular tissue engineering procedures. In addition, the findings of this study demonstrated that there is a method for producing multi-structured vascular patches as well as other types of biomaterials that are functional.

5.5 Concluding Remark and Future Perspectives

Chitosan is a biopolymer that occurs abundantly in nature and is represented by a union comprising physical characteristics, mechanical features, and biological mechanisms, and can be found in different practical forms. Thus, it is of great importance not only in the biomedical domain, but also for tissue engineering, cell culture, bioimaging, drug distribution, and development of tools for fast in vitro and in vivo diagnostics and the range can be extended to biotechnology, food industry, antimicrobial activities, etc. There are numerous moieties that are both reactive and function within a number of functional groups that can be used to modify the C polymer. To perform a functionalization or to composite the C, some reactions like addition, bonding to microscopic molecules, and even crosslinking can be used. For the functionalization of C, substitution reactions like thiolation and phosphorylation are the best reactions to use, while one can take advantage of the numerous crosslinking agents available in order to perform a crosslinking reaction without any difficulty. Other reactions that C can be involved are carboxylation with alkyl carboxylate, methacrylic acid, and, among others, carboxylic acid. Chitosan was readily available in numerous ionic forms, including sulfated and cationic. In cell culture, drug transportation, gene shipment, bioimaging, tissue engineering, and anti-bacterial activity, the functionalized composites of C or its derivatives are superior to their unfunctionalized counterparts, according to a study of the relevant literature. Thus, it can be deduced that composite materials based on C and its derivatives that come in a wide range of functional forms are promising contenders for a multitude of employment. Many fields are beginning to investigate the potential of composite materials based on C and its derivatives, but there are still some issues to be resolved before such materials can meet practical requirements. Additionally, challenges in biological applications still exist, including those related to drug loading capacity, drug release effectiveness, duration of administration, and rate of material degradation.

References

1. S. Sethi, B.S. Saruchi, M. Kaith, N. Sharma Kaur, Cross-linked xanthan gum – starch hydrogels as promising materials for controlled drug delivery. Cellulose **27**, 4565–4589 (2020)
2. D. Vukajlovic, J. Parker, O. Bretcanu, K. Novakovic, Chitosan based polymer/ bioglass composites for tissue engineering applications. Mater. Sci. Eng. C **96**, 955–967 (2018)
3. P. Sharma, H. Mittal, R. Jindal, D. Jindal, S.M. Alhassan, Sustained delivery of atenolol drug using gum dammar crosslinked polyacrylamide and zirconium based biodegradable hydrogel composites. Coll. Surf. A Physicochem. Eng. Asp. **62**, 136–145 (2019)
4. K.V. Harish Prashanth, R.N. Tharanathan, Chitin/chitosan: Modifications and their unlimited application potential-an overview. Trends Food Sci. Twechnol. **18**, 117–131 (2007)
5. B.R. Rizeq, N.N. Younes, K. Rasool, G.K. Nasrallah, Synthesis, bioapplications, and toxicity evaluation of chitosan-based nanoparticles. Int. J. Mol. Sci. **20**, 1–24 (2019)
6. S. Kou, L. Peters, M. Mucalo, Chitosan: a review of sources and preparation methods. Int. J. Biol. Macromol. **169**, 85–94 (2021)
7. A.O. Silva, R.S. Cunha, D. Hotza, R.A.F. Machado, Chitosan as a matrix of nanocomposites: a review on nanostructures, processes, properties, and applications. Carbohydr. Polym. **272**, 118472 (2021)
8. A. Baranwal, A. Kumar, A. Priyadharshini, G.S. Oggu, I. Bhatnagar, A. Srivastava, P. Chandra, Chitosan: an undisputed bio-fabrication material for tissue engineering and bio-sensing applications. Int. J. Biol. Macromol. **110**, 110–123 (2018)
9. S. Kumar, S.S. Kesharwani, B. Kuppast, M.A. Bakkari, H. Tummala, Pathogen-mimicking vaccine delivery system designed with a bioactive polymer (inulin acetate) for robust humoral and cellular immune responses. J. Control. Release. **261**, 263–274 (2017)
10. S. Kumar, S.S. Kesharwani, B. Kuppast, M. Rajput, M. Ali Bakkari, H. Tummala, Discovery of inulin acetate as a novel immune-active polymer and vaccine adjuvant: synthesis, material characterization, and biological evaluation as a toll-like receptor-4 agonist. J. Mater. Chem. B. **4**, 7950–7960 (2016)
11. M. Dash, F. Chiellini, R.M. Ottenbrite, E. Chiellini, Chitosan - a versatile semi-synthetic polymer in biomedical applications. Prog. Polym. Sci. **36**, 981e1014; Different deacetylated chitosans. J. Agric. Food Chem. **51**, 4930e4934 (2011)
12. N. Islam, I. Dmour, M.O. Taha, Degradability of chitosan micro/nanoparticles for pulmonary drug delivery. Heliyon **5**, e01684 (2019)
13. W.Y. Cheah, P.L. Show, I.S. Ng, G.Y. Lin, C.Y. Chiu, Y.K. Chang, Antibacterial activity of quaternized chitosan modified nanofiber membrane. Int. J. Biol. Macromol. **126**, 569–577 (2019)
14. W. Wang, Q. Meng, Q. Li, J. Liu, M. Zhou, Z. Jin, K. Zhao, Chitosan derivatives and their application in biomedicine. Int. J. Mol. Sci. **21**(2), 487 (2020)
15. T.C. Yang, C.C. Chou, C.F. Li, Antibacterial activity of N-alkylated disaccharide chitosan derivatives. Int. J. Food Microbiol. **97**, 237–245 (2005)
16. Z. Shariatinia, Carboxymethyl chitosan: properties and biomedical applications. Int. J. Biol. Macromol. **120**, 1406–1419 (2018)
17. E.D. Freitas, C.F. Moura Jr., J. Kerwald, M.M. Beppu, An overview of current knowledge on the properties, synthesis and applications of quaternary chitosan derivatives. Polymers (Basel) **12**, 2878 (2020)
18. C.K.S. Pillai, W. Paul, S.P. Sharma, Chitin and chitosan polymers: chemistry, solubility and fiber formation. Prog. Polym. Sci. **34**, 641–678 (2009)

19. B. Xia, Y. Deng, Y. Lv, G. Chen, Stem cell recruitment based on scaffold features for bone tissue engineering. Biomater. Sci. **9**, 1189–1203 (2021)
20. H. Tan, C.R. Chu, K.A. Payne, K.G. Marra, Injectable in situ forming biodegradable chitosan-hyaluronic acid based hydrogels for cartilage tissue engineering. Biomaterials **30**, 2499–2506 (2009)
21. A. Chenite, M. Buschmann, D. Wang, C. Chaput, N. Kandani, Rheological characterization of thermogelling chitosan/glycerol-phosphate solutions. Carbohydr. Polym. **46**, 39–47 (2001)
22. L. Upadhyaya, J. Singh, V. Agarwal, R.P. Tewari, Biomedical applications of carboxymethyl chitosans. Carbohydr. Polym. **91**, 452–466 (2013)
23. H. Yilmaz Atay, Antibacterial activity of chitosan-based systems, in *Functional Chitosan*, ed. by S. Jana, S. Jana. (Springer, Singapore, 2019)
24. C. Xu, S. Guan, S. Wang, W. Gong, T. Liu, X. Ma, C. Sun, Biodegradable and electroconductive poly(3,4-ethylenedioxythiophene)/carboxymethyl chitosan hydrogels for neural tissue engineering. Mater. Sci. Eng. C Mater. Biol. Appl. **84**, 32–43 (2018)
25. G. Paradossi, E. Chiessi, M. Venanzi, B. Pispisa, A. Palleschi, Branched-chain analogues of linear polysaccharides: a spectroscopic and conformational investigation of chitosan derivatives. Int. J. Biol. Macromol. **2**, 73–80 (1992)
26. M. Heuser, C. Rivera, C. Nunez, G. Cardenas, Chitosan–copper paint types as antifouling. J. Chil. Chem. Soc. **54**, 448–453 (2009)
27. D.T. Vo, S. Sabrina, C.K. Lee, Silver deposited carboxymethyl chitosan-grafted magnetic nanoparticles as dual action deliverable antimicrobial materials. Mater. Sci. Eng. C **73**, 544–551 (2017)
28. D. Jaikumar, K.M. Sajesh, S. Soumya, T.R. Nimal, K.P. Chennazhi, S.V. Nair, R. Jayakumar, Injectable alginate-O-carboxymethyl chitosan/nano fibrin composite hydrogels for adipose tissue engineering. Int. J. Biol. Macromol. **74**, 318–326 (2015)
29. M.P. Diamond, M.L. Freeman, Clinical implications of postsurgical adhesions. Hum. Reprod. Update **7**, 567–576 (2001)
30. R.P. den Broek, C. Strik, Y. Issa, R.P. Bleichrodt, H. van Goor, Adhesiolysis-related morbidity in abdominal surgery. Ann. Surg. **258**, 98–106 (2013)
31. D.J. Costain, R. Kennedy, C. Ciona, V.C. McAlister, T.D. Lee, Prevention of postsurgical adhesions with N, O-carboxymethyl chitosan: examination of the most efficacious preparation and the effect of N, O-carboxymethyl chitosan on postsurgical healing. Surgery **121**, 314–319 (1997)
32. M. Ways T. Mohammed, W.M. Lau, V.V. Khutoryanskiy, Chitosan and its derivatives for application in mucoadhesive drug delivery systems. Polymers **10**, 267 (2018)
33. A.D. Kulkarni, H.M. Patel, S.J. Surana, Y.H. Vanjari, V.S. Belgamwar, C.V. Pardeshi, N, N, N-Trimethyl chitosan: an advanced polymer with myriad of opportunities in nanomedicine. Carbohydr. Polym. **157**, 875–902 (2017)
34. S.A. Sreenivas, K.V. Pai, Thiolated chitosans: novel polymers for mucoadhesive drug delivery–a review. Trop. J. Pharm. Res. **7**, 1077–1088 (2008)
35. M. Rahimi, R. Ahmadi, H.S. Kafil, V. Shafiei-Irannejad, A novel bioactive quaternized chitosan and its silver-containing nanocomposites as a potent antimicrobial wound dressing: structural and biological properties. Mater. Sci. Eng. C. **101**, 360–369 (2019)
36. X. Huang, C. Xu, Y. Li, H. Cheng, X. Wang, R. Sun, Quaternized chitosan-stabilized copper sulfide nanoparticles for cancer therapy. Mater. Sci. Eng. C. **96**, 129–137 (2019)
37. H. Xue, L. Hu, Y. Xiong, X. Zhu, C. Wei, F. Cao, W. Zhou, Y. Sun, Y. Endo, M. Liu, Y. Liu, Quaternized chitosan-Matrigel-polyacrylamide hydrogels as wound dressing for wound repair and regeneration. Carbohydr. Polym. **226**, 115302 (2019)
38. L. Wei, Y. Mi, J. Zhang, Q. Li, F. Dong, Z. Guo, Evaluation of quaternary ammonium chitosan derivatives differing in the length of alkyl side-chain: synthesis and antifungal activity. Int. J. Biol. Macromol. **129**, 1127–1132 (2019)
39. H. Li, Z. Zhang, X. Bao, G. Xu, P. Yao, Fatty acid and quaternary ammonium modified chitosan nanoparticles for insulin delivery. Colloids Surf. B. **170**, 136–143 (2018)

40. Y. Zhou, H. Yang, X. Liu, J. Mao, S. Gu, W. Xu, Potential of quaternization-functionalized chitosan fiber for wound dressing. Int. J. Biol. Macromol. **52**, 327–332 (2013)
41. Y. Qin, R. Xing, S. Liu, K. Li, L. Hu, H. Yu, X. Chen, P. Li, Synthesis of chitosan derivative with diethyldithiocarbamate and its antifungal activity. Int. J. Biol. Macromol. **65**, 369–374 (2014)
42. P.J. Park, J.Y. Je, S.K. Kim, Angiotensin I converting enzyme (ACE) inhibitory activity of hetero-chitooligosaccharides prepared from partially different deacetylated chitosans. J. Agric. Food Chem. **51**, 4930–4934 (2003)
43. Z.J. Qian, T.K. Eom, B.M. Ryu, S.K. Kim, Angiotensin I-converting enzyme inhibitory activity of sulfated chitooligosaccharides with different molecular weights. J. Chitin Chitosan. **15**, 75–79 (2010)
44. P.S. Bakshi, D. Selvakumar, K. Kadirvelu, N.S. Kumar, Chitosan as an environment-friendly biomaterial – a review on recent modifications and applications. Int. J. Biol. Macromol. **150**, 1072–1083 (2020)
45. R.K. Kankala, C.G. Liu, D.Y. Yang, S.B. Wang, A.Z. Chen, Ultrasmall platinum nanoparticles enable deep tumor penetration and synergistic therapeutic abilities through free radical species-assisted catalysis to combat cancer multidrug resistance. Chem. Eng. J. **383**, 123138 (2020)
46. M. Agarwal, D.P. Nagar, N. Srivastava, M.K. Agarwal, Chitosan nanoparticles-based drug Delivery: an update. Int. J. Adv. Multidiscip. Res. **2**, 1–13 (2015)
47. C. Liu, Y. Tan, C. Liu, X. Chen, L. Yu, Preparations, characterizations and applications of chitosan-based nanoparticles. J. Ocean Univ. China **6**, 237–243 (2007)
48. J.H. Adair, M.P. Parette, E.I. Altinoğlu, M. Kester, Nanoparticulate alternatives for drug delivery. ACS Nano **4**, 4967–4970 (2010); S.-M. Lee, S.T. Nguyen, Smart nanoscale drug delivery platforms from stimuli-responsive polymers and liposomes. Macromolecules **46**, 9169–9180 (2013)
49. V. Valtchev, L. Tosheva, Porous nanosized particles: preparation, properties, and applications. Chem. Rev. **113**, 6734–6760 (2013)
50. A. Narmani, J. Mohammadnejad, K. Yavari, Synthesis and evaluation of polyethylene glycol- and folic acid-conjugated polyamidoamine G4 dendrimer as nanocarrier. J. Drug Deliv. Sci. Technol. **50**, 278–286 (2019)
51. M. Rajabzadeh-Khosroshahi, M. Pourmadadi, F. Yazdian, H. Rashedi, M. Navaei-Nigjeh, B. Rasekh, Chitosan/agarose/graphitic carbon nitride nanocomposite as an efficient pH-sensitive drug delivery system for anticancer curcumin releasing. J. Drug Deliv. Sci. Technol. **74**, 103443 (2022)
52. K. Mortezaee, A. Narmani, M. Salehi, H. Bagheri, B. Farhood, H. Haghi-Aminjan, M. Najafi, Synergic effects of nanoparticles-mediated hyperthermia in radiotherapy/chemotherapy of cancer. Life Sci. **269**, 119020 (2021)
53. A. Narmani, S.M. Jafari, Chitosan-based nano delivery systems for cancer therapy: recent advances. Carbohydr. Polym. **272**, 118464 (2021)
54. A. Bernkop-Schnürch, D. Guggi, Y. Pinter, Thiolated chitosans: development and in vitro evaluation of a mucoadhesive, permeation enhancing oral drug delivery system. J. Contr. Release **94**, 177–186 (2004)
55. S. Sudhakar, S.V. Chandran, N. Selvamurugan, R.A. Nazeer, Biodistribution and pharmacokinetics of thiolated chitosan nanoparticles for oral delivery of insulin in vivo. Int. J. Biol. Macromol. **150**, 281–288 (2020)
56. Z.R. Zad, S.S.H. Davarani, A. Taheri, Y. Bide, A yolk-shell Fe3O4@PA-Ni@Pd/Chitosan nanocomposite-modified carbon ionic liquid electrode as a new sensor for the sensitive determination of fluconazole in pharmaceutical preparations and biological fluids. J. Mol. Liq. **253**, 233–240 (2018)
57. W.A. Adeosun, A.M. Asiri, H.M. Marwani, Fabrication of conductive polypyrrole doped chitosan thin film for sensitive detection of sulfite in real food and biological samples. Electroanalysis **32**, 1725–1736 (2020)

58. W. Yu, Y. Tang, Y. Sang, W. Liu, S. Wang, X. Wang, Preparation of a carboxylated single-walled carbon-nanotube-chitosan functional layer and its application to a molecularly imprinted electrochemical sensor to quantify semicarbazide. Food Chem. **333**, 127524 (2020)
59. W.T. Wahyuni, Z. Arif, N.S. Maryam, Development of graphene-chitosan composite modified screen-printed carbon electrode and its application for detection of rutin. IOP Conf. Ser. Earth Environ. Sci. 299 (2019)
60. Y. Wang, Y. Xiong, J. Qu, J. Qu, S. Li, Selective sensing of hydroquinone and catechol based on multiwalled carbon nanotubes/polydopamine/gold nanoparticles composites. Sens. Actuators B Chem. **223**, 501–508 (2016)
61. A. Hassanein, N. Salahuddin, A. Matsuda, G. Kawamura, M. Elfiky, Fabrication of biosensor based on Chitosan-ZnO/Polypyrrole nanocomposite modified carbon paste electrode for electroanalytical application. Mater. Sci. Eng. C **80**, 494–501 (2017)
62. S.J. Kim, S.J. Park, S.I. Kim, Swelling behavior of interpenetrating polymer network hydrogels composed of poly(vinyl alcohol) and chitosan. React. Funct. Polym. **55**, 53–59 (2003)
63. X. Shen, F. Ju, G. Li, L. Ma, Smartphone-based electrochemical potentiostat detection system using pedot: Pss/chitosan/graphene modified screen-printed electrodes for dopamine detection. Sensors **20**, 2781 (2020)
64. A. Diouf, M. Moufid, D. Bouyahya, L. Österlund, N. El Bari, B. Bouchikhi, An electrochemical sensor based on chitosan capped with gold nanoparticles combined with a voltammetric electronic tongue for quantitative aspirin detection in human physiological fluids and tablets. Mater. Sci. Eng. C **110**, 110665 (2020)
65. H. KarimiMaleh, S. Rostami, V.K. Gupta, M. Fouladgar, Evaluation of ZnO nanoparticle ionic liquid composite as voltammetric sensing of isoprenaline in the presence of aspirin for liquid phase determination. J. Mol. Liq. **201**, 102–107 (2015)
66. S.M. Patil, S.R. Sataraddi, A.M. Bagoji, R.M. Pathan, S.T. Nandibewoor, Electrochemical behavior of graphene-based sensors on the redox mechanism of aspirin. Electroanalysis **26**, 831–839 (2014)
67. B. Lou, U. Rajaji, S. Chen, T. Chen, A Simple sonochemical assisted synthesis of porous NiMoO4/chitosan nanocomposite for electrochemical sensing of amlodipine in pharmaceutical formulation and human serum. Ultrason. Sonochemistry **64**, 104827 (2020)
68. Z.Z. Stoiljković, M.L. Avramov Ivić, S.D. Petrović, D.Z. Mijin, S.I. Stevanović, U.C. Lačnjevac, A.D. Marinković, Voltammetric and square-wave anodic stripping determination of amlodipine besylate on a gold electrode. Int. J. Electrochem. Sci. **7**, 2288–2303 (2012)
69. A. Mohammadi, A.B. Moghaddam, K. Eilkhanizadeh, E. Alikhani, S. Mozaffari, T. Yavari, Electro-oxidation and simultaneous determination of amlodipine and atorvastatin in commercial tablets using carbon nanotube modified electrode. Micro Nano Lett. **8**, 413–417 (2013)
70. S. Akhter, W.J. Basirun, Y. Alias, M.R. Johan, S. Bagheri, M. Shalauddin, M. Ladan, N.S. Anuar, Enhanced amperometric detection of paracetamol by immobilized cobalt ion on functionalized MWCNTs - Chitosan thin film. Anal. Biochem. **551**, 29–36 (2018)
71. K.R. Reddy, P.K. Brahman, L. Suresh, Fabrication of high-performance disposable screen printed electrochemical sensor for ciprofloxacin sensing in biological samples. Meas. J. Int. Meas. Confed. **127**, 175–186 (2018)
72. R. Zou, S. Shan, L. Huang, Z. Chen, T. Lawson, M. Lin, L. Yan, Y. Liu, High-performance intraocular biosensors from chitosan-functionalized nitrogen-containing graphene for the detection of glucose. ACS Biomater. Sci. Eng. **6**, 673–679 (2020)
73. I.-Y. Jeon, Y.-R. Shin, G.-J. Sohn, H.-J. Choi, S.-Y. Bae, J. Mahmood, S.-M. Jung, J.-M. Seo, M.-J. Kim, D.W. Chang, L. Dai, J.-B. Baek, Edge-carboxylated graphene nanosheets via ball milling. Proc. Natl. Acad. Sci. USA **109**, 5588–5593 (2012)
74. W. Zhang, X. Li, R. Zou, H. Wu, H. Shi, S. Yu, Y. Liu, Multifunctional glucose biosensors from Fe3O4 nanoparticles modified chitosan/graphene nanocomposites. Sci. Rep. **511129**, 1–9 (2017)
75. H. Kim, K. Lee, S.I. Woo, Y. Jung, On the mechanism of enhanced oxygen reduction reactions in nitrogen-doped graphene nanoribbons. Phys. Chem. Chem. Phys. **13**, 17505–17510 (2011)

76. A.L.M. Reddy, A. Srivastava, S.R. Gowda, H. Gullapalli, M. Dubey, P.M. Ajayan, Synthesis of nitrogen-doped graphene films for lithium battery application. ACS Nano **4**, 6337–6342 (2010)
77. K. Saha, S.S. Agasti, C. Kim, X. Li, V.M. Rotello, Gold nanoparticles in chemical and biological sensing. Chem. Rev. **112**, 2739–2779 (2012)
78. H. Bayley, P.S. Cremer, Stochastic sensors inspired by biology. Nature **413**, 226–230 (2001)
79. I. Moldoveanu, R.-I. Stefan-van Staden, J.F. van Staden, Chitosan based diamond paste stochastic microsensors modified with gold nanoparticles detect hepatitis C virus core antigen. Electroanalysis **27**, 1842–1846 (2015)
80. R.-I. Stefan-van Staden, I. Popa-Tudor, C. Ionescu-Tirgoviste, R.-A. Stoica, Molecular recognition and determination of interleukins 1β, 6, 12, and 17 in whole blood from diabetic patients. Anal. Lett. **53**(13), 2021–2033 (2020)
81. C. Cioates Negut, R.-I. Stefan-van Staden, M. Badulescu, B. Bita, Disposable stochastic sensors obtained by nanolayer deposition of copper, graphene, and copper-graphene composite on silk for the determination of isocitrate dehydrogenases 1 and 2. Anal. Bioanal. Chem. **414**, 1797–1807 (2022)
82. R.-I. Stefan-van Staden, R.M. Ilie, F. Pogacean, S.M. Pruneanu, Needle stochastic sensors for on-site fast recognition and quantification of biomarkers for gastric cancer in biological samples. New J. Chem. **44**, 20203–20211 (2020)
83. D.-C. Gheorghe, R.M. Ilie-Mihai, R.-I. Stefan-van Staden, A. Lungu-Moscalu, J.F. van Staden, Fast screening method for early diagnosis of gastric cancer based on utilization of chitosan–S-doped graphene-based needle stochastic sensors. J. Pharm. Biomed. Anal. **214**, 114725 (2022)
84. R.-I. Stefan-van Staden, R.-M. Ilie-Mihai, D.-C. Gheorghe, I.M. Bogea, M. Badulescu, 2D disposable stochastic sensors for molecular recognition and quantification of maspin in biological samples. Microchim. Acta **189**, 101 (2022)
85. I. Moldoveanu, R.-I. Stefan-van Staden, Pattern recognition of HER-2 in whole blood samples using stochastic microsensors. ECS J. Solid State Sci. **4**, S3067–S3070 (2015)
86. R.-I. Stefan-van Staden, D.-C. Gheorghe, V. Jinga, C.S. Sima, M. Geanta, Fast screening of whole blood and tumor tissue for bladder cancer biomarkers using stochastic needle sensors. Sensors **20**, 2420 (2020)
87. G. Turnbull, J. Clarke, F. Picard, P. Riches, L. Jia, F. Han, B. Li, W. Shu, 3D bioactive composite scaffolds for bone tissue engineering. Bioact. Mater. **3**, 278–314 (2018)
88. A. Di Martino, L. Liverani, A. Rainer, G. Salvatore, M. Trombetta, V. Denaro, Electrospun scaffolds for bone tissue engineering. Musculoskelet. Surg. **95**, 69–80 (2011)
89. H. Qu, H. Fu, Z. Hana, Y. Sun, Biomaterials for bone tissue engineering scaffolds: a review. RSC Adv. **9**, 26252–26262 (2019)
90. R. Khajavi, M. Abbasipour, A. Bahador, Electrospun biodegradable nanofibers scaffolds for bone tissue engineering. J. Appl. Polym. Sci. **133**, 42883 (2016)
91. M. Witzler, D. Büchner, S.H. Shoushrah, P. Babczyk, J. Baranova, S. Witzleben, E. Tobiasch, M. Schulze, Polysaccharide-based systems for targeted stem cell differentiation and bone regeneration. Biomolecules **9**, 840 (2019)
92. D.P. Bhattarai, L.E. Aguilar, C.H. Park, C.S. Kim, A review on properties of natural and synthetic based electrospun fibrous materials for bone tissue engineering. Membranes **8**, 62 (2018)
93. C.-H. Fang, Y.-W. Lin, F.-H. Lin, J.-S. Sun, Y.-H. Chao, H.-Y. Lin, Z.-C. Chang, Biomimetic synthesis of nanocrystalline hydroxyapatite composites: therapeutic potential and effects on bone regeneration. Int. J. Mol. Sci. **20**, 6002 (2019)
94. Z. Xu, N. Wang, P. Liu, Y. Sun, Y. Wang, F. Fei, S. Zhang, J. Zheng, B. Han, Poly(Dopamine) coating on 3D-printed poly-lactic-Co-glycolic acid/β-tricalcium phosphate scaffolds for bone tissue engineering. Molecules **24**, 4397 (2019)
95. G. Narayanan, N. Vernekar, E.L. Kuyinu, C.T. Laurencin, Poly (Lactic Acid)-based biomaterials for orthopedic regenerative engineering. Adv. Drug Deliv. Rev. **107**, 247–276 (2016)

96. F. Shahidi, R. Abuzaytoun, Chitin, chitosan, and co-products: chemistry, production, applications, and health effects. Adv. Food Nutr. Res. **49**, 93–135 (2005)
97. R. Logith Kumar, A. Keshav Narayan, S. Dhivya, A. Chawla, S. Saravanan, N. Selvamurugan, A review of chitosan and its derivatives in bone tissue engineering. Carbohydr. Polym. **151**, 172–188 (2016)
98. S. Saravanan, R.S. Leena, N. Selvamurugan, Chitosan based biocomposite scaffolds for bone tissue engineering. Int. J. Biol. Macromol. **93**, 1354–1365 (2016)
99. D. Cassan, A. Becker, B. Glasmacher, Y. Roger, A. Hoffmann, T.R. Gengenbach, C.D. Easton, R. Hänsch, H. Menzel, Blending chitosan-g-poly(caprolactone) with poly(caprolactone) by electrospinning to produce functional fiber mats for tissue engineering applications. J. Appl. Polym. Sci. **94**, 48650 (2019)
100. K.V. Chan, M. Asadian, I. Onyshchenko, H. Declercq, R. Morent, N. De Geyter, Biocompatibility of cyclopropylamine-based plasma polymers deposited at sub-atmospheric pressure on poly (ε-caprolactone) nanofiber meshes. Nanomaterials **9**, 1215 (2019)
101. M.A. Fernandez-Yague, S.A. Abbah, L. McNamara, D.I. Zeugolis, A. Pandit, M.J. Biggs, Biomimetic approaches in bone tissue engineering: integrating biological and physicomechanical strategies. Adv. Drug Deliv. Rev. 84, 1–29 (2014)
102. L.L. Hench, I. Thompson, Twenty-first century challenges for biomaterials. J. R. Soc. Interface **7**, S379–S391 (2010)
103. S. Saravanan, D.K. Sameera, A. Moorthi, N. Selvamurugan, Chitosan scaffolds containing chicken feather keratin nanoparticles for bone tissue engineering. Int. J. Biol. Macromol. **62**, 481–486 (2013)
104. N. Saranya, S. Saravanan, A. Moorthi, B. Ramyakrishna, N. Selvamurugan, Enhanced osteoblast adhesion on polymeric nano-scaffolds for bone tissue engineering. J. Biomed. Nanotechnol. **7**, 238–244 (2011)
105. R. Niranjan, C. Koushik, S. Saravanan, A. Moorthi, M. Vairamani, N. Selvamurugan, A novel injectable temperature-sensitive zinc doped chitosan/β-glycerophosphate hydrogel for bone tissue engineering. Int. J. Biol. Macromol. **54**, 24–29 (2013)
106. T. Jiang, R. James, S.G. Kumbar, C.T. Laurencin, Chitosan as a biomaterial: structure, properties, and applications in tissue engineering and drug delivery. Biomed. Mater. Eng. **5**, 91–113 (2014)
107. K. Madhumathi, K.T. Shalumon, V.D. Rani, H. Tamura, T. Furuike, N. Selvamurugan, R. Jayakumar, Wet chemical synthesis of chitosan hydrogel–hydroxyapatite composite membranes for tissue engineering applications. Int. J. Biol. Macromol. **45**, 12–15 (2009)
108. M. Peter, N.S. Binulal, S.V. Nair, N. Selvamurugan, H. Tamura, R. Jayakumar, Novel biodegradable chitosan–gelatin/nano-bioactive glass ceramic composite scaffolds for alveolar bone tissue engineering. J. Chem. Eng. **158**, 353–361 (2010)
109. M. Peter, N. Ganesh, N. Selvamurugan, S.V. Nair, T. Furuike, H. Tamura, R. Jayakumar, Preparation and characterization of chitosan– gelatin/nanohydroxyapatite composite scaffolds for tissue engineering applications. Carbohydr. Polym. **80**, 687–694 (2010)
110. S. Pattnaik, S. Nethala, A. Tripathi, S. Saravanan, A. Moorthi, N. Selvamurugan, Chitosan scaffolds containing silicon dioxide and zirconia nanoparticles for bone tissue engineering. Int. J. Biol. Macromol. **49**, 1167–1172 (2011)
111. H. Park, B. Choi, J. Nguyen, J. Fan, S. Shafi, P. Klokkevold, M. Lee, Anionic carbohydrate-containing chitosan scaffolds for bone regeneration. Carbohydr. Polym. **97**, 587–596 (2013)
112. J.P. Kumar, L. Lakshmi, V. Jyothsna, D.R. Balaji, S. Saravanan, A. Moorthi, N. Selvamurugan, Synthesis and characterization of diopside particles and their suitability along with chitosan matrix XE "Chitosan matrix" for bone tissue engineering in vitro and in vivo. J. Biomed. Nanotech. **10**, 970–981 (2014)
113. B.G. Zhang, D.E. Myers, G.G. Wallace, M. Brandt, P.F. Choong, Bioactive coatings for orthopaedic implants—recent trends in development of implant coatings. Int. J. Mol. Sci. **15**, 11878–11921 (2014)
114. J. Radwan-Pragłowska, Ł. Janus, M. Piątkowski, D. Bogdał, D. Matysek, 3D hierarchical, nanostructured chitosan/PLA/HA scaffolds doped with TiO2/Au/Pt NPs with tunable properties for guided bone tissue engineering. Polymers **12**, 792 (2020)

115. N.G. Rim, A. Yih, P. Hsi, Y.J. Wang, Y.H. Zhang, J.Y. Wong, Micropatterned cell sheets as structural building blocks for biomimetic vascular patches. Biomaterials **181**, 126–139 (2018)
116. J. Zhang, C. Liu, F.L. Feng, D.W. Wang, S.S. Lu, G. Wei, H. Mo, T. Qiao, A PCPU nanoparticle/PU/decellularized scaffold composite vascular patch: synergistically optimized overall performance promotes endothelialization. Colloid Surface B **160**, 192–200 (2017)
117. D. Cherepanov, T.G.K. Bentley, W. Hsiao, P. Xiang, F. O′Neill, Y. Qian, N. Yurgin, D. Beenhouwer, Real-world cardiovascular disease burden in patients with atherosclerotic cardiovascular disease: a comprehensive systematic literature review. Curr. Med. Res. Opin. **34**, 459–473 (2018)
118. H.L. Bai, H.D. Hu, J.M. Guo, M.Y. Ige, T. Wang, T. Isaji, T. Kudze, H.Y. Liu, B. Yatsula, T. Hashimoto, Y. Xing, A. Dardik, Polyester vascular patches acquire arterial or venous identity depending on their environment. J. Biomed. Mater. Res. A **105**, 3422–3431 (2017)
119. P. Chantawong, T. Tanaka, A. Uemura, K. Shimada, A. Higuchi, H. Tajiri, K. Sakura, T. Murakami, Y. Nakazawa, R. Tanaka, Silk fibroin-Pellethane® cardiovascular patches: effect of silk fibroin concentration on vascular remodeling in rat model. J. Mater. Sci.-Mater. Med. **28**, 191 (2017)
120. J. Zhang, Y. Wang, C. Liu, F.L. Feng, D.W. Wang, H. Mo, L. Si, G. Wei, J. Shen, Polyurethane/polyurethane nanoparticle-modified expanded poly(tetrafluoroethylene) vascular patches promote endothelialization. J. Biomed. Mater. Res. A **106**, 2131–2140 (2018)
121. Y. Hu, W.H. Dan, S.B. Xiong, Y. Kang, A. Dhinakar, J. Wu, Z.P. Gu, Development of collagen/polydopamine complexed matrix as mechanically enhanced and highly biocompatible semi-natural tissue engineering scaffold. Acta Biomater. **47**, 135–148 (2017)
122. Q.L. Zhao, H.Q. Cui, J. Wang, H.X. Chen, Y.L. Wang, L.D. Zhang, X.M. Du, M. Wang, Regulation effects of biomimetic hybrid scaffolds on vascular endothelium remodeling. ACS Appl. Mater. Interfaces **10**, 23583–23594 (2018)
123. X. Li, J.Y. Tang, L.H. Bao, L. Chen, F.F. Hong, Performance improvements of the BNC tubes from unique double-silicone-tube bioreactors by introducing chitosan and heparin for application as small-diameter artificial blood vessels. Carbohyd. Polym. **178**, 394–405 (2017)
124. E.D. Boland, J.A. Matthews, K.J. Pawlowski, D.G. Simpson, G.E. Wnek, G.L. Bowlin, Electrospinning collagen and elastin: preliminary vascular tissue engineering. Front Biosci. **9**, 1422–1432 (2004)
125. B.H. Walpoth, G.L. Bowlin, The daunting quest for a small diameter vascular graft. Expert Rev. Med. Dev. **2**, 647–651 (2005)
126. A.A. Ignatius, L.E. Claes, In vitro biocompatibility of bioresorbable polymers: poly (L, DL-lactide) and poly (L-lactide-co-glycolide). Biomaterials **17**, 831–839 (1996)
127. P. Zamiri, Y. Kuang, U. Sharma, The biocompatibility of rapidly degrading polymeric stents in porcine carotid arteries. Biomaterials **31**, 7847–7855 (2010)
128. J.E. McBane, S. Sharifpoor, K.H. Cai, R.S. Labow, J.P. Santerre, Biodegradation and in vivo biocompatibility of a degradable, polar/hydrophobic/ionic polyurethane for tissue engineering applications. Biomaterials **32**, 6034–6044 (2011)
129. X. Kong, W. Xu, Biodegradation and biocompatibility of a degradable chitosan vascular prosthesis. Int. J. Clin. Exp. Med. **8**, 3498–3505 (2015)
130. W.H. Zhou, Z.J. Jia, P. Xiong, J.L. Yan, M. Li, Y. Cheng, Y.F. Zheng, Novel pH responsive tobramycin-embedded micelles in nanostructured multilayer-coatings of chitosan/heparin with efficient and sustained antibacterial properties. Mater. Sci. Eng. C-Mater. **90**, 693–705 (2018)
131. L. Hoang, S. Islam, A. Hindenburg, Utilization rates of enoxaparin and heparin in deep venous thrombosis prophylaxis after education and electronic order change at a single institution: a quality improvement study. J. Thromb. Thrombolys **46**, 502–506 (2018)
132. I. Wijesekara, R. Pangestuti, S.K. Kim, Biological activities and potential health benefits of sulfated polysaccharides derived from marine algae. Carbohyd. Polym. **84**, 14–21 (2011)
133. G.K. Kumar, C.H. Su, P. Velusamy, Ciprofloxacin loaded genipin cross-linked chitosan/heparin nanoparticles for drug delivery application. Mater. Lett. **180**, 119–122 (2016)

134. R.P. Singh, H.V. Gangadharappa, K. Mruthunjaya, Phytosome complexed with chitosan for gingerol delivery in the treatment of respiratory infection: in vitro and in vivo evaluation. Eur. J. Pharm. Sci. **122**, 214–229 (2018)
135. L. Vannozzi, L. Ricotti, C. Filippeschi, S. Sartini, V. Coviello, V. Piazza, P. Pingue, C.L. Motta, P. Dario, A. Menciassi, Nanostructured ultra-thin patches for ultrasound-modulated delivery of anti-restenotic drug. Int. J. Nanomed. **11**, 69–92 (2016)
136. G. Kerch, M. Chausson, S. Gautier, R.M. Meri, J. Zicans, E. Jakobsons, M. Joner, Heparin-like polyelectrolyte multilayer coatings based on fungal sulfated chitosan decrease platelet adhesion due to the increased hydration and reduced stiffness. Biomater. Tissue Technol. **1**, 1–5 (2017)
137. H.D.M. Follmann, A.F. Naves, A.F. Martins, O. Félix, G. Decher, E.C. Muniz, R. Silva, Advanced fibroblast proliferation inhibition for biocompatible coating by electrostatic layer-by-layer assemblies of heparin and chitosan derivatives. J. Colloid Interface Sci. **474**, 9–17 (2016)
138. J. Zhang, D. Wang, X. Jiang, L. He, L. Fu, Y. Zhao, Y. Wang, H. Mo, J. Shen, Multistructured vascular patches constructed via layer-by-layer self-assembly of heparin and chitosan for vascular tissue engineering applications. J. Chem. Eng. **370**, 1057–1067 (2019)

Chapter 6
Chitosan Nanoparticles: A Potential Biomedical Device

Karolina Jafernik, Magdalena Anna Malinowska, Katarzyna Bialik-Wąs, Marta Sharafan, Halina Ekiert, Elżbieta Sikora, and Agnieszka Szopa

Abstract Chitosan nanoparticles (chitosan-based nanoparticles; chitosan nanostructures; ChNPs) constitute a very interesting and promising group of bio-based compounds, which have attracted a lot of attention in the last decades. They are more and more commonly used in various biomedical devices, especially in cancer diagnostics (fluorescent endoscopic diagnostics, detecting cancer cells), wound dressings, as the glucose detection sensor and the histamine biosensor, in bone tissue engineering and dentistry, which are presented in details in this chapter. Such a variety of application possibilities is mainly due to the properties of chitosan, which is characterized by high biocompatibility, biodegradability, non-toxicity, as well as the great potential as nanocarriers encapsulating active substances and providing a controlled release process. This chapter presents an overview of ChNPs preparation methods, mainly: reversed micelles, emulsification and crosslinking, SCASA, spray

K. Jafernik · H. Ekiert · A. Szopa (✉)
Chair and Department of Pharmaceutical Botany, Medical College, Jagiellonian University, Medyczna 9, 30-688 Kraków, Poland
e-mail: a.szopa@uj.edu.pl

K. Jafernik
e-mail: karolina.jafernik@doctoral.uj.edu.pl

H. Ekiert
e-mail: halina.ekiert@uj.edu.pl

M. A. Malinowska (✉) · K. Bialik-Wąs · M. Sharafan · E. Sikora
Faculty of Chemical Engineering and Technology, Institute of Organic Chemistry and Technology, Cracow University of Technology, Warszawska 24, 31-155 Kraków, Poland
e-mail: magdalena.malinowska@pk.edu.pl

K. Bialik-Wąs
e-mail: katarzyna.bialik-was@pk.edu.pl

M. Sharafan
e-mail: marta.sharafan@doktorant.pk.edu.pl

E. Sikora
e-mail: elzbieta.sikora@pk.edu.pl

S. K. Swain and A. Biswal (eds.), *Chitosan Nanocomposites*,
Biological and Medical Physics, Biomedical Engineering,
https://doi.org/10.1007/978-981-19-9646-7_6

drying, phase inversion precipitation, ionic gelation, and emulsion-droplet coalescence. Offering a lot of benefits, nanotechnology for medical and biomedical science has become the foundation for the development and improvement of human life. The safety of nanoparticles which can be toxic to the environment, organisms, and cells, had to be taken into account in their safety assessment in the biomedical fields. According to this, the overall safety of these materials is also evaluated.

6.1 Introduction

Nanoscience and nanotechnology are fields that have led to many important breakthroughs in therapeutic areas that can be used in medicine. They are based on nanoparticles (NPs), which are particles between 10 and 100 nm in size that occur naturally or are made in a laboratory. They have a wide range of applications, mainly due to the easier control, more detailed orientation, and a higher surface-to-volume ratio compared to microparticles. In the case of chitosan nanoparticles (chitosan-based nanoparticles; chitosan nanostructures; ChNPs) the mean particle size can range between 92 and 800 nm. This variation in particle distribution is primarily due to the preparation methods. ChNPs are relatively simple and easy to produce. In general, we can categorize various chitosan NPs preparation methods, which are presented in Fig. 6.1 [1].

When the reversed micelles (microemulsion) method, based on covalent crosslinking is used, the average size of ChNPs size will be below 100 nm [1, 2]. If the nano-precipitation method is used, the average particle size of ChNPs becomes much larger and varies between 600 and 800 nm [1, 3, 4]. Based on the required application and needs, researchers can design chitosan NPs with different basic characteristics, such as size, stability, drug loading capacity, and retention time. However, everyone uses the same biopolymer —chitosan, which is one of the most functional, naturally occurring polymer with a very high biodegradability and biocompatibility. This has been confirmed by a number of studies. Chitosan belongs to polysaccharides and it can be easily obtained by deacetylating chitin, which is the main component of crustacean shells and wall fungus cells. The chemical structure of chitosan is reflected in its properties, mainly in its functionality as a carrier. An extremely important here is the presence of a primary amine at the C-2 position of glucosamine residues [5–7]. From the point of view of the chemical structure, chitosan is a copolymer consisting of β-(1 → 4)-2-acetamido-D-glucose and β-(1 → 4)-2-amino-D-glucose units (Fig. 6.2) in different proportions and in random sequences [1, 8].

It is observed that scientists focus primarily on the use of ChNPs as biomedical dressings, which have a strong antimicrobial effect. It influences faster wound healing, but also allows substances to penetrate into the skin, reducing inflammatory reactions. Dressings based on ChNPs, in addition to accelerating the regeneration of difficult-to-heal wounds, are also designed to maintain homeostasis and reduce bleeding. ChNPs are very often combined with other antimicrobial NPs, such as silver, gold, or copper, which increases antifungal and antibacterial activity.

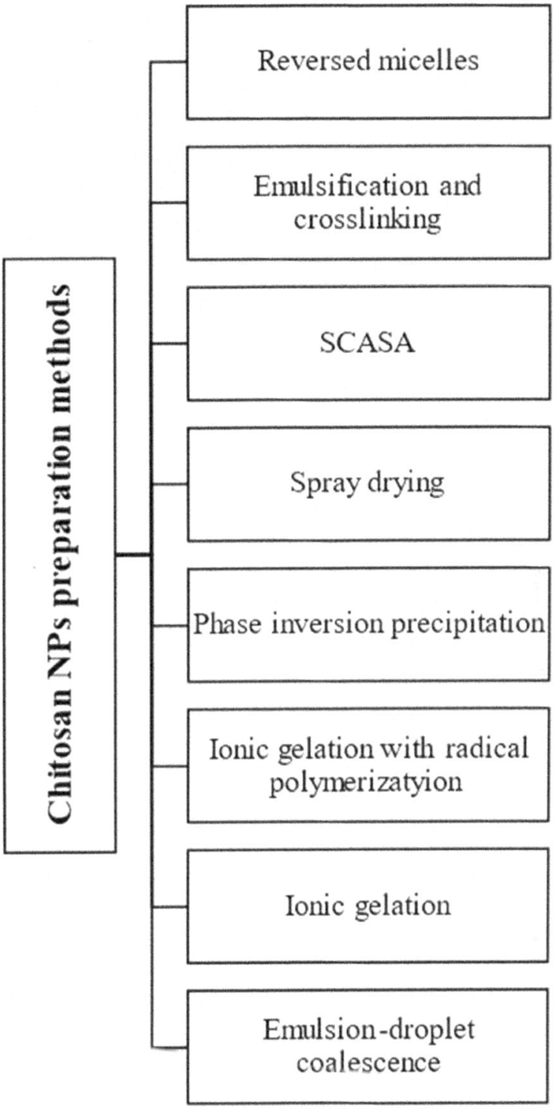

Fig. 6.1 Preparatory methods of Chitosan nanoparticles

ChNPs have also found their application in tissue engineering, especially in bone tissue regeneration. Numerous studies confirm the beneficial effect on the regeneration of osteoblasts and supporting other bone-forming mechanisms. Another area in which ChNPs are used is anti-cancer diagnostics. Nanostructures are used as carriers that bind to cancer cells, influencing their uptake, and thus aid in faster diagnosis. ChNPs were also used to develop glucose and histamine biosensors. Also, dentistry widely uses ChNPs, using them primarily for their antimicrobial activity [11–17].

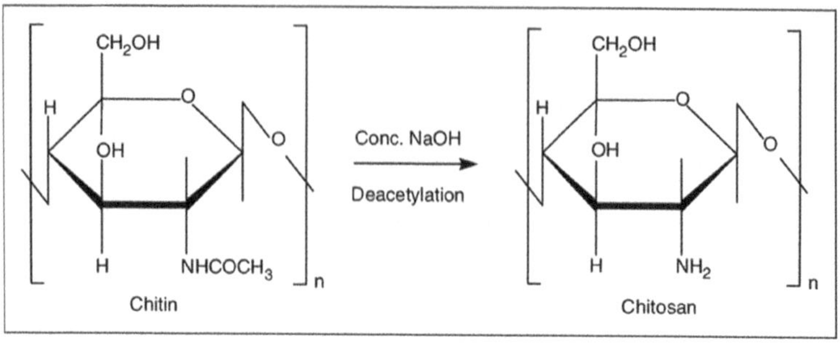

Fig. 6.2 The chemical structure of chitin and chitosan [9, 10]

Despite the many benefits and advantages of chitosan NPs, their safety assessment in medical applications should be also considered. Nowadays, scientists are aware of the danger, undesirable effects, and harmfulness, which can have components in the form of NPs. Studies on the toxicity of ChNPs are still ongoing, but their results are not sufficient yet. However, there are indications of their harmful effects at in vitro and in vivo levels. Therefore, this issue concerning the safety of chitosan NPs is also discussed in this chapter [1, 18].

The discussed topic is extremely important due to the growing need to explore new possibilities of using natural materials in medicine (as dressings, drug carriers, or diagnostic tools). Chitosan is a material that is known, tolerated by the body, and has been tested for safety. Chitosan nanoparticles open up completely new possibilities in human health and constitute an innovative solution to many existing questions in the field of medicine and pharmacy. The presented data will allow us to summarize the information gathered so far regarding the use of this material. Significant attention should be given to the enormous potential of chitosan nanoparticles, which in the coming years will be developed to be used in numerous fields of life science.

6.2 Methods of ChNPs Fabrication

As chitosan nanoparticles (ChNPs) are the object of various studies, several synthetic methods for preparation of these materials have been tested and developed. The methods take into account factors such as the particle size, stability, and retention time, as well as their loading capacity. Most common, chitosan nanoparticles can be obtained by processes such as precipitation and emulsification, but also ionic or covalent crosslinking. The methods can be also be combined. It is difficult to indicate which method has the most advantages. However, observations show that the methods that are as quick as possible, are the most valuable in terms of the production of chitosan nanoparticles for health purposes. These include techniques such as spray drying, ionic gelation, and self-assembly [1]. Figure 6.1 presents the overall view of the most usable and significant methods of obtaining chitosan NPs, which are shortly described below.

6.2.1 *Reversed Micelles*

The reversed micelles (microemulsion) method is the technique that engages covalent crosslinking. In this process, water-in-oil reverse micelle structures are formed [2]. The aqueous phase contains chitosan and glutaraldehyde. These components are mixed with the organic layer, usually a lipophilic surfactant and an organic solvent. Sodium bis (2-ethylhexyl) sulfosuccinate (AOT) and cetyltrimethylammonium bromide (CTAB) are usually used as surfactants, in turn, n-hexane is applied as the organic phase [19]. The core of the micelle contains chitosan, and this plays the role of a scaffold where ChNPs are formed during crosslinking process. The isolation of NPs takes place in three steps: first is the precipitation of surfactant with $CaCl_2$, second is the dialysis for the elimination of unreacted substances, and the last one is freeze-drying. NPs characterized by a size below 100 nm are obtained by this process. This makes them applicable in many fields of medicine, owing to their high loading capacity and their ability to sustained release. To avoid toxic substances, the preparation procedure can be modified by the use of non-harmful solvents and crosslinkers [3].

6.2.2 *Emulsification and Crosslinking*

Emulsification and crosslinking are the oldest methods for the obtaining of ChNPs. This technique uses the amino group of chitosan and the aldehyde group of a crosslinking agent [20]. The aqueous phase of the emulsion contains chitosan, and the oil phase includes toluene, glutaraldehyde, Span 80, and the crosslinker. Both phases are combined while intensive mixing, where the ChNPs are formed after crosslinking. The separation of the ChNPs can be done by using various methods, e.g., centrifugation, multiple washing steps, or freeze-drying. It is possible to obtain ChNPs characterized by uniform size. On the other hand, due to glutaraldehyde toxicity, as well as its drug integrity issues, this method is no longer used [1].

6.2.3 *SCASA*

Supercritical-CO_2-assisted solubilization and atomization (SCASA) method which is characterized by the safety of use. This "green" technique is a method that does not include acids or harmful organic solvents [21]. The process is based on water and supercritical CO_2. The dissolution of chitosan in water is possible through the acidifying effect of pressurized CO_2. The disadvantage of this method is the relatively high size of the ChNPs and the longtime of chitosan dissolution which takes up to 48 h. After this step, the chitosan solution is fed to a fluidized bed by a spraying nozzle which leads to its atomization. In effect, ChNPs are formed in the drying process. They are collected by a filter on top of the fluidized bed [22].

6.2.4 Spray Drying

Spray drying is the process where ChNPs are formed by passing chitosan through the aqueous acetic acid solution via a nozzle. The air temperatures range from 120 °C to 160 °C [23]. As a result, the mixture of chitosan microparticles and NPs is obtained, thus the low-size particles fraction needs to be isolated by other methods. The ChNPs sizes range from 95 to 358 nm, they are homogeneous and characterized by high zeta potential. Adjusting the concentration of the chitosan solution, the size of the nozzle, and the molecular weight of the chitosan allows for the control on ChNPs properties [24].

6.2.5 Phase Inversion Precipitation

The phase inversion precipitation technique is based on the combination of two methods: emulsification and precipitation. The principle of this method is the composition of the O/W (oil in water) type emulsion. The organic phase which can be dichloromethane and acetone and an aqueous solution of chitosan are combined in the presence of a stabilizer, e.g., poloxamer. After a high-pressure homogenization, nanometer-sized well-dispersed emulsion droplets are obtained. In the last step, methylene chloride is separated from the emulsion by low pressure evaporation at room temperature. This leads to acetone diffusing out of the droplets and the precipitation of ChNPs which is characterized by low sizes and high zeta potential [4].

6.2.6 Ionic Gelation with Radical Polymerization

The radical polymerization method combined with ionic gelation is also used commonly. This process consists of chitosan gelation and acrylic acid polymerization. Acrylic acid can be replaced by methacrylic acid. The polymerization agent used is usually potassium persulphate. The process lasts 6 h and runs at 60–70 °C [25]. There are some reagents that needs to be removed after the reaction. Usually, they are separated by washing them with water or through dialysis. The substances which can be loaded into these chitosan forms can be peptides (e.g., silk), serum albumin, and insulin. They are active substances dedicated mostly for oral route [26]. Currently, a limited number of applications are available, possibly due to the long production process.

6.2.7 *Ionic Gelation*

The ionic gelation technique is one of the most preferred preparation methods for bio-based NPs. This method uses a crosslinking reaction that takes place between the chitosan amino groups having a positive charge and the negatively charged polyanionic groups (e.g., sodium tripolyphosphate) [27]. Chitosan is dissolved in an aqueous acidic solution and then the aqueous solution of sodium tripolyphosphate is added during intensive stirring. Anionic molecules diffuse into the mixture of positively charged chitosan molecules. Moreover, crosslinking process starts leading to ChNPs formation. After centrifugation and washing processes, chitosan NPs can be obtained by drying or freeze-drying. The advantage of this method is safe, there are no harmful crosslinkers or toxic solvents used. The process can be carried out at room temperature and the final nanoparticle size can be adjusted by changing the chitosan/sodium tripolyphosphate ratio. That directly affects drug encapsulation efficiency and delivery [28].

6.2.8 *Emulsion—Droplet Coalescence*

The desolvation process, also called the emulsion-droplet coalescence method is based on the coalescence process of two emulsions, both W/O types. This induces precipitation of chitosan nanoparticles. The precipitation agent used in this process (e.g., NaOH) decreases the solubility of chitosan in an aqueous phase, which leads to the formation of small chitosan droplets. The process requires the application of high-speed homogenization. The continuous phase consists of liquid paraffin and sorbitan sesquioleate. The whole process of ChNPs formation includes the following steps: centrifugation, washing, and freeze-drying [29]. The particle size ranged from 600 to 800 nm and they are characterized by the ability of encapsulation of hydrophobic active substances. Importantly, the droplet coalescence method is preferable because of the usage of various organic solvents [30].

6.3 Application of ChNPs in Fluorescent Endoscopic Diagnostics

ChNPs have enormous potential to use them in the diagnosis of many diseases, including cancer. The methods can be divided into several different groups according to the apparatuses applied and the materials used. A lot of studies focus on the application of ChNPs in endoscopic diagnostics of fluorescent cancer of the colon and oral cavity. Chitosan coupled with active compounds is highly effective in capturing cancer cells, which results in faster diagnosis.

Yang et al. have carried out a series of studies focused on the use of ChNPs in the diagnosis of colorectal cancer through fluorescent endoscopy. In the early stages of work, they designed oral NPs for encapsulating 5-aminolevulinic acid (5-ALA) to mend disclosure of colon tumor cells in vivo. The primary goal was for the NPs to be resistant to bacterial uptake in the gastrointestinal tract, which sincerely distorts the conclusions of endoscopic observations. Chitosan was used for the study and blended with sodium tripolyphosphate (STPP) and 5-ALA to assemble ChNPs and ChNPs weighted with 5-aminolevulinic acid (CNA) by adding other pH amounts and concentrations of 5-ALA solution. Fluorescence microscopy examination presented that CNA can be absorbed by Caco-2 colon tumor cells but presented no absorption by *Escherichia coli*. The conclusion suggests that CNA may eliminate the control of normal gut flora and provides a suitable device for fluorescent endoscopic identification of colorectal tumor cells tested in vivo conditions, bidding researchers to further research on ChNPs [11].

Yang et al. in the second stage conducted research on the conjunction of ChNPs with folic acid for use in endoscopic examinations in the early detection of colorectal tumor. In the first research, they combined folic acid with chitosan, while maintaining a high affection for colon tumor cells caused by overexpression of the folate receptor. ChNPs conjugated with folic acid were developed as a suitable 5-aminolevulinic acid carrier to increase the identification of colorectal-type tumor cells in vivo *conditions*. The HT29 and Caco-2 colorectal tumor cell lines were exposed to the activity of the resulting molecule. It was found that cancer cells were easier to pick up, prompting the research team to continue their research [31].

In Yang et al. study, folic acid molecules were conjugated with ChNPs and combined with alginate to create high-throughput NPs for the photodynamic detection of colorectal tumor. ChNPs incorporated into alginate and conjugated with folic acid have been found to be the finest vectors for the different delivery of 5-ALA to the colon for fluorescent endoscopic disclosure. The resulting ChNPs were easily absorbed by colon tumor cells via endocytosis via the folic acid receptor. The loaded 5-ALA was then discharged in the lysosome, which was promoted by the cut-down intensity of interest between chitosan and 5-ALA by the deprotonated alginate, producing in higher intracellular accretion of PpIX (protoporphyrin IX) for photodynamic detection. The research confirmed the effectiveness of ChNPs for fluorescent endoscopic detection [32].

Yang and other researchers, in addition to work on the endoscopic diagnosis of colorectal cancer, also focused on research related to the diagnosis of oral tumor, a subtype of neck and head cancer. A high-throughput ChNPs for the photodynamic detection of oral cancer was designed. The resulting ChNPs contained succinate-modified chitosan (SCHI) conjugated to folic acid to produce ChNPs with great drug loading efficiency and advanced drug release in cell lysosomes. An experiment was also performed where the resulting complex of ChNPs was loaded with 5-aminolevulinic acid (5-ALA). Studies have shown that there are no significant differences between the effects of ChNPs loaded with 5-ALA acid and ChNPs without 5-ALA. The NPs have been shown to be readily absorbed by oral tumor cells through endocytosis via the folate receptor. The release of loaded 5-ALA in the lysosome was promoted by the decreased concentration of attraction between ChNPs and 5-ALA by deprotonated succinate particle, resulting in higher agglomeration of intracellular protoporphyrin IX (PpIX) which was used for photodynamic detection. The results obtained present that ChNPs embedded with N-succinyl chitosan. Importantly, if conjugated with folic acid, they are a great vector exhibiting the potential for the oral delivery of 5-aminolevulinic acid for endoscopic fluorescence detection [33].

6.4 Use of ChNPs in Medical Dressings

ChNPs are characterized by broad antimicrobial action against both fungi and bacteria, which has been confirmed by a large number of clinical trials both in vitro and in vivo. Biomedical functions of medical dressing include, in particular, antimicrobial coatings that promote wound healing, prevent infection, and reduce the rate of infectious diseases. This has attracted great interest among researchers in ChNPs, which they use to create medical dressings those not only support faster wound healing, but also act as occlusion and support the delivery of drugs. ChNPs are most often supported by antimicrobial elements such as silver, gold, copper, or zinc, as well as with other polymers (e.g., hyaluronic acid), which, like chitosan, show high biocompatibility and biodegradability [35–41].

Researchers are focusing primarily on creating a dressing that, in addition to its antimicrobial action, will contain an active ingredient that will result in faster tissue reconstruction, Lin and his team have designed a NPs containing curcumin, acting as an active ingredient with anti-inflammatory, antimicrobial wound healing functions, combined with tripolar nanocomplex containing poly-γ-glutamic acid, chitosan and pluronic using ion gel technology. The nanocomplex decreased inflammation and infection of bacteria during wound regeneration. Additionally, in vitro analyzes showed a controlled release of curcumin in a prompted model of skin tissue. In vivo studies present that chitosan wound dressing containing chitosan/poly-γ-glutamic acid/Pluronic/curcumin NPs stimulated neo collagen regeneration and reconstruction of tissue [36].

ChNPs are also used in wound dressings. Sharaf and his team designed a smart dressing that allows the release of the drug to accelerate healing. The bandage was made of natural materials—ChNPs combined with propolis with cellulose acetate nanofibers were used. Research on the cytotoxicity of propolis and chitosan/propolis NPs was thoroughly investigated and showed no abnormalities. In addition, a cell viability test was carried out—chitosan/propolis NPs introduced into cellulose nanostructures showed cell growth that the level of 89.46%. In vivo analysis implied that after 21 days of treatment with loaded nanostructures in an albino mouse model, repair of air follicles, epithelial cells, and sebaceous glands in the skin of a burn wound took place [41].

Many solutions are sought in the case of burn wounds. El-Feky et al. designed a dressing closed with ChNPs containing silver sulfadiazine for controlled release into a burn wound to control bacterial growth. The dressing was defined in terms of physical functions, in addition to the characterization, FTIR, RTG, SEM, and in vitro release were used. The dressing turned out to be effective in inhibiting the growth of Gram-negative and Gram-positive bacteria as well as *Candida* fungi strains on the wound which was infected [42].

Biranje and other scientists set out to develop a biodegradable dressing that would accelerate wound healing in necrotic tissue. For this purpose, they used ChNPs, which were folded into a porous dressing by means of lyophilization. The designed dressing was studied for biodegradability, morphology, pore volume, porosity, and surface area. Bigger surface and porosity of the dressing promoted its limited biodegradation through enzymatic action. In vitro analyzes using dermal fibroblasts (HDF) confirmed the safety of the dressing for wound healing functions. An in vitro ELISA based on human thrombin-antithrombin was also performed, which presented accelerated hemostatic activity through greater thrombin stimulation and stable blood clot formation. Blood in contact with the dressing contained twice the level of thrombin-antithrombin. The conclusions indicate the huge potential of the advanced dressing to remove necrotic tissues and accelerate hemostasis for efficient and fast wound healing [38].

Rahimi et al. designed a series of quaternized ChNPs layers blended with great potential silver NPs for wound dressing applications. The membranes were prepared by the following methods: scanning electron microscopy (SEM), Fourier transform infrared spectroscopy (FTIR), X-ray diffraction (XRD), element mapping analysis (MAP), and energy dispersive X-ray spectroscopy (EDX). More parameters such as swelling ratio, membrane biocompatibility, and blood clotting activity were also tested. In addition, the antimicrobial activity of the assembled samples was tested for: *E. coli, S. aureus, P. aeruginosa, C. albicans,* and multi-drug resistant *P. aeruginosa* as a drug-resistant strain of bacteria. The conclusion presented that the quaternary layer of chitosan showed a strong antibacterial effect against all strains. Moreover, all nanocomposites showed strong antimicrobial activity. Overall, the results of cell proliferation and attachment by MTT and DAPI staining on HFFF cells demonstrated the cytocompatibility of nanoparticle membranes. These conclusions indicate that engineered bioactive quaternary nanocomposite chitosan membranes can be used as wound dressings [48].

Tien et al. designed biocompatible fiber scaffolds based on deacetylated chitosan, manufactured using high performance blow spinning. Scanning electron microscopy analysis showed that the chitosan nanofiber structures had ultra-thin and continuous fibers, characterized by the size between 300 and 1200 nm. They had also porous structures, strongly interconnected (between 30 and 75% porosity). Imitating particular characteristics of the extracellular skin tissue matrix. After the treatment of the nanofibers after spinning with an aqueous solution of K_2CO_3, a fibrous structure with high chitosan concentration was created. Interestingly, it remained in its fibrous scaffold integrity for cell culture. The results of the chitosan nanofiber scaffold mechanical properties held various conditions (dry and wet) proved that these materials are characterized with strength and durability suitable for dressing implementation. It is worth to point out, that the wet scaffold underwent significant elastic deformation during tensile such that the elongation at break increased dramatically to 44% of its original length, showing the morphology of the wavy fibers near the break point. Culturing normal human skin fibroblast cells on scaffolds for 1 day to 2 weeks period days showed that the structures are highly compatible and provide an appropriate platform for cell adhesion, viability, and proliferation. The secretion profiles of wound healing proteins into the cell culture medium have shown that chitosan fibers are promising scaffolds for wound healing applications. Overall, the dense, high-porosity fibrous network of chitosan nanofiber structures and their mechanical properties shows that they can be used to design and manufacture new materials that imitate the cuticle layer of natural leather [49].

Cai et al. used magnetic Fe_3O_4 NPs to improve the antimicrobial and mechanical properties of composite chitosan/gelatin (CS/GE) nanofiber membranes. Homogeneous Fe_3O_4/CS/GE nanostructures have been successfully subjected to electrospinning. The incorporation of Fe_3O_4 NPs resulted in a significant improvement in mechanical characterizations. The optimal mechanical performance was demonstrated on nanofiber membranes with Fe_3O_4/CS/GE, achieving a 155% increase in Young's modulus, a 128% increase in tensile strength, and a 100% increase in toughness with CS/GE. The outstanding mechanical reinforcement can be explained by the effective filler dispersion and filler-matrix interactions, which ensure efficient load transfer from the CS/GE matrix to the Fe3O4 nanofillers. Additionally, thanks to Fe_3O_4 NP supplementation, the zones of inhibition of *E. coli* and *S. aureus* were significantly expanded. Overall, nanofiber membranes made of Fe_3O_4/CS/GE composite with developed mechanical functions as well as antibacterial activity can be proposed as a dressing material [51].

Table 6.1 Biomedical dressing using chitosan nanoparticles and their roles in wound healing

Composition of the dressing	Biological activity	References
Tripolier nanocomposite with chitosan, poly-γ-glutamic acid, and Pluronic combined with curcumin nanomolecules	• Faster tissue reconstruction • Reduce inflammation • Reduce bacterion infection • Promote neocolagen regeneration	[36]
Chitosan and propolis nanoparticles with cellulose acetate nanofibers	• Show cell viability at the level of 89.46% • Repair of hair follicles, epithelial cells, and sebaceous glands in the skin of a burn wound	[41]
Chitosan nanoparticles containing silver sulfadiazine	• Inhibiting the growth of Gram-positive and Gram-negative bacteria • Antibacterial activity against *Candida* strains	[42]
Chitosan nanoparticles	• Hemostatic activity • Stable blood clot formation • Rapid wound healing	[38]
Fiber nanoscaffolds with deacetylated chitosan	• Carrying out cell culture (human skin fibroblasts) on nanoscaffolds (1–14 days) has shown that the scaffolds are very compatible • Nanoscaffolds provide an appropriate platform for cell adhesion, viability, and proliferation	[49]
Fe_3O_4 nanoparticles with chitosan and gelatin nanofiber membranes	• Antibacterial activity	[51]

Table 6.1 presents the summary of the ChNPs application in biomedical wound dressings. Chitosan nanoparticles have a significant potential as safe and active material that can be loaded with various therapeutics according to the activity needed. Importantly, it ensures the controlled releasing process of the healing agents.

6.5 ChNPs as Components of Sensors Used in Biomedical Engineering

As biomedical engineering is gaining importance as a novel treatment tool, it focuses on the advantages of biology, medicine, and technology. The application of chitosan is a prime example of how these areas can be combined together in the service of human health.

6.5.1 Application of ChNPs in a Glucose Detection Sensor

Wang et al. focused on the improvement of a new nonenzymatic glucose sensor, which works by electroplating copper-cobalt dendritic nanostructures (Cu-Co NS) on a glassy carbon electrode (GCE), which was altered with nanostructures of reduced graphene oxide and chitosan (RGO- CHIT). The sensor's electrochemical behavior and electrocatalytic parameters for glucose oxidation were assessed using cyclic voltammograms, chronoamperometry, and amperometric method. Analyzed to monometallic Cu or Co NS sensors, the bimetallic Cu-Co NS sensor has great electrocatalytic activity against glucose oxidation. The effect of electrodeposition time and the scale of Cu^{2+} to Co^{2+} in the electrodeposition solution, the electrocatalytic effect of the Cu-Co NSs sensor was investigated in detail. The best catalytic activity against glucose oxidation can be achieved under advanced conditions: a settling time of 2600 s and a molar ratio of Cu^{2+} to Co^{2+} of 2:1. The catalytic current density is linear to the glucose concentration in the scope 0.015–6.95 mM with a sensitivity of 1921 $\mu A\ cm^{-2}\ mM^{-1}$ and a detection limit of 10 μM. Excellent catalytic activity, good sensitivity, and high stability indicate that the newly established dendritic Cu-Co NSs/RGO-CHIT/GCE sensor is a promising tool for use as glucose sensors [16].

6.5.2 ChNPs as an Element Used in the Histamine Biosensor

ChNPs were used in the histamine biosensor project. Nontipichet et al. developed a histamine biosensor based on a screen-printed prussian blue (PB) carbon electrode deposited on multi-wall carbon nanotunnels coated with a macroporous layer of compound cryogel with ChNPs (CS-AuNPs Cry). Due to its great specific surface area and conductivity, CS-AuNPs Cry has been tested to be a great auxiliary material for immobilization with diamine oxidase (DAO). PB acted as a redox mediator, promoting electron relocation between hydrogen peroxide and the electrode surface. The PB reduction current was measured as DAO catalyzed oxidation of histamine releasing hydrogen peroxide. The expected biosensor showed two linear areas: 2.50–125.0 $\mu mol\ L^{-1}$ and 125.0–400.0 $\mu mol\ L^{-1}$. The detection check was 1.81 $\mu mol\ L^{-1}$. The reproducibility was great (RSD = 5.46%), the operational stability was good, the long-term stability was high and the selectivity was excellent. The biosensor determined the histamine levels in fish and shrimp with solid recoveries and the results accessed definite with the conclusions accessed in the ELISA test [52].

6.6 ChNPs in Bone Tissue Engineering

Scientists are also developing methods to use ChNPs in the reconstruction of bone tissue. The researchers focused on how ChNPs affect bone. Moradikhah and others conducted an experiment to confirm the beneficial effects of CHN on bone tissue. To this end, they prepared an alendronate-conjugated ChNPs using a microfluidic cross-linked device. The study showed that ChNPs approximately enhanced the osteogenic emission of hA-MSC and may be a suitable compartment for bone engineering nanoscaffolds [53].

Other researchers at the head of Saravanan focused on designing nanoscaffolds that could be used in bone engineering. For this purpose, they isolated keratin from chicken feathers, and then synthesized keratin NPs, which were coupled with a chitosan matrix that acts as a scaffold. A lot of studies have been imposed using SEM, FTIR, and XRD analyses. The nanoscaffolds in the range had a porous architecture that assist in cell infiltration and tissue in growth. The NPs of keratin interacted with the chitosan matrix and did not change the semi-crystalline nature of the chitosan nanoscaffolds. The biodegradation and adsorption of the nanoscaffolds proteins was significantly increased after the addition of keratin NPs. The scaffolds were further construct not to be cytotoxic to human osteoblastic cells [54].

Ullah et al. conducted an experiment in which they introduced zinc oxide NPs into 3D porous chitosan-collagen structures and considered the effect of incorporating ZnO NPs on the microstructure, biodegradation, mechanical properties, and cytocompatibility of three-dimensional porous nanoscaffolds. The effects of incorporation of ZnO NPs in the form of three-dimensional porous chitosan-collagen scaffolds were investigated by mechanical and swelling tests, and the effect on the morphology of the scaffolds was examined microscopically. Cytocompatibility and biodegradation tests investigated the effect of the incorporation of zinc oxide NPs on the ability of the scaffolds to be used in tissue engineering. The mean pore size and the swelling scale of the scaffolds decreased after the introduction of ZnO NPs, however, the porosity, tensile modulus, and the rate of biodegradation increased after the introduction of zinc oxide NPs. In vitro culture of human fibroblasts and keratinocytes presented that zinc oxide NPs facilitate cell adhesion, proliferation, and infiltration of 3D porous nanostructures with chitosan and collagen [55].

ChNPs are also being tested for their use in cartilage restoration as a biocompatible hydrogel scaffold. Li and other scientists conducted a study to evaluate the ability of silk fibroin hydrogel scaffolds with the addition of ChNPs to improve cartilage defects of the knee joint. In the present study, ChNPs systems incorporated with transforming growth factor-β1 (TGF-β1) and silk fibroin incorporated with bone morphogenetic protein-2 were created. The nanocomplexes were developed and characterized in terms of size distribution, zeta potential, morphology, and TGF-β1 and BMP-2 release. Bone marrow stromal cells (BMSC) were co-cultured with the nanocomplexes to assess chondrogenesis in vitro using the Kit-8 cell count assay followed by in vivo evaluations in a rabbit knee cartilage defect model. The constructed complex system was well characterized and presented agreeable biocompatibility. In the presence of nanocomplexes, they showed normal cell morphology and increased chondrogenic capacity both in vitro and in vivo, as evidenced by promoting cell viability and mitigating cartilage damage. Thus, the hydrogel advanced in the present study cooperates with the chondrogenic capacity of BMSC both in vivo and in vitro by releasing TGF-β1 and BMP-2, thus offering a new therapeutic method for repairing joints cartilage defects in the knee joints [56].

Lin et al. designed a scaffold with nanofiber containing chitosan-stabilized bovine serum albumin (BSA) NPs for the delivery of abaloparatide and aspirin (ASA). Chitosan-stabilized BSA NPs simulated as a release barrier for encapsulated abaloparatide. Polymer nanofibers were made by electrospinning from a complex of NPs loaded with abaloparatide, ASA, poly (ε-caprolactone) (PCL), and nanohydroxyapatite (n-HA). Scaffolds with NPs and nanofibers were characterized in terms of their structure, surface hydrophilicity, degradation, morphology, and drug release efficiency. In vitro osteogenesis and in vitro cell adhesion, viability, and proliferation were determined to check their osteoinductive activity. The results presented that the drugs were successfully confined to nanoscaffolds. Most of the ASA was released within 7 days, while abaloparatide was released as well as 30 days. The scaffolds with nanofibers loaded with the two drugs increased the proliferation and osteogenic differentiation of osteoblasts. These findings indicate that electrospun nanofibers with chitosan-stabilized BSA NPs may be effective in bone tissue engineering [57].

Wang et al. formulated advanced electrospun nanofibers made up of chitosan/polyethylene oxide combined with NPs of silver. Silver NPs were synthesized in situ after ultraviolet (UV) with $AgNO_3$ as a precursor and chitosan/polyethylene oxide as a reducing and protecting agent, appropriately. The obtained fibers presented a homogeneous morphology with silver NPs distributed throughout the fiber. Moreover, the fibers presented some tensile strength and high antibacterial action against Gram-positive (*S. aureus*) and Gram-negative (*E. coli*) bacteria. The sustained release of silver NPs from nanofibers can take more than 72 h. Silver with chitosan and polyethylene oxide nanofibers presented great cytocompatibility [40].

Correia et al. in their research proposed combining chitosan (CH) containing bioactive glass NPs (BG-NPs) to produce CH/BG-NPs scaffolds which uses chitosan's shape memory properties and the biomineralization ability of BG-NP for bone regeneration methods. The addition of sol–gel-prepared BG-NPs to the polymer CH matrix enhanced the bioactivity of the nanocomplex scaffold, as evidenced by the precipitation of the bone-like apatite layer upon immersion in simulated body fluid (SBF). Shape memory tests were shown by immersing the patterns in different compositions of water/ethanol mixtures. Dehydration with ethanol allows the fixation of the temporary shape of the deformed scaffold, which, after absorbing water, regains its original geometry. It was presented that the scaffolds have great shape memory properties, characterized by a recovery rate of 87.5% for CH and 89.9% for CH/BG-NP, and a durability factor of 97.2% for CH and 98.2% for CH/BG-NP (for 30% compressive deformation). The usefulness of designs has demonstrated an excellent geometric fit of the previously compressed scaffold in the bone defect. The conclusions indicate that the developed CH/BG-NPs nanocomposite scaffolds have a high potential to purpose in bone tissue engineering [58].

6.7 ChNPs in Dentistry

ChNPs are also used in dentistry mainly due to their strong antimicrobial activity. Mousavi et al. focused their research on the use of ChNPs to reduce the growth of strains of bacteria and fungi on the surface of flexible linings of tissue conditioners in full dentures. The research was carried out by growing bacteria and fungi, and then the obtained suspensions were bred for 24 and 48 h, appropriately, with a tissue conditioner with ChNPs. The growth amount of the bacterial and fungal microorganisms was measured in the culture medium using a spectrophotometer. Various concentrations of ChNPs have been presented to inhibit the growth of microorganisms within 24 and 48 h, suggesting that adding ChNPs to tissue nutrients will reduce microbial growth [59].

After the effective antimicrobial activity of ChNPs, Mousavi and other researchers decided to extend their research on the inhibition of fungal and bacterial growth on full dentures. For this purpose, they conducted studies on the antifungal and antibacterial activity of silver NPs, zinc oxide, and chitosan on tissue conditioners in full dentures. After 24 and 48 h, the growth of three bacteria and one fungus strain was examined in six different concentrations of Ag, ZnO, and ChNPs. The NPs were synthesized by optical sequestration and validated by scanning electron microscopy, X-ray diffraction, and infrared (FTIR) diffraction. The studies presented that *C. albicans* growth was inhibited at a concentration of 2.5%. However, inhibition of growth of *S. mutans, Enterococcus faecalis,* and *P. aeruginosa* was 5% at both 24 and 48 h. It was also found that the optimal concentration of NPs for *C. albicans* is 1.25% for both times. On the other hand, the optimal concentration of NPs for *S. mutans, E. faecalis,* and *P. aeruginosa* was 2.5% for both timescales. The conclusions suggest that the complex with Ag, ZnO, and ChNPs inhibited the growth of microorganisms in tissue conditioning preparations [60].

Scientists are also focusing on the use of ChNPs for root canal treatment. Xiong et al. focused on a study to consider the effect of dentin pre-treatment in root canals with EDC crosslinking agent (carbodiimide) in combination with ChNPs on the bonding quality of fibrous posts in root canals in terms of bond strength, nanoleakage, quantitative assessment of collagen degradation, and degree of adhesive conversion. It was presented that the bonding of ChNPs to dentin in an EDC connection significantly reduced the degeneration of dentin collagen and enhanced the stability of the adhesive bond without compromising the adhesive polymerization [61].

ChNPs also proved to be an auxiliary factor in bone formation when placing a dental implant in the accompanying osteoporosis, which adversely affects the osseointegration of dental implants. Takanche et al. conducted a study to overcome implant failure and facilitate osseointegration of dental implants by c-myb in ovariectomy-induced osteoporosis (OVX). C-myb is a transcription factor and promotes bone formation. Plasmid DNA/c-myb was conjugated to chitosan-gold NPs (Ch-GNP/c-myb). The resulting nanocomplex was supposed to promote osteogenesis and inhibit osteoclast genesis in MC-3T3 E1 cells. In vivo conclusions in rat jaws presented that titanium implants coated with the formed nanocomposite expanded the volume and density of the newly formed bone and the osseointegration of the dental implant with the bone in a CT scan after OVX-induced osteoporosis. The immunohistochemical analysis presented expanded expression of c-myb and upregulation of bone morphogenic proteins osteoprotegerin and EphB4, as well as downregulation of RANKL by Ch-GNP/c-myb coated titanium implants. Hematoxylin and eosin staining expressed new bone formation by Ch-GNP/c-myb coated Ti implants. The obtained conclusions demonstrate that the c-myb determined by Ch-GNP helps the osseointegration of the dental implant even in the state of osteoporosis. C-myb can be used to protect the assimilation of dental implants and the therapy of age-related bone destruction [62]. Figure 6.3 represents the summary of the ChNPs application in the biomedical field.

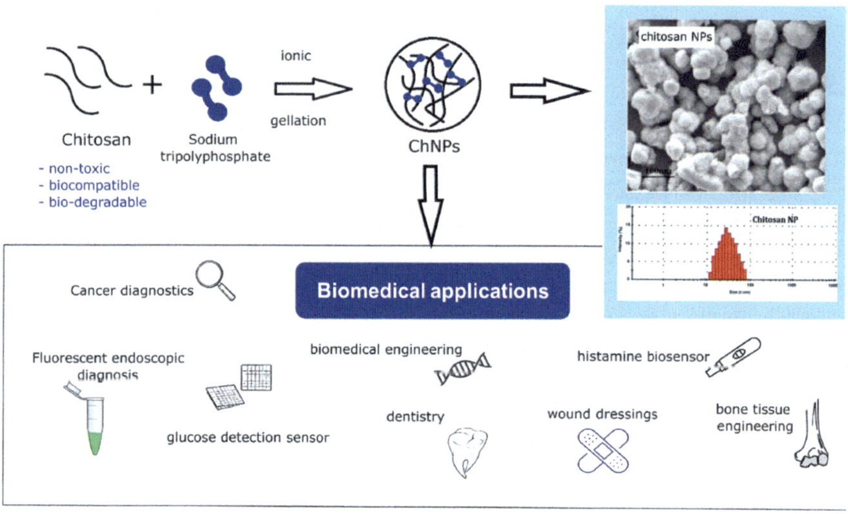

Fig. 6.3 The various applications of ChNPs in the field of medicine

6.8 Safety Assessment of ChNPs in Medical Applications

Nanotechnology combines multiple disciplines, such as medicine, chemistry, biology, biotechnology, physics, material engineering, modern electronics, and many others. From the point of view of biomedical applications, nanotechnology should also be related directly to nanotoxicology. It turns out, that ChNPs can be toxic and the cytotoxicity effect during in vitro tests, was observed [63, 64]. It was proved that the surface area, surface charge, average size, and concentration of ChNPs play an important part in their toxicity. In these studies, the zebrafish model was used. It occurs that in the case of low concentrations of NPs, apparent toxic effects were not observed. However, increasing ChNPs concentrations caused death and malformation of zebrafish embryos. When a concentration of 40 mg/L for the 200 nm ChNPs was used, the mortality was almost 100%. We can conclude that the toxicity of ChNPs is concentration, dose, and size dependent. Additionally, the genotoxic effect and skin irritation were confirmed. Despite that, ChNPs had an impact on the survival rate of mice during in vivo studies, and there are number of benefits to using it in biomedicine [65]. ChNPs are promising and relevant components for biomedical devices because of their biocompatibility and biodegradability [66].

Table 6.2 represents examples of chitosan nanostructures applications in cancer diagnosis, tissue engineering, dentistry, and other medical fields, showing the enormous potential of this biomaterial for human health. The table also summarizes the mechanism of action of these structures.

6.9 Conclusions

To sum up, chitosan nanoparticles (ChNPs) are relevant in the preparation and design of new biomedical devices. Due to their properties and possibilities for drug delivery, they can be used in different biomedical applications, especially in endoscopic diagnostics of fluorescent cancer of the colon and oral cavity. Moreover, ChNPs can be used in the production of medical dressings, in tissue engineering, and as components of sensors and biosensors. It turns out that ChNPs are most often incorporated with different components, for example with 5-aminolevulinic acid (5-ALA), folic acid, silver sulfadiazine, or transforming growth factor-β1 (TGF-β1). Its excellent biocompatibility and its polymer cationic character makes them more and more useful for medical application. Thus, ChNPs enriched with different substances will play a wide role in various fields of the medical industry, especially cancer diagnosis and are a potent tool in human health modern technologies.

Table 6.2 Chitosan nanostructures, their applications and mechanism of action in medical devices

Use structures	Use in medicine	Mechanism of action	References
Alginate and chitosan nanoparticles conjugated with folic acid as vectors to deliver specific 5-aminolevulinic acid	Fluorescent endoscopy	• Nanoparticles are easily absorbed by colorectal cancer cells through endocytosis via the folic acid receptor • Release of 5-aminolevulinic acid in the lysosome, thanks to the reduced intensity of attraction between chitosan and 5-aminolevulinic acid by deprotonated alginate molecules—higher intracellular accumulation of protoporphyrin IX for photodynamic detection	[11]
Succinate-modified chitosan (SCHI) conjugated with folic acid, the resulting complex of nanoparticles loaded with 5-aminolevulinic acid (5-ALA)	Fluorescent endoscopy	• Nanomolecules have been easily absorbed by oral cancer cells through endocytosis via the folic acid receptor • Release of loaded 5-ALA in lysosomes, due to the reduced intensity of 5-ALA and chitosan attraction by deprotonated succinate structure, which resulted in increased accumulation of intracellular protoporphyrin IX necessary for photodynamic detection	[31]
Cobalt and copper nanostructures modified with nanocomposites of reduced graphene oxide and chitosan	Nonenzymatic glucose sensor	• Compared with Cu or Co NS monometallic sensors, the Cu-Co NS bimetallic sensor has good electrocatalytic activity toward glucose oxidation • Excellent catalytic activity, high sensitivity, and great stability have been demonstrated	[16]
Alendorate-conjugated chitosan nanoparticles Chitosan and keratin nanoparticles	Nanoscaffolds in bone tissue engineering	• Widely strengthened the osteogenic differentiation of hA-MSC	[53]

(continued)

Table 6.2 (continued)

Use structures	Use in medicine	Mechanism of action	References
	nanoscaffolds in bone tissue engineering	• Biodegradation and adsorption of the scaffold proteins was widely increased • The nanoscaffolds not to be cytotoxic to human osteoblastic cells	[54]
Hydrogel with silk fibroin and chitosan nanoparticles and TGF-β1 i BMP-2 Chitosan-stabilized serum albumin nanoparticles loaded with abaloparatide, ASA, poly (ε-caprolactone) and nanohydroxyapatite electrospun nanofibers chitosan/polyethylene oxide combined with silver nanoparticles	nanoscaffolds in bone tissue engineering	• Promoted the chondrogenic - capacity of BMSC both in vivo and in vitro by releasing TGF-β1 and BMP-2	[56]
	Nanoscaffolds in bone tissue engineering	• The scaffolds with nanofibers loaded with the two drugs increased the osteogenic differentiation of osteoblasts proliferation and	[57]
	Nanoscaffolds in bone tissue engineering	• Antibacterial action against Gram-positive (*S. aureus*) and Gram-negative (*E. coli*) bacteria	[40]
Nanocomposite with chitosan and glass nanoparticles Chitosan nanoparticles Chitosan nanoparticles	nanoscaffolds in bone tissue engineering	scaffolds have great shape memory properties, characterized by a recovery rate a durability factor	[58]
	Dentistry	• Antibacterial activity	[59]
	Dentistry	• Widely reduced the degradation of dentin collagen and enhanced the stability of the adhesive bond without compromising the adhesive polymerization	[61]

(continued)

Table 6.2 (continued)

Use structures	Use in medicine	Mechanism of action	References
Chitosan-gold nanoparticles	Dentistry	• Was supposed to cooperate osteogenesis and inhibit osteoclastogenesis in MC-3T3 E1 cells • In vivo results in rat jaws showed that titanium implants coated with the formed nanocomposite increased the density and the density of the newly formed bone and the osseointegration of the dental implant with the bone in a CT scan after OVX-induced osteoporosis • Increased expression of c-myb and upregulation of bone morphogenic proteins like osteoprotegerin and EphB4, as well as downregulation of RANKL by Ch-GNP/c-myb coated Titanium implants	[62]

References

1. M. Yanat, K. Schro, Preparation methods and applications of chitosan nanoparticles; with an outlook toward reinforcement of biodegradable packaging 161 (2021). https://doi.org/10.1016/j.reactfunctpolym.2021.104849
2. M. Kafshgari, M. Khorram, Preparation of alginate and chitosan nanoparticles using a new reverse micellar system. Iran Polym. Petrochem. Inst. 99–107 (2012). https://doi.org/10.1007/s13726-011-0010-1
3. A. Grenha, Chitosan nanoparticles : a survey of preparation methods. J. Drug Targenting **20**, 291–300 (2012). https://doi.org/10.3109/1061186X.2011.654121
4. M.H. ElShabouri, Positively charged nanoparticles for improving the oral bioavailability of cyclosporin-A. Int. J. Pharm. **249**, 101–108 (2002)
5. K. Lewandowska, M. Szulc, Rheological and film-forming properties of chitosan composites. Int. J. Mol. Sci. **23**, 8763 (2022). https://doi.org/10.3390/IJMS23158763
6. K. Nagpal, S.K. Singh, D.N. Mishra, Chitosan nanoparticles: a promising system in novel drug delivery. Chem. Pharm. Bull. (Tokyo) **58**, 1423–1430 (2010)
7. S. Naskar, K. Kuotsu, S. Sharma, Chitosan-based nanoparticles as drug delivery systems: a review on two decades of research. J. Drug Target **27**, 379–393 (2018)
8. R.A.A. Muzzarelli, Chitosan-based dietary foods. Carbohydr. Polym. **29**, 309–316 (1996)
9. A.A. Zaki, Dielectric and optical properties of chitosan-Pb and chitosan-Bi nanocomposites. J. Mater. Sci.: Mater. Electron. **32**, 3603–3611 (2021). https://doi.org/10.1007/s10854-020-05107-7
10. J. Dutta, S. Tripathi, P.K. Dutta, F. Science, Progress in antimicrobial activities of chitin, chitosan and its oligosaccharides: a systematic study needs for food applications. Food Sci. Technol. Int. 18 (2012). https://doi.org/10.1177/1082013211399195
11. S.J. Yang, M.J. Shieh, F.H. Lin et al., Colorectal cancer cell detection by 5-aminolaevulinic acid-loaded chitosan nano-particles. Cancer Lett. **273**, 210–220 (2009). https://doi.org/10.1016/J.CANLET.2008.08.014
12. I. Karakurt, K. Ozaltin, H. Pištěková et al., Effect of saccharides coating on antibacterial potential and drug loading and releasing capability of plasma treated polylactic acid films. Int. J. Mol. Sci. **23**, 8821 (2022). https://doi.org/10.3390/IJMS23158821
13. P. Hu, Q. Lei, S. Duan et al., In-situ formable dextran/chitosan-based hydrogels functionalized with collagen and EGF for diabetic wounds healing. Biomater. Adv. 136. https://doi.org/10.1016/J.BIOADV.2022.212773
14. A. Ragamin, K.B. Fieten, R.A. Tupker et al., The effectiveness of antibacterial therapeutic clothing based on silver or chitosan as compared with non-antibacterial therapeutic clothing in patients with moderate to severe atopic dermatitis (ABC trial): study protocol for a pragmatic randomized controlled trial. Trials 22 (2021). https://doi.org/10.1186/S13063-021-05836-Y
15. S.H. Kang, D. Han, S. Kim et al., Hemostasis pad combined with compression device after transradial coronary procedures: a randomized controlled trial. PLoS One 12 (2017). https://doi.org/10.1371/JOURNAL.PONE.0181099
16. L. Wang, Y. Zhen, X.L. Lu et al., Dendritic copper-cobalt nanostructures/reduced graphene oxide-chitosan modified glassy carbon electrode for glucose sensing. Sens Actuators B Chem. **195**, 1–7 (2014)
17. X. Lin, J. Liu, F. Zhou et al., Poly(2-hydroxyethyl methacrylate-co-quaternary ammonium salt chitosan) hydrogel: a potential contact lens material with tear protein deposition resistance and antimicrobial activity. Biomater. Adv. **136**, 212787 (2022). https://doi.org/10.1016/J.BIOADV.2022.212787
18. G. Oberdorster, Safety assessment for nanotechnology and nanomedicine : concepts of nanotoxicology. J. Int. Med. 89–105 (2009). https://doi.org/10.1111/j.1365-2796.2009.02187.x
19. T. Banerjee, S. Mitra, A. Kumar, Preparation, characterization and biodistribution of ultrafine chitosan nanoparticles. Int. J. Pharm. **243**, 93–105 (2002)

20. Y. Ohya, M. Shiratani, H. Kobayashi, Behavior of 5-fluorouracil from chitosan-gel nanospheres immobilizing 5-fluorouracil coated with polysaccharides and their cell specific cytotoxicity. J. Macromol. Sci. 37–41 (2012). https://doi.org/10.1080/10601329409349743
21. N. Hijazi, Moigne N Le, Rodier E, et al (2018) Biocomposite films based on poly (lactic acid) and chitosan nanoparticles: elaboration , microstructural and thermal characterization. Polym. Eng. Sci. 1–11. https://doi.org/10.1002/pen.24983
22. N. Hijazi, E. Rodier, J. Letourneau et al., Chitosan nanoparticles generation using CO2 assisted processes. J. Supercrit. Fluids **95**, 118–128 (2014). https://doi.org/10.1016/j.supflu.2014.08.003
23. E. Basaran, E. Yenilmez, M.S. Berkman et al., Chitosan nanoparticles for ocular delivery of cyclosporine A. J. Microencapsul. **2048**, 1–9 (2013). https://doi.org/10.3109/02652048.2013.805839
24. L.T.K. Ngan, S. Wang, P.M. Luong et al., Preparation of chitosan nanoparticles by spray drying, and their antibacterial activity. Res. Chem. Int. **40**, 2165–2175 (2014). https://doi.org/10.1007/s11164-014-1594-9
25. S. Sajeesh, C.P. Sharma, Novel pH responsive polymethacrylic acid – chitosan – polyethylene glycol nanoparticles for oral peptide delivery, in *Biomedical Technology Wing, Sree Chitra Tirunal Institute for Medical Sciences and Technology*, pp 298–305 (2005)
26. S. Sajeesh, C.P. Sharma, Cyclodextrin – insulin complex encapsulated polymethacrylic acid based nanoparticles for oral insulin delivery. Int. J. Pharm. **325**, 147–154 (2006). https://doi.org/10.1016/j.ijpharm.2006.06.019
27. W. Fan, W. Yan, Z. Xu, H. Ni, Formation mechanism of monodisperse, low molecular weight chitosan nanoparticles by ionic gelation technique. Colloids Surf. B Biointerfaces **90**, 21–27 (2012). https://doi.org/10.1016/j.colsurfb.2011.09.042
28. E.M.A. Hejjaji, A.M. Smith, G.A. Morris, Evaluation of the mucoadhesive properties of chitosan nanoparticles prepared using different chitosan to tripolyphosphate (CS: TPP) ratios. Int. J. Biol. Macromol. **120**, 1610–1617 (2018). https://doi.org/10.1016/j.ijbiomac.2018.09.185
29. H. Tokumitsu, H. Ichikawa, Y. Fukumori, Chitosan-gadopentetic acid complex nanoparticles for gadolinium neutron-capture therapy of cancer: preparation by novel emulsion-droplet coalescence technique and characterization. Pharm. Res. **16**, 1830–1835 (1999). https://doi.org/10.1023/A:1018995124527
30. H. Ichikawa, H. Tokumitsu, M. Miyamoto, Nanoparticles for neutron capture therapy of cancer, in *Nanotechnologies for the Life Sciences* (2006), pp. 104–105
31. S.-J. Yang, F.-H. Lin, K.-C. Tsai et al., Folic acid-conjugated chitosan nanoparticles enhanced protoporphyrin IX accumulation in colorectal cancer cells. Bioconjug. Chem. **21**, 679–689 (2010)
32. S.-J. Yang, F.-H. Lin, H.-M. Tsai et al., Alginate-folic acid-modified chitosan nanoparticles for photodynamic detection of intestinal neoplasms. Biomaterials **32**, 2174–2182 (2011)
33. S.J. Yang, C.F. Lin, M.L. Kuo, C.T. Tan, Photodynamic detection of oral cancers with high-performance chitosan-based nanoparticles. Biomacromol **14**, 3183–3191 (2013). https://doi.org/10.1021/BM400820S
34. I.-C. Sun, C.-H. Ahn, K. Kim, S. Emelianov, Photoacoustic imaging of cancer cells with glycol-chitosan-coated gold nanoparticles as contrast agents. J. Biomed. Opt. **24**, 1 (2019). https://doi.org/10.1117/1.JBO.24.12.121903
35. A. Ładniak, M. Jurak, Physicochemical AW-C and SA, 2021 undefined physicochemical characteristics of chitosan-TiO2 biomaterial , in *Wettability and Biocompatibility*, vol. 2. Elsevier
36. Y.H. Lin, J.H. Lin, Y.S. Hong, Development of chitosan/poly-γ-glutamic acid/pluronic/curcumin nanoparticles in chitosan dressings for wound regeneration. J. Biomed. Mater. Res. B Appl. Biomater. **105**, 81–90 (2017). https://doi.org/10.1002/JBM.B.33394
37. A. Mohandas, B.S. Anisha, K.P. Chennazhi, R. Jayakumar, Chitosan-hyaluronic acid/VEGF loaded fibrin nanoparticles composite sponges for enhancing angiogenesis in wounds. Colloids Surf. B Biointerfaces **127**, 105–113 (2015). https://doi.org/10.1016/J.COLSURFB.2015.01.024

38. S.S. Biranje, P.v. Madiwale, K.C. Patankar et al., Hemostasis and anti-necrotic activity of wound-healing dressing containing chitosan nanoparticles. Int. J. Biol. Macromol. **121**, 936–946 (2019). https://doi.org/10.1016/J.IJBIOMAC.2018.10.125
39. N. Masood, R. Ahmed, M. Tariq et al., Silver nanoparticle impregnated chitosan-PEG hydrogel enhances wound healing in diabetes induced rabbits. Int. J. Pharm. **559**, 23–36 (2019). https://doi.org/10.1016/J.IJPHARM.2019.01.019
40. X. Wang, F. Cheng, J. Gao, L. Wang, Antibacterial wound dressing from chitosan/polyethylene oxide nanofibers mats embedded with silver nanoparticles. J. Biomater. Appl. **29**, 1086–1095 (2015). https://doi.org/10.1177/0885328214554665
41. S.M. Sharaf, S.E.D. AlMoftyl, E.S.M. El-Sayed et al., Deacetylated cellulose acetate nanofibrous dressing loaded with chitosan/propolis nanoparticles for the effective treatment of burn wounds. Int. J. Biol. Macromol. **193**, 2029–2037 (2021). https://doi.org/10.1016/J.IJBIOMAC.2021.11.034
42. G.S. ElFeky, S.S. Sharaf, A. el Shafei, A.A. Hegazy, Using chitosan nanoparticles as drug carriers for the development of a silver sulfadiazine wound dressing. Carbohydr. Polym. **158**, 11–19 (2017). https://doi.org/10.1016/J.CARBPOL.2016.11.054
43. R.A. Farghali, A.M. Fekry, R.A. Ahmed, H.K.A. Elhakim, Corrosion resistance of Ti modified by chitosan-gold nanoparticles for orthopedic implantation. Int. J. Biol. Macromol. **79**, 787–799 (2015). https://doi.org/10.1016/J.IJBIOMAC.2015.04.078
44. M. Ishihara, V.Q. Nguyen, Y. Mori et al., Adsorption of silver nanoparticles onto different surface structures of chitin/chitosan and correlations with antimicrobial activities. Int. J. Mol. Sci. **16**, 13973–13988 (2015). https://doi.org/10.3390/IJMS160613973
45. P. Zhou, Z. Xia, C. Qi et al., Construction of chitosan/Ag nanocomposite sponges and their properties. Int. J. Biol. Macromol. **192**, 272–277 (2021). https://doi.org/10.1016/J.IJBIOMAC.2021.10.011
46. B. Anisha, D. Sankar, A. Mohandas et al., Chitosan-hyaluronan/nano chondroitin sulfate ternary composite sponges for medical use. Carbohydr. Polym. **92**, 1470–1476 (2013)
47. A. Bal-Ozturk, O. KaralYilmaz, Z.P. Akguner et al., Sponge-like chitosan-based nanostructured antibacterial material as a topical hemostat. J. Appl. Polym. Sci. **136**, 47522 (2019). https://doi.org/10.1002/APP.47522
48. M. Rahimi, R. Ahmadi, H. SamadiKafil, V. Shafiei-Irannejad, A novel bioactive quaternized chitosan and its silver-containing nanocomposites as a potent antimicrobial wound dressing: structural and biological properties. Mater. Sci. Eng. C **101**, 360–369 (2019). https://doi.org/10.1016/J.MSEC.2019.03.092
49. N.D. Tien, T. Geng, C.A. Heyward et al., Solution blow spinning of highly deacetylated chitosan nanofiber scaffolds for dermal wound healing. Biomater. Adv. 137 (2022). https://doi.org/10.1016/J.BIOADV.2022.212871
50. A. Regiel-Futyra, M. Kus-Lis, V. Sebastian et al., Development of noncytotoxic chitosan−gold nanocomposites as efficient antibacterial materials (2014). https://doi.org/10.1021/am508094e
51. N. Cai, C. Li, C. Han et al., Tailoring mechanical and antibacterial properties of chitosan/gelatin nanofiber membranes with Fe3O4 nanoparticles for potential wound dressing application. Appl. Surf. Sci. **369**, 492–500 (2016). https://doi.org/10.1016/J.APSUSC.2016.02.053
52. N. Nontipichet, S. Khumngern, J. Choosang et al., An enzymatic histamine biosensor based on a screen-printed carbon electrode modified with a chitosan-gold nanoparticles composite cryogel on Prussian blue-coated multi-walled carbon nanotubes. Food Chem. 364 (2021). https://doi.org/10.1016/J.FOODCHEM.2021.130396
53. F. Moradikhah, M. Doosti-Telgerd, I. Shabani et al., Microfluidic fabrication of alendronate-loaded chitosan nanoparticles for enhanced osteogenic differentiation of stem cells. Life Sci. 254 (2020). https://doi.org/10.1016/J.LFS.2020.117768
54. S. Saravanan, D.K. Sameera, A. Moorthi, N. Selvamurugan, Chitosan scaffolds containing chicken feather keratin nanoparticles for bone tissue engineering. Int. J. Biol. Macromol. **62**, 481–486 (2013). https://doi.org/10.1016/J.IJBIOMAC.2013.09.034
55. S. Ullah, I. Zainol, R.H. Idrus, Incorporation of zinc oxide nanoparticles into chitosan-collagen 3D porous scaffolds: Effect on morphology, mechanical properties and cytocompatibility of

3D porous scaffolds. Int. J. Biol. Macromol. **104**, 1020–1029 (2017). https://doi.org/10.1016/J.IJBIOMAC.2017.06.080
56. Y. Li, Y. Liu, Q. Guo, Silk fibroin hydrogel scaffolds incorporated with chitosan nanoparticles repair articular cartilage defects by regulating TGF-β1 and BMP-2. Arthritis Res. Ther. 23 (2021). https://doi.org/10.1186/S13075-020-02382-X
57. P. Lin, W. Zhang, D. Chen et al., Electrospun nanofibers containing chitosan-stabilized bovine serum albumin nanoparticles for bone regeneration. Colloids Surf. B Biointerfaces 217 (2022). https://doi.org/10.1016/J.COLSURFB.2022.112680
58. C.O. Correia, Á.J. Leite, J.F. Mano, Chitosan/bioactive glass nanoparticles scaffolds with shape memory properties. Carbohydr. Polym. **123**, 39–45 (2015). https://doi.org/10.1016/J.CARBPOL.2014.12.076
59. S.A. Mousavi, R. Ghotaslou, S. Kordi et al., Antibacterial and antifungal effects of chitosan nanoparticles on tissue conditioners of complete dentures. Int. J. Biol. Macromol. **118**, 881–885 (2018). https://doi.org/10.1016/J.IJBIOMAC.2018.06.151
60. S.A. Mousavi, R. Ghotaslou, A. Khorramdel et al., Antibacterial and antifungal impacts of combined silver, zinc oxide, and chitosan nanoparticles within tissue conditioners of complete dentures in vitro. Ir. J. Med. Sci. **189**, 1343–1350 (2020). https://doi.org/10.1007/S11845-020-02243-1
61. J. Xiong, L. Shen, Q. Jiang, A. Kishen, Effect of crosslinked chitosan nanoparticles on the bonding quality of fiber posts in root canals. J. Adhes. Dent. **22**, 321–330 (2020). https://doi.org/10.3290/J.JAD.A44555
62. J.S. Takanche, J.E. Kim, J.S. Kim et al., Chitosan-gold nanoparticles mediated gene delivery of c-myb facilitates osseointegration of dental implants in ovariectomized rat. Artif. Cells Nanomed. Biotechnol. **46**, S807–S817 (2018). https://doi.org/10.1080/21691401.2018.1513940
63. L. Qi, Z. Xu, In vivo antitumor activity of chitosan nanoparticles. Bioorg. Med. Chem. Lett. **16**, 4243–4245 (2006). https://doi.org/10.1016/j.bmcl.2006.05.078
64. Y. Hu, Toxicity evaluation of biodegradable chitosan nanoparticles using a zebrafish embryo model. Int. J. Nanomedicine **6**, 3351–3359 (2011)
65. M. Elgadir, S. Uddin, S. Ferdous et al., Science direct impact of chitosan composites and chitosan nanoparticle composites on various drug delivery systems: a review. J. Food Drug Anal. 1–11 (2014). https://doi.org/10.1016/j.jfda.2014.10.008
66. M.A. Malinowska, K. Billet, S. Drouet, T. Munsch, M. Unlubayir, D. Tungmunnithum, N. Giglioli-Guivarc'h, C. Hano, A. Lanoue, Grape cane extracts as multifunctional rejuvenating cosmetic ingredient: evaluation of sirtuin activity, tyrosinase inhibition and bioavailability potential. Mol. **25**, 2203 (2020). https://doi.org/10.3390/molecules25092203

Chapter 7
Antibacterial Response of Nanostructured Chitosan Hybrid Materials

Vellaichamy Mageshwaran, Palanisamy Sivasubramanian, Praveen Kumar, and Yalavarthi Nagaraju

Abstract In recent years, nanotechnology has gained much interest due to its potential application in different sectors. Among the different nanomaterials, chitosan nanoparticles (CS-NPs) and their derivatives were found to have antibacterial properties. Bio- and nanohybrid materials based on chitosan have been developed for corrosion prevention, antimicrobial packaging, and also the removal of harmful metal ions. The different nanostructured chitosan hybrid materials having potential antibacterial application are silver-chitosan, lipid-chitosan, poly (lactic acid)-chitosan, gold-chitosan and silica-chitosan hybrid. Although CS-NPs exhibit a wide range of antibacterial activity, their inhibitory efficiency differs with respect to gram-negative (G−) and gram-positive (G +) bacteria. The difference in antibacterial action of CS-NPs is due to variations in the chemistry of the cell wall of G (+) and G (–) bacteria. The various mechanism of antibacterial activity of hybrid CS-NPs is chelation of ions or nutrients, binding with cell wall and cell membrane thereby disrupts the stability of the cell and more importantly interruption of DNA/RNA or protein synthesis. The various factors influencing antibacterial activity of CS-NPs are target bacteria (G (+) or G (−) bacteria), the rate of bacterial growth, molecular weight, concentration, zeta-potential, level of acetylation and pH. Overall, this review focuses on different CS-NPs, their mechanism of action, factors influencing antibacterial activity against bacteria that cause diseases in plants and animals. In addition, the modern procedures and applications of CS-NPs are covered in this chapter.

V. Mageshwaran · Y. Nagaraju
Microbial Technology Laboratory, ICAR-National Bureau of Agriculturally Important Microorganisms, Uttar Pradesh, Kushmaur, Mau Nath Bhanjan 275103, India

P. Sivasubramanian (✉)
Department of Mechanical Engineering, Dilkap Research Institute of Engineering and Management Studies, At Village Mamdapur, Post Neral, Tal- Karjat, Dist Raigad 410101, India
e-mail: sivaresearch948@gmail.com

P. Kumar
Department of Mechanical Engineering, Sant Longowal Institute of Engineering and Technology, Longowal, Sangrur, Punjab, India

S. K. Swain and A. Biswal (eds.), *Chitosan Nanocomposites*,
Biological and Medical Physics, Biomedical Engineering,
https://doi.org/10.1007/978-981-19-9646-7_7

7.1 Introduction

Antimicrobial materials are gaining importance in terms of providing antimicrobial finishing in the products which can provide clean, safe and protective environment. They have been researched for potential usage in a range of settings, including homes, businesses, labs and healthcare facilities [1, 2]. The primary purpose of these materials is to sterilise the medical amenities and apparatus to reduce the number of fatalities caused by hospital-transmitted infections that may spread through linens and clothes. An appropriate environment must be created to prevent the spread of infectious illnesses caused by fungus or viruses, [3]. Today, it is still important to ensure the products' microbiological safety while keeping their nutritive and organoleptic qualities [4]. To expand the margin of safety and quality, another alternative is to utilise antimicrobial packaging. This packaging may lengthen the product's shelf life by preventing the development of microbes on its surface [5]. Therefore, antibacterial and antimicrobial systems are crucial for laboratory, residential, marine and certain industrial uses in addition to the hospital and healthcare settings [2].

Chitosan is a naturally occurring, inexpensive biopolymer. Chitosan is extracted from shrimp and crab shells waste. The process of extraction of chitosan and their characteristic property is outlined in Fig. 7.1. Due to its novel features for a variety of applications, chitosan is gaining interest as a nanocomposite's matrix [7]. The emphasis is on presenting alternative methods for producing embedded or coated nanoparticles, shaping methods that have been used (3D printing, electrospinning) and new uses for nanocomposite materials in areas such as corrosion prevention, tissue engineering, wastewater treatment and medicine. There are several evaluations of chitosan as a single substance and its derivatives for various uses. There is, however, no research on chitosan as a nanocomposite matrix that explains the potential for nanomaterial additions, the interaction of the attached species, and the potential applications following the methods to mix chitosan with nanostructures [8].

One of the most cutting-edge research areas is the creation of improved materials through the interaction of macromolecules with inorganic species. By employing this method, the synthesised hybrid materials can customise characteristics on length scales ranging from atomic to microscopic and macroscopic. Recent scientific progress has allowed for bio- and nanohybrid materials based on chitosan have been developed for corrosion prevention, antimicrobial packaging, and also the removal of harmful metal ions [9]. Chitosan has been combined with a number of polymers, including polycarbonate, poly (methyl-methacrylate), polysulfide, polystyrene and polyaniline. Their usage in a variety of technological domains as biosensors, antimicrobial, membrane technology, and packaging, textile industry, and the biomedical materials, is made possible by improved material qualities and bioactive capabilities. The range of application of CS-NPs in animal and plant system is illustrated in Fig. 7.2. Several approaches have been examined to better understand the antibacterial properties of chitosan and also its derivatives. These methods range from the initial solution used in agriculture to film structures in the food industry to widely utilised medicinal nanostructure materials.

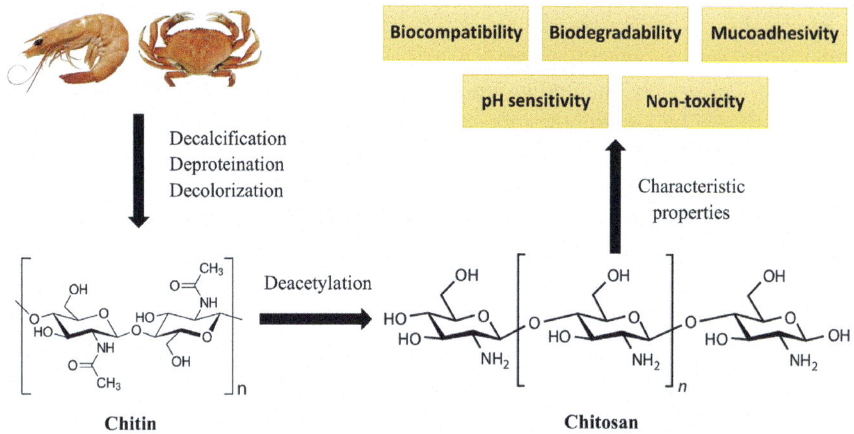

Fig. 7.1 Extraction of chitin from natural resources and chemical transformation of chitin into chitosan. Reproduced with permission from Elsevier [6]

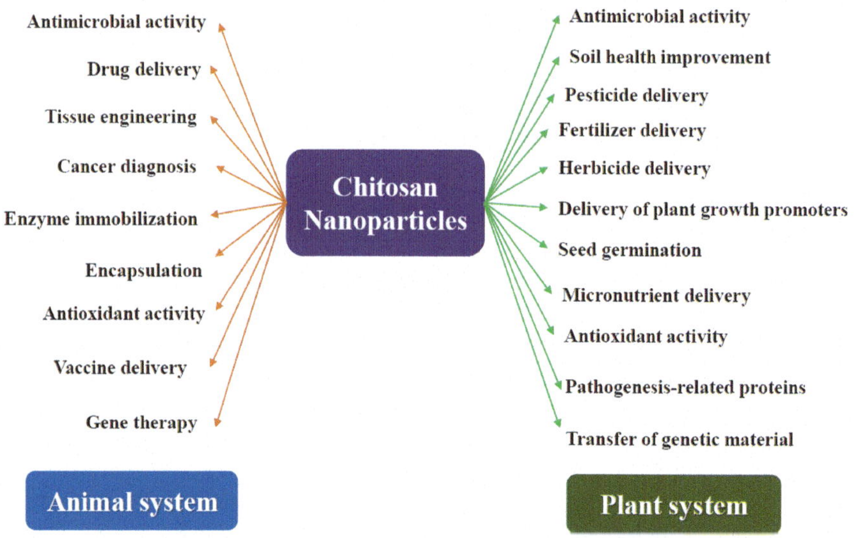

Fig. 7.2 Applications of chitosan nanoparticles. Reproduced with permission from MDPI [10]

7.2 Nanostructured Chitosan Hybrid Materials

NPs' physical characteristics have changed significantly from their initial form due to their tiny size. CS-NPs share the benefits of NPs, like increased surface area, quantum size effects and their small size [11]. Several techniques, including ionic gelation [12], reverse micellar approach, microemulsion, emulsion droplet coalescence and spray drying, were used to create CS-NPs. The most successful technique was found to be physical crosslinking by ionic gelation. The ion gelation technique makes use of the electrostatic contact between a negatively charged group of tripolyphosphate (TPP) and a positively charged group of CS. Size as well as surface charge of NPs might be altered by varying the ratio of CS to TPP. Additionally, since no chemical is involved in the crosslinking process, the negative side effects are diminished. The antibacterial activity of several metal ions with CS complexes ($CSZn^{2+}$ NPs, CS-Fe^{2+}NPs, CS-Ag + NPs, CS-Mn^{2+}NPs and CS-Cu^{2+}NPs,) has recently improved [13]. In addition, hybrid CS-NPs containing lysozyme, protamine, oleic acid-grafted chitosan oligosaccharide (CSO-OA) NPs, curcumin NPs and essential oil, among others, have all been studied to see how effective they are against microorganisms. In comparison to CS alone, polycationic CS-NPs with high density of surface charger can converse with bacteria more effectively. Additionally, CS-NPs damage the bacterial cell walls and membranes, which causes internal chemicals to leak out and cause the bacteria to die. The size of NPs is influenced by several variables, which also have an impact on how well they inhibit bacteria. These comprise the kind of bacteria, stage of development, concentration, pH, zeta-potential, MW and DA [14].

7.3 Preparation of Hybrid Chitosan Material

7.3.1 Silver-Chitosan Hybrid

A method is briefly described here for producing silver nanospheres (AgNS). The first step was to combine 55 ml of distilled water, 1 mM AgNO3 in 10 ml and 30 ml of 5 mM Na_3Cit. Then, a pale-yellow dispersion was produced by adding 8 mM $NaBH_4$ in 1 ml dropwise and liquid when stirring for 10 min at room temperature. In this procedure, a flat-bottomed flask containing 20 mM CTAC in 50 ml and 0.1 M AA in 5 ml was heated for 10 min at 60 °C. The pH was then brought down to 3.1 using 0.1 M HCl. The reaction was then allowed to occur for 4 h under natural light by adding 10 mM CF_3COOAg in 500 L and 4.29 mM $FeCl_3$ in 1 ml. The result was a yellow-green dispersion. After making the dispersions, nanoparticle glues were extracted from the supernatants using micropipettes. The centrifugation times for each system were 20 min at the proper speed (8000 rpm for AgNS & 4000 rpm for AgNC). Due to the AgNS's tiny size and the fact that smaller nanoparticles require more effort to be centrifuged than bigger ones, before centrifuging, the dispersion must be made more concentrated by heating. To prevent this behaviour, which is typical of AgNC nano inks, Ag nanoparticles must be utilised right away and maintained carefully. This is especially true when the dispersion is concentrated, as it is when they are found in nano inks. Aluminium cans that had been cleaned, chopped and manually polished using a polish made of aluminium oxide were used to create the Al substrates. The dried active region of the nanoparticle substrates was then drop cast with 0.125% w/vv of chitosan in an acetic acid solution of 1% in 4 L and allowed to dry [15].

7.3.2 Lipid-Chitosan Hybrid

The method used to create the nanoparticles was designed by Sonvico et al. [16]. In 92 ml of deionized water with 0.1% acetic acid, 10 mg of chitosan was dissolved. With constant stirring, cisplatin was dissolved in the same solution. Pure ethanol was used to dissolve the lipid (25 mg/ml). The medication solution was then gradually mixed with drops of the ethanolic solution. Nanoparticles were created using ionic gelation. Before being lyophilized, the lipid and chitosan nanoparticles were centrifuged for 30 min at 10,000 rpm. Six distinct formulations were tested for surface charge, size, entrapment effectiveness and drug loading, with the lipid: chitosan ratio varying from 5:1 to 60:1. Fluorescent dyes were added to the chitosan solution drop by drop, with Rhodamine 123 as well as Rhodamine-PE diluted in ethanol with lipid for cell absorption experiments [17].

7.3.3 Poly (Lactic Acid)-Chitosan Hybrid

A fine dispersion of about 100 mg of PLA was made in 10 ml of dichloromethane. This mixture was quickly added to 10 ml of 1% acetic acid solution that also included 200 mg of PEO and 40 mg of CS. After 15 min of sonication to create an emulsion, the liquid was vigorously agitated until all of the organic solvents had been evaporated. The nanoparticles were then lyophilized after being precipitated by water. Lamivudine-loaded PLA/CS nanoparticle preparation, Hydrophilic drugs were encapsulated using the w/o/w emulsion method. An aqueous drug solution (2 ml) was initially added to a polymer solution to make a w/o emulsion (100 mg of PLA dissolved in 10 ml dichloromethane). The 1% acetic acid solution containing 40 mg of chitosan and 200 mg of PEO was quickly added to 10 ml of the w/o emulsion. The liquid was sonicated for 15 min to produce an emulsion, and then it was aggressively churned. The organic solvent was stirred until it had completely evaporated. As a result of the incorporation of water, the nanoparticles precipitated and were then lyophilized [18].

7.3.4 Gold-Chitosan Hybrid

The process described below was utilised to create the gold nanospheres (AuNS). 2.5 ml of 10% w/v Na_3Cit were added after heating 50 ml of 0.25 mM HAuCl4 to boiling. The combination was then allowed to react for 20 min to give off a ruby red colour. Finally, a modification work done by Ye et al. was employed to produce gold nanorods (AuNR). In this process, a seed dispersion is made by combining, in that order, 0.5 mM $HAuCl_4$ in 1 ml, 6 mM $NaBH_4$ in 200 M, along with 0.2 M CTAB in 1 ml while constantly stirring at 1000 rpm. Another 30 min of preparation time is allotted before the system is put to use. The growth solution, on the other hand, was made by adding 0.990 g of CTAB and 15 ml of warm water, mixing them together, and letting them cool to ambient temperature. In 10 ml of warm water, 0.0961 g of 3MSA was dissolved, and the mixture was then allowed to react with 0.025 g of NaOH until the solution became completely clear. 0.6 ml of 4 mM $AgNO_3$ was added to the liquid, which was then chilled before the reaction could begin. 25 ml of 1 mM $HAuCl_4$ was then added to the dispersion, which resulted in a colourless mixture. The mixture was then given 15 min to react before 0.1 ml of 0.064 M AA was added. Once 80 l of the Au seed dispersion had been added, the growth fluid was again rapidly agitated. The system was allowed to mature for 12 h while not being stirred after 1 min, eventually turning dark blue. Following the creation of the dispersions, nanoparticle pastes were created by centrifuging each system at the appropriate speed (8000 rpm for AuNR and 6000 rpm for AuNS) for 20 min and extracting the supernatant with a micropipette. While water was added to the other dispersions to a final volume (40 l for AuNS), the wet pellet formed in the case of the AuNR was employed as the paste. The pastes were preserved until they were needed after being wrapped in aluminium foil. 4 L of the paste was applied to a polished Au substrate for AuNS substrates, and it was allowed to cure under normal humidity (60%) conditions. After that, the dried active region of the nanoparticle substrates was drop cast with 4 L of 0.125% w/v chitosan in 1% v/v acetic acid solution [19].

7.3.5 Silica-Chitosan Hybrid

The mineral silica is made up of silicon and oxygen, the two substances that make up most of the earth's crust. Following changes from what was previously reported that it was determined to be possible to produce hybrid materials by combining OCS and $nSiO_2$ in an equal mass ratio (1:1) throughout the synthesis process. While beaker B had 0.30 g of CMC that had been dissolved in distilled water, beaker A had 1.00 g of $nSiO_2$ that had been diluted in 6.65 ml of 1 M NaOH. The CMC solution was added to beaker A and agitated for two hours. $OCS2/nSiO_2$, $OCS1/nSiO_2$, along with $OCS3/nSiO_2$ were created by progressively adding dropwise the obtained solutions of OCS with various molecular weights (OCS1, OCS2, and OCS3) into the mixture in beaker. The pH of these combinations was later adjusted to 7 using 1 M HCl. After a three-hour stir, they were left at room temperature. In the area of resistance to the *Phytophthora infestans* fungus, the study found that nanohybrid materials performed better than their component parts individually, with a concentration of 800 mg/L for an $OCS2/nSiO_2$ hybrid material being the minimum concentration at which the material entirely prevented *Phytophthora infestans* growth. This finding not only produces a brand-new, eco-friendly substance with special synergistic effects that can replace harmful agrochemicals now in use, but it also establishes a new research base for further study in environmentally friendly agricultural applications [20].

7.4 Mechanisms of Antibacterial Activity

Depending on where the antimicrobial effects of chitosan are most concentrated, its mode of action against bacteria can be classified as either intracellular, extracellular or both (Fig. 7.3). Chitosan has been found to exhibit antibacterial properties in all three of these locations. Due to its frequent inability to pass through cell walls and membranes, high-MW chitosan may function as a chelator of significant metals, inhibit the extracellular uptake of nutrients and affect cell permeability. These are just a few examples of high-MW chitosan's potential antimicrobial effects. However, low-MW chitosan affects intracellular functions such as protein/RNA/DNA synthesis and mitochondrial activity. Numerous medications contain chitosan and its derivatives, which play an ever-increasing role in antimicrobials against gram-positive, gram-negative and fungal pathogens. Since the first notion by Allan and Hardwiger [8] regarding the substance's broad-spectrum antibacterial activity and huge economic potential, scientists have been studying chitosan as well as its derivatives extensively due to its antibacterial properties. Investigation on antifungal activity of CS-NPs against foodborne as well as soilborne pathogenic fungus in food sector and agriculture, respectively had started 20 years ago [9]. Chitosan and its derivatives antimicrobial activity was explored for several microorganisms like algae, fungi, bacteria such as *Escherichia coli*, *Staphylococcus aureus*, *Candida albicans* and *Actinobacillus actinomycetemcomitans* [7]. However, its uses as an antimicrobial agent are limited by its high molecular weight and limited positive charge.

By transforming primary amine groups into quaternary salts with a continuing positive charge, chitosan becomes more active. Recent comparative studies have examined the mechanism of action of the antibacterial chitosan. Due to modifications in the structure of their membranes, various bacteria have varying sensitivity levels to antibacterial compounds like chitosan, and antibacterial compounds with a positive charge have the potential to harm cell membranes. The impact that chitosan has on the cell membrane of bacteria was discovered via research conducted utilising electron microscopy on many bacterial strains. As a result, permeability of the cell membrane and cytoplasmic component leakage emerge from the interaction between the positive charge glucosamine amine groups (NH^{3+}) as well as the negative-charged bacterial cell membranes, which ultimately results in cell death. More negatively charged forms of *Staphylococcus aureus* were shown to be more susceptible to chitosan, according to another study. Other hypothesis includes the reduced mRNA activity brought on by chitosan's intranuclear binding to the target microorganism's DNA. In research using a confocal laser microscope to examine *Escherichia coli*, chitooligomers found inside the cells suggested that DNA was being down regulated [21]. Mg^{2+} and Ca^{2+}, two cationic ions crucial to bacterial metabolism, may be chelated, may also contribute to chitosan's antibacterial properties. Another theory is that chitosan acts as a barrier, keeping bacteria from getting nutrients and oxygen.

However, microbes are thought to be resistant to chitosan, which poses a severe threat to its use in medicine and hampers the mode of action due to its congruent

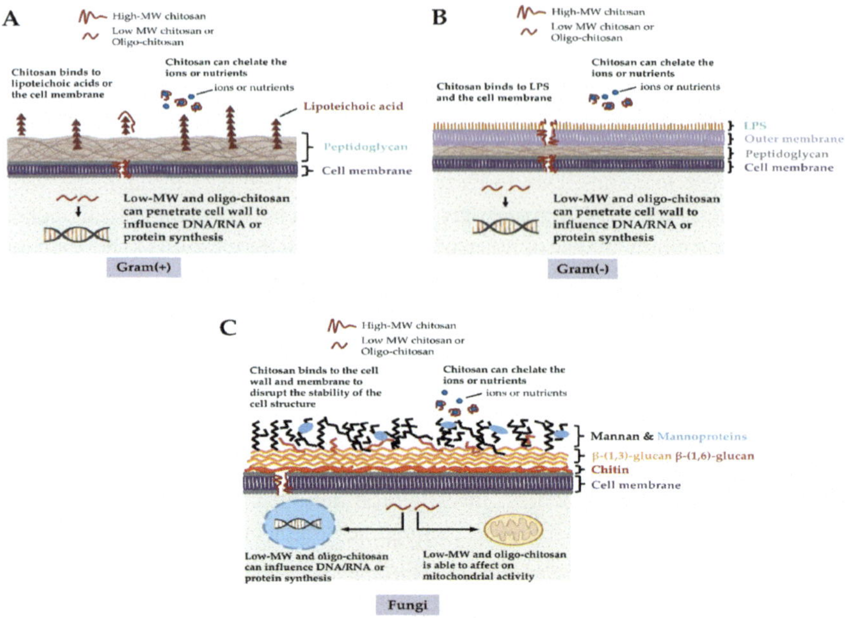

Fig. 7.3 Mechanism of antimicrobial activity of CS-NPs

molecular size. A reduction in membrane negative charge is consistent with increased chitosan resistance among *Salmonella typhimurium* mutations. The spinnable coating of polymer/metal nanocomposites was found to be an excellent material to control the bacteria. They recommend a biostatic coating made of a polymer nanocomposite containing stabilised copper/silver/gold nanoparticles. There are clear linkages between the material's properties and its biological effects. Chitosan primarily creates bio-nanocomposites for various uses, including tissue engineering, medication delivery, packing, decontamination, gas sensors and cultural heritage preservation. Chitosan, along with silver nanoparticles, exhibits extensive antimicrobial activity. The positively charged chitosan-based nanoparticles (CS-NPs) interact with the negatively charged cell surface of bacteria resulting in the membrane lysis, cytoplasmic discharge, inactivation of enzymes, metabolites, and chelation behaviour of nanoparticles, leading to the death of bacterial cells. It was observed in the *E. coli*, *Salmonella typhi*, *Klebsiella pneumoniae*, *Pseudomonas aeruginosa* and *Enterobacter cloacae*. Interaction of AgNPs with internal compounds like phosphorous and proteins and sulphur in the cell wall leads to the inhibition of cell respiration as well as cell wall division. AgNPs binging with the small subunit of ribosomes (the 30S) prevents protein translation in *E. coli*. The mode of action in a variable with the cell wall composition of bacteria (Fig. 7.3).

7.4.1 Gram-Negative Bacteria

The outer lipopolysaccharide membrane covers a thinner peptidoglycan layer that makes up the gram-negative cell wall. Since LPS is often connected to phosphorylated groups, gram-negative bacteria are more negatively charged than gram-positive bacteria. When the ambient pH is below 6.5, more negatively charged cell surfaces enable cationic chitosan to attach to phospholipids. Chitosan was formerly thought to be more effective against gram-negative bacteria; however, recent research has shown that gram-positive bacteria are actually more sensitive to chitosan.

The most common antibacterial activity of chitosan is said to include attachment to bacterial cell walls having a negative charge, causing cell disruption and changing the permeability of membranes. Additionally, binding to DNA prevents DNA replication, which results in cell death [22]. It has been hypothesised that chitosan acts as a chelating agent, willingly binding to trace metal ions to form poisons and inhibiting microbial development [23]. For chitosan to be effective against bacteria, its polycationic structure is required. Because the pKa of chitosan is greater than pH of the environment as well as its derivatives, there is a lot of electrostatic communication between the mostly anionic parts and the polycationic structure of the microorganisms' surface, like cell surface proteins and Gram-negative lipopolysaccharides. This contributes significantly to the antimicrobial effect [7]. Because some derivatives' grafted groups may alter the chitosan pKa and lead to protonation at a higher pH value, the polycationic structure arises needlessly under acidic circumstances [24]. As chitosan's density of positive charges increases, as it does in the case of chitosan metal complex and quaternized chitosan, its antibacterial activity will increase proportionately [25]. On the other hand, the associated antibacterial activity will be diminished or lost if the polycationic characteristic of chitosan is eliminated or reversed. Numerous amino groups may boost antibacterial activity, and in addition to protonation, the number of amino groups connected to C-2 on chitosan backbone plays a vital role in electrostatic interaction. Therefore, compared to a molecule with a lesser DD, native chitosan with a more significant DD has a more notable inhibitory impact. The presence of two positively charged sites on asparagine N-conjugated chitosan oligosaccharide has also been observed to interact strongly with the carboxyl-negative charges on bacterial cell walls [26]. The second attempt to increase the number of amino groups by replacing amino with formamidine produced guanidinylated chitosan, which was more antibacterial than chitosan [27, 28].

The polymer composites supplemented with Ag^+ nanoparticles demonstrated an effective inhibitory region against *S. aureus* and *E. coli*. The interaction between silver and the thiol groups in bacterial proteins is thought to be the root of silver's antibacterial function, according to Ylmaz atay et al. [29]. Due to gram-negative bacteria's thin cell wall layer, the silver nanoparticles made from chitosan exhibit more inhibitory action against G(-) bacteria than G(+) bacteria. The interaction between the cationic charge of the CS-NPs and the anionic charge of the cell wall, which results in cell lysis and prevents the uptake of vital nutrients, is what gives CS-NPs their antibacterial properties.

Instead, high molecular weight (HMW) water-soluble and solid chitosan, especially bigger size nanoparticles, interact with the cell surface and alter the cell's permeability or form an impenetrable barrier around the cell, which hinders the movement of essential solutes into the cell [30]. The same microbial species might exhibit noticeable changes in mode of action based on the particles of chitosan that have two distinct dimensions, according to experiments done on treatment of *E. coli* with oleoyl-chitosan nanoparticles (OCNP) and CM [31]. The chitosan microspheres displayed a number of different states on the cell surfaces: several were still intact, while others were oozing interior contents, and some had already ruptured, leaving just the membranes. According to these findings, CMs destroy bacteria by preventing them from sticking to the microspheres surface [32]. Chitosan acts on surface of the cell rather than full-contact mode which usually happens in a liquid state [2].

7.4.2 Gram-Positive Bacteria

Chitosan's cationic nature enables it to demonstrate exceptional inhibitory action against various microbes, including bacteria, fungus and trypanosomes [33]. Positively charged chitosan molecules may interact with negatively charged microbial cell membranes, altering the permeability of the cell wall, and enabling internal contents to leak. However, the antibacterial properties of chitosan can be impacted by variables including molecular weight, deacetylation level and positive charge concentration. Chitosan has been treated with sulfonate or quaternary ammonium groups in various studies to boost its antimicrobial action. Antibacterial herbs or enzymes have also been added to chitosan-based beads or nanoparticles. A naturally occurring cationic antimicrobial peptide (CAP) called protamine mainly comprises fundamental arginine residues. The broad-spectrum antibacterial actions of protamine can kill both gram-positive and gram-negative bacteria. Chitosan has been used with protamine to deliver heparin, DNA, siRNA, and insulin for intensive insulin therapy and gene therapy [34].

Teichoic acids found in gram-positive bacteria have a similar negatively charged structure as a result of the presence of phosphate groups throughout their composition. Although *Staphylococcus aureus* with the teichoic acid synthesis pathway ablation has increased resistance to chitosan, this shows that chitosan's mode of action is more complicated than simple electrostatic interactions. Also, gram-positive bacteria have thicker cell walls than gram-negative bacteria, making it more difficult for chitosan to connect to the cell membrane directly. The cell wall can be penetrated by certain chitosan oligomers (5 kDa), which can then affect protein or DNA synthesis. It's interesting to note that studies have shown that chitosan, which has a molecular weight of 50 kDa, may penetrate cell walls and block DNA transcription.

7.4.3 Antifungal Activity

Similar to bacteria, it's thought that chitosan exerts fungistatic rather than fungicidal effects on fungi, which might indicate changes in regulatory structure in both the host and the fungus. The prevention of spore germination, elongation of the germ tube, and radial growth has typically been demonstrated to be successful using chitosan. Most of the study has been on yeasts and moulds connected to the degradation of plants and food. For these, chitosan stimulates chitinases, which influence vesicular–arbuscular mycorrhizal fungi, entomopathogenic fungi, biotrophic and necrotrophic mycoparasites, and plant tissue [35, 36]. Due to their branching and unbranded polymers and strong cell wall, which prevents cell wall lysis, *C. albicans* requires larger doses of CS-NPs for fungicidal efficacy. N-acetylation, which all of a sudden prevents amine groups from changing into protonated forms, dramatically reduces chitosan's ability to suppress the growth of fungi.

Numerous fungi that cause disease in both plants and people can be killed by chitosan. Chitosan's interactions with cell walls and membranes are primarily responsible for its antifungal activities. MICs (Minimum inhibitory concentrations) of chitosan against fungi differs, depending on factors such as solvent pH, molecular weight (MW), degree of deacetylation (DDA), and the kind of fungus being targeted. It's probable that higher quantities of unsaturated fatty acids on the cell membrane positively correlate with chitosan susceptibility because they promote greater membrane fluidity and provide a stronger negative charge on the cell membrane. For instance, the quantity of unsaturated fatty acids on the cell membranes explains the differences in characteristics between chitosan-sensitive and chitosan-resistant *Neurospora crassa* strains. These findings might help explain why Candida species, such as *Candida albicans*, *Candida tropicalis* and others, exhibit strikingly different susceptibilities to the same chitosan, at least partly. Compared to *C. albicans*, *C. tropicalis* showed an increase in sensitivity to certain chitosans of more than 1,000-fold. Like low-MW chitosan inhibits DNA/RNA and protein production by penetrating cell walls and surfaces and having antifungal effects outside of cells, this ability is a key feature of low-MW chitosan.

7.5 Factors Affecting the Action

Several variables, which may be divided into three broad groups: environmental factors, chitosan-related factors, and factors influencing different types of microbes, might alter the antimicrobial action of chitosan [37]. Species, cell age, intrinsic variables and physical properties all have a role in the varying antibacterial efficacy of chitosan-based nanoparticles.

7.5.1 Gram-Negative and Gram-Positive Bacterial Cells

Chitosan is known to inhibit bacterial and fungal species in very diverse ways. Due to their dissimilar cellular compositions, Gram-positive and also Gram-negative bacteria need distinct antimicrobial activities that have to be carried out through complex mechanisms. Cell membranes of Gram-positive bacteria are made of peptidoglycans and teichoic acid, which provide a platform for the action of enzymes bound to the cell surface, in contrast to the hydrophilic lipopolysaccharides of Gram-negative bacteria, which protect the cell from hydrophobic toxins including macromolecule penetration. Thus, chitosan seems to have a greater effect in inhibiting G(−) than on G (+) bacteria [7]. It is not easy to determine which individual molecules of the membrane chitosan affects. It was predicted by Young, Kohle and Kauss [38] that the presence of chitosan in the plant cell membrane would cause it to interact with polygalacturonate, hence increasing the permeability of the plant cell wall. Additionally, some studies on bacteria have hypothesised that chitosan electrostatically interacts with one of the following components of amino acids found in the membrane of Gram-positive bacteria, a variety of lipopolysaccharides on the outer membrane of Gram-negative bacteria, which could affect the integrity of cell membrane and infiltration rate. In, *Pseudomonas aeruginosa* membrane receptors bind chitosan derivatives [39]. Teichoic acid, a negatively charged component of Gram-positive bacteria, electrostatically interacted with other molecules, including the polycationic chitosan macromolecule [40]. The relatively weak antibacterial activity of chitosan on gram-negative species supports these experiments' findings since chitosan binding to LPS does not compromise the integrity of cell membranes.

7.5.2 pH

Environmental pH is one of the primary elements stated above that affects the antimicrobial properties of chitosan and its analogs. Acidic pH, as opposed to pH higher than 6.5, can greatly dissolve chitosan, creating an appropriate environment for its action. This is supported by several results, including the greatly lowered antibacterial activity of chitosan at pH 7, which is thought to be caused by the deprotonation of amine groups and reduced ability to dissolve. It's possible that this is because of the high rate at which amino groups in alkaline or neutral circumstances get uncharged [41]. However, this problem has been solved with newly developed chitosan derivatives showing excessive water solubility. Multiple functional groups, including carboxyalkyl, quaternary ammoniumyl, extended alkyl chains, guaindinyl and thiol, may be introduced into chitosan to change its properties [42].

7.5.3 *Temperature*

Chitosan's activity could also be affected by the temperature at which it is kept. According to a study, the chitosan solution held at 4 °C (within 15 weeks) inhibited *Listeria monocytogenes*, *Salmonella enterica*, *Staphylococcus aureus* and *E. coli* more efficiently. Low-weight chitosan's structural characteristics also affect how easily it attaches to the cell surface, with short chains having ionic interaction, easier mobility and attraction [36].

7.5.4 *Molecular Weight*

Chitosan's molecular weight also influences the functionality of the compound, i.e., lower the molecular weight, higher the toxicity and vice versa. In addition to chitosan's size, the target organism's molecular weight significantly impacts chitosan's antibacterial efficacy. Since molecular weights less than 470 kDa inhibited Gram-positive in addition to Gram-negative bacteria growing, *E. coli* was exclusively inhibited by 470 kDa less than molecular weights [43]. Likewise, chitosan's antibacterial activity was increased by decreased acetylation levels [36]. Amino groups are created during the deacetylation process by removing acetyl groups from chitin, and they are useful in determining the level of deacetylation that has occurred [2].

7.6 Applications

7.6.1 *Agriculture*

The potential application of CS-NPs in agriculture as pesticides or herbicides which can enhance the yield and food quality are just a few of the numerous uses for chitosan that have received in-depth literature reviews [44]. When storing fresh fruits like strawberries, jujubes, loquats and longans, compounds based on nano-chitosan or chitosan mixed simultaneously with other nanoparticles were used as a preservative. Furthermore, because of its cationic nature, non-toxicity, biodegradability and adsorption capabilities, chitosan may be used alone or in conjunction with other substances to create slow-release fertilisers as an encapsulating agent [45].

Materials derived from naturally occurring substances including chitosan, cellulose and activated carbon, are attractive solutions with cheap costs and excellent energy efficiency when looking for sustainable wastewater treatment and water purification methods. Chitosan's crosslinked structure contains a lot of hydroxyl and amino groups, which have been widely used in its microstructures, nanoparticles and nanocomposites as absorbents to get rid of various inorganic and organic contaminants [23]. Since chitosan-based nanoparticles have higher surface areas than their microparticles, they are more able to adsorb dyes, insecticides, heavy metal ions and phosphate including Hg(II), Pb(II), Cr(III), Cd(II), Cr(VI), Co(II), Cu(II), Ni(II) and rare earth metals [46]. Various chitosan nanoparticles are added to the ultrafiltration and nanofiltration membranes in several modern membrane-based wastewater treatments to minimise the need for chemical oxygen, colour, metal ions, and to improve the antifouling capabilities [47].

Chitosan-based nanoparticles were also exploited in the salinity alleviation; in one study, the salt stress in maize was regulated by using the nitric oxide (NO) releasing chitosan nanoparticles [48]. It has been shown that giving chitosan-polyvinyl alcohol hydrogels to tomato plants under salt stress stimulates jasmonic acid (JA) and superoxide dismutase (SOD) gene expression [49]. By boosting the synthesis of phenolic compounds, osmoregulators, and crop output as well as by encouraging gas exchange in leaves, nanochitosans can lessen the impacts of drought stress.

When applied topically to pearl millet under drought stress, nano-chitosan emulsion has been found to improve plant water status by reducing stomatal conductance and transpiration [50]. It has been demonstrated that S-nitrosoglutathione (NO donor) nano-chitosan may help sugarcane plants recover from drought stress [51]. Wheat plants with low water content were treated with chitosan nanoparticles at a dosage of 90 ppm, which enhanced the plants' biochemical and physiological traits. A hybrid maize species vulnerable to drought stress was shown to be more tolerant of water stress when N-succinyl chitosan and NO-dicarboxymethylated chitosan derivatives were applied topically, according to Rabelo et al. [52].

7.6.2 Biomedical Fields

Numerous biomedical applications, such as drug delivery, tissue engineering, wound healing, gene therapy, and bioimaging and biosensing, have made extensive use of nanostructured hybrid materials. The primary factors influencing the choice of these materials are the increased precision and effectiveness in medical diagnosis, monitoring, and subsequent therapy. Polymer nanocomposites are now being actively exploited to provide diverse bioactive compounds for medicine delivery and wound healing. Many biomedical applications of polymeric matrices use either synthetic polymers such as poly (-caprolactone), acrylic, polylactides, and tyrosine-based polymers, or natural polymers such as alginate, cellulose, and chitosan, or a combination of the two. Depending on the intended usage, these nanocomposites are quickly processed into different forms. For biomedical usage, nanocomposite hydrogels, films, and membranes are frequently strongly advocated. Due to its elastic biological and physical characteristics, distinctive adaptability and broad range of applications, nanocomposite hydrogels have attracted the attention of material scientists [53]. The antimicrobial spectrum of CS-NPs against potential clinical pathogens has been illustrated in Fig. 7.4.

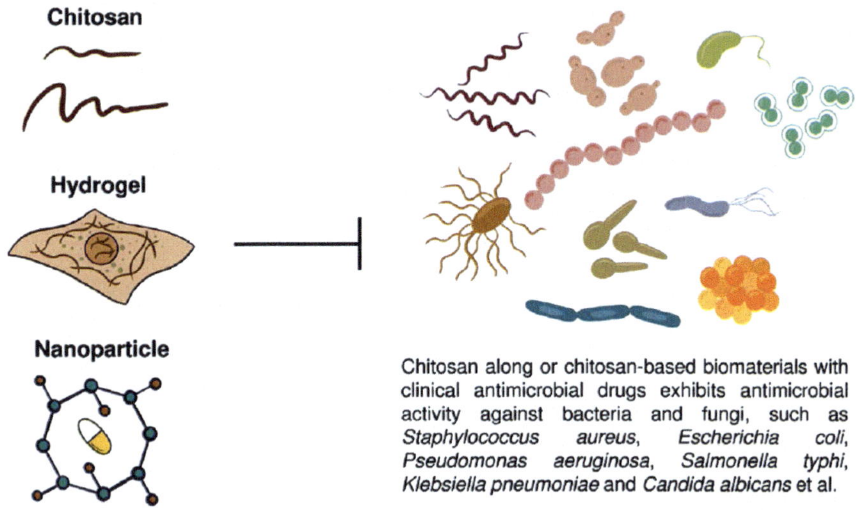

Fig. 7.4 Antimicrobial spectrum of CS-NPs

Owing to its high affinity for absorbing water, porous shape, lack of toxicity, and, most importantly, its capacity to imitate biological processes, the hydrogel is generally considered an attractive choice for creating tissue engineering scaffolds [53, 54]. But the absence of mechanical and optical qualities in virgin matrix hydrogels limits their use [55, 56]. By adding adhesion sites and growth agents to support cellular development, the incorporation of nanoparticles overcomes the drawbacks of composite hydrogels [57]. Additionally, using nanoparticles enhances the thermal characteristics, stimulates stem cell development, and enhances the self-healing nature, which permits the creation of implantable scaffolds for tissue engineering. The application of nanoparticles increases their applicability in drug administration by allowing for more control over the rate of crosslinking and porosity [58]. Nanoparticles may also immediately impact how the hydrogel reacts to outside stimuli like electromagnetic radiation, which enables the introduction of orthogonal drugs.

They were also used for wound dressing, and antibiotics are typically needed to prevent microbial infection using conventional wound dressing materials. The addition of nanoparticles gave the hydrogels angiogenic, antipathogenic and bioadhesive characteristics [59]. According to the information in the previous section, including nanoparticles can enhance cell attachment's mechanical strength and sites, enabling high-resolution prints and post-removal cell survival of the nanocomposite hydrogels [60].

7.6.3 *Textiles*

Ionic gelation is an appealing method for producing chitosan nanoparticles because it enables the repeatable creation of NPs with tiny diameters and restricted population distribution in a benign environment (such as water) without the need for any harmful crosslinking agents or organic solvents. Chitosan NPs are applied in the textile industry, where the utilisation is explored for enhancing the braking strength, shrink-proof, wrinkle resistance properties and textile dyeing [61]. The CS-NPs are known for the interaction of various fabrics and dyes with reduced toxicity and provide a larger surface area.

7.7 Conclusion and Future Perspective

The potential for antibacterial uses of natural biopolymers like CS is enormous. It is possible to create strong, highly active CS-NPs for a range of industrial applications by applying the appropriate NP synthesis techniques. When used biologically, CS-NPs, either by themselves or in intermingling with other elements, inhibits both G(+) and G(-) microorganisms. Numerous factors, such as the chemical makeup of CS, concentration, pH and particle size, predispose to this bioactivity. In order to establish solutions and alternatives to the challenges that both the medical and agricultural professions face, in situ research assessments must be completed. The majority of the features of in vitro investigations have been made so far. The mechanisms of action of CS-NPs against bacteria are not yet fully explored, necessitating continued research efforts. Additionally, it is crucial to keep tracking and evaluating the toxicity associated with CS-NPs as antibacterial agent, as well as to offer direction for the laws and practices surrounding their usage and applications. To fill the information gap regarding the antibacterial activity of CS-NPs, the following kinds of investigations must be conducted: (1) Examining the mechanisms by which CS-NPs exert their antibacterial effects; (2) figuring out why G (-) bacteria are less susceptible to CS-NPs than G + bacteria; (3) discovering the mechanism by which medium MW-CS outperforms high MW-CS in killing bacteria and (4) exploring the toxicity and method of synthesis for CS-NPs as well as hybrid CS-NPs that incorporate metal ions. These investigations are helpful for animal and plant research, as well as the creation of a new class of antibacterial drugs.

References

1. A. Ali, N.M. Noh, M.A. Mustafa, Antimicrobial activity of chitosan enriched with lemongrass oil against anthracnose of bell pepper. Food Packag. Shelf Life **3**, 56–61 (2015)
2. M. Hosseinnejad, S.M. Jafari, Evaluation of different factors affecting antimicrobial properties of chitosan. Int. J. Biol. Macromol. **85**, 467–475 (2016)

3. H.Y. Atay, E. Çelik, Investigations of antibacterial activity of chitosan in the polymeric composite coatings. Prog. Org. Coatings **102**, 194–200 (2017)
4. R. Severino, K.D. Vu, F. Donsì, S. Salmieri, G. Ferrari, M. Lacroix, Antimicrobial effects of different combined non-thermal treatments against Listeria monocytogenes in broccoli florets. J. Food Eng. **124**, 1–10 (2014)
5. D.S. Cha, M.S. Chinnan, Biopolymer-based antimicrobial packaging: a review. Crit. Rev. Food Sci. Nutr. **44**, 223–237 (2004)
6. S.A. Qamar, M. Ashiq, M. Jahangeer, A. Riasat, M. Bilal, Chitosan-based hybrid materials as adsorbents for textile dyes–a review. Case Stud. Chem. Environ. Eng. **2**, 100021 (2020)
7. M. Kong, X.G. Chen, K. Xing, H.J. Park, Antimicrobial properties of chitosan and mode of action: a state of the art review. Int. J. Food Microbiol. **144**, 51–63 (2010)
8. C.R. Allan, L.A. Hadwiger, The fungicidal effect of chitosan on fungi of varying cell wall composition. Exp. Mycol. **3**, 285–287 (1979)
9. E.I. Rabea, M.E.-T. Badawy, C.V. Stevens, G. Smagghe, W. Steurbaut, Chitosan as antimicrobial agent: applications and mode of action. Biomacromol **4**, 1457–1465 (2003)
10. M. Chandrasekaran, K.D. Kim, S.C. Chun, Antibacterial activity of chitosan nanoparticles: a review. Processes **8**, 1173 (2020)
11. N.A.S. Rozman, W.Y. Tong, C.R. Leong, W.N. Tan, M.A. Hasanolbasori, S.Z. Abdullah, Potential antimicrobial applications of chitosan nanoparticles (ChNP) (2019)
12. P. Calvo, C. Remuñan-López, J.L. Vila-Jato, M.J. Alonso, Chitosan and chitosan/ethylene oxide-propylene oxide block copolymer nanoparticles as novel carriers for proteins and vaccines. Pharm Res **14**, 1431–1436 (1997)
13. H. Katas, C.S. Lim, A.Y.H.N. Azlan, F. Buang, M.F.M. Busra, Antibacterial activity of biosynthesized gold nanoparticles using biomolecules from Lignosus rhinocerotis and chitosan. Saudi. Pharm. J. **27**, 283–292 (2019)
14. S.S. Davis, L. Illum, Sustained release chitosan microspheres prepared by novel spray drying methods. J. Microencapsul. **16**, 343–355 (1999)
15. O.E. Eremina, O.O. Kapitanova, E.A. Goodilin, I.A. Veselova, Silver-chitosan nanocomposite as a plasmonic platform for SERS sensing of polyaromatic sulfur heterocycles in oil fuel. Nanotechnology **31**, 225503 (2020)
16. F. Sonvico, A. Cagnani, A. Rossi, S. Motta, M.T. Di Bari, F. Cavatorta, M.J. Alonso, A. Deriu, P. Colombo, Formation of self-organized nanoparticles by lecithin/chitosan ionic interaction. Int. J. Pharm. **324**, 67–73 (2006)
17. M.M. Khan, A. Madni, V. Torchilin, N. Filipczak, J. Pan, N. Tahir, H. Shah, Lipid-chitosan hybrid nanoparticles for controlled delivery of cisplatin. Drug Deliv. **26**, 765–772 (2019)
18. A. Dev, N.S. Binulal, A. Anitha, S.V. Nair, T. Furuike, H. Tamura, R. Jayakumar, Preparation of poly (lactic acid)/chitosan nanoparticles for anti-HIV drug delivery applications. Carbohydr. Polym. **80**, 833–838 (2010)
19. C. Puente, M. Sánchez-Domínguez, C.L. Brosseau, I. López, Silver-chitosan and gold-chitosan substrates for surface-enhanced Raman spectroscopy (SERS): effect of nanoparticle morphology on SERS performance. Mater Chem. Phys. **260**, 124107 (2021)
20. T.V. Podust, T.V. Kulik, B.B. Palyanytsya, V.M. Gun'Ko, A. Tóth, L. Mikhalovska, A. Menyhárd, K. László, Chitosan-nanosilica hybrid materials: preparation and properties. Appl. Surf. Sci. **320**, 563–569 (2014)
21. X. Fei Liu, Y. Lin Guan, D. Zhi Yang, Z. Li, K. De Yao, Antibacterial action of chitosan and carboxymethylated chitosan. J. Appl. Polym. Sci. **79**, 1324–1335 (2001)
22. A. Nagy, A. Harrison, S. Sabbani, R.S. Munson Jr., P.K. Dutta, W.J. Waldman, Silver nanoparticles embedded in zeolite membranes: release of silver ions and mechanism of antibacterial action. Int. J. Nanomedicine **6**, 1833 (2011)
23. K. Divya, S. Vijayan, T.K. George, M.S. Jisha, Antimicrobial properties of chitosan nanoparticles: mode of action and factors affecting activity. Fibers Polym. **18**, 221–230 (2017)
24. T.-C. Yang, C.-C. Chou, C.-F. Li, Antibacterial activity of N-alkylated disaccharide chitosan derivatives. Int. J. Food Microbiol. **97**, 237–245 (2005)

25. Y. Xie, X. Liu, Q. Chen, Synthesis and characterization of water-soluble chitosan derivate and its antibacterial activity. Carbohydr Polym. **69**, 142–147 (2007)
26. Y.-J. Jeon, P.-J. Park, S.-K. Kim, Antimicrobial effect of chitooligosaccharides produced by bioreactor. Carbohydr. Polym. **44**, 71–76 (2001)
27. Y. Hu, Y. Du, J. Yang, J.F. Kennedy, X. Wang, L. Wang, Synthesis, characterization and antibacterial activity of guanidinylated chitosan. Carbohydr. Polym. **67**, 66–72 (2007). https://doi.org/10.1016/j.carbpol.2006.04.015
28. Y. Hu, Y. Du, J. Yang, Y. Tang, J. Li, X. Wang, Self-aggregation and antibacterial activity of N-acylated chitosan. Polymer (Guildf) **48**, 3098–3106 (2007)
29. H. Yilmaz Atay, İ Yaşa, E. Çelik, Antibacterial polymeric coatings with synthesized silver nanoparticles. Synth. React. Inorg. Met. - Org. Nano - Met. Chem. **45**, 784–798 (2015)
30. J.L. Leuba, P. Stossel, Chitosan and other polyamines: antifungal activity and interaction with biological membranes. In: Chitin in Nature and Technology (Springer, 1986), pp. 215–222
31. K. Xing, X.G. Chen, C.S. Liu, D.S. Cha, H.J. Park, Oleoyl-chitosan nanoparticles inhibits Escherichia coli and Staphylococcus aureus by damaging the cell membrane and putative binding to extracellular or intracellular targets. Int. J. Food Microbiol. **132**, 127–133 (2009)
32. M. Kong, X.G. Chen, C.S. Liu, C.G. Liu, X.H. Meng, Antibacterial mechanism of chitosan microspheres in a solid dispersing system against E. coli. Colloids Surf. B Biointerfaces **65**, 197–202 (2008)
33. M. Sathiyabama, R. Parthasarathy, Biological preparation of chitosan nanoparticles and its in vitro antifungal efficacy against some phytopathogenic fungi. Carbohydr. Polym. **151**, 321–325 (2016)
34. J. Sheng, H. He, L. Han, J. Qin, S. Chen, G. Ru, R. Li, P. Yang, J. Wang, V.C. Yang, Enhancing insulin oral absorption by using mucoadhesive nanoparticles loaded with LMWP-linked insulin conjugates. J. Control. Release **233**, 181–190 (2016)
35. A. El Ghaouth, J. Arul, J. Grenier, A. Asselin, Antifungal activity of chitosan on two postharvest pathogens of strawberry fruits. Phytopathology **82**, 398–402 (1992)
36. R.C. Goy, D. de Britto, O.B.G. Assis, A review of the antimicrobial activity of chitosan. Polímeros **19**, 241–247 (2009)
37. G. Kravanja, M. Primožič, Ž Knez, M. Leitgeb, Chitosan-based (Nano) materials for novel biomedical applications. Molecules **24**, 1960 (2019)
38. D.H. Young, H. Köhle, H. Kauss, Effect of chitosan on membrane permeability of suspension-cultured Glycine max and Phaseolus vulgaris cells. Plant Physiol. **70**, 1449–1454 (1982)
39. M. Morimoto, H. Saimoto, H. Usui, Y. Okamoto, S. Minami, Y. Shigemasa, Biological activities of carbohydrate-branched chitosan derivatives. Biomacromol **2**, 1133–1136 (2001)
40. D. Raafat, K. Von Bargen, A. Haas, H.-G. Sahl, Insights into the mode of action of chitosan as an antibacterial compound. Appl. Environ. Microbiol. **74**, 3764–3773 (2008)
41. K. Aiedeh, M.O. Taha, Synthesis of iron-crosslinked chitosan succinate and iron-crosslinked hydroxamated chitosan succinate and their in vitro evaluation as potential matrix materials for oral theophylline sustained-release beads. Eur. J. Pharm. Sci. **13**, 159–168 (2001)
42. P. Sahariah, M. Másson, Antimicrobial chitosan and chitosan derivatives: a review of the structure–activity relationship. Biomacromol **18**, 3846–3868 (2017)
43. T. Takahashi, M. Imai, I. Suzuki, Water permeability of chitosan membrane involved in deacetylation degree control. Biochem. Eng. J. **36**, 43–48 (2007)
44. R.V. Kumaraswamy, S. Kumari, R.C. Choudhary, A. Pal, R. Raliya, P. Biswas, V. Saharan, Engineered chitosan based nanomaterials: Bioactivities, mechanisms and perspectives in plant protection and growth. Int. J. Biol. Macromol. **113**, 494–506 (2018)
45. P.L. Kashyap, X. Xiang, P. Heiden, Chitosan nanoparticle based delivery systems for sustainable agriculture. Int. J. Biol. Macromol. **77**, 36–51 (2015)
46. H. Zhang, X. Tan, T. Qiu, L. Zhou, R. Li, Z. Deng, A novel and biocompatible Fe3O4 loaded chitosan polyelectrolyte nanoparticles for the removal of Cd2+ ion. Int. J. Biol. Macromol. **141**, 1165–1174 (2019)
47. S.K. Ramachandran, A. Gangasalam, Reduction of chemical oxygen demand and color from the rice mill wastewater by chitosan/2 (5 H)-furanone-incorporated ultrafiltration membrane system. Sep. Sci. Technol. **54**, 409–425 (2019)

48. H.C. Oliveira, B.C.R. Gomes, M.T. Pelegrino, A.B. Seabra, Nitric oxide-releasing chitosan nanoparticles alleviate the effects of salt stress in maize plants. Nitric Oxide **61**, 10–19 (2016)
49. H. Hernández-Hernández, A. Juárez-Maldonado, A. Benavides-Mendoza, H. Ortega-Ortiz, G. Cadenas-Pliego, D. Sánchez-Aspeytia, S. González-Morales, Chitosan-PVA and copper nanoparticles improve growth and overexpress the SOD and JA genes in tomato plants under salt stress. Agronomy **8**, 175 (2018)
50. M. Priyaadharshini, N. Sritharan, A. Senthil, S. Marimuthu, Physiological studies on effect of chitosan nanoemulsion in pearl millet under drought condition. J. Pharmacogn. Phytochem. **8**, 3304–3307 (2019)
51. N.M. Silveira, A.B. Seabra, F.C.C. Marcos, M.T. Pelegrino, E.C. Machado, R.V. Ribeiro, Encapsulation of S-nitrosoglutathione into chitosan nanoparticles improves drought tolerance of sugarcane plants. Nitric Oxide **84**, 38–44 (2019)
52. V.M. Rabêlo, P.C. Magalhães, L.A. Bressanin, D.T. Carvalho, C.O. dos Reis, D. Karam, A.C. Doriguetto, M.H. dos Santos, P.R.S. dos Santos Filho, T.C. de Souza, The foliar application of a mixture of semisynthetic chitosan derivatives induces tolerance to water deficit in maize, improving the antioxidant system and increasing photosynthesis and grain yield. Sci. Rep. **9**, 1–13 (2019)
53. M.R. Sakr, M.T. Bassuoni, Silane and methyl-methacrylate based nanocomposites as coatings for concrete exposed to salt solutions and cyclic environments. Cem. Concr. Compos. **115**, 103841 (2021)
54. K. Wang, Y. Hao, Y. Wang, J. Chen, L. Mao, Y. Deng, J. Chen, S. Yuan, T. Zhang, J. Ren, Functional hydrogels and their application in drug delivery, biosensors, and tissue engineering. Int. J. Polym. Sci. (2019)
55. Z. Wang, J. Zhao, W. Tang, L. Hu, X. Chen, Y. Su, C. Zou, J. Wang, W.W. Lu, W. Zhen, Multifunctional nanoengineered hydrogels consisting of black phosphorus nanosheets upregulate bone formation. Small **15**, 1901560 (2019)
56. K.-C. Cheng, C.-F. Huang, Y. Wei, S. Hsu, Novel chitosan–cellulose nanofiber self-healing hydrogels to correlate self-healing properties of hydrogels with neural regeneration effects. NPG Asia Mater. **11**, 1–17 (2019)
57. N. Wang, M. Ma, Y. Luo, T. Liu, P. Zhou, S. Qi, Y. Xu, H. Chen, Mesoporous silica nanoparticles-reinforced hydrogel scaffold together with pinacidil loading to improve stem cell adhesion. ChemNanoMat **4**, 631–641 (2018)
58. A.K. Gaharwar, C.P. Rivera, C.-J. Wu, G. Schmidt, Transparent, elastomeric and tough hydrogels from poly (ethylene glycol) and silicate nanoparticles. Acta Biomater. **7**, 4139–4148 (2011)
59. R.G. Patel, A. Purwada, L. Cerchietti, G. Inghirami, A. Melnick, A.K. Gaharwar, A. Singh, Microscale bioadhesive hydrogel arrays for cell engineering applications. Cell. Mol. Bioeng. **7**, 394–408 (2014)
60. Z. Deng, T. Hu, Q. Lei, J. He, P.X. Ma, B. Guo, Stimuli-responsive conductive nanocomposite hydrogels with high stretchability, self-healing, adhesiveness, and 3D printability for human motion sensing. ACS Appl. Mater. Interfaces **11**, 6796–6808 (2019)
61. S.W. Ali, S. Rajendran, M. Joshi, Synthesis and characterization of chitosan and silver loaded chitosan nanoparticles for bioactive polyester. Carbohydr. Polym. **83**, 438–446 (2011)

Chapter 8
Chitosan-Based Nanocomposites as Efficient Wound Dressing Materials

Lei Nie, Qianqian Wei, Jingyu Li, Chenxu Zhao, Peng Ding, Oseweuba Valentine Okoro, and Amin Shavandi

Abstract Wound dressings play a key role in wound healing with conventional wound dressings typically based on synthetic polymers. These synthetic polymers however present several risks such as poor biocompatible and toxicity, which are capable of hindering cell proliferation, cell growth and limiting wound closure. These synthetic polymers may also be characterized by their high cost. In response to these limitations, the use of naturally sourced polymers such as chitosan was proposed. This is because chitosan is non-toxic, biocompatible and possesses inherent biological properties (i.e. antibacterial effects, antioxidant ability, etc.), which can promote wound healing. In this chapter, different types of chitosan-based nanocomposites that could be used in the fabrication of wound healing dressings, such as hydrogels, cryogel, membranes, gauze, etc. are explored. Furthermore, the main properties of chitosan-based nanocomposites are summarized, and the specific biomedical applications of chitosan-based nanocomposites as wound dressings are discussed. In conclusion, future outlooks of chitosan-based nanocomposites, to facilitate the efficient design of new-generation wound dressings for wound healing, are also included.

L. Nie (✉) · Q. Wei · J. Li · C. Zhao · P. Ding (✉) · O. V. Okoro · A. Shavandi (✉)
College of Life Sciences, Xinyang Normal University, Xinyang 464000, China
e-mail: nielei@xynu.edu.cn

P. Ding
e-mail: Pengding@xynu.edu.cn

A. Shavandi
e-mail: amin.shavandi@ulb.be

L. Nie
Université Libre de Bruxelles (ULB), École Polytechnique de Bruxelles, 3BIO-BioMatter, Avenue F.D. Roosevelt, 50-CP 165/61, 1050 Brussels, Belgium

S. K. Swain and A. Biswal (eds.), *Chitosan Nanocomposites*,
Biological and Medical Physics, Biomedical Engineering,
https://doi.org/10.1007/978-981-19-9646-7_8

8.1 Introduction

The skin is the initial barrier that serves to protect the human body from external issues or stimuli and maintaining the human body fluids, electrolytes, and nutrients [1–3]. When wounds appear after trauma, burns, surgeries, and chronic diseases, this barrier is destroyed [4]. It can self-repair spontaneously, but it is not enough when facing bigger open wounds, serious burns, and chronic wounds. The wound healing process incorporates diverse enzymes, growth factors, and cytokines, which can synergistically modulate relevant cell activities to promote the wound closed and tissue repair [5]. The wound healing processes include the stages, of homeostasis, inflammation, proliferation, and remodeling, which occur successively [6].

A common approach to wound treatment is wound dressing with different traditional dressings like bandages and gauze employed as passive physical barriers for wounds. The advanced dressings act as temporary skin alternatives to prevent or treat infection, manage exudate release, maintain humidity and actively promote wound healing, even reducing scar formation [7]. Wound dressings may be fabricated from biocompatible natural biopolymers such as polysaccharides, proteoglycans, and proteins [8]. Polysaccharides mainly include heparin, hyaluronic acid, cellulose, alginates, chitosan etc. Among them, chitosan (CS), which is an amino polysaccharide and its derivatives have been used extensively employed in tissue engineering applications (i.e., wound healing) due to their favorable properties such as biocompatibility, antibacterial activity, hemostatic effects, etc. [9–12]. Previous researches demonstrated that CS is suitable for wound treatment because of its bioactive abilities including promoting collagen deposition, fibroblast proliferation, wound contraction, and tissue repair [13–16]. In addition, CS-based wound dressings enhance wound healing while also limiting scar formation and healing time. Cai et al. [7] have synthesized hydroxybutyl chitosans (HBCSs) and prepared sponge-like HBCSs dressings using freeze-drying, they confirmed that HBCSs promote fibroblast growth, and could enable a reduction in the recovery period of wounds compared to CS, when a rat model was employed. These favorable effects of the HBCS dressing were due to its capacity to enhance epithelialization, collagen deposition, and neovascularization.

The solubility of CS in water limits its wound healing application. Chemical modification of CS is the dominant route to broaden the wound healing use of CS [17]. Therefore, introducing other functional groups with antibacterial and antioxidant abilities to CS is a predominant method to fabricate chitosan-based wound healing materials. To obtain different types of multifunctional chitosan-based materials, the methods of its preparation including emulsion-droplet coalescence, cross-linking, spray drying, ionic gelation, nanoprecipitation have been used. On the other hand, CS-based nanocomposites are considered as useful tools to enhance the wound healing process due to its drug loading property and high surface area to volume ratio. In addition, CS-based nanocomposites could also influence collagen deposition and skin tissue regeneration, and the healing progress could also be greatly influenced due to nanoparticles could penetrate the wound [18, 19].

This chapter therefore discusses CS-based nanocomposites for wound healing applications with particular emphasis on suitable properties that make them desirable for wound healing.

8.2 Types of the CS-Based Nanocomposites

8.2.1 Hydrogel

Hydrogel is a three-dimensional (3D) network structure composed of polymer chains that are hydrophilic, such that the structure can retain a substantial amount of water. The polymers are either natural or synthetic and may be crosslinked via physical, chemical, or polymerization strategies [20–22]. Hydrogels as dressings have been of great importance because of their capacity to enhance wound healing via the provision of a moist environment that reduces the wound-related pains. The hydrogels also serve to facilitate wound exudate removal for a cleaner microenvironment and gas change. Relying on dynamic covalent linkage, multiple supramolecular interaction containing host–guest interaction, van der Waals Force, hydrogen-bonding, hydrophobic interaction, electrostatic interaction, and metal–ligand coordination have been adopted to form hydrogels to be applied on wounds. Additionally, one-pot and three-dimensional printing (3D printing) technology are usual strategies to fabricate hydrogel dressings [23–25].

CS-based nanocomposites hydrogels have been identified as important materials for wound dressings due to its multi-properties that promote wound healing. The nanocomposite hydrogels based on different nanostructures were considered a promising approach to prepare hydrogels with appropriate properties for wound healing. Four major groups of nanomaterials are used for fabricating nanocomposite hydrogels, namely, metal, nanofiber, carbon, and mineral (Fig. 8.1A). At the same time, the tissue adhesiveness properties, antioxidant properties, antibacterial properties, and mechanical properties of CS-based nanocomposites hydrogels needs to be considered for effective wound healing (Fig. 8.1B) [26]. Wen et al. [27] fabricated a self-healing carboxymethyl cellulose (CMC) hydrogel using a two-step process in combination with molding and acidification processes. The obtained CMC hydrogel could self-heal, autonomously, under ambient temperature in the absence of external stimulus, and the healed hydrogels could be stretched with ~80% of healing efficiency after the compressive strength (~2.5 MPa). Yu et al. [11] prepared a hydrogel with favorable antimicrobial properties while also having the capacity to sustained drug release performance through mixing N-glycosylated chitosan and polyvinyl alcohol (PVA) with ofloxacin. The study showed that when the ofloxacin-loaded hydrogel was applied to the full-thickness wound on a rat model, a good wound healing ability was observed.

CS-based nanocomposites hydrogels as effective wound dressings provide a new vision of wound healing that promotes effective skin regeneration. Based on

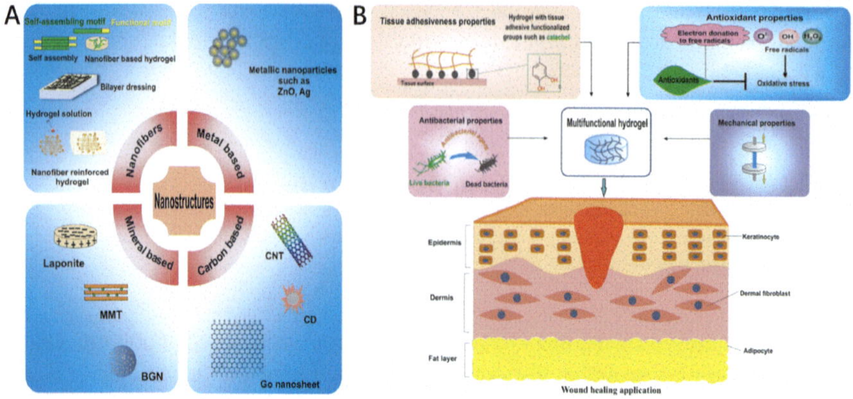

Fig. 8.1 **A** Nanofiber, metal, carbon, and mineral could be used for nanocomposite hydrogels fabrication for wound healing. **B** useful properties of multifunctional wound healing hydrogels. Reproduced with permission from Elsevier [26]

the multi-functionalities, CS-based nanocomposite hydrogels displayed potential impact as wound healing dressings for wound healing application. However, there are still challenges in applying CS-based nanocomposites hydrogels in the future, such as in combination with wound monitoring tools, and taking advantage of other biomacromolecules such as DNA and keratin.

8.2.2 Cryogel

Cryogels with inherent interconnected microporous structure allow the free flow of water across the pore structure and possess rapid shape recovery characteristic via squeezing out or absorbing water. Cryogels are a highly promising biomaterial [28, 29] and own hemorrhage ability, antibacterial properties to prevent infections and promote wound healing. Additionally, due to the abilities of high-water absorption and other functions such as hemostasis, antibacterial or antibiofilm, cryogels are suitable for treating wounds due to burns, which may lead to bacterial infection and exudate release at the early stage of healing process [30].

Li et al. [31] developed a series of cryogel dressings with antibacterial and antioxidant properties that could adhere to tissues. These dressings were based on polydopamine (PDA) and quaternized chitosan (QCS) to enhance wound healing. The prepared QCS/PDA cryogel exhibited multiple properties such as promotion of surface hemostasis, favorable antioxidant activity, favorable antibacterial performance, etc. In a full-thickness skin defect model, the cryogels showed an improved healing effect compared to chitosan sponge and Tegaderm film. In another research,

Zhao et al. [32] developed a cryogel, composed of glycidyl methacrylate functionalized quaternized chitosan (QCSG) and carbon nanotubes (CNT), for noncompressible hemorrhage and wound healing. The cryogel had favorable properties such as shape memory hemostatic, conductive, fast recovery, etc. (Fig. 8.2). The developed QCSG/CNT cryogel showed improved blood-clotting ability, favorable blood cell and improved platelet adhesion and activation compared to gelatin sponge and gauze. In addition, QCSG/CNT cryogel displayed favorable hemostatic ability in rabbit liver defect lethal noncompressible hemorrhage model.

Using chitosan and polydopamine (PDA), Zhao et al. [33] also designed cryogels (dry) with injectable, blood-absorbing, biodegradability, antioxidant, photo-thermal

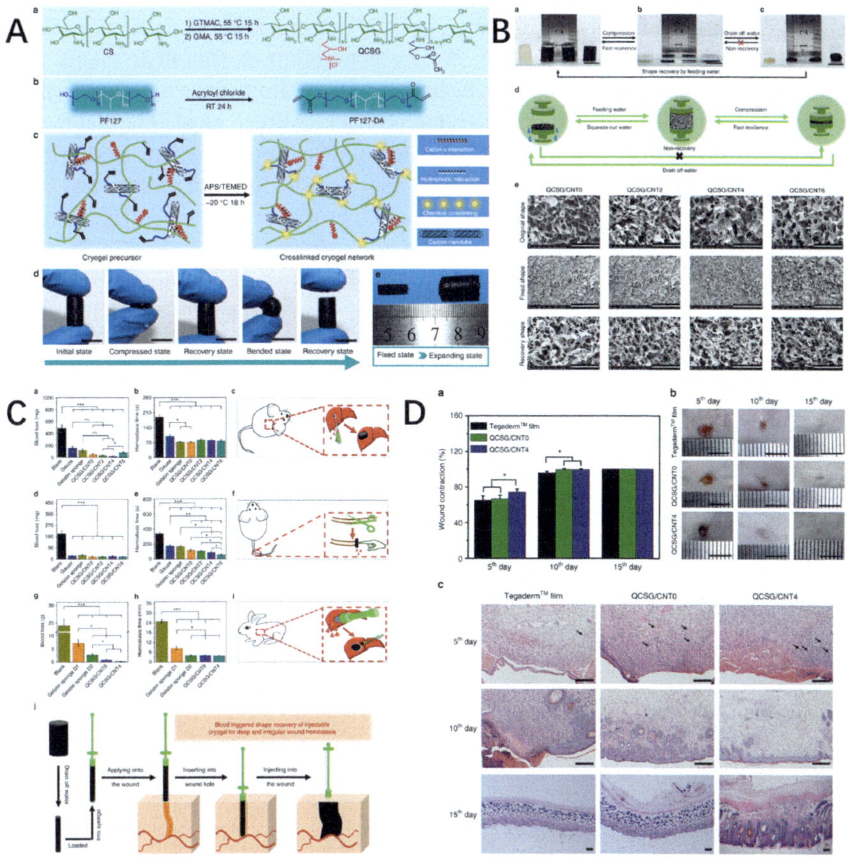

Fig. 8.2 **A** Illustration of glycidyl methacrylate functionalized quaternized chitosan (QCSG) and carbon nanotubes (CNT) cryogel dressings synthesis. **B** Shape memory properties of the QCSG/CNT cryogel. **C** In vivo hemostatic capacity evaluation of the cryogel, including blood loss and hemostatic time evaluation. **D** In vivo wound healing performance of the cryogels assessment using photographs and histomorphological images. Reproduced with permission from Springer Nature [32]

and antibacterial properties, and blood clotting properties. The cryogel dressings also showed properties such as improved blood-clotting ability, favorable blood cell, and improved platelet adhesion and activation compared to gelatin sponge and gauze.

8.2.3 Membranes and Nanofiber Membranes

Membranes and nanofiber membranes have potential as wound dressings due to the characteristics of biodegradation, low cost, harmless to life and environment [34]. Especially, electrospun nanofiber membranes possesses high porosity with excellent pore-interconnectivity, which promote the removal of fluids from wounds, and the invasion of exogenous microorganisms could be inhibited due to the high porosity and surface area (Fig. 8.3) [35, 36]. A new generation of nanofiber membranes as wound dressings to efficiently enhance wound healing process is to combine nanomaterials, such as nanocomposite membranes. Alireza et al. [37] fabricated cerium-doped bioactive glass (Ce-BG)-loaded chitosan/polyethylene oxide (PEO) nanofibrous membranes. The results showed that the swelling and mechanical properties of chitosan/PEO membranes were increased by the increase of Ce-BG/chitosan (w/w) ratio, and the mechanical properties of chitosan/PEO containing Ce-BG were similar to skin. The data displayed that the length of chitosan/PEO containing Ce-BG was 20% shorter when braking skin tissue, than the required length which increased in length by 28.6% after broken.

He et al. [38] prepared multifunctional nanofibrous membranes as wound dressings via electrospinning quaternized chitosan-graft-polyaniline (QCSP) and poly(ε-caprolactone) polymer solutions. The nanofibrous wound dressings possess electroactivity, antioxidant, antibacterial and biocompatibility properties. In vivo mouse

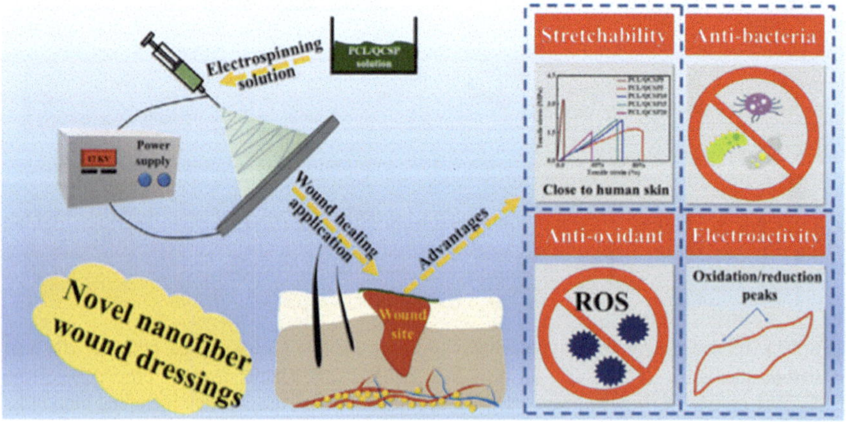

Fig. 8.3 Schematic diagram of production and advantages of the nanofiber wound dressing. Reproduced with permission from Elsevier [36]

full-thickness defect model exhibited the prepared wound dressings significantly effect on accelerating the healing process than commercial dressing (Tegaderm™ Film). In addition, the dressings including QCSP showed favorable collagen deposition, granulation tissue thickness, and angiogenesis properties, for enhanced wound healing. In another study, Liu et al. [39] prepared a nanofibrous dressing via chitosan electrospun nanofibrous material containing Cur@beta-CD/AgNPs nanoparticles composed of silver and curcumin. The dressings loaded with dual antibacterial agents and possessed synergic effects on effective antibacterial activity and wound healing. During the wound healing process, the nanocomposite dressings enhanced collagen deposition. Moreover, the chitosan nanocomposite nanofibers dressings presented antibacterial and anti-scarring abilities.

8.2.4 Foam

Foam dressings possess porous and other desirable physiochemical properties which can support cell proliferation and growth as well as skin wound remodeling. Foam dressings could provide a moist and incubator-like microenvironment, which is crucial for wound healing and minimizes scar formation. Especially, bilayered foam wound dressings composed of microporous top layer and highly porous sponge-like sublayer that helps to reduce dehydration, as well as bacterial penetration [40]. Thermoplastic polyurethanes (TPU) have been extensively studied for the fabrication of polyurethanes-based wound dressings due to the favorable wound healing properties of flexibility, biocompatibility, and permeability [41, 42]. Aleksandra et al. [43] produced nanocomposites forms based on TPU and ZnO using thermally induced phase separation method (TIPS). The prepared nanocomposite forms were found to present antibacterial effects on Gram-positive and Gram-negative bacteria, confirming that flexible TPU/ZnO nanocomposites foams are potential for wound dressings applications. Wang et al. [44] designed the multi-functional bionanocomposite foams using nanofibrillated cellulose (NFC) to reinforce a chitosan matrix. The density of prepared nanocomposite foams could be controlled by altering the NFC/chitosan (w/w) ratio and solid content. The prepared nanocomposite NFC/chitosan foams possessed good biocompatibility.

8.2.5 Sponge

The chitosan based sponge is an efficient wound dressing material due to its porosity, favorable swelling capacity and antibacterial activity, and rapid hemostatic ability [46]. Wu et al. [47] prepared a series of ampicillin-grafted chitosan sponges via a green and facile method without any leaching. The sponges exhibited excellent biocompatibility with HEK293 cells and good antibacterial properties. At the same time, the sponges showed accelerating wound healing efficiency in vivo model experiments. Hu et al. [48] produced a nanocomposite sponge by physically mixing hydroxybutyl chitosan with chitosan to form sponge dressing via vacuum freeze-drying. The resulting sponge showed high porosity (~85%) for enhanced water absorption, and lower blood-clotting index than sponge made from pure chitosan. In addition, in vivo evaluation in Sprague–Dawley (SD) rats confirmed that epithelial cells could attach to the composite sponge and penetrate it, and thus could promote wound healing.

8.2.6 Gauze

Cotton gauze is characterized by the intrinsic properties of cotton fibers that contribute to its usefulness in wound treatment. Gauze refers to a wide range of wound dressing products which are categorized based on the fabric construction or material composition. Usually, gauze wound dressings are categorized into woven and non-woven gauze dressing [49]. Before the 1960s, gauze dressings were used for most wounds via absorbing wound exudate and allowing the wound to form an eschar. Considering that modern wound dressings require more functions than conventional cotton gauze, such as antibacterial and moisture maintaining capabilities, and so on. Abbasipour et al. [50] prepared a new cotton gauze sample by impregnating gauze with chitosan/Ag/ZnO nanocomposite liquid to achieve modern wound dressings properties using the dip, dry, and cure method. The prepared gauze dressing has increased drying time and water absorbency by 78% and 38% respectively. In addition, the gauze dressings presented excellent antibacterial efficiency (96% for *Escherichia coli (E. coli)* and 99% for *staphylococcus aureus (S. aureus)*). Another research developed chitosan-based nanocomposite dressings based on salicyl-imine-chitosan Schiff base. The cotton gauze fabrics were modified with Quat 188 as cationizing agent first, then silver nanoparticles (AgNPs) and oxytetracyline hydrochloride drug were loaded to obtain the cationized cotton gauze fabrics/AgNPs/drug. The chitosan was then reacted with salicylaldehyde (SA) via Schiff base reaction to produce salicyl-imine-chitosan biopolymer, and coated onto fabrics/AgNPs/drug. The results confirmed that the prepared gauze dressings with improved swelling rate capacity and antimicrobial ability could be efficiently used as wound healing dressings [51].

8.3 Properties of the Chitosan-Based Nanocomposites as Efficient Wound Dressings

As efficient wound dressings, chitosan-based nanocomposites can promote the physical isolation extrudate and the creation of a moist environment, and some basic

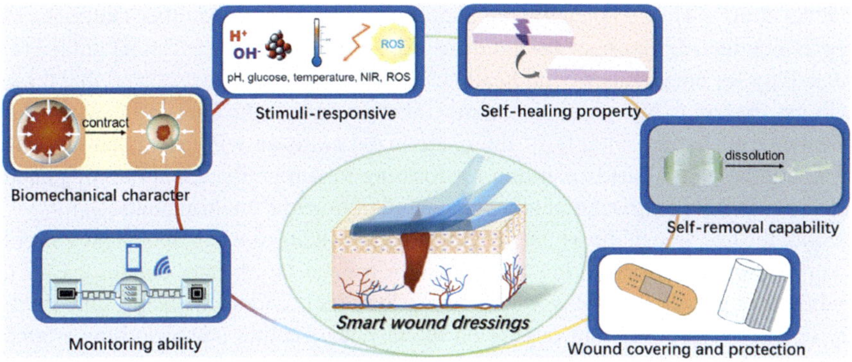

Fig. 8.4 Major functions of smart wound dressings Reproduced with permission from Elsevier [52]

properties should be proposed, including, biocompatibility, antibacterial properties, hemostasis, and so on. However, with the increase of clinical requirements for wound dressings, some advanced functions were discovered, including self-healing, adhesive and hemostatic, stimulus-responsive, anti-oxidative and anti-inflammatory, conductive, etc. Recently, the researchers developed the smart wound dressings that can use smart materials or inbuilt sensors to respond to the wound condition or external environment. Different smart wound dressings were developed, such as stimuli-responsive wound dressing, dynamic wound dressing, self-removable wound dressing, monitoring wound dressing, and biomechanical wound dressing (Fig. 8.4) [52]. In this section, some basic properties and advanced functions of chitosan-based nanocomposites as wound dressings will be summarized.

8.3.1 Basic Properties

Because the wound dressings directly contact wound tissue, the good biocompatibility and non-toxicity are the basic and essential properties for wound dressings. The biocompatibility of dressings is attributed to the biocompatible materials. Due to the excellent biocompatibility of chitosan, chitosan-based nanocomposites could display good biocompatibility. Bacterial infection hinders wound healing via increasing inflammatory reactions for slower wound healing. To reduce the risk of wound infections, the wound dressings should possess favorable antibacterial property. Currently, the materials such as antibiotic and other antibacterial drugs, cationic polymer, inherent antibacterial peptides, inorganic metals and metal oxides, and the strategies of photothermal antibacterial and photodynamic antibacterial, were introduced into dressings to achieve the antibacterial property [53]. Chitosan-based nanocomposites possess endogenous antimicrobial activity against microorganisms.

Zhao et al. [54], used quaternized chitosan-g-polyaniline and other materials, to prepare conductive hydrogel wound dressings characterized by excellent antibacterial activity for cutaneous wound healing. The experiment results showed that these hydrogel dressings had great candidates for treating full-thickness skin wound. It contributed to the advantages of the quaternized chitosan with good solubility in physiological conditions benefitting for forming in situ antibacterial dressings and excellent inherent antibacterial activity even in non-acidic environments.

Since the first step of the wound healing process is hemostasis, hemorrhage control is very important for reducing blood loss in trauma emergency conditions [55]. Studies have proved that the wound dressing hydrogels endow the hemostatic property in two ways, which are relying on the physical sealing and absorbing wound extract to enrich coagulation factors [56]. Many studies have found that some cationic polymers, anion, metal, silicon-based materials and natural polyphenols such as tea polyphenols, can effectively hemostasis. In most situations, hemostasis is often used by combining other properties, such as adhesiveness. For bioadhesive and hemostasis wound dressing, the hemostasis is provided by sealing the wound which limits the risk of wound infection by reducing contact with the external environment [53].

It was known that reactive oxygen species (ROS) that are responsible for cellular oxidant/antioxidant imbalance are abundant in the damaged wound. Moreover, overproduction of ROS will also destroy the fibroblast activity, which leads to the diabetic organ damage, infection, and slow tissue regeneration and wound healing [54, 57–59]. Dressings with an antioxidant ability play an important role in eliminating and controlling inflammation and are favorable to wound healing. Adding antioxidant materials, such as curcumin, dopamine, some conductive polymers, honey, and so on, into dressings for eliminating free radicals has gradually become an important strategy.

8.3.2 Advanced Functions

In the last few decades, several wound dressings with potential commercial and clinical applications have been developed. These wound dressing applications are however limited, for instance, the current wound dressings are usually applied in a one-for-all manner, and are unable to respond to wound environment changes. Based on such situations, more and more wound dressings with smart, efficient, advanced functions were developed [60].

The acceleration of healing using wound dressings may occur in several steps such as by triggering and enhancing several or one stage of wound healing, and promoting wound contraction [61]. For instance the design of a hydrogel with adhesive properties composed of a thermosensitive matrix that can respond to temperature changes was reported in the literature [62]. The thermosensitive poly(N-isopropylacrylamide) (PNIPAM) provides the contraction behavior. Additionally, carbodiimide-mediated reactions and the encapsulated silver nanoparticles endow the adhesiveness, and favorable mechanical and antibacterial properties. During the healing process, pH, temperature, glucose level, oxygen lever, and many other factors could affect wound healing, thus, the efficient wound dressings that can respond to these variations can positively alter the environment (wound) and thus promote efficient wound healing. Zhao et al. [63] developed a chitosan based nanocomposite dressing that could trigger glucose release based on the Schiff-base and phenylpyruvate ester reaction. In this work, phenylboronate was used to modify chitosan, and then reacted with PVA, at the same time, the reactions involving chitosan employed benzaldehyde-capped PEG as crosslinker. The prepared hydrogel displayed pH- and glucose-responsive properties, and the hydrogel as wound dressing could also facilitate the regulation of the level of glucose in the blood which provides an alternative approach for disease therapy in combination with wound healing.

Motional wounds occur in regions associated with motional stress (i.e., stretching, rotation, etc.) and require tailored wound dressings such as self-healing wound dressing. For instance, a self-healing dressing based on the quaternized chitosan that was crosslinked with benzaldehyde-terminated Pluronic F127 (PF127-CHO), was developed in the literature [64]. The prepared wound dressing was useful for motional skin defects subjected to consistent and significant stretch stress. In addition, the prepared dressing is degradable, and thus limits the discomfort of changing dressings and possible impairment of the repair. Self-removal wound dressing could also avoid these drawbacks. These self-removals wound dressings may incorporate thermosensitive polymers or light-triggered dissolutions. For instance, a wound dressing that can be UV triggered was designed and was based on glycol chitosan and PEG-4-(3-(1-(N-hydroxysuccinimidyl carbonic ester)ethyl)-4-nitrophenoxy)butanoate (PNN) [65]. In the study, crosslinking was achieved via the reaction between the amino group and the N-hydroxysuccinimide carbonate in Glycol chitosan and PNN respectively under the action of UV light [65]. Even, the prepared dressing can be partially removed to retain its pattern while shielding with a patterned tinfoil.

Recently, the wound dressing that can monitor wound conditions, such as glucose level change monitoring, was developed [66, 67]. Wound dressings with monitoring ability are desirable in infection control. Wound site infections may lead to the variations in conditions (i.e. pH), with these changes enabling the monitoring of the wound site for treatments in the early stage.

8.4 Chitosan-Based Nanocomposites as Wound Dressings

8.4.1 Treatment of Acute Wound

The "acute wound" usually refers to the type of wound characterized by the introduction of injury or damage and a rapid repair process [53]. But it also should be given enough attention because it can develop to chronic wound due to delay treatment.

8.4.2 *Treatment of Chronic Wound*

In the healing process, chronic wound may be formed due to many internal and external factors of the wound, such as pressure ulcers, infections, diabetes, frequent movement of motion wounds (joint, elbows, wrists, nape, and knees), which can hinder wound healing [64, 68–72].

8.4.2.1 Diabetic Foot Ulcers

For diabetic patients, the vascular lesions around the lower extremities result in poor blood supply to foot wounds, contribute to the poor healing of diabetic foot wounds [73]. The treatment of the diabetic foot ulcers is difficult, bringing a very considerable health care burden for patients and social health care systems.

8.4.2.2 Infected Wound

Infected wound is one of the typical chronic wounds caused by microbial infection, which is a significant obstacle to wound healing, finally leading to long-term unrepaired wound, non-functional scar, and other complications, even death in severe cases [74]. Traditional treatment usually uses a large number of antibiotics, finally leading serious drug-resistance, which threats the human life and increases the burden of treating. Therefore, there are many researchers focusing on the research of substitutes for antibiotics and more advanced biomaterials for infected wound.

Liang et al., using chitosan, carbon nanotubes (CNT), and gelatin-grafted-dopamine, prepared several antibacterial, conductive, injectable, antioxidant composite hydrogel dressings to enhance the infected wounds healing [74]. And they also used glycidyl methacrylate functionalized quaternized chitosan and gelatin methacrylate and prepared conductive injectable multi-antibacterial hydrogel dressings suitable for infectious skin tissue defect repair [75].

8.4.2.3 Burns

In daily life, due to the duration exposure and the total body area exposed to the thermal source, burns vary. Moreover, burns are a common hazard in the military [76]. It is therefore necessary to find an effective substance to promote and hasten burn wounds healing. During the treatment of burns, the clinician's goal is to provide comfort to the patient, speed up proper wound healing in order to minimal scarring, and minimize antimicrobial infection to short healing time [77].

Cherng et al. [76] using chitosan and gold nanoparticles (AuNPs), prepared a series of wound dressings. They clarified that chitosan alone or AuNP-coated chitosan dressings represented a prospective therapeutic substance for burn wound care because of their excellent hemorrhage and regulation of the expression of genes- and cytokines/chemokines-related.

8.4.3 Scarless Wound Healing

A difference exists in the wound healing processes between adult and fetal in the extracellular matrix, inflammatory response, cellular mediators, and gene expression profiles. While human embryonic skin wound repair is occurring without scar formation, the adult suffers from slow healing and scars, which is the result of abnormal wound-healing response after skin injury and may bring significant physical and psychological impact for adult patients [2, 78–80]. The process of wound healing and scarring formation involves many molecular, biological, and mechanical factors [78]. Although some current clinical strategies have a little effect in remedying for scarring, such as surgical excision, thermal laser treatment, gene therapy, drug application, and so on, it is difficult to eliminate scars completely [78, 81]. As a preferred wound dressing, except accelerating wound closure and promoting wound healing, it's better that the dressing can reduce the scar formation [54].

8.4.4 Wound Monitoring

During the wound healing process, some parameters are dynamically changing, e.g., pH, humidity, temperature, glucose, and uric acid [82, 83]. Then, it is essential to monitor wound for better wound management and treatment. Monitoring the pH value changes of a wound can get many useful informations about the wound healing. The pH of healthy skin is approximately 5–5.5, slightly acidic. On contrary, because of the existing of different types of enzymes and bacteria, the infected wound's pH ranges from 7 to 8.5, neutral or slightly basic [84]. Although different pH values may benefit wound healing at different healing stages, the mechanism of the dependence of wound healing on pH remains unclear.

Real time monitoring the temperature of a wound is crucial since local temperature helps to predict the risk of infection. Wound repair is hindered when the temperature of a wound is <33 °C. Due to local congestion and inflammation, the local temperature of acute wounds may be >37 °C. Otherwise, it is regarded as a signal of infection when occurs a sudden increase in wound temperature [85]. In a word, monitoring wound not only simply reflects the wound status but also achieves early diagnosis, infection diagnosis, on-demand drug delivery and therapy, even digital remote diagnosis and treatment.

8.4.5 Personalized Wound Dressings

Although varieties of wound dressings have been developed, yet there does not exist a single dressing that is fully available for treatment of all wounds. Otherwise, treatment of certain wound types and patients requires patient-specific antibiotics or antimicrobial agents, even patient-specific bioactive materials. These problems increase the difficulty of dressing efficiency applied to wounds. Hence, it is necessary to achieve personalized dressings for different wound types to enhance healing [4]. An idea and emerging strategy to prepare personalized dressings with low cost, high precision, customization and effectively is 3D printing. It can combine a variety of compositions and bioactive agents to manufacture personalized and multifunctional wound dressings.

Alizadehgiashi et al. [4], prepared 3D-printed multicomponent biocomposite hydrogel wound dressings by mixing chitosan methacrylamide, cellulose nanocrystals and some bioactive agents like proteins, small molecules, or nanoparticles. These dressings not only can selectively load bioactive agents for controlled release at the wound site but also can exhibit varied physiological responses in vivo which is potential for personalized treatment of various wound types.

Considering the complex burns circumstances, it is vital to provide optimal treatment for burn victims who need personalized medicine and dressings. Teoh et al. [86] used synthesized chitosan methacrylate as said material with great printability, biodegradability, and biocompatible properties to prepare personalized and photo-crosslinkable hydrogel wound dressings for the treatment of thermal burns by a 3D printer. The results showed that these dressings loaded various drugs relevant to the treatment of burns and contained different dosages did not significantly influence printability. Moreover, the dressings incorporated antimicrobial agents to enhance its antibacterial capabilities. In vivo models, the personalized 3D printing hydrogel wound dressings containing relevant burns drugs and antimicrobial agents have shown great wound healing results with no side effects in the whole healing process.

8.5 Summary

Skin is a barrier to protect human body. Once the skin faces serious defects, especially the extensive full-thickness defect needing longer time to repair, the wound will impact people's life and health severely. Over the past decade, a range of chitosan-based nanocomposites as dressings with desirable properties are developed, and some of them are expected to expand the use in wound healing in combination with disease therapy. Although the currently available dressings can provide an excellent environment to promote wound repair, they still exist some problems limiting their widely applications. For example, some dressings need surgical or mechanical separation from the wound and this process will cause the second injury of the newly formed tissues, leading to wound infection and delaying healing time. And the dressings also need add more function to meet wound treatment increasing demand. Overall, the emerging challenges will be resolved via the use of multi-disciplinary technologies, including nanotechnologies, polymer, tissue engineering, drug delivery, and so on. It is believed that the exploration of chitosan-based nanocomposites as efficient wound dressing materials will be sustained in the near future, and more dressings would promote transition to clinical trials from laboratory work.

Acknowledgements The authors acknowledge the support from the Nanhu Scholars Program for Young Scholars of XYNU.

Conflicts of Competing Interest The authors declare no competing financial interest.

References

1. Y. Yang, Y. Liang, J. Chen, X. Duan, B. Guo, Mussel-inspired adhesive antioxidant antibacterial hemostatic composite hydrogel wound dressing via photo-polymerization for infected skin wound healing. Bioact. Mater. **8**, 341–354 (2022)
2. M. Li, Y.P. Liang, J.H. He, H.L. Zhang, B.L. Guo, Two-pronged strategy of biomechanically active and biochemically multifunctional hydrogel wound dressing to accelerate wound closure and wound healing. Chem. Mater. **32**(23), 9937–9953 (2020)
3. R. Xu, G. Luo, H. Xia, W. He, J. Zhao, B. Liu, J. Tan, J. Zhou, D. Liu, Y. Wang, Z. Yao, R. Zhan, S. Yang, J. Wu, Novel bilayer wound dressing composed of silicone rubber with particular micropores enhanced wound re-epithelialization and contraction. Biomaterials **40**, 1–11 (2015)
4. M. Alizadehgiashi, C.R. Nemr, M. Chekini, D. Pinto Ramos, N. Mittal, S.U. Ahmed, N. Khuu, S.O. Kelley, E. Kumacheva, Multifunctional 3D-printed wound dressings, ACS Nano. (2021)
5. B.K. Sun, Z. Siprashvili, P.A. Khavari, Advances in skin grafting and treatment of cutaneous wounds. Science **346**(6212), 941–945 (2014)
6. I. De Luca, P. Pedram, A. Moeini, P. Cerruti, G. Peluso, A. Di Salle, N. Germann, Nanotechnology development for formulating essential oils in wound dressing materials to promote the wound-healing process: a review. Appl. Sci. **11**(4), 1713 (2021)
7. Y. Cai, Z. Zhong, C. He, H. Xia, Q. Hu, Y. Wang, Q. Ye, J. Zhou, Homogeneously synthesized hydroxybutyl chitosans in alkali/urea aqueous solutions as potential wound dressings. ACS Appl. Bio. Mater. **2**(10), 4291–4302 (2019)
8. S. Homaeigohar, A.R. Boccaccini, Antibacterial biohybrid nanofibers for wound dressings. Acta Biomater. **107**, 25–49 (2020)
9. Y. Cai, J. Cao, C. Xu, J. Zhou, Thermo responsive behaviors and bioactivities of hydroxybutyl chitosans prepared in alkali/urea aqueous solutions. Carbohydr. Polym. **215**, 90–98 (2019)
10. M. Rinaudo, Chitin and chitosan: properties and applications. Prog. Polym. Sci. **31**(7), 603–632 (2006)
11. Y. Yi, T. Huang, Y.X. Xu, J.F. Mei, Y.L. Zhang, X.D. Wang, J.S. Chen, G.Q. Ying, Preparation and properties of N-glycosylated chitosan/polyvinyl alcohol hydrogels for use in wound dressings, J. Appl. Biomater. & Funct. Mater. **20** (2022)
12. W. Peng, D. Li, K.L. Dai, Y.X. Wang, P. Song, H.R. Li, P. Tang, Z.Y. Zhang, Z.Y. Li, Y.C. Zhou, C.C. Zhou, Recent progress of collagen, chitosan, alginate and other hydrogels in skin repair and wound dressing applications. Int. J. Biol. Macromol. **208**, 400–408 (2022)
13. G.I. Howling, P.W. Dettmar, P.A. Goddard, F.C. Hampson, M. Dornish, E.J. Wood, The effect of chitin and chitosan on the proliferation of human skin fibroblasts and keratinocytes in vitro. Biomaterials **22**(22), 2959–2966 (2001)

14. R.-N. Chen, G.-M. Wang, C.-H. Chen, H.-O. Ho, M.-T. Sheu, Development of N,O-(carboxymethyl)chitosan/collagen matrixes as a wound dressing. Biomacromol. **7**(4), 1058–1064 (2006)
15. S. Liu, Y. Zhao, H. Wei, L. Nie, P. Ding, H. Sun, Y. Guo, T. Chen, O.V. Okoro, A. Shavandi, Injectable hydrogels based on silk fibroin peptide grafted hydroxypropyl chitosan and oxidized microcrystalline cellulose for scarless wound healing, Colloids Surf. A: Phys. Chem. Eng. Asp. (2022) 129062
16. H. Jafari, C. Delporte, K.V. Bernaerts, G. De Leener, M. Luhmer, L. Nie, A. Shavandi, Development of marine oligosaccharides for potential wound healing biomaterials engineering. Chem. Eng. J. Adv. **7**, 100113 (2021)
17. J. Cao, J. You, L. Zhang, J. Zhou, Homogeneous synthesis and characterization of chitosan ethers prepared in aqueous alkali/urea solutions. Carbohyd. Polym. **185**, 138–144 (2018)
18. S. Hamdan, I. Pastar, S. Drakulich, E. Dikici, M. Tomic-Canic, S. Deo, S. Daunert, Nanotechnology-driven therapeutic interventions in wound healing: potential uses and applications. ACS Cent. Sci. **3**(3), 163–175 (2017)
19. P. Zarrintaj, A.S. Moghaddam, S. Manouchehri, Z. Atoufi, A. Amiri, M.A. Amirkhani, M.A. Nilforoushzadeh, M.R. Saeb, M.R. Hamblin, M. Mozafari, Can regenerative medicine and nanotechnology combine to heal wounds? The search for the ideal wound dressing. Nanomedicine **12**(19), 2403–2422 (2017)
20. P. Zou, J. Suo, L. Nie, S. Feng, Temperature-sensitive biodegradable mixed star-shaped block copolymers hydrogels for an injection application. Polymer **53**(6), 1245–1257 (2012)
21. L. Nie, P. Zou, S. Feng, J. Suo, Temperature-sensitive star-shaped block copolymers hydrogels for an injection application: phase transition behavior and biocompatibility. J. Mater. Sci.—Mater. Med. **24**(3), 689–700 (2013)
22. L. Nie, X. Li, Z. Wang, K. Hu, R. Cai, P. Li, Y. Han, M. Sun, H. Yuan, J. Suo, In vitro biomineralization on poly (vinyl alcohol)/biphasic calcium phosphate hydrogels. Bioinspired, Biomim. Nanobiomaterials **9**(2), 122–128 (2020)
23. A. Shavandi, S. Hosseini, O.V. Okoro, L. Nie, F.E. Babadi, F. Melchels, 3D bioprinting of lignocellulosic biomaterials, Adv. Healthc. Mater. **9**(24) (2020)
24. M. Mirzaei, O.V. Okoro, L. Nie, D.F.S. Petri, A. Shavandi, Protein-Based 3D Biofabrication of Biomaterials, Bioeng.-Basel **8**(4) (2021)
25. Y.T. Han, Q.Q. Wei, P.B. Chang, K.H. Hu, O.V. Okoro, A. Shavandi, L. Nie, Three-dimensional printing of hydroxyapatite composites for biomedical application, Crystals **11**(4) (2021)
26. N. Asadi, H. Pazoki-Toroudi, A.R. Del Bakhshayesh, A. Akbarzadeh, S. Davaran, N. Annabi, Multifunctional hydrogels for wound healing: Special focus on biomacromolecular based hydrogels. Int. J. Biol. Macromol. **170**, 728–750 (2021)
27. W.J. Zheng, J. Gao, Z. Wei, J. Zhou, Y.M. Chen, Facile fabrication of self-healing carboxymethyl cellulose hydrogels. Eur. Polymer J. **72**, 514–522 (2015)
28. Y. Huang, X. Zhao, Z. Zhang, Y. Liang, Z. Yin, B. Chen, L. Bai, Y. Han, B. Guo, Degradable gelatin-based IPN cryogel hemostat for rapidly stopping deep noncompressible hemorrhage and simultaneously improving wound healing. Chem. Mater. **32**(15), 6595–6610 (2020)
29. C. Wang, Y. Liang, Y. Huang, M. Li, B. Guo, Porous photothermal antibacterial antioxidant dual–crosslinked cryogel based on hyaluronic acid/ polydopamine for non-compressible hemostasis and infectious wound repair. J. Mater. Sci. Technol. **121**, 207–219 (2022)
30. Y. Huang, L. Bai, Y. Yang, Z. Yin, B. Guo, Biodegradable gelatin/silver nanoparticle composite cryogel with excellent antibacterial and antibiofilm activity and hemostasis for Pseudomonas aeruginosa-infected burn wound healing. J. Colloid. Interface Sci. **608**(Pt 3), 2278–2289 (2022)
31. M. Li, Z. Zhang, Y. Liang, J. He, B. Guo, Multifunctional tissue-adhesive cryogel wound dressing for rapid nonpressing surface hemorrhage and wound repair. ACS Appl. Mater. Interfaces **12**(32), 35856–35872 (2020)
32. X. Zhao, B. Guo, H. Wu, Y. Liang, P.X. Ma, Injectable antibacterial conductive nanocomposite cryogels with rapid shape recovery for noncompressible hemorrhage and wound healing. Nat. Commun. **9**(1), 2784 (2018)

33. X. Zhao, Y. Liang, B. Guo, Z. Yin, D. Zhu, Y. Han, Injectable dry cryogels with excellent blood-sucking expansion and blood clotting to cease hemorrhage for lethal deep-wounds, coagulopathy and tissue regeneration, Chem. Eng. J. **403** (2021)
34. H.K. Noh, S.W. Lee, J.-M. Kim, J.-E. Oh, K.-H. Kim, C.-P. Chung, S.-C. Choi, W.H. Park, B.-M. Min, Electrospinning of chitin nanofibers: degradation behavior and cellular response to normal human keratinocytes and fibroblasts. Biomaterials **27**(21), 3934–3944 (2006)
35. X. Liu, T. Lin, J. Fang, G. Yao, H. Zhao, M. Dodson, X. Wang, In vivo wound healing and antibacterial performances of electrospun nanofibre membranes. J. Biomed. Mater. Res., Part A **94**(2), 499–508 (2010)
36. J. He, Y. Liang, M. Shi, B. Guo, Anti-oxidant electroactive and antibacterial nanofibrous wound dressings based on poly (ε-caprolactone)/quaternized chitosan-graft-polyaniline for full-thickness skin wound healing. Chem. Eng. J. **385**, 123464 (2020)
37. A. Saatchi, A.R. Arani, A. Moghanian, M. Mozafari, Cerium-doped bioactive glass-loaded chitosan/polyethylene oxide nanofiber with elevated antibacterial properties as a potential wound dressing. Ceram. Int. **47**(7), 9447–9461 (2021)
38. J. He, Y. Liang, M. Shi, B. Guo, Anti-oxidant electroactive and antibacterial nanofibrous wound dressings based on poly(ε-caprolactone)/quaternized chitosan-graft-polyaniline for full-thickness skin wound healing, Chem. Eng. J. **385** (2020)
39. C.L. Liu, Y.H. Zhu, X.J. Lun, H. Sheng, A.Q. Yan, Effects of wound dressing based on the combination of silver@curcumin nanoparticles and electrospun chitosan nanofibers on wound healing. Bioengineered **13**(2), 4328–4339 (2022)
40. S.M. Lee, I.K. Park, Y.S. Kim, H.J. Kim, H. Moon, S. Mueller, Y.-I. Jeong, Physical, morphological, and wound healing properties of a polyurethane foam-film dressing. Biomater. Res. **20**(1), 1–11 (2016)
41. M. Sahraro, H. Yeganeh, M. Sorayya, Guanidine hydrochloride embedded polyurethanes as antimicrobial and absorptive wound dressing membranes with promising cytocompatibility. Mater. Sci. Eng., C **59**, 1025–1037 (2016)
42. X. Liu, Y. Niu, K.C. Chen, S. Chen, Rapid hemostatic and mild polyurethane-urea foam wound dressing for promoting wound healing. Mater. Sci. Eng., C **71**, 289–297 (2017)
43. A. Bužarovska, S. Dinescu, A.D. Lazar, M. Serban, G.G. Pircalabioru, M. Costache, C. Gualandi, L. Avérous, Nanocomposite foams based on flexible biobased thermoplastic polyurethane and ZnO nanoparticles as potential wound dressing materials. Mater. Sci. Eng., C **104**, 109893 (2019)
44. Y. Wang, K. Uetani, S. Liu, X. Zhang, Y. Wang, P. Lu, T. Wei, Z. Fan, J. Shen, H. Yu, Multifunctional bionanocomposite foams with a chitosan matrix reinforced by nanofibrillated cellulose. ChemNanoMat **3**(2), 98–108 (2017)
45. X. Huang, Y. Sun, J. Nie, W. Lu, L. Yang, Z. Zhang, H. Yin, Z. Wang, Q. Hu, Using absorbable chitosan hemostatic sponges as a promising surgical dressing. Int. J. Biol. Macromol. **75**, 322–329 (2015)
46. L.W. Chan, C.H. Kim, X. Wang, S.H. Pun, N.J. White, T.H. Kim, PolySTAT-modified chitosan gauzes for improved hemostasis in external hemorrhage. Acta Biomater. **31**, 178–185 (2016)
47. J.M. Wu, C. Su, L. Jiang, S. Ye, X.F. Liu, W. Shao, Green and facile preparation of chitosan sponges as potential wound dressings. ACS Sustain. Chem. & Eng. **6**(7), 9145–9152 (2018)
48. S. Hu, S. Bi, D. Yan, Z. Zhou, G. Sun, X. Cheng, X. Chen, Preparation of composite hydroxybutyl chitosan sponge and its role in promoting wound healing. Carbohyd. Polym. **184**, 154–163 (2018)
49. V.J. Jones, The use of gauze: will it ever change? Int. Wound J. **3**(2), 79–88 (2006)
50. M. Abbasipour, M. Mirjalili, R. Khajavi, M.M. Majidi, Coated cotton gauze with Ag/ZnO/chitosan nanocomposite as a modern wound dressing. J. Eng. Fibers Fabr. **9**(1), 124–130 (2014)
51. A. Montaser, M. Rehan, W. El-Senousy, S. Zaghloul, Designing strategy for coating cotton gauze fabrics and its application in wound healing. Carbohyd. Polym. **244**, 116479 (2020)
52. R. Dong, B. Guo, Smart wound dressings for wound healing. Nano Today **41**, 101290 (2021)

53. Y.P. Liang, J.H. He, B.L. Guo, Functional hydrogels as wound dressing to enhance wound healing. ACS Nano **15**(8), 12687–12722 (2021)
54. X. Zhao, H. Wu, B.L. Guo, R.N. Dong, Y.S. Qiu, P.X. Ma, Antibacterial anti-oxidant electroactive injectable hydrogel as self-healing wound dressing with hemostasis and adhesiveness for cutaneous wound healing. Biomaterials **122**, 34–47 (2017)
55. J. Zhu, H. Han, F. Li, X. Wang, J. Yu, X. Qin, D. Wu, Peptide-functionalized amino acid-derived pseudoprotein-based hydrogel with hemorrhage control and antibacterial activity for wound healing. Chem. Mater. **31**(12), 4436–4450 (2019)
56. J. Zhu, F. Li, X. Wang, J. Yu, D. Wu, Hyaluronic acid and polyethylene glycol hybrid hydrogel encapsulating nanogel with hemostasis and sustainable antibacterial property for wound healing. ACS Appl. Mater. Interfaces. **10**(16), 13304–13316 (2018)
57. H. Cui, L. Cui, P. Zhang, Y. Huang, Y. Wei, X. Chen, In situ electroactive and antioxidant supramolecular hydrogel based on cyclodextrin/copolymer inclusion for tissue engineering repair. Macromol. Biosci. **14**(3), 440–450 (2014)
58. W. Chen, X. Shen, Y. Hu, K. Xu, Q. Ran, Y. Yu, L. Dai, Z. Yuan, L. Huang, T. Shen, Surface functionalization of titanium implants with chitosan-catechol conjugate for suppression of ROS-induced cells damage and improvement of osteogenesis. Biomaterials **114**, 82–96 (2017)
59. C. Dunnill, T. Patton, J. Brennan, J. Barrett, M. Dryden, J. Cooke, D. Leaper, N.T. Georgopoulos, Reactive oxygen species (ROS) and wound healing: the functional role of ROS and emerging ROS-modulating technologies for augmentation of the healing process. Int. Wound J. **14**(1), 89–96 (2017)
60. A. Benedetto, J. Staidle, J. Schoenfeld, E. Benedetto, P. Benedetto, Comparing the use of a novel antibiotic-free film-forming topical wound dressing versus a topical triple antibiotic in dermatologic surgical procedures including Mohs micrographic surgery. J. Eur. Acad. Dermatol. Venereol. **35**(1), 247–255 (2021)
61. E. Brauer, E. Lippens, O. Klein, G. Nebrich, S. Schreivogel, G. Korus, G.N. Duda, A. Petersen, Collagen fibrils mechanically contribute to tissue contraction in an in vitro wound healing scenario. Adv. Sci. **6**(9), 1801780 (2019)
62. S. Blacklow, J. Li, B. Freedman, M. Zeidi, C. Chen, D. Mooney, Bioinspired mechanically active adhesive dressings to accelerate wound closure, Sci. Adv. **5**(7) (2019) eaaw3963
63. L. Zhao, L. Niu, H. Liang, H. Tan, C. Liu, F. Zhu, pH and glucose dual-responsive injectable hydrogels with insulin and fibroblasts as bioactive dressings for diabetic wound healing. ACS Appl. Mater. Interfaces. **9**(43), 37563–37574 (2017)
64. J. Qu, X. Zhao, Y. Liang, T. Zhang, P.X. Ma, B. Guo, Antibacterial adhesive injectable hydrogels with rapid self-healing, extensibility and compressibility as wound dressing for joints skin wound healing. Biomaterials **183**, 185–199 (2018)
65. H. Wu, Z. Qin, X. Yu, J. Li, H. Lv, X. Yang, On-demand removable hydrogels based on photolabile cross-linkings as wound dressing materials. J. Mater. Chem. B **7**(37), 5669–5676 (2019)
66. C. Withycombe, K.J. Purdy, S.E. Maddocks, Micro-management: curbing chronic wound infection. Mol. Oral Microbiol. **32**(4), 263–274 (2017)
67. T. Yu, M. Gao, P. Yang, D. Liu, D. Wang, F. Song, X. Zhang, Y. Liu, Insulin promotes macrophage phenotype transition through PI3K/Akt and PPAR-γ signaling during diabetic wound healing. J. Cell. Physiol. **234**(4), 4217–4231 (2019)
68. M. Li, Y. Liang, Y. Liang, G. Pan, B. Guo, Injectable stretchable self-healing dual dynamic network hydrogel as adhesive anti-oxidant wound dressing for photothermal clearance of bacteria and promoting wound healing of MRSA infected motion wounds, Chem. Eng. J. **427** (2022)
69. S. Li, L. Wang, W. Zheng, G. Yang, X. Jiang, Rapid fabrication of self-healing, conductive, and injectable gel as dressings for healing wounds in stretchable parts of the body. Adv. Func. Mater. **30**(31), 2002370 (2020)
70. C. Ge, R. Wu, Y. Chong, G. Fang, X. Jiang, Y. Pan, C. Chen, J.-J. Yin, Synthesis of Pt hollow nanodendrites with enhanced peroxidase-like activity against bacterial infections: implication for wound healing. Adv. Func. Mater. **28**(28), 1801484 (2018)

71. Y. Mori, G. Nakagami, A. Kitamura, T. Minematsu, M. Kinoshita, H. Suga, M. Kurita, C. Hayashi, A. Kawasaki, H. Sanada, Effectiveness of biofilm-based wound care system on wound healing in chronic wounds. Wound Repair and Regeneration **27**(5), 540–547 (2019)
72. J. Xiao, S. Chen, J. Yi, H.F. Zhang, G.A. Ameer, A cooperative copper metal-organic framework-hydrogel system improves wound healing in diabetes. Adv. Func. Mater. **27**(1), 1604872 (2017)
73. D.G. Armstrong, A.J.M. Boulton, S.A. Bus, Diabetic foot ulcers and their recurrence. N. Engl. J. Med. **376**(24), 2367–2375 (2017)
74. Y.P. Liang, X. Zhao, T.L. Hu, Y. Han, B.L. Guo, Mussel-inspired, antibacterial, conductive, antioxidant, injectable composite hydrogel wound dressing to promote the regeneration of infected skin. J. Colloid Interface Sci. **556**, 514–528 (2019)
75. Y. Liang, B. Chen, M. Li, J. He, Z. Yin, B. Guo, Injectable antimicrobial conductive hydrogels for wound disinfection and infectious wound healing. Biomacromol **21**(5), 1841–1852 (2020)
76. J.H. Cherng, C.A.J. Lin, C.C. Liu, J.Z. Yeh, G.Y. Fan, H.D. Tsai, C.F. Chung, S.D. Hsu, Hemostasis and anti-inflammatory abilities of AuNPs-coated chitosan dressing for burn wounds, J. Pers. Med. **12**(7) (2022)
77. Q. Jiang, Z.-H. Chen, S.-B. Wang, X.-D. Chen, Comparative effectiveness of different wound dressings for patients with partial-thickness burns: study protocol of a systematic review and a Bayesian framework network meta-analysis. BMJ Open **7**(3), e013289 (2017)
78. A. El Ayadi, J.W. Jay, A. Prasai, Current Approaches Targeting the Wound Healing Phases to Attenuate Fibrosis and Scarring, Int. J. Mol. Sci. **21**(3) (2020)
79. A.E. Stoica, A.M. Grumezescu, A.O. Hermenean, E. Andronescu, B.S. Vasile, Scar-free healing: current concepts and future perspectives, Nanomaterials **10**(11) (2020)
80. P. Martin, J. Lewis, Actin cables and epidermal movement in embryonic wound healing. Nature **360**(6400), 179–183 (1992)
81. J. Wang, J. Lin, L. Chen, L. Deng, W. Cui, Endogenous electric fields-coupled electrospun short fiber via collecting wound exudation, Adv. Mater. **34** (2022) 2108325
82. R. Dong, B. Guo, Smart wound dressings for wound healing, Nano Today **41** (2021)
83. G. Xu, Y. Lu, C. Cheng, X. Li, J. Xu, Z. Liu, J. Liu, G. Liu, Z. Shi, Z. Chen, F. Zhang, Y. Jia, D. Xu, W. Yuan, Z. Cui, S.S. Low, Q. Liu, Battery-free and wireless smart wound dressing for wound infection monitoring and electrically controlled on-demand drug delivery, Adv. Funct. Mater. **31**(26) (2021)
84. T.R. Dargaville, B.L. Farrugia, J.A. Broadbent, S. Pace, Z. Upton, N.H. Voelcker, Sensors and imaging for wound healing: a review. Biosens. Bioelectron. **41**, 30–42 (2013)
85. Z. Tu, M. Chen, M. Wang, Z. Shao, X. Jiang, K. Wang, Z. Yao, S. Yang, X. Zhang, W. Gao, Engineering bioactive M2 macrophage-polarized anti-inflammatory, antioxidant, and antibacterial scaffolds for rapid angiogenesis and diabetic wound repair. Adv. Func. Mater. **31**(30), 2100924 (2021)
86. J.H. Teoh, A. Mozhi, V. Sunil, S.M. Tay, J. Fuh, C.H. Wang, 3D printing personalized, photocrosslinkable hydrogel wound dressings for the treatment of thermal burns, Adv. Funct. Mater. **31**(48) (2021)

Chapter 9
Chitosan-Based Hybrid Dressing Materials for Treatment of Diabetic Wounds

Feng Cheng, Jiliang Dai, and Hongbin Li

Abstract Diabetic wounds are not easy to repair mainly attributed to their complicated wound situation which can lead to the injure wound further infection and even death of patients. Chitosan-based hybrid dressing materials have recently attracted considerable attention as excellent potential biomedical materials with excellent hydrophilicity, absorbability, anti-inflammatory, and antibacterial properties, and substantial contribution to tissue repairment and regeneration for treating diabetes wounds and accelerating wound healing. In this chapter, we analyze the preparation of the chitosan-based hybrid dressing materials, as well as the applications of these wound dressing toward diabetes wounds. It is believed that chitosan-based hybrid dressing materials will play an essential role in the field of treating diabetic wounds due to their unique performances, such as excellent biocompatibility for increasing the angiogenesis and collagen deposition, cell proliferation, and promoting the aggregation of growth factors, as well as the antibacterial performance for inhibiting the wound inflammation surround the injured wound. The present chapter reveals an overall idea regarding the introduction to chitosan-based hybrid dressing materials for the treatment of diabetic wounds.

9.1 Introduction

Health has always been the heart of people's concerns. Of all chronic diseases, diabetes mellitus (DM) has been proven to be a metabolic chronic disease which can cause a great pain and discomfort to patients mainly attributed to the hyperglycemia

F. Cheng · H. Li (✉)
MIIT Key Laboratory of Critical Materials Technology for New Energy Conversion and Storage, School of Chemistry and Chemical Engineering, Harbin Institute of Technology, Harbin 150001, PR China
e-mail: lhb987258@163.com

J. Dai · H. Li
College of Light Industry and Textile, Qiqihar University, Qiqihar, Heilongjiang 161000, PR China

S. K. Swain and A. Biswal (eds.), *Chitosan Nanocomposites*,
Biological and Medical Physics, Biomedical Engineering,
https://doi.org/10.1007/978-981-19-9646-7_9

(high levels of glucose in the blood) caused by defective insulin secretion, insulin action, or both [1–4]. For the DM patients, the long-term insulin deficiency can further affect patient's organs or tissues and form many complications to threaten people health, such as neuropathy [5–7], cardiovascular diseases [8–10], nephropathy [11, 12], ocular diseases, and ulcerations [13–15]. Among those the complications of DM, the diabetic foot ulcers (DFUs) mainly attributed to the chronic non-healing foot ulcerations is considered to be a major health impact [16, 17], which can significantly increase the risk of amputation and result from hospitalization of diabetes patients [18, 19].

Human skin acts as the main defense organ is to be protected from any damage by the external environment (such as the pollutants, bacteria, and sharps) in our daily life [20–22]. The regular wound injury is always easy to repair and regeneration through using the regular wound dressing (such as the pure cotton fabric) to cover the wound site to avoid contact with the external environment and protect from wound infection [23, 24]. However, the wound treating of the chronic wound result from DM mainly includes the following processes: (i) cleaning of the injured wound site, (ii) debridement of necrotic tissue, (iii) using the appropriate wound dressings or antibiotics for decreasing the injured wound infection, compression, and (iv) hyperbaric oxygen therapy [25–28]. Moreover, the nonhealing of the DM wound mainly due to the synergistic effect between extracellular matrix (ECM) degrading protease enzymes and disruption of proliferating cells, which can further damage the regeneration of the new ECM due to the accumulation of macrophages and inflammatory factors to enhance inflammation situation during wound healing of DM [29–32]. Therefore, the wound repair and regeneration of the DM need more time than the common wound mainly due to the poorly healing wounds easy to pathogen invasion, bacterial infection and formation biofilm, and further injury to the surrounding healthy tissue [33–35].

In order to address the non healing issue of the DM wound, the nature polysaccharides (such as cellulose, chitosan and their modified products, alginate, starch, and other natural polymer materials [36, 37]), as biological polymers, have been widely researched and modified by various technology to design and fabricate the multifunctional wound dressing for improving wound healing in chronic wounds [38, 39]. Among those natural polymeric materials, chitosan, as a polysaccharide, obtained from the shells of crustaceans that has recently received extensive attention as a biomedical material, which has been widely selected to design and develop the wound dressing with antibacterial performance, hemostasis ability, anti-injection and wound healing for various wound repair and regeneration (such as wound injured, wound injection, DFU, and other complications) [40–43]. Moreover, due to the absorbability, biocompatibility, low immunogenicity, and ease of production or modification of chitosan-based biomaterials, many chitosan-based hybrid dressing materials has been designed and developed for the treatment of DM wounds via various strategies [44, 45]. Most significantly, chitosan can obviously improve the stem cell differentiation and growth factors release surrounding the injured wound area during the progress of the wound repair and regeneration, especially synergy with other polymer biomedical materials [46–48]. Due to the insolublity in water, hydrophilic modification of chitosan is important to improve the biodegradability of the chitosan-based hybrid wound dressing. Therefore, in order to overcome that above-mentioned issues, many modified chitosan-based functional materials are further to be developed for treating the diabetic wound, such as succinyl chitosan [49], zwitterionized chitosan [50], carboxymethyl chitosan [51], N-carboxymethyl chitosan, 5-methyl pyrrolidinone chitosan, and N-succinyl chitosan [52]. Therefore, these chitosan-based hybrid materials show high potential in designing various wound dressings for chronic wound repair and regeneration.

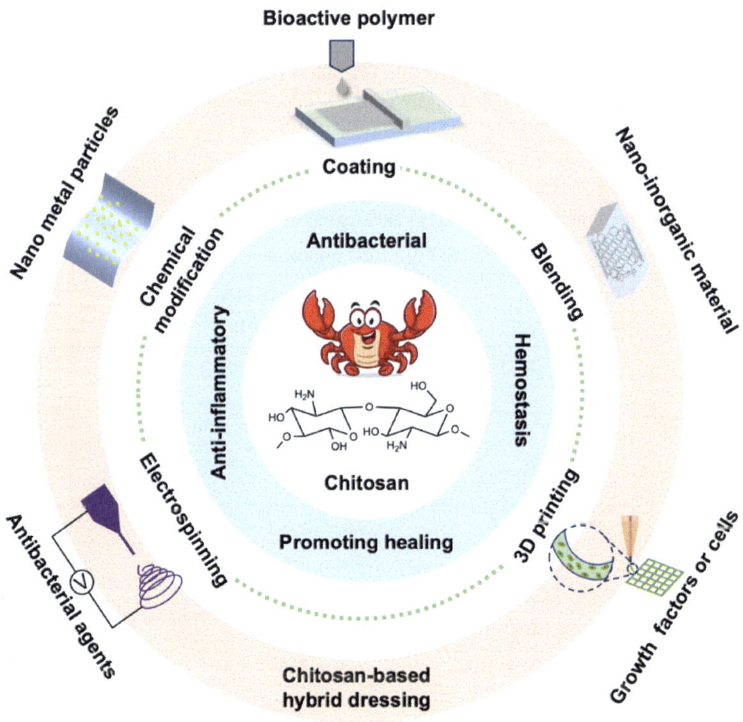

Fig. 9.1 Fabrication of the chitosan-based hybrid materials via various technologies

With the development of medical processing technology, the wound dressings with multi-functional properties are widely applied in clinical trials and wound repair [53–55]. According to the actual condition of the wound, the traditional wound dressings usually include gauze, bandages, cotton, and tulle used in the surgery operation [56–58]. Nowadays, chitosan-based hybrid dressing materials can be designed and manufactured in various structure (for examples, films, foams, hydrogels, scaffolds, and fibers) to treat the chronic wounds [59–61]. Although dry wound dressings (cotton fabrics and gauzes) have been used to cover the initial wound site for a long time, however, the traditional dry dressing materials were convenient to adhere to the injured wound site. Due to the weak ability to maintain moist environment around the wound and even finally cause secondary traumas when need remove from the dressing material from the wound area [62–64]. Especially for DM wound, the non healing characteristics makes the wound maintain a wet state for a long time which not conducive for using dry wound dressing [65, 66]. Based on the current efforts for treating the DM wound repair and regeneration, chitosan-based hybrid dressing materials hold huge potential to improve the wound nonhealing issues in the real clinical operation [67, 68]. This chapter provides key insights into the design and fabrication the multi-functional chitosan-based hybrid wound dressing materials for DM wound repair and regeneration. Furthermore, this chapter focuses on the different chitosan-based hybrid materials reported till date as well as their equipped manufacturing strategy for DM wound repair and regeneration (Fig. 9.1 and Table 9.1).

Table 9.1 Summary of the chitosan-based hybrid materials and their related applications

Method	Base material	Composition	References
Chemical modification	CS	GMA + SBMA	[50]
Chemical modification	CMC	CMC-g-PVPI	[51]
Blending	Chitosan hydrogel	Chitosan hydrogel + carbon nanotubes	[75]
Electrospinning	PVA	PVA + ZnO + chitosan	[88]
3D printing	CS	D-(+) raffinose pentahydrate + Chitosan	[78]
Coating	CS	Insulin + chitosan	[100]
Blending	CS	Psi + GLP-1 + DPP4	[104]
Blending-electrospinning	CS + PVA	CS + PVA + ANE	[108]
Blending	MPC	MPC + NT	[52]
Blending	CS + HA	CS + HA + nAg	[116]
Blending	CS	CS + AuNPs + insulin	[117]
Blending	Chitosan hydrogel	α-LA + CS	[118]
Blending	Chitosan hydrogel	CS + EGF + Ag^{+}	[124]
Blending	Alginate-chitosan	Alginate-chitosan + beta-TC-6	[131]

9.2 Fabrication of Chitosan-Based Hybrid Dressing Materials

Although the chitosan-based hybrid dressing materials have been considered as an effective medical product for wound hemostasis, but the non healing performance and ease to infection of the DM wounds also are a vital issue that need to be addressed [69, 70]. Currently, the chitosan materials have been confirmed that it can inhibit the bacteria proliferation and avoid wound bacterial infections mainly attributed to the cell membranes of bacterial cells destroyed by electrostatic interactions which could further prevent bacteria from absorbing nutrients from the wound tissue [71–74]. However, the anti-bacterial abilities of pure chitosan are limited due to the complicated wound of DM. Therefore, to enhance the antibacterial properties of chitosan-based wound dressing materials, many bioactive materials has been selected to modify chitosan to produce chitosan-based hybrid wound dressing. This is achieved by using chemical modification [50], blending [75], coating [76], electrospinning [77], 3D printing [78, 79] for developing the multi-functional wound dressing materials to meet the requirements of the different patients. The above mentioned approach further made it possible to design and produce the chitosan-based hybrid wound dressing by incorporating the functional agent, growth factors and cells into the wound dressing during or after synthesis.

9.2.1 Chemical Modification

Firstly, the chitosan derivatives are widely selected to repair chronic wounds due to their excellent biodegradability, antimicrobial performance, and substantial contribution to tissue regeneration [65, 80]. To get the desired performance of chitosan-based wound dressing, chitosan should be functionalized by chemical modification. In Lin et al. work, they designed and developed chitosan-based soft membranes wound dressing by using zwitterionic copolymers of glycidyl methacrylate (GMA) units and sulfobetaine methacrylate (SBMA) segments for treating DM wounds [50]. Moreover, Yu et al. also designed and prepared a carboxymethyl chitosan (CMC)-grafted polyvinylpyrrolidoneiodine (CMC-g-PVPI) microspheres for treating the DM wounds [51]. In their work, the CMC was grafted with PVPI to fabricate a CMC-g-PVPI microspheres which can have high shearing dispersion into the CMC-g-PVPI complex hydrogel. In addition, their in vivo results demonstrated that the obtained chitosan-based hybrid wound dressing exhibits accelerated wound healing efficiency due to the presence of the microspheres on the DM model, which further indicates that the CMC-g-PVPI has an excellent ability to promote the repair and regeneration of DM wound.

9.2.2 Coating

Coating strategy is always considered a relatively easy method to design and develop a special functional coating system containing multi-bioactive materials for producing the antimicrobial, anti-inflammatory and improving healing wound dressing [81, 82]. Prakash et al. designed and fabricated a bamboo fiber-based wound dressing through a coating solution system which contain the leave extracts of *Aloe vera*, piper betel, and neem leaf [83]. The coating strategy is an effective technology to build the multifunctional coating system via mixture the different bioactive medical materials. However, the coating strategy also has an unavoidable disadvantage which is the unstable characteristics on/into the surface of the traditional wound dressing.

9.2.3 Blending

Unlike coating strategy, in order to overcome the unstable performance of the coating solution, the chitosan-based materials (powders, hydrogels, and dressings) were usually complexed with bioactive molecules, antimicrobial agent or other function materials to form the chitosan-based hybrid wound dressing for wound healing, which is a relatively simpler strategy [84, 85]. Kittana et al. designed and fabricated single-wall or multi-wall carbon nanotubes blended with chitosan-based hydrogel dressing for treating the impaired wound healing in diabetic patients, and further to resolve the relevantly health problems, such as, bedsores, and extensive burns [75].

9.2.4 Electrospinning

Due to the lack of ability of most antibacterial materials or bioactive medical materials for the surface chemical groups modification of the wound dressing, the impregnation by coating or blend maybe result in a lower quality drug loading efficiency. On the contrary, the electrospinning strategy may be considered as an appropriate method mainly attributed to its unique preparation process for effective drug loading [86, 87]. Ahmed et al. designed and fabricated nanofiber mats wound dressings with antibacterial and healing performance which were composed of chitosan, polyvinyl alcohol (PVA), and zinc oxide (ZnO) for faster healing of DM wounds [88]. In their work, the in vivo wound healing results demonstrated that the obtained chitosan/PVA/ZnO nanofibrous membranes can significantly promote wound healing than the chitosan/PVA nanofibrous wound dressing without ZnO nanoparticles in the

DM wound model. However, the green electrospinning solution with low cytotoxicity also shows some serious issues need to be further overcome in future.

9.2.5 3D Printing

Due to the poor coverage performance of the traditional wound dressing (such as cotton fabric, film), especially the planner structure for deep wounds, the three-d imensional (3D) structure are important to treat the irregular or complicated wounds. 3D printing, as the famous additive manufacture technology, also play an important role to design and develop the wound dressing with irregular structure and size to match the real injured wound condition with various wound depths [89, 90]. In Intini et al. work, the 3D chitosan scaffolds were fabricated using an extrusion-based 3D printer [78], which can effectively maintain the structural distribution of the chitosan-based scaffolds. Their research results demonstrated that the designed and developed chitosan-based scaffolds can be an appropriate wound dressing to promote the wound tissue regeneration than a commercial product at in vivo tests on rat DM models, which further confirms that chitosan-based scaffolds have a great potential for treating the chronic wounds.

9.3 Chitosan-Based Hybrid Dressing Materials for Treatment of Diabetic Wounds

As mentioned above, chitosan-based wound dressing is being developed for treating diabetic wounds via various technologies, mainly due to their excellently anti-inflammatory performance during the healing process [91]. In addition, the wet condition of the DM wound site obtained from the wound dressing is very important to accelerate wound healing and facilitate cell proliferation around the injured wound area [92]. As we all known, the chitosan, as excellent functional biomaterials, has gained attention as a tissue regeneration scaffold for DM wound healing [93, 94]. In addition, the $-NH_2-$ and $-OH$ groups of chitosan can provide multi-sites for reacting with other active functional groups of other bioactive medical materials (such as the bioactive and therapeutic agents, nanoparticles, growth factors and the matched cells from the wound site), which further renders chitosan as a potential candidate for designing and developing the multi-functional wound dressing via various chemical routes and other technologies [95, 96].

9.3.1 Chitosan-Based Hybrid Dressing Materials Containing Anti-Inflammatory Agents for Wound Healing

With the development of chitosan-based therapies, this multifaceted biomolecule has become popular as an anti-diabetic drug and a highly effective drug carrier. Moreover, the chitosan loaded drugs can easily adhere to wound site and release various types of macromolecules for wound repair and regeneration [97–99].

Erel and their team designed and fabricated an insulin-loaded chitosan nanoparticles system encapsulated into the inner phase of a water-in-oil microemulsion system for DM wound repair and regeneration [100]. The obtained optimal insulin-loaded chitosan nanoparticles improved the absorption of insulin in the gastrointestinal tract in in vivo diabetic rats through oral administration. Therefore, the microemulsion system may be considered as a potential platform to encapsulate the functionalized nanoparticles for improving the adverse effects of subcutaneous insulin administration. Moreover, the insulin-loaded chitosan nanoparticles could maintain a relatively long dosing time after oral administration, which may reduce the frequency of drug administration in diabetic patients. However, the intestine's pH value can accelerate the degradation of chitosan which can enhance the release of insulin in the intestine and lead have a negative impact on the efficacy of the drug.

In addition to the insulin loading, the glucagon-like peptide-1 (GLP-1) with antidiabetic performance has been confirmed that induces secreting insulin by L cells in the gut that is a form of incretin for treating DM wounds [101–103]. Thereby, chitosan-based nano/micro materials were selected to stimulate L cell for releasing of GLP-1. In the study conducted by Shrestha et al., they designed and developed a dual protein-drug delivery system for treating DM wounds [104]. To treat the type 2 DM, the mucoadhesive porous silicon (PSi) nanoparticles encapsulated into a pH-responsive polymeric chitosan-based nano-system with an antidiabetic peptide, GLP-1, and the enzyme inhibitor, dipeptidyl peptidase-4 (DPP4). The nanosystem successfully exhibited huge clinical potential for treating DM with a peptide through controlled delivery, improved micro-adhesion, and higher intestinal permeability. Their results demonstrated that the nanosystem with multi-drug loaded chitosan-based nanoparticles exhibited excellent intestinal permeability of GLP-1 in in vitro intestinal epithelium model which was mainly due to the permeation enhancer effect of chitosan and inhibition of GLP-1 degradation by the DPP4 inhibitor. The results further also confirmed that the co-therapeutic nanosystem via combining various peptides and bioactive materials exhibited huge potential as a novel and multipurpose strategy for DM wound repair and regeneration.

9.3.2 Chitosan-Based Hybrid Dressing Materials Containing Bioactive Materials for Wound Healing

For diabetic wounds, the complexity of wounds and complications seriously affect their healing efficiency, therefore, the design and manufacture of the multi-functional chitosan-based hybrid wound dressing is necessary to treat chronic wounds in DM patients through cell-material interactions [105–107]. Zhang et al. developed a chitosan-polyvinyl alcohol (CS-PVA) nanofibrous loaded a bioactive ingredient of chinese medicine Pulsatilla, anemoside B4 (ANE) with different contents through the blend-electrospinning strategy for repairing the DM wound [108]. Their results demonstrated that the developed CS-PVA-ANE composite nanofibrous materials have multifunctional properties including the excellent water absorption, bionic elastic mechanical performances, and controlled releasing of ANE bioactive materials, which also exhibited the obviously hemostatic ability to treat the DM wound. Moreover, the obtained CS-PVA-ANE nanofiber mats also demonstrated that the inflammation of the wounds can be effectively controlled and the wound healing time also can be shortened, which shows the excellent wound repair and regeneration performance and makes an ideal wound dressing for treating DM wounds.

In addition, DFUs are a serious complication of DM which characterize as chronic, non-healing [109, 110]. Moura et al. designed and developed a chitosan derivative (5-methyl pyrrolidinone chitosan (MPC))-based hybrid wound dressing for the controlled delivery of neurotensin (NT) which can effectively modulate the inflammation in DM wounds repair and regeneration [52]. The results demonstrated that the MPC and NT-loaded MPC scaffolds can obviously decrease the wound inflammatory and improve wound healing efficiency during the early stages of the wound repair which mainly attributed to the potential healing effect of NT on DM wounds. Therefore, their study provides an effective strategy by combining the chitosan derivatives and NT to fabricate a biomedical scaffold for DM wound repair and regeneration through regulating the inflammatory response and stimulating re-epithelialization. However, the efficacy of the NT-loaded MPC wound dressing, as a therapy for DFUs, requires further clinical inflammation.

9.3.3 Chitosan-Based Hybrid Dressing Materials Containing Metal Nanoparticles for Wound Healing

Currently, loading metal nanoparticles have been considered as an effective method to formulate wound dressings with antibacterial activity to treat the DM and improve the wound healing effects [111–113]. Among these, the silver nanoparticles (AgNPs), gold nanoparticles (AuNPs), zinc nanoparticles (ZnO nanoparticles), and titanium nanoparticles are mostly selected to combine with other biomedical materials for promoting the wound healing in DM wounds [114, 115].

AgNPs, as a broad-spectrum antimicrobial material, have received extensive attention for the injured wound repair and infection prevention. Anisha et al. manufactured an antimicrobial wound dressing with a flexible and porous structure which contain chitosan, hyaluronic acid (HA) and nano silver (nAg) for treating DFUs wound infection [116]. The results demonstrated that the nanocomposite sponges wound dressing have excellent antibacterial performance against the regular bacterial infection from the DM wounds. Therefore, these nanocomposite sponges can be used as a potential biomedical product for treating the DFUs wound infection by adding the nAg. However, the wound healing process and complications of DM wound are extremely complex, and only relying on antibacterial function is insufficient to completely repair the DM wounds. In addition, the AuNPs also have the excellent antibacterial performance for wound repair attributed to their nanoscale efficacy, high reactivity to living cells, and good cellular penetration. Bhumkar et al. developed novel carriers using insulin loaded chitosan reduced AuNPs for releasing insulin to repair the DM wounds [117]. The results implied that the novel drug carrier obviously improved pharmacodynamic activity through oral and nasal administration. Thus, chitosan incorporated AuNPs drug carrier can be considered as a potential biomedical material to reduce the blood glucose levels for the DM patients. However, the excretion, accumulation, and toxicity of nanoparticles need to be addressed in future for proper treatment procedures.

In addition, Li et al. developed a chitosan-based hybrid hydrogel with metal organic frameworks wound dressing loaded α-lipoic acid (α-LA) for treating DM wound [118]. In their work, they have confirmed that the chitosan-based hybrid hydrogel wound dressing can effectively improve cell proliferation and migration. In addition, chitosan-based hybrid hydrogel wound dressing loaded with α-LA have higher wound healing efficiency than the hydrogel without α-LA. The results further demonstrated that the chitosan-based hybrid hydrogel wound dressing with special targeted drugs or bioactive compounds may be considered an appropriate method to design and produce the multi-functional wound dressing for the various complications of DM.

9.3.4 Chitosan-Based Hybrid Dressing Materials Containing Growth Factors for Wound Healing

DFUs, as a common DM complication, is a hard-to-heal skin wound with uncontrolled deterioration rate mainly attributed to injured metabolism and weak immune responses [119, 120]. To effectively repair the DM wound, the biomaterials incorporated with growth factors have fetched widely attention to solve the problem in the wound healing process [121–123].

To promote the efficiency of DM wound healing, Lee et al. designed and developed a chitosan-based hybrid hydrogel wound dressing incorporating silver ions (Ag^+) and nanoparticle-encapsulated epidermal growth factor (EGF) [124]. Their results demonstrated that the chitosan-based hybrid hydrogel wound dressing exhibit an excellent antibacterial performance duo to the constituent Ag^+, while the encapsulated EGF can obviously promote NIH/3T3 cell proliferation. Therefore, the obtained chitosan-based hybrid hydrogel wound dressing with sustained antibacterial and cell growth promotion performance can maintain a significantly promote wound healing in the DFUs wound. However, more in-depth questions need to be addressed in further studies.

9.3.5 Chitosan-Based Hybrid Dressing Materials Containing Cells for Enhanced Wound Healing

In type 1 DM, the autoimmune disorder is caused by the selective destruction of islet cells by the body's own immune cells [125, 126]. The ideal way to repair diabetes is to remove and clean the damaged debris in DM wounds, but it can lead to immune suppression, leaving patients susceptible to infections. Transplantation of allogeneic insulin-secreting cells encased in biocompatible microcapsules have shown promising results, thus eliminating the need for immunosuppression [127, 128].

Currently, the microfluidics technology has been considered an effective strategy to encapsulate the relative cells into the hydrogel micro-devices for wound healing which mainly attributed to its controlled size, shape, and reproducibility [129, 130]. In the work conducted by Bansal et al., an alginate-chitosan microcapsules loading insulin-secreting beta-TC-6 cells was designed and fabricated for treating Type 1 DM [131]. Moreover, their work demonstrated that the chitosan-based microencapsulated beta-TC-6 cells allograft can reduce blood glucose levels in the DM mice model for 5 weeks without immunosuppression. Importantly, the viability of microencapsulated β-TC-6 cells in vitro was highly correlated with the maintenance of normoglycemia in vivo. The results implied that the obtained transplantation of microencapsulated beta-TC-6 cells can be considered as a promising alternative in the treatment of Type 1 DM with tolerable immune response. However, the in vivo service life of the chitosan-based graft need to further addressed in future.

9.4 Conclusion

Chitosan-based hybrid wound dressing materials are a new class of biomedical materials with potential applications because of their wound repair and regeneration properties. Out of different fabricate strategies, the designing of hybrid materials for DM wound repair and regeneration has achieved a special attention of researchers with challenge for solving the hard-to-heal issues of DM wounds. In summary, chitosan-based hybrid materials provide the possibility to develop the wound dressings for treating diabetic wounds, but this is only the front end of the development. With further developments of advanced processing technology, we believe that chitosan-based hybrid medical dressings can be used effectively for repairing and regenerating the complex wound for further solving the issue of diabetes wound nonhealing.

Acknowledgements This work is supported by grants from the Natural Science Foundation of Heilongjiang Province of China (Grant No. LH2021B032), Heilongjiang Provincial Universities Basal Research Foundation-Youth Innovation Talent Project (Grant No. 145109210) and Natural Science Foundation of China (Grant No. 22005077).

References

1. C.A. Alarcon-Ruiz, B. Pantoja-Torres, M. Guarnizo-Poma, H. Lazaro-Alcantara, S. Paico-Palacios, M.G. Moscoso-Porras, V. Benites-Zapata, I. Resistance, Vitamin b12 associated to total body fat in healthy non-obese adults according to insulin resistance status. *Metabolism* ***116*** (2021)
2. M.C. Petersen, G.I. Shulman, Mechanisms of insulin action and insulin resistance. Physiol Rev. **98**, 2133–2223 (2018)
3. C.E. Mendez, C. Eiler, R.J. Walker, L.E. Egede, Insulin therapy for insulin resistant patients—harm or benefit? *Diabetes* **67**, 1577 (2018)
4. Q. Sun, Q. Yang, H. Xu, J. Xue, C. Chen, X. Yang, X. Gao, Q. Liu, Mir-149 negative regulation of mafa is involved in the arsenite-induced dysfunction of insulin synthesis and secretion in pancreatic beta cells. Toxicol. Sci. **167**, 116–125 (2019)
5. C.H. Gibbons, Treatment-induced neuropathy of diabetes. Curr. Diab. Rep. **17**, 127 (2017)
6. S.L. Ennis, M.P. Galea, D.N. O'Neal, M.J. Dodson, Peripheral neuropathy in the hands of people with diabetes mellitus. Diabetes Res. Clin. Pract. **119**, 23–31 (2016)
7. P.S. Ng, P.J. Dyck, R.S. Laughlin, P. Thapa, M.V. Pinto, P.J.B. Dyck, Lumbosacral radiculoplexus neuropathy: Incidence and the association with diabetes mellitus. Neurology **92**, e1188–e1194 (2019)
8. R. Almourani, B. Chinnakotla, R. Patel, L.R. Kurukulasuriya, J. Sowers, Diabetes and cardiovascular disease: An update. Curr. Diab. Rep. **19**, 161 (2019)
9. A.M. Schmidt, Diabetes mellitus and cardiovascular disease. Arterioscler. Thromb. Vasc. Biol. **39**, 558–568 (2019)
10. N.O. Stitziel, J.E. Kanter, K.E. Bornfeldt, Emerging targets for cardiovascular disease prevention in diabetes. Trends Mol. Med. **26**, 744–757 (2020)
11. J. Wang, F. Liu, R. Kong, X. Han, Association between globulin and diabetic nephropathy in type2 diabetes mellitus patients: a cross-sectional study. Front. Endocrinol. (Lausanne). **13**, 890273 (2022)
12. S. Prabhakar, H. Deshmukh, D. Cooper, Fabp1: a potential biomarker for nephropathy in type II diabetes. FASEB J. **36** (2022)
13. F. Jansson Sigfrids, E.H. Dahlstrom, C. Forsblom, N. Sandholm, V. Harjutsalo, M.R. Taskinen, P.H. Groop, Remnant cholesterol predicts progression of diabetic nephropathy and retinopathy in type 1 diabetes. J. Intern. Med. **290**, 632–645 (2021)
14. I.D. Jesus Ascencio-Montiel, E. JesÚS Zitle-GarcÍA, C. Alberto Reyes-MartÍNez, M. Altagracia GarcÍA-Lopez, Diabetes as leading cause of permanent occupational disability due to ocular diseases in the mexican institute of social security. Diabetes **68**, 1562 (2019)

15. M. Markoulli, J. Flanagan, S.S. Tummanapalli, J. Wu, M. Willcox, The impact of diabetes on corneal nerve morphology and ocular surface integrity. Ocul. Surf. **16**, 45–57 (2018)
16. D. Dayya, O.J. O'Neill, T.B. Huedo-Medina, N. Habib, J. Moore, K. Iyer, Debridement of diabetic foot ulcers. Adv Wound Care (New Rochelle) (2021)
17. W. Chai, Y. Wang, H. Zheng, S. Yue, Y. Liu, Y. Wu, X. Li, The profile of microbiological pathogens in diabetic foot ulcers. Front. Med. (Lausanne) **8**, 656467 (2021)
18. Z. Schafer, A. Mathisen, K. Svendsen, S. Engberg, T. Rolighed Thomsen, K. Kirketerp-Moller, Toward machine-learning-based decision support in diabetes care: a risk stratification study on diabetic foot ulcer and amputation. Front. Med. (Lausanne) **7**, 601602 (2020)
19. P. Pourghaderi, K. Yuquimpo, S. Sukpraprut-Braaten, D. Waggoner, Retrospective clinical study of lower extremity amputation in patients with diabetes mellitus and peripheral arterial disease: predictors of postoperative outcomes. J. Am. Coll.E Surg. **225** (2017)
20. S.E. Vidal Yucha, K.A. Tamamoto, D.L. Kaplan, The importance of the neuro-immuno-cutaneous system on human skin equivalent design. Cell Prolif. **52**, e12677 (2019)
21. A. Veves, V. Falanga, D.G. Armstrong, M.L. Sabolinski, A.D.F. Ulcer, S, Graftskin, a human skin equivalent, is effective in the management of noninfected neuropathic diabetic foot ulcers: A prospective randomized multicenter clinical trial. Diabetes Care **24**, 290–295 (2001)
22. Y. Zhao, Z. Li, S. Song, K. Yang, H. Liu, Z. Yang, J. Wang, B. Yang, Q. Lin, Skin-inspired antibacterial conductive hydrogels for epidermal sensors and diabetic foot wound dressings. *Adv. Funct. Materials* **29** (2019)
23. H. Xue, L. Hu, Y. Xiong, X. Zhu, C. Wei, F. Cao, W. Zhou, Y. Sun, Y. Endo, M. Liu, Y. Liu, J. Liu, A. Abududilibaier, L. Chen, C. Yan, B. Mi, G. Liu, Quaternized chitosan-matrigel-polyacrylamide hydrogels as wound dressing for wound repair and regeneration. Carbohydr. Polym. **226**, 115302 (2019)
24. G.C. Gurtner, S. Werner, Y. Barrandon, M.T. Longaker, Wound repair and regeneration. Nature **453**, 314–321 (2008)
25. P.J. Buch, Y. Chai, E.D. Goluch, Treating polymicrobial infections in chronic diabetic wounds. Clin. Microbiol. Rev. **32** (2019)
26. M.H. Otaviano, M. Salles, T.H. Ching, J.L. Dettoni, I.G.S. Coulibaly, E.T. Fukunaga, M.A. Gamba, J.C. Moraes, Topical oxygen jet therapy (tojt) for treating infected chronic surgical wounds. Braz. J. Infect. Dis. **25**, 101547 (2021)
27. J.G. Powers, C. Higham, K. Broussard, T.J. Phillips, Wound healing and treating wounds: chronic wound care and management. J Am Acad Dermatol. **74**, 607–25; quiz 25–6 (2016)
28. M. Song, J. Zong, L. Zou, Z. Fu, J. Liu, S. Wang, Biological debridement combined with stem cell therapy will be a convenient and efficient method for treating chronic wounds in the future. Med. Hypotheses **166** (2022)
29. C. Cangemi, V. Skov, M.K. Poulsen, J. Funder, W.O. Twal, M.A. Gall, V. Hjortdal, M.L. Jespersen, T.A. Kruse, J. Aagard, H.H. Parving, S. Knudsen, P.F. Hoilund-Carlsen, P. Rossing, J.E. Henriksen, W.S. Argraves, L.M. Rasmussen, Fibulin-1 is a marker for arterial extracellular matrix alterations in type 2 diabetes. Clin Chem. **57** 1556-65 (2011)
30. A. MØLler, A.S. Bojer, F. Genovese, M.H. SØRensen, P. GÆDe, M.A. Karsdal, P.L. Madsen, D.G. Rasmussen, Association of extracellular matrix remodeling biomarkers with complications of type 2 diabetes. Diabetes **71**, 1153-p (2022)
31. P.K. Barman, T.J. Koh, Macrophage dysregulation and impaired skin wound healing in diabetes. Front. Cell Dev. Biol. **8**, 528 (2020)
32. J.A. Knipper, X. Ding, S.A. Eming, Diabetes impedes the epigenetic switch of macrophages into repair mode. Immunity **51**, 199–201 (2019)
33. J.B. Cole, J.C. Florez, Genetics of diabetes mellitus and diabetes complications. Nat. Rev. Nephrol. **16**, 377–390 (2020)
34. W.B. Horton, E.J. Barrett, Microvascular dysfunction in diabetes mellitus and cardiometabolic disease. Endocr. Rev. **42**, 29–55 (2021)
35. A.A. Oktay, H.K. Akturk, K. Esenboga, F. Javed, N.M. Polin, E. Jahangir, Pathophysiology and prevention of heart disease in diabetes mellitus. Curr. Probl. Cardiol. **43**, 68–110 (2018)

36. W.K. Abdelbasset, S.M. Elkholi, K.A. Ismail, H.S. Al-Ghamdi, S. Mironov, H.S.H. Ridha, M.S. Maashi, L. Thangavelu, T. Mahmudiono, Y.F. Mustafa, Mequinol-loaded carboxymethyl cellulose/chitosan electrospun wound dressing as a potential candidate to treat diabetic wounds. Cellulose **29**, 7863–7881 (2022)
37. Y. Yang, Y. Zhang, Y. Yan, Q. Ji, Y. Dai, S. Jin, Y. Li, J. Chen, L. Teng, A sponge-like double-layer wound dressing with chitosan and decellularized bovine amniotic membrane for promoting diabetic wound healing. Polymers (Basel) **12** (2020)
38. C. Chen, P. Zhou, C. Huang, R. Zeng, L. Yang, Z. Han, Y. Qu, C. Zhang, Photothermal-promoted multi-functional dual network polysaccharide hydrogel adhesive for infected and susceptible wound healing. Carbohydr. Polym. **273**, 118557 (2021)
39. Y. Ren, A. Aierken, L. Zhao, Z. Lin, J. Jiang, B. Li, J. Wang, J. Hua, Q. Tu, Huc-mscs lyophilized powder loaded polysaccharide ulvan driven functional hydrogel for chronic diabetic wound healing. Carbohydr. Polym. **288**, 119404 (2022)
40. H.-T. Lin, A. Venault, Y. Chang, Zwitterionized chitosan based soft membranes for diabetic wound healing. J Membr. Sci. **591** (2019)
41. E. Nydrioti, B. Saramago, A.P. Serro, Polyhexanide and chlorhexidine loaded chitosan wound dressings. Annals of Medicine. **53**, S119-S (2021)
42. M.M. Sheir, M.M.A. Nasra, O.Y. Abdallah, Chitosan alginate nanoparticles as a platform for the treatment of diabetic and non-diabetic pressure ulcers: formulation and in vitro/in vivo evaluation. Int. J. Pharm. **607**, 120963 (2021)
43. T. Shen, K. Dai, Y. Yu, J. Wang, C. Liu, Sulfated chitosan rescues dysfunctional macrophages and accelerates wound healing in diabetic mice. Acta Biomater. **117**, 192–203 (2020)
44. T. Jin, T. Liu, E. Lam, A. Moores, Chitin and chitosan on the nanoscale. Nanoscale Horiz. **6**, 505–542 (2021)
45. Z. Shariatinia, Pharmaceutical applications of chitosan. Adv. Colloid. Interface Sci. **263**, 131–194 (2019)
46. K.E. Lee, D.H. Choi, C. Joo, S.W. Kang, K.M. Huh, Y.S. Park, Octanoyl glycol chitosan enhances the proliferation and differentiation of tonsil-derived mesenchymal stem cells. Carbohydr. Polym. **264**, 117992 (2021)
47. A. Shamosi, D. Mehrabani, M. Azami, S. Ebrahimi-Barough, V. Siavashi, H. Ghanbari, E. Sharifi, R. Roozafzoon, J. Ai, Differentiation of human endometrial stem cells into endothelial-like cells on gelatin/chitosan/bioglass nanofibrous scaffolds. Artif. Cells Nanomed. Biotechnol. **45**, 163–173 (2017)
48. P. Sanchez-Cid, M. Jimenez-Rosado, J.F. Rubio-Valle, A. Romero, F.J. Ostos, M. Rafii-El-Idrissi Benhnia, V. Perez-Puyana, Biocompatible and thermoresistant hydrogels based on collagen and chitosan. Polymers (Basel) **14** (2022)
49. D. Huber, G. Tegl, A.O. Mensah, B. Beer, M. Baumann, N. Borth, C. Sygmund, R. Ludwig, G.M. Guebitz, A dual-enzyme hydrogen peroxide generation machinery in hydrogels supports antimicrobial wound treatment. ACS Appl. Mater. Interfaces 9, 15307-16 (2017)
50. H.T. Lin, A. Venault, Y. Chang, Zwitterionized chitosan based soft membranes for diabetic wound healing. J. Membr. Sci. **591**, 117319 (2019)
51. J. Yu, P. Wang, M. Yin, K. Zhang, X. Wang, B. Han, Carboxymethyl chitosan-grafted polyvinylpyrrolidone-iodine microspheres for promoting the healing of chronic wounds. Bioengineered **13**, 8735–8746 (2022)
52. L.I.F. Moura, A.M.A. Dias, E.C. Leal, L. Carvalho, H.C. de Sousa, E. Carvalho, Chitosan-based dressings loaded with neurotensin—an efficient strategy to improve early diabetic wound healing. Acta Biomater. **10**, 843–857 (2014)
53. V. Brumberg, T. Astrelina, T. Malivanova, A. Samoilov, Modern wound dressings: hydrogel dressings. Biomedicines **9** (2021)
54. C. Merzougui, F. Miao, Z. Liao, L. Wang, Y. Wei, D. Huang, Electrospun nanofibers with antibacterial properties for wound dressings. J. Biomater. Sci. Polym. Ed.1–19 (2022)
55. D. Wang, N. Zhang, G. Meng, J. He, F. Wu, The effect of form of carboxymethyl-chitosan dressings on biological properties in wound healing. Colloid. Surf. B Biointerfaces **194**, 111191 (2020)

56. C. Yang, G. Liu, J. Chen, B. Zeng, T. Shen, D. Qiu, C. Huang, L. Li, D. Chen, J. Chen, Z. Mu, H. Deng, X. Cai, Chitosan and polyhexamethylene guanidine dual-functionalized cotton gauze as a versatile bandage for the management of chronic wounds. Carbohydr. Polym. **282**, 119130 (2022)
57. C. Gibbs, T. Orth, M. Gerkovich, E. Heitmann, M. Parrish, G. Lu, Traditional dressing compared with an external negative pressure system in preventing wound complications. Obstet. Gynecol. **123** (2014)
58. L. Long, W. Liu, C. Hu, L. Yang, Y. Wang, Construction of multifunctional wound dressings with their application in chronic wound treatment. Biomater. Sci. **10**, 4058–4076 (2022)
59. M. Georgescu, M. Chifiriuc, L. Marutescu, I. Gheorghe, V. Lazar, A. Bolocan, S. Bertesteanu, Bioactive wound dressings for the management of chronic wounds. Curr. Org. Chem. **21**, 53–63 (2016)
60. A. Moeini, P. Pedram, P. Makvandi, M. Malinconico, G. Gomez d'Ayala, Wound healing and antimicrobial effect of active secondary metabolites in chitosan-based wound dressings: a review. Carbohydr. Polym. **233**, 115839 (2020)
61. B. Vigani, S. Rossi, G. Sandri, M.C. Bonferoni, C.M. Caramella, F. Ferrari, Hyaluronic acid and chitosan-based nanosystems: A new dressing generation for wound care. Expert Opin. Drug Deliv. **16**, 715–740 (2019)
62. H. Li, F. Cheng, X. Wei, X. Yi, S. Tang, Z. Wang, Y.S. Zhang, J. He, Y. Huang, Injectable, self-healing, antibacterial, and hemostatic n,o-carboxymethyl chitosan/oxidized chondroitin sulfate composite hydrogel for wound dressing. Mater. Sci. & Eng., C. Mater. Biol. Appl. **118**, 111324 (2021)
63. M.E. Aljghami, S. Saboor, S. Amini-Nik, Emerging innovative wound dressings. Ann. Biomed. Eng. **47**, 659–675 (2019)
64. G. Dabiri, E. Damstetter, T. Phillips, Choosing a wound dressing based on common wound characteristics. Adv. Wound Care (New Rochelle) **5**, 32–41 (2016)
65. Y. Kong, X. Tang, Y. Zhao, X. Chen, K. Yao, L. Zhang, Q. Han, L. Zhang, J. Ling, Y. Wang, Y. Yang, Degradable tough chitosan dressing for skin wound recovery. Nanotechnol. Rev. **9**, 1576–1585 (2020)
66. H. Wang, Z. Xu, M. Zhao, G. Liu, J. Wu, Advances of hydrogel dressings in diabetic wounds. Biomater. Sci. **9**, 1530–1546 (2021)
67. G. Cavallaro, S. Micciulla, L. Chiappisi, G. Lazzara, Chitosan-based smart hybrid materials: a physico-chemical perspective. J. Mater. Chem. B **9**, 594–611 (2021)
68. D. Mukherjee, M. Azamthulla, S. Santhosh, G. Dath, A. Ghosh, R. Natholia, J. Anbu, B.V. Teja, K.M. Muzammil, Development and characterization of chitosan-based hydrogels as wound dressing materials. J. Drug Deliv. Sci. Technol. **46**, 498–510 (2018)
69. F. Cao, B. Zeng, Y. Zhu, F. Yu, M. Wang, X. Song, X. Cheng, L. Chen, X. Wang, Porous zno modified silk sutures with dual light defined antibacterial, healing promotion and controlled self-degradation capabilities. Biomater. Sci. **8**, 250–255 (2020)
70. X. Zhang, Z. Yin, Y. Guo, H. Huang, J. Zhou, L. Wang, J. Bai, Z. Fan, A facile and large-scale synthesis of a pva/chitosan/collagen hydrogel for wound healing. New J. Chem. **44**, 20776–20784 (2020)
71. S. Hoshino, Y. Yoshida, S. Tanimura, Y. Yamauchi, T. Noritomi, Y. Yamashita, A study of the efficacy of antibacterial sutures for surgical site infection: a retrospective controlled trial. Int. Surg. **98**, 129–132 (2013)
72. T. Baygar, Characterization of silk sutures coated with propolis and biogenic silver nanoparticles (agnps); an eco-friendly solution with wound healing potential against surgical site infections (ssis). Turk. J. Med. Sci. **50**, 258–266 (2020)
73. A.R. Franco, E. Palma Kimmerling, C. Silva, F.J. Rodrigues, I.B. Leonor, R.L. Reis, D.L. Kaplan, Silk-based antimicrobial polymers as a new platform to design drug-free materials to impede microbial infections. Macromol. Biosci. **18**, e1800262 (2018)
74. S. Zhang, X. Liu, H. Wang, J. Peng, K.K. Wong, Silver nanoparticle-coated suture effectively reduces inflammation and improves mechanical strength at intestinal anastomosis in mice. J. Pediatr. Surg. **49**, 606–613 (2014)

75. N. Kittana, M. Assali, H. Abu-Rass, S. Lutz, R. Hindawi, L. Ghannam, M. Zakarneh, A. Mousa, Enhancement of wound healing by single-wall/multi-wall carbon nanotubes complexed with chitosan. Int. J. Nanomedicine **13**, 7195–7206 (2018)
76. C. He, T. Ye, W. Teng, Z. Fang, W.S. Ruan, G. Liu, H. Chen, J. Sun, L. Hui, F. Sheng, D. Pan, C. Yang, Y. Zheng, M.B. Luo, K. Yao, B. Wang, Bioinspired shear-flow-driven layer-by-layer in situ self-assembly. ACS Nano **13**, 1910–1922 (2019)
77. J. Yin, L. Xu, A. Ahmed, Batch preparation and characterization of electrospun porous polylactic acid-based nanofiber membranes for antibacterial wound dressing. Adv. Fiber Mater. **4**, 832–844 (2022)
78. C. Intini, L. Elviri, J. Cabral, S. Mros, C. Bergonzi, A. Bianchera, L. Flammini, P. Govoni, E. Barocelli, R. Bettini, M. McConnell, 3D-printed chitosan-based scaffolds: An in vitro study of human skin cell growth and an in-vivo wound healing evaluation in experimental diabetes in rats. Carbohydr. Polym. **199**, 593–602 (2018)
79. M. Alizadehgiashi, C.R. Nemr, M. Chekini, D. Pinto Ramos, N. Mittal, S.U. Ahmed, N. Khuu, S.O. Kelley, E. Kumacheva, Multifunctional 3d-printed wound dressings. ACS Nano (2021)
80. R. Jayakumar, M. Prabaharan, S.V. Nair, H. Tamura, Novel chitin and chitosan nanofibers in biomedical applications. Biotechnol. Adv. **28**, 142–150 (2010)
81. H.M. Fahmy, A.A. Aly, A. Abou-Okeil, A non-woven fabric wound dressing containing layer - by - layer deposited hyaluronic acid and chitosan. Int. J. Biol. Macromol. **114**, 929–934 (2018)
82. B. Venkatrajah, V.V. Malathy, B. Elayarajah, R. Rajendran, R. Rammohan, Synthesis of carboxymethyl chitosan and coating on wound dressing gauze for wound healing. Pak. J. Biol. Sci. **16**, 1438–1448 (2013)
83. C. Prakash, N. Sukumar, P. Ramesh, S.K.S. Kumar, Development and characterization of wound dressing material coated with natural extracts of curcumin, aloe vera and chitosan solution enhanced with rhegf (regen-dtm). J. Nat. Fibers **18**, 2019–2032 (2020)
84. A. Oryan, S. Sahvieh, Effectiveness of chitosan scaffold in skin, bone and cartilage healing. Int. J. Biol. Macromol. **104**, 1003–1011 (2017)
85. V. Perumal, T. Manickam, K.S. Bang, P. Velmurugan, B.T. Oh, Antidiabetic potential of bioactive molecules coated chitosan nanoparticles in experimental rats. Int. J. Biol. Macromol. **92**, 63–69 (2016)
86. A. Celebioglu, A.F. Saporito, T. Uyar, Green electrospinning of chitosan/pectin nanofibrous films by the incorporation of cyclodextrin/curcumin inclusion complexes: Ph-responsive release and hydrogel features. ACS Sustain. Chem. Eng. **10**, 4758–4769 (2022)
87. M. Rahmati, D.K. Mills, A.M. Urbanska, M.R. Saeb, J.R. Venugopal, S. Ramakrishna, M. Mozafari, Electrospinning for tissue engineering applications. Progress in Materials Science Prog. Mater. Sci. **117** (2021)
88. R. Ahmed, M. Tariq, I. Ali, R. Asghar, P. Noorunnisa Khanam, R. Augustine, A. Hasan, Novel electrospun chitosan/polyvinyl alcohol/zinc oxide nanofibrous mats with antibacterial and antioxidant properties for diabetic wound healing. Int. J. Biol. Macromol. **120**, 385–393 (2018)
89. T. Hu, G.-P. Wu, H. Bu, H. Zhang, W.-X. Li, K. Song, G.-B. Jiang, An injectable, adhesive, and self-healable composite hydrogel wound dressing with excellent antibacterial activity. Chem. Eng. J. 450 (2022)
90. Y. Liang, M. Li, Y. Yang, L. Qiao, H. Xu, B. Guo, Ph/glucose dual responsive metformin release hydrogel dressings with adhesion and self-healing via dual-dynamic bonding for athletic diabetic foot wound healing. ACS Nano **16**, 3194–3207 (2022)
91. Y. Qian, Y. Zheng, J. Jin, X. Wu, K. Xu, M. Dai, Q. Niu, H. Zheng, X. He, J. Shen, Immunoregulation in diabetic wound repair with a photoenhanced glycyrrhizic acid hydrogel scaffold. Adv. Mater. **34**, e2200521 (2022)
92. W.C. Lin, C.C. Lien, H.J. Yeh, C.M. Yu, S.H. Hsu, Bacterial cellulose and bacterial cellulose-chitosan membranes for wound dressing applications. Carbohydr. Polym. **94**, 603–611 (2013)
93. H. Li, F. Cheng, X. Wei, X. Yi, S. Tang, Z. Wang, Y.S. Zhang, J. He, Y. Huang, Injectable, self-healing, antibacterial, and hemostatic n, o-carboxymethyl chitosan/oxidized chondroitin

sulfate composite hydrogel for wound dressing. Mater. Sci. Eng. C Mater. Biol. Appl. **118**, 111324 (2021)

94. Q. Zhang, Y. Chen, P. Wei, Y. Zhong, C. Chen, J. Cai, Extremely strong and tough chitosan films mediated by unique hydrated chitosan crystal structures. Mater. Today **51**, 27–38 (2021)
95. Y. Liu, H. Niu, C. Wang, X. Yang, W. Li, Y. Zhang, X. Ma, Y. Xu, P. Zheng, J. Wang, K. Dai, Bio-inspired, bio-degradable adenosine 5'-diphosphate-modified hyaluronic acid coordinated hydrophobic undecanal-modified chitosan for hemostasis and wound healing. Bioact. Mater. **17**, 162–177 (2022)
96. N. Nguyen, Z.-H. Lin, S.R. Barman, C. Korupalli, J.-Y. Cheng, N.-X. Song, Y. Chang, F.-L. Mi, H.-L. Song, H.-W. Sung, Y.-J. Lin, Engineering an integrated electroactive dressing to accelerate wound healing and monitor noninvasively progress of healing. Nano Energy **99** (2022)
97. J.R. Hilton, D.T. Williams, B. Beuker, D.R. Miller, K.G. Harding, Wound dressings in diabetic foot disease. Clin. Infect. Dis. **39**(Suppl 2), S100–S103 (2004)
98. R. Jayakumar, M. Prabaharan, S.V. Nair, S. Tokura, H. Tamura, N. Selvamurugan, Novel carboxymethyl derivatives of chitin and chitosan materials and their biomedical applications. Prog. Mater. Sci. **55**, 675–709 (2010)
99. X. Yang, J.L. Guo, J. Han, R.J. Si, P.P. Liu, Z.R. Zhang, A.M. Wang, J. Zhang, Chitosan hydrogel encapsulated with Il-37 peptide promotes deep tissue injury healing in a mouse model. Mil. Med. Res. **7**, 20 (2020)
100. G. Erel, M. Kotmakçı, H. Akbaba, S.S. Karadagli, A.G. Kantarcı, Nanoencapsulated chitosan nanoparticles in emulsion-based oral delivery system: In vitro and in vivo evaluation of insulin loaded formulation. J. Drug Deliv. Sci. Technol. **36**, 161–167 (2016)
101. T.D. Muller, B. Finan, S.R. Bloom, D. D'Alessio, D.J. Drucker, P.R. Flatt, A. Fritsche, F. Gribble, H.J. Grill, J.F. Habener, J.J. Holst, W. Langhans, J.J. Meier, M.A. Nauck, D. Perez-Tilve, A. Pocai, F. Reimann, D.A. Sandoval, T.W. Schwartz, R.J. Seeley, K. Stemmer, M. Tang-Christensen, S.C. Woods, R.D. DiMarchi, M.H. Tschop, Glucagon-like peptide 1 (glp-1). Mol. Metab. **30**, 72–130 (2019)
102. D.J. Drucker, The cardiovascular biology of glucagon-like peptide-1. Cell Metab. **24**, 15–30 (2016)
103. T.A. Lutz, E. Osto, Glucagon-like peptide-1, glucagon-like peptide-2, and lipid metabolism. Curr. Opin. Lipidol. **27**, 257–263 (2016)
104. N. Shrestha, M.A. Shahbazi, F. Araujo, E. Makila, J. Raula, E.I. Kauppinen, J. Salonen, B. Sarmento, J. Hirvonen, H.A. Santos, Multistage ph-responsive mucoadhesive nanocarriers prepared by aerosol flow reactor technology: a controlled dual protein-drug delivery system. Biomaterials **68**, 9–20 (2015)
105. B. Hu, M. Gao, K.O. Boakye-Yiadom, W. Ho, W. Yu, X. Xu, X.Q. Zhang, An intrinsically bioactive hydrogel with on-demand drug release behaviors for diabetic wound healing. Bioact. Mater. **6**, 4592–4606 (2021)
106. Y. Li, R. Fu, Z. Duan, C. Zhu, D. Fan, Injectable hydrogel based on defect-rich multi-nanozymes for diabetic wound healing via an oxygen self-supplying cascade reaction. Small **18**, e2200165 (2022)
107. Z. Shao, T. Yin, J. Jiang, Y. He, T. Xiang, S. Zhou, Wound microenvironment self-adaptive hydrogel with efficient angiogenesis for promoting diabetic wound healing. Bioact. Mater. **20**, 561–573 (2023)
108. H. Zhang, M. Zhang, X. Wang, M. Zhang, X. Wang, Y. Li, Z. Cui, X. Chen, Y. Han, W. Zhao, Electrospun multifunctional nanofibrous mats loaded with bioactive anemoside b4 for accelerated wound healing in diabetic mice. Drug Deliv. **29**, 174–185 (2022)
109. G. Theocharidis, H. Yuk, H. Roh, L. Wang, I. Mezghani, J. Wu, A. Kafanas, M. Contreras, B. Sumpio, Z. Li, E. Wang, L. Chen, C.F. Guo, N. Jayaswal, X.L. Katopodi, N. Kalavros, C.S. Nabzdyk, I.S. Vlachos, A. Veves, X. Zhao, A strain-programmed patch for the healing of diabetic wounds. Nat. Biomed. Eng. (2022)
110. Z. Zhu, L. Wang, Y. Peng, X.Chu, L. Zhou, Y. Jin, H. Guo, Q. Gao, J. Yang, X. Wang, Z. Long, Y. Ge, S. Lu, B. Wang, Continuous self-oxygenated double-layered hydrogel under natural

light for real-time infection monitoring, enhanced photodynamic therapy, and hypoxia relief in refractory diabetic wounds healing. Adv. Funct. Mater. **32** (2022)
111. L. Yang, F. Liang, X. Zhang, Y. Jiang, F. Duan, L. Li, F. Ren, Remodeling microenvironment based on mofs-hydrogel hybrid system for improving diabetic wound healing. Chem. Eng. J. **427** (2022)
112. C. Liu, Y. Zhu, X. Lun, H. Sheng, A. Yan, Effects of wound dressing based on the combination of silver@curcumin nanoparticles and electrospun chitosan nanofibers on wound healing. Bioengineered **13**, 4328–4339 (2022)
113. M. Zhang, G. Wang, D. Wang, Y. Zheng, Y. Li, W. Meng, X. Zhang, F. Du, S. Lee, Ag@mof-loaded chitosan nanoparticle and polyvinyl alcohol/sodium alginate/chitosan bilayer dressing for wound healing applications. Int. J. Biol. Macromol. **175**, 481–494 (2021)
114. J. Chen, Y. Liu, G. Cheng, J. Guo, S. Du, J. Qiu, C. Wang, C. Li, X. Yang, T. Chen, Z. Chen, Tailored hydrogel delivering niobium carbide boosts ros-scavenging and antimicrobial activities for diabetic wound healing. Small **18**, e2201300 (2022)
115. B. Sun, F. Wu, X. Wang, Q. Song, Z. Ye, M. Mohammadniaei, M. Zhang, X. Chu, S. Xi, N. Zhou, W. Wang, C. Yao, J. Shen, An optimally designed engineering exosome-reductive cof integrated nanoagent for synergistically enhanced diabetic fester wound healing. Small **18**, e2200895 (2022)
116. B.S. Anisha, R. Biswas, K.P. Chennazhi, R. Jayakumar, Chitosan-hyaluronic acid/nano silver composite sponges for drug resistant bacteria infected diabetic wounds. Int. J. Biol. Macromol. **62**, 310–320 (2013)
117. D.R. Bhumkar, H.M. Joshi, M. Sastry, V.B. Pokharkar, Chitosan reduced gold nanoparticles as novel carriers for transmucosal delivery of insulin. Pharm. Res. **24**, 1415–1426 (2007)
118. Q. Li, K. Liu, T. Jiang, S. Ren, Y. Kang, W. Li, H. Yao, X. Yang, H. Dai, Z. Chen, Injectable and self-healing chitosan-based hydrogel with mof-loaded α-lipoic acid promotes diabetic wound healing. Mater. Sci. Eng.: C **131**, 112519 (2021)
119. M. Chang, T.T. Nguyen, Strategy for treatment of infected diabetic foot ulcers. Acc. Chem. Res. **54**, 1080–1093 (2021)
120. A.S. MacLeod, Bad "staph" in the wound environment of diabetic foot ulcers. Cell Host Microbe. **25**, 638–640 (2019)
121. S. Barrientos, O. Stojadinovic, M.S. Golinko, H. Brem, M. Tomic-Canic, Growth factors and cytokines in wound healing. Wound Repair Regeneration **16**, 585–601 (2010)
122. H.S. Kim, H.S. Yoo, In vitro and in vivo epidermal growth factor gene therapy for diabetic ulcers with electrospun fibrous meshes. Acta Biomater. **9**, 7371–7380 (2013)
123. M.W. Tsang, W.K. Wong, C.S. Hung, K.M. Lai, W. Tang, E.Y. Cheung, G. Kam, L. Leung, C.W. Chan, C.M. Chu, E.K. Lam, Human epidermal growth factor enhances healing of diabetic foot ulcers. Diabetes Care **26**, 1856–1861 (2003)
124. Y.H. Lee, Y.L. Hong, T.L. Wu, Novel silver and nanoparticle-encapsulated growth factor co-loaded chitosan composite hydrogel with sustained antimicrobility and promoted biological properties for diabetic wound healing. Mater. Sci. Eng. C Mater. Biol. Appl. **118**, 111385 (2021)
125. F.Z. Syed, Type 1 diabetes mellitus. Ann. Intern. Med. **175**, ITC33–ITC48 (2022)
126. H.M. Manukumar, J. Shiva Kumar, B. Chandrasekhar, S. Raghava, S. Umesha, Evidences for diabetes and insulin mimetic activity of medicinal plants: Present status and future prospects. Crit. Rev. Food Sci. Nutr. **57**, 2712–2729 (2017)
127. A. Jeyagaran, C.E. Lu, A. Zbinden, A.L. Birkenfeld, S.Y. Brucker, S.L. Layland, Type 1 diabetes and engineering enhanced islet transplantation. Adv. Drug Deliv. Rev. 114481 (2022)
128. M.J.V. White, P.S. Briquez, D.A.V. White, J.A. Hubbell, Vegf-a, pdgf-bb and hb-egf engineered for promiscuous super affinity to the extracellular matrix improve wound healing in a model of type 1 diabetes. NPJ Regen. Med. **6**, 76 (2021)
129. X. Liu, Y. Yu, D. Liu, J. Li, J. Sun, Q. Wei, Y. Zhao, S.J. Pandol, L. Li, Porous microcapsules encapsulating β cells generated by microfluidic electrospray technology for diabetes treatment. NPG Asia Mater. **14**, (2022)

130. K. Duan, G. Ghosh, J.F. Lo, Optimizing multiplexed detections of diabetes antibodies via quantitative microfluidic droplet array. Small **13** (2017)
131. A. Bansal, S. D'Sa, M.J. D'Souza, Biofabrication of microcapsules encapsulating beta-tc-6 cells via scalable device and in-vivo evaluation in type 1 diabetic mice. Int. J. Pharm. **572**, 118830 (2019)

Chapter 10
Nanostructured Chitosan Hybrid Materials for Tissue Engineering Applications

Dhiman Santra and Mitali Sarkar

Abstract Chitosan, a natural polysaccharide, is derived from chitin, and has unique antimicrobial activity, along with biodegradability, and biocompatibility. On chemical modification chitosan develops additional functional properties that are utilized in fabrication of a variety of biomaterials for biomedical applications, drug delivery, regenerative medicine, tissue engineering etc. In recent times there is an enormous development in the synthesis of chitosan-containing scaffolds, in form of gel, sponge, particle, film, fiber, and net. The possible applications of such scaffolds as a component for drug delivery applications, particularly in tissue repair and regeneration are getting prominence. The long-term therapeutic use, drug release for tissue fixation together with regeneration makes the chitosan hybrid materials more fascinating for future research. The present chapter highlights systematic findings in regard to the fabrication and utility of nanostructured chitosan-containing scaffolds in various biomedical applications emphasizing on the tissue engineering.

10.1 Introduction

Tissue engineering (TE), an interdisciplinary science involving biological sciences and engineering, is employed to build tissues that maintain, restore, or enhance tissue function. TE, in biomedical research, is a promising field, aimed to highlight the correlation between structure and function in cells and tissues, along with the development of suitable materials, as biological substitutes, having potential to replace and repair the diseased or damaged tissues as well as to function as vehicles for delivery of drugs, to cells or cellular biomolecules and also the nonbiological apparatus (ex. Stents, materials coated with biomaterials to interact with cells) to interact with cells

D. Santra · M. Sarkar (✉)
Department of Chemistry, University of Kalyani, Kalyani, West Bengal 741235, India
e-mail: msarkar@klyuniv.ac.in

S. K. Swain and A. Biswal (eds.), *Chitosan Nanocomposites*,
Biological and Medical Physics, Biomedical Engineering,
https://doi.org/10.1007/978-981-19-9646-7_10

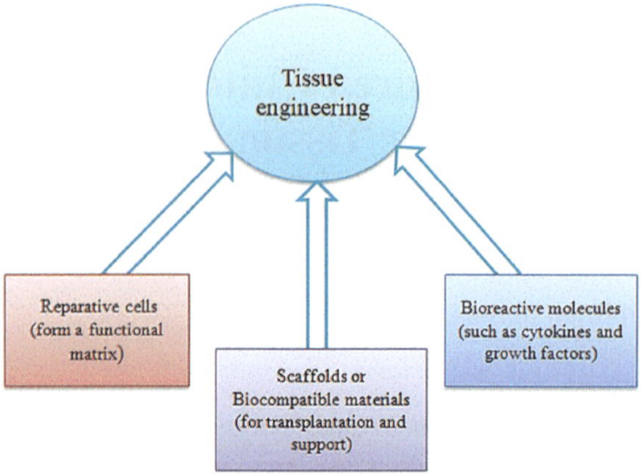

Fig. 10.1 Components in tissue engineering

[1]. In the current state of affairs, TE provides a promising trust to patients suffering from chronic or critical injuries [2]. A tissue can be engineered based on the three main aspects of molecules (Fig. 10.1), viz.

(1) reparative, mostly embryonic stem cell, which promotes functional tissue formation,
(2) scaffold or biocompatible or bioactive material which promotes support for cells growth and proliferation, and
(3) bioactive molecules (cytokine and growth factor, etc.), which support and choreograph desired tissue formation [3].

Scaffolds or synthetic biomaterials need to have architectural similarity to the native tissues [4], in terms of geometry and morphology, in repairing and replacing damaged or diseased tissues and for suitable medical applications [5]. Various natural materials, viz., chitosan, alginate, gelatin, silk fibroin, collagen, hydroxyapatite, polyethylene glycol, etc., are used to engineer synthetic biomaterials for TE applications [6]. However, limitations, viz. average mechanical properties, risk of infection, local acidic environments, and uncontrolled degradation of the engineered materials, prompted to design and fabricate hybrid bio composite materials, employing two or more biopolymers, organic, inorganic, and metallic nanocomposites for enhancement of biological as well as physico-chemical properties. Nanotechnology provides creative and outstanding opportunities for the use of nanocomposites in several biomedical applications viz. tissue engineering, vaccination, drug delivery, cancer therapy, and other medicinal aspects [7–9]. The advantages of the nanostructured hybrid biocomposite materials involve high colloidal stability, hydrophilicity, dispersibility, flexibility, biocompatibility, improved surface reactivity, easy cellular uptake, and the capacity to stimulate cells, enhancing adhesion and differentiation [1, 8].

Among the different biopolymers, chitosan, the second abundant natural biopolymer on earth, is produced commonly from chitin deacetylation. Chitosan, as the precursor of the nanostructured hybrid materials, has received great attention in biomedical fields due to its easy availability, abundance, and unique properties *viz.* non-toxicity, biocompatibility, biodegradability, hydrophilicity, mucoadhesiveness, antimicrobial, anti-inflammatory, antioxidant, and anticancer, activities [10, 11]. Moreover, chitosan, on degradation, produces amino sugar, which is harmless and absorbed through human tissues. Thus, chitosan finds potential applications in a wide range of biomedical fields, prominently in drug delivery, TE, surgical thread, etc. The poor solubility and stability issues of chitosan in water, however, can be modified in chitosan hybrid or derivative materials, i.e., composites, following electrostatic interactions, chemical cross-linking, metal ion coordination, and chemical modifications [12–14]. Chitosan composites show enhanced biological as well as physico-chemical properties than the parent chitosan. Chitosan along with its hybrids or derivatives, in different forms viz. sponge, hydrogel, thin film, nanoparticle, and microsphere, are thus found suitable to build desired nanostructured hybrid materials for use in various biomedical applications. Specifically, chitosan nanocomposites can improve the therapeutic efficacy and pharmacokinetic behavior of traditional drugs while enhancing their inherent biological properties to regenerate or repair tissues [15].

Several review articles on chitosan materials for biomedical applications though available in literature, a systemic review on nanostructure chitosan hybrid materials for tissue engineering applications is lacking. The chapter is, therefore, presented on the chitosan nanostructure and nanocomposite materials for tissue engineering applications.

10.2 Chitosan Structure

Chitosan is obtained from chitin deacetylation via chemical, biological, or combined chemical and biological methods. This linear polysaccharide chitosan is composed of (1→4)-2-acetamido-2-deoxy-β-D-glucan (N-acetyl D-glucosamine) and (1→4)-2-amino-2-deoxy-β-D-glucan (D-glucosamine) units [16] (Fig. 10.2). Chitosan, though not naturally abundant is developed easily from the partial deacetylation of chitin, which is found in shells of crabs, tortoise, shrimps, lobsters, etc. In fact, DD, the degree of deacetylation, determines the average molecular weight (M.Wt) of chitosan and hence influences the nature and quality of the polymer. The lower DD provides higher M.Wt with higher mechanical strength, higher chemical stability, and lower solubility in common solvents [8]. Generally, two common methods were used to produce chitosan with varying DD of chitin [6], viz. heterogeneous and homogeneous deacetylation. In the first case solid phase chitin and in the last case pre-swollen chitin was used in an aqueous medium under vacuum. However, these methods require long time and the use of concentrated alkali solutions, which necessitates the development of alternative soft methods. Such methods include thermo-mechanical process at low

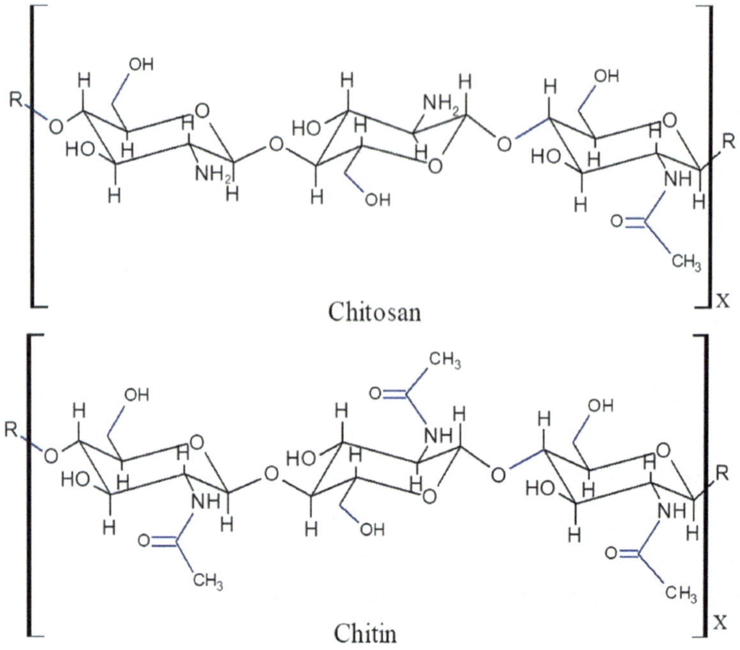

Fig. 10.2 Structure of Chitosan and Chitin

alkali concentrations using a cascade reactor, microwave-assisted dielectric heating, flash treatment by saturated steam, enzymatic preparation, γ–irradiation technique, etc. [6, 17–22].

10.3 Chitosan Nanostructure Properties

Chitosan has been progressively more investigated in building low-cost, eco-friendly, sustainable, and renewable nanocomposite. It is considered as a very effective biomaterial in TE because of its unique properties, such as

(i) rate of biodegradability related to the new tissue formation rate,
(ii) biocompatibility with tissues,
(iii) non-toxicity as well as mucoadhesive nature,
(iv) mechanical strength,
(v) adequate porosity together with suitable morphology for cellular uptake and transcellular transport, (within, across materials, and the host environment), of metabolites, nutrients, gases, and signal molecules [6].

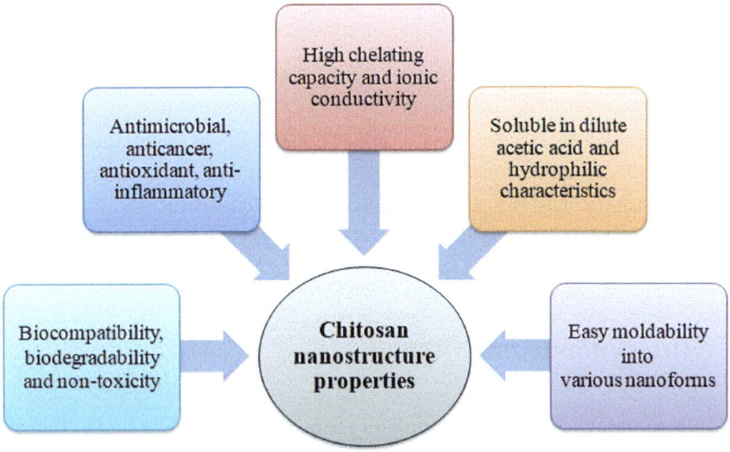

Fig. 10.3 Biological and physico-chemical properties of chitosan nanostructure

Moreover, moldability into different forms, drug carrying capacity, compatibility with a wide variety of delivery materials, cationic nature, formation ability of interconnected porous structure, and electrostatic interactions with proteoglycans, anionic species, e.g., glycosaminoglycans etc., [6], as well as its unique biological properties has raised its significance in biomedical engineering [5, 23]. The structural aspects of chitosan, together with the presence of active functional groups viz. amino ($-NH_2$), hydroxyl ($-OH$), and carboxyl ($-COOH$), endorse different intra- and inter-molecular hydrogen bond formation and allow embedding of nanostructured fillers. Chitosan with different DD and M.Wt have low toxicity, comparable to that of chitosan nanoparticles [24, 25]. The prominent biological aspects of chitosan are its antimicrobial, anti-inflammatory, antioxidant, and anticancer activities (Fig. 10.3).

10.3.1 Antimicrobial Activity

Chitosan, in different forms (solution, film and composite), is used as an antimicrobial agent/material against different target organisms, e.g., yeasts, bacteria, algae, and fungi and in experiments involving in-vivo and in-vitro interactions [26]. Generally, electrostatic interaction between positively charged amine groups (NH_3^+) of glucosamine and the negatively charged microbial membranes, results in the release of intracellular constituents leading to cell death [23]. Other possibility includes DNA-chitosan binding which inhibits mRNA as it penetrates the microorganism's nuclei. Several factors, viz. ionic strength, pH, time, and temperature of solution together with concentration, molecular weight, DD, and chelating capacity of chitosan, affect antimicrobial activity toward different types of microorganisms [10].

10.3.2 Antioxidant Activity

The antioxidant behavior of chitosan can further be improved in chitosan nanocomposites or derivatives. The amine and hydroxyl groups in chitosan monomer (Fig. 10.2) react with free radicals [11, 27]. Thus, chitosan and its derivatives were shown to develop scavenging ability, in-vitro, of the active oxygen free radicals. The low molecular weight (LMW) chitosan molecules have more ability to eliminate free radicals compared to that of the high molecular weight (HMW) chitosan molecules.

10.3.3 Anti-inflammatory Activity

The anti-inflammatory property of chitosan is demonstrated to be due to the presence of charged moieties, that negatively regulates the pro-inflammatory reactions [28]. The anti-inflammatory activity of both the LMW (5.2 kDa) and HMW (115 kDa) chitosan was demonstrated [29]. The HMW chitosan inhibited the in-vitro synthesis of anti-inflammatory cytokine *IL-10* (Interleukin 10), the tumor necrosis factor alpha, induced by endotoxin. The synthesis of IL-10 was reported to be stimulated by both HMW and LMW chitosan in the blood serum of mice for the peroral introduction. It was further revealed that anti-inflammatory activity of chitosan was not dependent on M.Wt, but was influenced by structural features of the molecule [29]. Morphological and biochemical characteristics revealed that the effect of chitosan was comparable to that of a hormone anti-inflammatory drug, e.g., prednisolone.

10.3.4 Anticancer Activity

The present-day challenge for cure of cancer, adopting surgical resection, is, so far, to adopt an efficient and effective management protocol. Nano drug delivery by chitosan derivatives is one of the most up to date and prominent delivery systems in cancer therapy. The technology involves most promising target therapy with minimum systemic toxicity, induced cell necrosis, apoptosis, enhanced immunity and maximum cytotoxicity to the cancer and tumor cells [30, 31].

10.4 Chitosan Nanostructure

Chitosan may exhibit some drawbacks viz. low thermal stability, poor barrier and mechanical properties [32]. Polymer, organic, inorganic, and metallic nanocomposite can be applied, as additive to chitosan, to fabricate different kind of nanostructures such as nanofilm, nanofiber, nanoparticle, nanohydrogel, and nanonet (Fig. 10.4) to be used for tissue engineering. Nanoplate, nanoparticle, and nanofiber are the three types of nanofillers according to their geometry and particle length to thickness ratio (aspect ratio), the key factor toward enhancement of nanostructure properties [33]. These hybrid nanostructure materials having specific and desired functionalities show unique performance with improved biological properties toward tissue engineering [34]. An improvement in biological as well as physico-chemical properties of chitosan-based nanocomposites, even at a very low level of nanofillers (5–10 wt. %), in comparison to conventional composites (40–50 wt.% nanofillers level) was observed. The amount of chitosan in the composites is generally maintained high, yielding nanocomposites of good biocompatibility and bioactivity along with improvement of barrier and mechanical properties, transparency, and thermal stability [35]. Thus, different chitosan nanostructures are getting prominence in various biomedical applications prominently in TE.

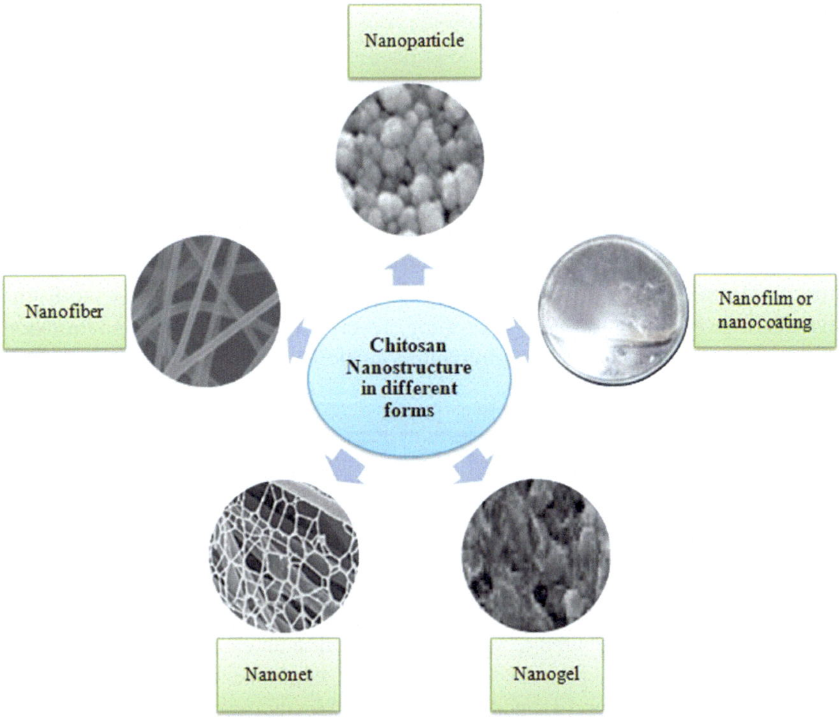

Fig. 10.4 Processibility of chitosan nanostructure into different forms

10.4.1 Nanofilm

A good permeability, unique antibacterial activity, and large surface area, make the chitosan-based films a potential material for wound dressing and an alternative to artificial skin [36]. Nanostructured thin films and multilayer coatings have been investigated for improved surface properties and are used in different medical devices. Nanostructures at the surface of the chitosan matrix enhance the leaching or release of the nano species [37]. Generally, two methods, viz Langmuir–Blodgett (LB) and the layer-by-layer (LBL) deposition technique are employed to synthesize chitosan thin films. The polycationic nature makes chitosan well-suited for the LBL process. The chitosan-based films are widely employed in wound dressings, bone engineering, drug delivery, and sensors [38].

10.4.2 Nanoparticle

Chitosan nanoparticles (CSNPs) are found to have excellent drug carrier potential along with biodegradability, biocompatibility, bioactivity, and non-toxicity [33]. The incorporation of therapeutic agents (anticancer drugs) into CSNPs might enhance chemical stability, bioavailability, and biocompatibility, modify the pharmacokinetics and pharmacodynamics of the therapeutic agent, enhance the efficacy of the active agents and reducing toxicity [39]. Nanomaterials possess the ability to enter into cytoplasmic space (high cellular uptake). Ionic gelation, covalent and ionic cross-linking, precipitation or coacervation, reverse micelles formation, emulsion-droplet coalescence, sieving, nanoprecipitation, etc., have been reported for CSNPs synthesis [40].

10.4.3 Nanofiber

Biopolymers turned into ultrafine nanofibers using electrospinning technique (ESP) is a simple, straight forward, cost-effective, and novel technique [41, 42]. In ESP when a high voltage is applied to a polymeric solution an electrically charged jet is created and the polymer fibers are formed [43, 44]. Nanofibers offer potential applications in drug delivery and related biomedical fields, due to their surface property, enhanced mass transfer, and cell attachment ability.

10.4.4 Nanogel

Currently, hydrogels, especially nanogels, are considered as potent biomaterials in the fields of TE, drug delivery, implants, etc. [45]. The nanogels size can be monitored within the range of few nanometers up to thousand nm. The nanogels surface properties facilitate multivalent conjugations producing a 'smart material' for stimulus responsiveness. Nanogels show cytotoxicity, biodegradability, biocompatibility, inflammatory response, cellular uptake, etc., along with unique mechanical resistance, degree of swelling, etc. In nanomedicine development, in-vivo biodistribution study and intracellular fate of nanogels are certainly the key requirements. Nanogels are popularly designed using several cross-linking protocols (covalent, ionic), the introduction of unstable bonds, which may cleave in physiological conditions [46].

The versatile use of some prominent chitosan-based nanostructure hybrid composites/scaffolds in the biomedical field are summarized in Table 10.1.

10.5 Tissue Engineering Application

Chitosan-based hybrid nanocomposites, with unique physico-chemical, mechanical as well as biological features have gained continuous importance in TE along with other different biomedical fields. The Scheme 10.1 demonstrates the efficacy of chitosan derived scaffolds in different branches of TE applications.

10.5.1 Skin Tissue Engineering

Skin damage affected by any chemical, thermal, or electrical stimuli may result in cutaneous complications leading to critical or chronic injuries in the worst situations. The injured/damaged skin tissue initiates healing by a natural repair process. A complex and overlapping process of skin regeneration proceed via (i) hemostasis, (ii) inflammation, (iii) proliferation, and (iv) tissue remodeling, the four phases [12]. It is a spontaneous and dynamic repair process involving cells, soluble factors, blood elements, and extracellular components [82]. Regeneration of skin requires dressings (medical) that ensures physical protection and healing of the wound, protection from microbial growth, and reduction of scar formation. However, successful skin tissue regeneration still remains a biomedical challenge and a major healthcare issue, especially for chronic cases of diabetic and ischemic wounds, which result in loss of functional ability, decreased quality of life as well as increased pain and infections. Therefore, an enormous effort is being made to develop new and efficient approaches that can efficiently promote full skin tissue recovery with minimum side effects and costs [23].

Table 10.1 Chitosan-based nanostructure scaffolds for biomedical applications

Sl no	Nanostructure	Composite/scaffold	Application	References
1	Particle	Titanium dioxide nano particle loaded chitosan–pectin	Wound healing	[47]
2		Chitosan–alginate	Drug therapy for infectious dermatologic treatment	[48]
3		ZnO/gentamicin–chitosan gel	Wound care	[49]
4		Ceramide incorporated poly (lactic-co-glycolic acid) NPs with chitosan coating	Atopic dermatitis	[50]
5		Dexamethasone sodium phosphate loaded Chitosan nanoparticles	Anti-inflammatory efficacy, ophthalmic therapeutics	[51]
6		NPs of Gelatin–chitosan–epigallocatechin gallate incorporated in hydrogel of poly (γ-glutamic acid)-gelatin	Regeneration of wound tissue	[52]
7		Curcumin loaded chitosan NPs immobilized on scaffolds of collagen–alginate	Healing of diabetic wounds	[53]
8		Gold nanoparticles coated with chitosan, poly (lactic-co-glycolic acid), and poly (N-isopropyl acryl amide)-co-(2-dimethylamino ethyl methacrylate)	Antibacterial efficacy; wound healing	[54]
9		NPs of Chitosan-5-fluorouracil	Skin regeneration	[55]
10		NPs of Chatoyant/poly-γ-glutei acid/plutonic/cur cumin	Regeneration of neo collagen; reconstruction of tissues	[56]
11		NPs of Chitosan-capped silver	Antioxidant efficacy; wound repair/regeneration	[57]
12		Composite sponge dressing of Chitosan loaded on iturin-AgNPs	Antibacterial efficacy in-vitro; wound healing in-vivo	[58]
13		Silver chitosan nanocomposites	Antifungal efficacy	[59]
14		Carvacrol and Eugenol encapsulated in chitosan NPs	Antimicrobial efficacy against microbial contaminants in cosmetic products	[60]

(continued)

Table 10.1 (continued)

Sl no	Nanostructure	Composite/scaffold	Application	References
15	Capsule, particle	Bi-layered self-assembly of hyaluronic acid and chitosan	Skin regeneration; wound healing	[61]
16	Particle, fiber	AgNPs incorporation into chitosan-based nanofibrous membranes	Antibacterial efficacy; wound healing	[62]
17		Silver NPs/chitosan oligosaccharide/poly (vinyl alcohol) nanofiber	Wound healing	[63]
18	Tube	Chitosan–halloysite nanotubes (HNTs) composite sponge	Burn wounds, chronic wounds, diabetic foot ulcers	[64]
19	Fiber	Nanofibrous Poly (3-hydroxybutyrate-co-3-hydroxyvalerate)/chitosan mats	Regeneration of skins	[65]
20		Biomimetic LBL nanofibrous matrices assembled by chitosan/collagen	Healing of wounds	[66]
21		Coating of nanofibrous scaffold of Poly (L-lactic acid)/collagen embedded on chitosan	Skin tissue engineering	[67]
22		Chitosan /polyethylene oxide/berberine nanofibers	Leishmania ulcer healing	[68]
23	Gel	Alginate-coated chitosan nanogels	Blood dendritic cells and intracellular vaccine delivery	[69]
24		Silica-modified chitosan/polyethylenimine nanogel	Gene carrier ability	[70]
25		Erythrocytes incorporated pravastatin–chitosan nanogels	Drug carriers in liver cancer treatment	[71]
26		pH-sensitive nanogels, methacrylated carboxymethyl chitosan co-polymerized with methacrylated ortho ester-based monomer	Anti-tumor efficacy; nano-carrier for drug delivery in cancer therapy	[72]
27		Diacrylamide containing ortho ester and methacrylated succinyl chitosan in aqueous solution	Improvement of drug in tumor tissues, enhanced therapeutic efficacy	[73]

(continued)

Table 10.1 (continued)

Sl no	Nanostructure	Composite/scaffold	Application	References
28		physical cross-linking of Pectin hydrogels and chitosan nanogels	Injectable material for biomedical applications	[74]
29		Probiotic Lysate-treated chitosan nanogel	Wound healing; nanomedicine	[75]
30	Film, whisker	Chitosan and eucalyptus wood cellulose nanowhiskers multilayer using LBL	Biomedical	[76]
31	Film	with and without cross-linking of alginate and chitosan multilayer thin films and/or impregnation with extracellular matrix proteins	Tunable model for cell adhesive and non- adhesive multilayered films	[77]
32		Free standing multilayer membranes of chitosan and dopamine-modified hyaluronic acid	Bone tissue engineering	[78]
33		Chitosan/silver nanocomposites films	Skin engineering	[79]
34		LBL coating of Alginate structures with chitosan and arginine–glycine–aspartic acid-impregnated with xanthan gum hydrogels	Vascular tissue engineering	[80]
35	Film, fiber, particle	Layer-by-layer deposition of chitosan nanoparticles coatings on poly(ε-caprolactone) nanofibers	Tissue engineering	[81]

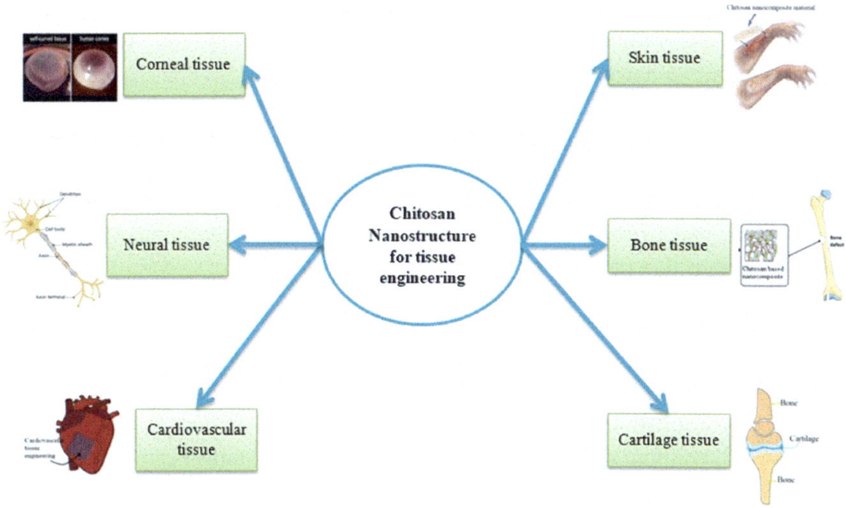

Scheme 10.1 Chitosan nanostructure for tissue engineering applications

Many biomaterials are being used as wound dressings in critical and chronic cases. The rate of wound recovery varies depending upon the nature and characteristics of the used biomaterials. Chitosan-based nanoparticles, nanocoatings, and nanogels have made a significant contribution in developing new scaffolds suitable for skin tissue engineering [23]. Nanomaterials, due to their nanometer size, and very large surface area compared with their volume, have pronounced ability to enter into cytoplasmic space [83].

Cai et al. [84] utilized magnetic Fe_3O_4 nanoparticles (FeNPs) to modify the antibacterial as well as mechanical properties of nanofiber membranes of chitosan (CS)/gelatin (GE) composite [84]. Uniform FeNPs/CS/GE composite nanofibers fabricated by electrospinning was found to gain antibacterial efficacy at ≥ 1 wt% optimum filler loading. Such membranes, with tailored antibacterial as well as mechanical properties, are used as potential wound dressing. Liu et al. [85] synthesized a scaffold using chitosan–silk–fibroin (CS-SF) emulsion and silver nanoparticles (AgNPs) following lyophilization technique and subsequent asymmetric coating by grafting stearic acid (SA) onto CS-SF-AgNP surface [85]. The designed scaffold CS-SF-Ag0.5/SA, used as dressing, has good structural channels, hygroscopicity, moisture retention, and adequate mechanical integrity, and exhibits antibacterial functionality, promoting wound healing. Rahimi et al. [86] fabricated an antimicrobial quaternized chitosan (QC) film and QCA1, QCA2, and QCA3 (QC derived nano-silver nanocomposites) containing 0.125, ~0.25, and ~0.5% w/w AgNPs to the QC, respectively, synthesized by casting method [86]. The parent QC and those three QC derivatives gained prominent antimicrobial efficacies against E. coli, S. aureus, P. aeruginosa, C. albicans, as well as a multi-drug resistant strain from P. aeruginosa. Blood clotting capability, the cell cytotoxicity on human foreskin fibroblasts, cell attachment assays, swelling ratio, and biocompatibility showed the synthesized scaffolds to possess high efficacy for antimicrobial wound dressing. Moxifloxacin (Mox) loaded silver nanoparticles (<100 nm), made via in-situ synthesis, were incorporated successfully by solvent casting technique within the colloidal composite of chitosan-sericin film dressing [87]. A sustained Mox release and cell permeation were indicated by the composite films. The promising antibacterial behavior of such composite films against the pathogenic bacteria viz. Gram-negative (A. baumannii, P. aeruginosa) and Gram-positive (S. epidermidis, S. aureus including Methicillin-resistant S aureus that promotes hospital-acquired skin infections) was observed via in-vitro antibacterial susceptibility studies. With an aim to accelerate wound healing in diabetic patients, Masood et al. [88] developed silver nanoparticle immobilized chitosan-poly ethylene glycol (PEG) hydrogel [88]. The formulated hydrogel indicated a slow and sustained release of AgNPs. The enhanced antibacterial and antioxidant properties of AgNPs loaded chitosan-PEG hydro gels, in comparison to parent chitosan-PEG, recommended the use of the nanogels for wounds healing in diabetic rabbits. A sponge-like nano-Ag/ZnO-loaded chitosan composite dressing material was developed [89], using lyophilization technique and subsequently modifying the chitosan sponge by incorporating Ag/ZnO nanocomposites. In-vitro antibacterial studies against drug-resistant pathogenic bacteria proved its high antibacterial efficacies. In-vivo studies proved that the chitosan-Ag/ZnO would have faster capability of wound healing with denser collagen deposition properties and more complete re-epithelialization in mice. Zinc oxide (ZnO) nanoparticles doped with potassium (K) containing porous hydrogels were developed following the freeze gelation technique [90]. CH-$K_{2.0}$, the ZnO NPs, containing 2% potassium (dopant), showed a higher degree of swelling than CLH-$K_{1.0}$ and CLH-$K_{0.5}$ (1.0% and 0.5% potassium dopant concentration, respectively). With an aim to investigate their angiogenic potential the synthesized hydrogels were rooted on the chick chorioallantoic membrane. The CLH-$K_{2.0}$ hydrogel stimulated greater angiogenesis compared to the other two, when blood vessels were allowed to grow inside the scaffold. The large extent of uniformly distributed pores, potent angiogenic ability, biodegradability, and biocompatibility make the scaffolds effective for skin tissue regeneration and wound healing. Ding et al. [91] developed a spongy dressing, the chemically cross-linked bilayer composite, for full-thickness dermal wound healing applications. The layers were composed of genipin cross-linked chitosan-Ag nanoparticles as the upper layer and genipin cross-linked chitosan partially oxidized Bletilla striata polysaccharide as the lower layer [91]. The lower layer enhanced cell proliferation while the upper layer suppressed the wound bacterial (L929 cell) proliferation. However, both the layers indicated more gas permeation ability and high-water retention capacity. The bilayer showed faster tissue rebuilding capability and improved antibacterial activity compared to commonly used medical gauze and hence could act as a potential dressing for healing and protection of wound. Andrographolide (AND) lipid nanocarriers enriched chitosan hyaluronic acid (HAL) composite sponge scaffold was synthesized by Sanad and Abdel-Bar 2017 [92]. The optimized lipid nanoparticle formulation, using solvent diffusion method, was incorporated successfully into the chitosan–HAL scaffold. The synthesized chitosan–HAL/NLC4 (nanostructured lipid carriers) scaffold was found to possess perfect porosity, reduced scar formation, swelling ratio-controlled drug release, and improved histological progress, making the scaffold effective for healing of wounds. The antioxidant as well as anti-inflammatory activity of chitosan and HAL in combination with AND were established. A mixture of chitosan-l-glutamic acid (CG) and silver nitrate ($AgNO_3$) was freeze dried to yield CG/AgNP. This was loaded with HAL solution to synthesize chitosan-l-glutamic acid/hyaluronic acid, a spongy composite. The composite possessed higher water retention capacity, swelling behavior, and good mechanical properties. In-vitro studies revealed that the composite effectively inhibited bacterial (S. aureus and E. coli) penetration and growth. Spongy composite with low AgNP concentrations was found to be non-toxic and hence safe to L929 cells, and promoted wound healing [93]. Chitosan/alginic acid/zinc oxide (CS/AA/ZnO) nanostructured hydrogel sponges were developed by Bal-Ozturk et al. [94], following freeze-drying technique and immobilizing ZnO nanoparticles (<100 nm) into polymer matrix using genipin (crosslinker). ZnO-incorporated hydrogel sponges showed good bacteriostatic effect on S. aureus [94]. The hemostatic efficacy and cytotoxicity of crosslinked nanostructured hydrogel sponges were evaluated *in-vivo and in-vitro* studies, respectively. The antibacterial property was found to increase with the increase of ZnO content in the polymer network. Kumari et al. [95] developed chitosan and gelatin-based scaffolds doped with copper NPs, for skin tissue engineering [95]. The scaffold incorporated with 0.1% copper nanoparticles performed the best for skin cell culture. Cell viability ensured the proliferation and growth of the skin cells over the scaffolds. Oxidative stress evaluation on the cells showed no production of reactive oxygen species (ROS) with the scaffolds. The scaffolds developed via copper NPs doping showed appropriate biological and physico-chemical properties for skin tissue engineering.

Vaccarin (VAC-NPs) hydrogel-loaded chitosan NPs were developed to evaluate its efficacy in a rat model in promoting the full-thickness dermal wound healing [96]. The encapsulation efficiency of VAC in NPs was estimated to be $51.7 \pm 1.7\%$. The VAC-NPs were estimated to possess uniform distribution and particle size, with no aggregation. The in-vitro biocompatibility of VAC-NPs was assessed employing human umbilical vein endothelial cells. In-vivo studies revealed that VAC-NPs promoted wound healing, following a mechanistic pathway through up-regulating a cytokine (interleukin, IL-1β) and platelet-derived growth factor. Radwan-Pragłowska et al. [97], developed some hybrid poly lactic acid (PLA)/chitosan nanofibers doped with metal NPs (ZnO, Fe_3O_4, and Au) and investigated the morphology by TEM analysis [97]. The prepared nanofibers were incorporated over the acylated chitosan porous layer so as to enable and promote the process of neovascularization. The top and the bottom layers were developed from PLA nanofibers, and the acylated chitosan, respectively. The hybrid materials, shown to have good water vapor permeability, conductivity, and biodegradation susceptibility, promote the healing process for wounds. The lack of bioactivity and cytotoxicity was observed using both primary cells (human dermal fibroblasts) and fibroblasts cell line (L929).

Curcumin (Cur) loaded chitosan nanoparticle (CSNPs) within poly (ε-caprolactone) and gelatin fiber mat were employed [98] to assess in-vivo wound healing capability. The incorporation of Cur containing CNs and Gela improved PCL fiber hydrophilicity, wettability, degradability, and decreased mechanical property. The cell attachment as well as proliferation on the PCL/Gela/NCs/Cur was found higher in comparison to PCL and PCL/Gela scaffolds. The potential of the scaffolds as a skin substitute to repair and regenerate the skin was evaluated in histological analysis using endometrial stem cells (EnSCs). In a full-thickness excisional animal model and desirable biocompatibility for the host immune system, human endogenous neural stem cells seeded scaffold showed wound healing ability. The scaffold could be recommended as a skin substitute for regenerative medicine applications and to repair injured skin.

10.5.2 Bone Tissue Engineering

Bone is a highly vascularized tissue consisting of 10% water, 30% matrix, and 60% mineral. It contains both

(a) inorganic substances (69%): hydroxyapatite (HAP) (99%), which is responsible for bone rigidity and hardness and
(b) organic substances (22%): collagen (Col) (90%) and proteoglycans, sialoproteins and glycoproteins, the non-Col structural proteins [99–101].

Osteoclasts, osteocytes, osteoblasts, and bone lining cells, the four types of active cells, responsible for mineralized bone matrix, regulate dynamic bone resorption, bone formation, and bone remodeling [101]. Bone transplanting for patient, especially elder one, is a big clinical challenge for the medical surgeons. Although bone has, in certain extent, its own healing ability, in cases of large defects, age-related factors, injuries after accidents and bone tumor, etc., resection prevents desirable bone regeneration. Generally, allograft, autograft, and xenograft are the common medical protocols to repair damaged bone tissues [102]. Among all, autograft is considered as a standard protocol by which cells from the injured patient are harvested to repair the damaged bone tissue [103]. In allografts, cells from other persons and in case of xenografts, cells from a different species, viz., pigs, are used to repair tissues. However, such grafting procedures have several limitations viz., poor availability, scar formation, mismatching, tissue rejection, and disease transmission [104]. Currently, bone tissue engineering mainly focuses on designing three-dimensional (3D) scaffolds which mimic the extra cellular matrix (ECM), providing mechanical assistance toward formation of new bones. Scaffolds should have osteoinductive, osteoconductive, and osteogenic properties to assist survival, adhesion, together with osteogenic cell migration. A more challenging factor is to arrest scaffold degradation after natural tissue formation [101, 105, 106]. Biomaterials, particularly, chitosan, due to their biodegradability, biocompatibility, and bioactivity have received great success to build suitable scaffolds. Chitosan nanocomposites have the ability to improve biological and/or physical properties of the nanocomposites and are the most versatile choice for bone engineering applications.

ZrO_2 nanoparticles doped chitosan polyvinyl alcohol hydroxyapatite (CTS-PVA-HAP) composites viz. (ZrCPH I–III) were fabricated for bone tissue engineering applications [107]. The tensile strength of ZrCPH I–III was improved than that of the composite CTS–PVA–HAP. However, improved tensile strength, strongest antimicrobial effect, and greatest osteoblastic MG-63 cell proliferation were demonstrated in ZrCPH III containing maximum extent of nano-HAP-ZrO_2 compared to ZrCPH I and ZrCPH II. The strongest antimicrobial effect and cytocompatibility of ZrCPH III and the maximum cell proliferation for human osteoblastic MG-63 cells were demonstrated. In order to use in the treatment of osteosarcoma chitosan nanoparticles loaded with copper (CuCSNPs) was synthesized by Ai et al. [108]. The recommended size range of CuCSNPs is <200 nm for drug delivery applications and cancer targeting. CuCSNPs showed a significantly improved anticancer effect than that of parent copper sulphate ($CuSO_4$). The studies revealed a higher mitochondrial ROS generation capacity of CuCNPs compared to that of the control. CuCSNPs were found to exhibit higher apoptotic activities viz. sixfold and twofold than the control and $CuSO_4$, respectively, as shown by Caspase 3/7 activity. Thus, the anticancer effect of copper was shown to improve effective therapy in osteosarcoma, by incorporating it in biocompatible chitosan nanoparticles. Zhou et al. [109] synthesized hydroxyapatite (HAP) and whitlockite (WH) hollow microspheres, following microwave-assisted hydrothermal method, and using an organic phosphorus source, e.g., creatine phosphate disodium salt [109]. The membranes, such as, WH/chitosan and HAP/chitosan composites were prepared following freeze-drying method, to study in-vitro biocompatibility was investigated. In a rat model bone regeneration in calvarial defects was studied using these porous scaffolds. WH/chitosan scaffold showed better biocompatibility for human mesenchymal stem cells, with enhanced proliferation and ability of osteogenic differentiation as compared to HAP/chitosan.

The porous WH/chitosan scaffold was suggested to have higher applicability in TE applications. Erbium-doped hydroxyapatite (eHAP)-chitosan, a fluorescent nanocomposite film, was prepared to employ erbium (III) ion as a dopant and following hydrothermal assisted precipitation method [110]. A strong antimicrobial activity was observed through contact inhibition against S. aureus and E. coli on agar plates using eHAP-incorporated chitosan films. Furthermore, an excellent biocompatibility was demonstrated for human lung fibroblast cells (WI-38), indicating proliferation and strong attachment with the chitosan films having poor cytotoxicity. Shakir et al. [111] synthesized Euryale ferox (EF)/ nano-hydroxyapatite (n-HAP)/ chitosan (CS) tricomposite scaffold. The n-HAP/CS/EF composite showed better antibacterial effect on different bacterial pathogens [111]. In simulated body fluid, the biomineralization of n-HAP/CS/EF was observed after 2 weeks, which was further confirmed by alizarin red S staining procedure. The osteoconductivity of the scaffold was assessed from osteogenic differentiation test (ALP assay).

The scaffold was suggested to be a promising candidate for use in bone defect reparation. Maji et al. [112] fabricated gelatin/carboxymethyl chitosan (CMC)/nano-hydroxyapatite matrix, a 3D scaffold, having macroporosity employing a combined technique viz, high stirring induced foaming and freeze-drying method [112]. The synthesized macroporous (SGC) scaffold showed high porosity, better water retention capacity, high mechanical strength, and greater pore size, than the non-macroporous (NGC) one. The wjhMSC-MT (Wharton's jelly Mesenchymal stem cells micro tissue) biological studies supported its viability, proliferation, differentiation, and high mineralization capacity on the SGC. The micro tissue seeded on the SGC showed a nearly up-regulated expression of collagen type I, Osteocalcin, and Runx2 expression upon osteogenic differentiation. A silk fibroin (SF) coating loaded with AgNPs/gentamicin (Gen) having desirable osteogenic aptitude and antibacterial activity was constructed, with an aim to get rid of poor osseointegration and titanium implants-associated infection [113].

A chitosan barrier layer was developed for the prebuilt SF-based films following spin-coating (SCS) and dip-coating (DCS). An improved hydrophilic and protein-absorbed performance was observed in chitosan barrier layer constructed by SCS. The SCS barrier multilayer coating, also, showed biofilm-inhibited and bacteria-trigged antibacterial activities. However, limited enhancement in chitosan barrier layer, constructed by DCS coating for antibacterial abilities, was observed. It was suggested that accelerated bactericides release due to collapse of chitosan barrier layer was triggered by pH lowering, induced by the bacteria attachment following proliferation. Moreover, the functional coatings indicated migration, enhanced initial adhesion, and preosteoblast MC3T3-E1 cell proliferation. This resulted in the subsequent production of alkaline phosphatase, i.e., accelerated osteoblast differentiation.

With a target to enhance the biological attributes, including antibacterial activity of titanium implants, Li et al. [114] prepared a micro-nanostructured HAP coating on titanium surface employing micro-arc oxidation, and subsequent loading of chitosan (as antibacterial agent), using dip-coating method, on the HAP surface [114]. The study in simulated body fluid revealed that HAP/CS composite coating enhanced HAP layer formation, spreading, cell proliferation, and adhesion. Inhibition of bacterial growth as well as improved biological and antibacterial properties were also observed. Thus, the HAP/CS coating was recommended for promising applications in orthopedics. Zhao et al. [115] constructed gadolinium phosphate/chitosan ($GdPO_4$/CS) scaffolds by lyophilization method. The macropores were developed by connecting the plate-like films. An enhancement of the osteoconductivity of the $GdPO_4$/CS scaffolds was observed for bone tissue growth via cell adhesion [115]. The released Gd^{3+} ions were found to be in safe concentration effecting no toxicity to rBMSCs (mesenchymal stem cells of rat bone marrow). Significantly, $GdPO_4$/CS scaffolds promoted osteogenic differentiation of rBMSCs through activated Smad/Runx2 signaling route. *In-vivo* studies showed that $GdPO_4$/CS scaffolds, compared to β-TCP (β-tricalcium phosphate)/CS scaffolds, significantly accelerated new bone regeneration and, thus, were suggested as a promising material for regeneration of bone. In-vitro and in-vivo osteogenic potential of chitosan scaffolds crosslinked with HAP-incorporated guanosine 5′-diphosphate (purine), with or without pyrophosphatase activity, was examined [116]. Cells embedded scaffolds significantly enhanced proliferation (up to three-fold), alkaline phosphatase activity (two-fold), and osterix expression (two-fold). Moreover, increase of calcium phosphate deposits (in-vitro) was observed. The chitosan scaffolds, having both hydroxyapatite and pyrophosphatase, when implanted in a mouse model at the fracture site showed increased callus formation. Thus, pyrophosphatase hydroxyl apatite composite scaffold may act as a promising scaffold to facilitate bone fracture healing.

10.5.3 Cartilage Tissue Engineering

Chondrocytes in the cartilage tissue produce type II collagen and chondroitin sulphate or aggrecan, the ECM proteins [117]. Aggrecan, a critical component for cartilage structure, provides cartilage and intervertebral disc the ability to resist compressive loads. The tensile strength of the tissue are governed by these proteins. Degradation of ECM, due to pathological situations like traumatic conditions, osteo, and rheumatoid arthritis may lead to loss of cartilaginous tissues. Self-repair of the injured articular cartilage is very challenging, as cartilage tissue is avascular, inhomogeneous, and of complex structure and functions under an extremely tough environment [118]. Currently, different clinical treatments involving autograft, mosaicplasty, micro facture, or autologous chondrocytes injection have although been employed, have some common limitations due to the structural and biomechanical mismatch of the newly repaired cartilage tissues with the native cartilage tissues and additional high risk of disease transmission [119].

In recent years, nanotechnology for cartilage tissue repair has gained great impetus. The nanocomposite-based scaffolds play the most significant role in cartilage and bone tissue engineering. Liu et al. [120] reported that chitosan may inactivate metal proteinases, responsible for collagen degradation, and interact with the growth factors for their retention at the healing site. Moreover, being immune and nontoxic, it is a safe player in the biomedical applications [120]. However, to overcome the difficulties such as low mechanical strength and poor elasticity, chitosan was hybridized using biopolymer, silk fibroin, nanocomposite, genipin, etc. Different glycosaminoglycans (GAGs) that are present in articular cartilage, due to the similarity in structure to chitosan, are suggested as suitable matrix for cartilage repair. Chitosan in complexation with silk fibroin has been used for TE due to its biological compatibility and excellent mechanical strength. Maleilated chitosan (MCS)/methacrylated silk fibroin (MSF) nanocomposite hydrogels were synthesized for applications of cartilage tissue repair [121]. The process follows photo crosslinking of MCS and MSF in aqueous solutions under UV light irradiation in presence of the photo initiator Darocur 2959. The nanocomposite with MSF content of 0.1%, showed compressive modulus in the range of that of articular cartilage, with a value of 0.32 ± 0.07 MPa for the present composite. Cell culture and in-vitro cytotoxic exhibited that the nanocomposites with TGF-β1 were biocompatible to articular chondrocytes of mouse with well-supported cell attachment. Thus, the scaffolds could be applied for cartilage tissue repair. Poly (3-Hydroxybutyrate)-Chitosan/Al_2O_3 nanocomposite scaffold, for use in cartilage tissue repair, was synthesized using alumina (Al_2O_3) nanowires and polyhydroxy butyrate-chitosan (PHB-CS) solution, following electro spinning technique [122]. The tensile strength of PHB and PHB-CS scaffolds was found to be significantly increased on addition of alumina. The cell viability and cell attachment studies exhibited that the rabbit chondrocyte cells spread on PHB-CS and PHB-CS/3% Al_2O_3 composite scaffolds were in a larger extent than PHB scaffolds. Improvement of mechanical and physical properties of chitosan-based scaffolds was achieved by graphene oxide (GO) nanoparticles incorporation via cross-linking [123]. An increased cell proliferation with augmentation of the percentage of GO was observed in a prolonged cultivation period of 14 days, on seeding of the human articular chondrocytes to the scaffolds. The study of the human articular chondrocyte indicated more spherical morphology of the cells with the cross-linked scaffolds during in vitro culture for 21 days. Nanofabrication using electrospinning was employed for the preparation of chitosan-gelatine/single-walled carbon nanotubes functionalized by COOH (SWNTs-COOH) electrospun nanocomposite scaffolds [124]. The porosity, morphology, mechanical properties, thermal behavior, chemical structure, wettability, biodegradability, as well as in vitro cell culture study with all the scaffolds were evaluated. The porosity percentage (>80%) of all scaffolds match with that of scaffolds suitable for use in TE. The presence of SWNTs-COOH did not show any adverse effect on cytocompatibility. Among all the scaffolds, the chitosan-gelatin/1wt% SWNTs-COOH was found to be effective in cartilage tissue engineering. Mirmusavi et al. [125] prepared a highly strengthened electrospun nanocomposite scaffolds viz. polycaprolactone (PCL)-chitosan/carboxyl-functionalized multi-walled carbon nanotubes (MWCNTs) using varying concentrations of MWCNTs. Morphology, porosity, hydrophilicity, tensile strength, biodegradation, bioactivity, and cell response of the samples were evaluated to assess the efficacy of the scaffolds in cartilage tissue engineering [125]. Scaffold containing MWCNTs showed more cell viability when chondrocytes were well cultured compared to the sample without MWCNTs. The PCL-chitosan/0.5wt% MWCNTs scaffold could be used in cartilage tissue engineering applications.

10.5.4 Cardiovascular Tissue Engineering

The world in the twenty first century is facing a challenging health problem of cardiovascular diseases. Cardiomyocytes death is mostly caused due to myocardial infarction, the most common cardiac disease. The end result of thrombosis, atherosclerosis or cardiac dysfunction, and eventual heart failure are now the most frequent and prevalent cases of serious concern. Currently, there is no standard medical procedure to repair the damaged myocardium other than use of pharmaceutical drugs that thin blood, and/or heart transplantation, to boost the heart pump functioning [126]. It is highly a necessity of the day to formulate protocols toward damaged cardiac tissue repair. A successful vascular graft which provides mechanical strength for the prevention of surface thrombosis and also maintains well-organized structures combining with ECM proteins can be achieved through TE.

Limongi et al. [127] demonstrated that surface modification of biomaterials on nanoscales enhances cardiac tissue repair by influencing cell adhesion, alignment, and differentiation [127]. Kalishwaralal et al. [128] prepared a composite film, the chitosan-selenium nanoparticle (SeNPs) film that was employed in the cardiac patches to generate electrical conductivity [128]. The H9C2 cells incorporated with in chitosan film SeNPs were found to be very well mature and proliferated. The synthesized film, as substrate, may develop improved cellular behavior via electrical stimulation, and mechanical strengthening and may therefore be used as an appropriate biocompatible film for engineered cardiac tissues. Chitosan scaffolds containing a conductive polymer viz. poly (3,4-ethylenedioxythiophene) polystyrene sulfonate (PEDOT: PSS), was designed by Abedi et al. [129]. PEDOT: PSS was added to the CS/PVA (polyvinyl alcohol) composites for fabrication of scaffolds having different PEDOT: PSS content (0.3, 0.6, and 1 wt%), using electrospinning technology. The mechanical and electrical properties, as well as cell viability and biocompatibility of the scaffolds, were examined. The scaffolds exhibited a significant increase in electrical conductivity, tensile and mechanical strength of heart tissues together with improvement of biocompatibility and cell viability. TiO_2 nanoparticles incorporated PEG/CS (PEGylated chitosan) composite hydrogel was developed as a cardiac material to increase functional activity in cardiac tissue repair [130]. The engineered TiO_2 nanoparticles are homogeneously dispersed and cross-linked with PEG/CS hydrogels. The adhesion of cardiomyocytes along with cell retention activity applying this hydrogel network was found to be increased. Chitosan matrix loaded with gold nanoparticles (GNPs), for increasing electrical coupling between adjacent mesenchymal stem cells, was studied by Baei et al. [131]. The hybrid hydrogel CS-GNP when seeded with mesenchymal stem cells (MSCs) the cell viability and growth rate were maintained. However, an enhanced cardiomyogenic differentiation of MSCs promoted by GNP's electrical properties harnessing was observed.

10.5.5 Neural Tissue Engineering

The complex network of nervous system consisting of nerves and neurons, transporting messages via signals to and from the spinal cord and brain to different organs of the body. The nerve system has a very important role in human biological system which helps to interact with physiological processes. Physiological and pathological complex changes happen in the damaged spinal cord tissue due to spinal cord injury (SCI). Damaged nerves can enforce severe consequences, and is hard to repair. Although implantation of allograft, autograft, and xenograft, have been used for repairing peripheral nerve gap, those have many drawbacks such as limited donor, the recipient and donor nerve tissue mismatch, transmission of infections, and thus need complex surgeries [132]. In recent years, TE is getting momentum due to great success in experimental works in neurological tissue repair [126]. Micro and nanostructured scaffolds play an important role to promote repair or regenerate by reconnecting axons. The scaffolds control the repair of damaged axons bridging the transplanted and host cells [133].

Habibizadeh et al. [133] developed some novel 3D scaffold nanocomposite through surface modification of polycaprolactone (PCL)/chitosan nanofiber/net (CS-NFN) using microlayer of alginate hydrogel, simultaneously utilizing unique features of both nanofibers and alginate hydrogel. NFN structure consisting of PCL and chitosan was developed employing electrospinning technology. Coating of PCL/chitosan nanofiber/net was made with alginate loaded neurotrophin-3 (NT-3) microlayer and conjunctiva mesenchymal stem cells (CJMSCs). Compared to the unmodified scaffold a higher proliferation of CJMSCs and lower inflammatory response was found due to nanofiber/net surface modification by microlayer of alginate hydrogel. A sustained release of NT-3 was reported from the scaffolds stimulated neuronal differentiation for up to 21 days and induced regeneration consequently in injured nerve. A differentiation of CJMSCs after 7 days into neuron-like cells was affected. An increase of nesting (a protein marker for detecting newly formed endothelial cells) expression by sixfold, microtubule-associated protein 2 (MAP-2) by 5.4-fold and β-tubulin III genes by 8.8-fold was evidenced from real time polymerase chain reaction (RT-PCR) analysis. The result recommended NT-3 and CJMSCs loaded 3D scaffolds an effective biomaterial in spinal cord injury recovery. Different matrices of chitosan (CS)-NPs were fabricated with different concentrations of gold (Au) nanoparticles (25, 50, and 100 ppm) [134] for in vivo and in vitro studies. In vitro experiments demonstrated better biocompatibility and colony formation. In mesenchymal stem cells (MSCs) matrix metalloproteinase activation was also observed for CS–Au 50 (chitosan embedded with 50 ppm of Au nanoparticles). On CS–Au 50 treatment monocyte conversion, platelet activation, and intracellular generation of ROS were decreased markedly. Moreover, RT-PCR analysis and immunostaining assays revealed CS–Au 50 induced β-Tubulin, glial fibrillary acidic protein (GFAP), and nestin protein expression in MSCs for neural differentiation. CS–Au 50 showed a marked endothelialization and anti-inflammatory abilities when treated in the subcutaneous implantation of a rat model. Nanocomposite membranes *viz.* chitosan/graphene membranes and poly (D, L-lactic-co-glycolic acid) (PLGA)/graphene, were fabricated by incorporating graphene nano-sheets (0.5–1.5 wt. %), to chitosan or PLGA matrix, and employing solution casting method. Chitosan 1.5% graphene (CS/1.5 G) membrane, among all, was found to have about 10^6 times higher electrical conductivity compared to that of neat polymers. A highest proliferation of PC12 cell line ($p < 0.05$) by MTT [3-(4, 5-di-methylthiazol-2-yl)-2, 5-diphenyltetrazolium bromide] assay after 72 h was evidenced [132]. Thus, the nanocomposite membrane was recommended for nerve tissue engineering. A conductive porous scaffold was prepared from chitosan/gelatin matrix by incorporation of polyaniline/graphene (PAG) conductive nanoparticles and used for peripheral nerve regeneration [135]. The mechanical and electrical properties of the scaffolds were found to be enhanced in PAG-incorporated chitosan/gelatin matrix. Among all the scaffolds, C2G6-PAG2.5 containing 2.5 wt. % PAG, yielded highest number of Schwann cells attachment, observed from in-vitro cytotoxicity study, the scaffold was recommended as suitable for the repair of nerve tissue. A polyglycerol sebacate/chitosan/gelatin nanofiber with an average diameter of 80 nm was prepared following electrospinning [136]. MTT assay of cell culture using PC12 cell line approved the capability of the scaffold for application in nerve tissue engineering. A simple and fast approach was employed for the fabrication of nanographene oxide/chitosan (NGO/CS) hydrogels where a crosslinker (lactic acid) was used [137]. The hydrogel was characterized for mechanical properties, swelling, and morphology. The adhesion along with proliferation rate of nerve cells in association with the synthesized hydrogel was ascertained. The study revealed pore structure change, higher mechanical strength of the hydrogel, and up to 20% growth of nerve cells in addition to NGO. Thus, functional hydrogel could be designed for nerve regeneration employing an appropriate concentration of NGO.

10.5.6 Corneal Tissue Engineering

The first lens of the eye is the cornea, which consists of three optically transparent cell layers from outside to the inside involving the epithelium, the endothelium, and the stroma [138]. Damage to any layer can lead to major visual problems. Corneal damage or dysfunction, such as inherent disorders, immune diseases, trauma, and aging results in ultimate vision impairment, due to the reduction of endothelium cell function and may need cornea transplantation. However, corneal transplantation has several drawbacks involving allograft tissue rejection, lack of donor corneas, high cost of surgeries, and poor long-term graft survival [139]. Scaffolds or membranes fabricated from gelatin, chitosan, genipin, polycaprolactone, etc., have great potential in corneal tissue engineering though have some limitations including poor mechanical strength. In order to overcome such limitations, scaffolds are prepared by blending polymers with chitosan. These scaffolds must have optical and mechanical properties similar to those of the cornea. Salehi et al., 2020 discussed in detail polycaprolactone-based biomaterials in corneal tissue engineering [139]. Corneal transplantation is applied mainly for common corneal diseases that suffers from tissue rejection by the host immune system along with a lack of suitable donors. Recently, TE, with its wide range of applicability has proved to be a good substitute for corneal transplantation. In order to facilitate corneal regeneration, different materials have been tested. Polycaprolactone (PCL), being readily available, biocompatible, biodegradable, easy to modify, and low-cost polymer finds promising applications in TE. Moreover, PCL withstands physical, chemical, and mechanical insults without significant property change Tayebi et al. [138] constructed, via the incorporation of chitosan nanoparticles (CSNPs) into chitosan/polycaprolactone (PCL) membranes, a transparent and biodegradable film for application in corneal endothelial cells culture [138]. Different ratios of CSNP/PCL, maintaining constant concentration of chitosan were taken and solvent casting method was applied for the film construction. With increase of the CSNP/PCL ratio, surface wettability and transparency were improved. CSNP/PCL 50/25, with the lowest water contact angle, was indicated to develop transparency comparable to that of human acellular corneal stroma. Flow cytometry, cell counting, SEM, as well as hematoxylin and eosin staining studies indicated appropriate scaffold attachment to human corneal endothelial cells and a compact monolayer formation. A hydroxypropyl chitosan–gelatin-derived scaffold was synthesized employing 1,4- butanediol diglycidyl ether as the crosslinker, and tested for use in corneal stroma tissue engineering. The scaffold exhibited 83–88% visible light transmission, permeation of glucose and common salt, high water content, and was suitable for keratocyte growth on its surface [140]. An ultrathin chitosan–poly (ethylene glycol) hydrogel film was synthesized for corneal tissue engineering [141]. The film was >95% optically transparent in the visible region having identical tensile stress and strain to those of human corneal tissue. Chitosan/polycaprolactone blended membranes were used to study corneal endothelial cell (CEC) differentiation mechanism [142]. The stable blended membranes were found to be acted as an effective cell culture system under extracellular matrix compositions.

10.6 Conclusion

Recent research studies implies the efficacy of chitosan as a smart, approachable, and promising biopolymer with wide-ranging functionalities due to brilliant physicochemical properties, low toxicity, biodegradability, brilliant biocompatibility, and multipurpose biological activity. Chitosan has definite interactions with cells, proteins, and living organisms. Hence, it fascinates significant interest in biomedical fields, including cell culture, drug delivery, tissue engineering, bioimaging, and many more promising applications such as antimicrobial and biotechnology. Nanostructured chitosan-containing scaffolds are gaining interest in recent years in tissue engineering application. Due to its unique mechanical properties of chitosan, chitosan can easily be transformed into different nanoforms and polymer, organic, inorganic, and metallic nanocomposite can be also applied, as additive to chitosan, to fabricate different kinds of nanostructures such as nanofilm, nanofiber, nanoparticle, nanohydrogel, and nanonet. As we studied other research reports on nanostructured chitosan derivatives for tissue engineering which highlights the greatest promise of such bio-scaffolds to achieve extensive improvements in long-term therapeutic use, drug release for tissue fixation, regeneration and creating fascinating nanostructure materials which improve tissue repair mechanism in human. Researchers anticipate the successful use of nanostructure chitosan derivative materials in different types of tissues such as skin, bone, cartilage, cardiovascular, corneal, and neural tissue. FeNPs/CS/GE membranes, nano-Ag/ZnO-loaded chitosan composite, CS/AA/ZnO nanostructured hydrogel, and VAC-CSNPs hydrogel were great development in skin regeneration. ZrO_2 nanoparticles doped CS-PVA-HAP composite showed great tensile strength in bone tissue engineering, CuCNPs showed a significant anticancer effect, eHAP incorporated chitosan films showed a strong antimicrobial activity, n-HAP/CS/EF composite showed a better antibacterial activity, these scaffolds were suggested to be a promising candidate for bone defects reparation. MCS/MSF nanocomposite hydrogels and Al_2O_3 nanowire added PHB and PHB-CS scaffolds showed excellent mechanical strength in cartilage tissue repair. Also, chitosan–gelatin/1wt% SWNTs-COOH and PCL-CS/0.5wt% MWCNTs scaffolds were used in cartilage tissue regeneration. Chitosan-SeNPs, CS-GNPs, and chitosan containing a conductive polymer viz PEDOT:PSS scaffolds were shown to exhibit a significant increase in electrical conductivity, tensile, and mechanical strength of heart tissues. CSNPs fabricated gold nanoparticles and PCL/CS-NFN with alginate scaffolds showed great improvement in nerve injury. Conductive PAG nanoparticles into a CS/gelatin matrix and NGO/CS hydrogel are recommended scaffolds for nerve regeneration. CSNP/PCL blended membranes were found an effective cell culture system under extracellular matrix compositions for corneal tissue repair.

The use of nanostructure chitosan derivatives is expected to have major implications for tissue engineering resulting in quality of the tissue outcomes, due to its natural similarity in chemistry, structure, and function of hosting environment. Despite the new technology, further research is needed to investigate the progression, mechanisms to improve cell-specific interactions, more quality tissue regeneration, and its long-term biocompatibility effects in the human body.

References

1. D. Vukajlovic, J. Parker, O. Bretcanu, K. Novakovic, Chitosan based polymer/bioglass composites for tissue engineering applications. Mater. Sci. Eng. **C96**, 955–967 (2019)
2. F.H. Zulkifli, F.S.J. Hussain, S.S. Zeyohannes, M.S.B.A. Rasad, M.M. Yusuff, A facile synthesis method of hydroxyethyl cellulose-silver nanoparticle scaffolds for skin tissue engineering applications. Mater. Sci. Eng. **C79**, 151–160 (2017)
3. N. De Isla, C. Huseltein, N. Jessel, A. Pinzano, V. Decot, J. Magdalou, D. Bensoussan, J.F. Stoltz, Introduction to tissue engineering and application for cartilage engineering. Bio-Med. Mater. Eng. **20**(3–4), 127–133 (2010)
4. E. García-Gareta, M.J. Coathup, G.W. Blunn, Osteoinduction of bone grafting materials for bone repair and regeneration. Bone **81**, 112–121 (2015)
5. C.H. Kim, S.J. Park, D.H. Yang, H.J. Chun, Chitosan for tissue engineering, in Novel biomaterials for regenerative medicine, ed. by H. Chun, K. Park, C.H. Kim, G. Khang. Adv. Exp. Med. Biol, vol. 1077 (2018), pp. 475–485
6. M.M. Islam, M. Shahruzzaman, S. Biswas, M.N. Sakib, T.U. Rashid, Chitosan based bioactive materials in tissue engineering applications-A review. Bioact. Mater. **5**(1), 164–183 (2020)
7. P. Aiello, S. Consalvi, G. Poce, A. Raguzzini, E. Toti, M. Palmery, M. Biava, M. Bernardi, M.A. Kamal, G. Perry, I. Peluso, Dietary flavonoids: nano delivery and nanoparticles for cancer therapy. Semin. Cancer Biol. **69**, 150–165 (2021)
8. A. Khan, K.A. Alamry, Recent advances of emerging green chitosan-based biomaterials with potential biomedical applications: a review. Carbohydr. Res. **506**, 108368 (2021)
9. R. Vecchione, V. Quagliariello, P. Giustetto, D. Calabria, A. Sathya, R. Marotta, M. Profeta, S. Nitti, N. Silvestri, T. Pellegrino, R.V. Iaffaioli, Oil/water nano-emulsion loaded with cobalt

ferrite oxide nanocubes for photo-acoustic and magnetic resonance dual imaging in cancer: In vitro and preclinical studies. Nanomed.: Nanotechnol. Biol. Med. **13**(1):275–286 (2017)
10. H. Yilmaz Atay, Antibacterial activity of chitosan-based systems, in *Functional Chitosan*. (Springer, Singapore, 2019), pp.457–489
11. M.E. Abd El-Hack, M.T. El-Saadony, M.E. Shafi, N.M. Zabermawi, M. Arif, G.E. Batiha, A.F. Khafaga, Y.M. Abd El-Hakim, A. Al-Sagheer, Antimicrobial and antioxidant properties of chitosan and its derivatives and their applications: a review. Int. J. Biol. Macromol. **164**, 2726–2744 (2020)
12. Y. Xia, D. Wang, D. Liu, J. Su, Y. Jin, D. Wang, B. Han, Z. Jiang, B. Liu, Applications of chitosan and its derivatives in skin and soft tissue diseases. Front. Bioeng. Biotechnol. **10**, 894667 (2022)
13. I. Shahid Ul, B.S. Butola, Recent advances in Chitosan polysaccharide and its derivatives in antimicrobial modification of textile materials. Int. J. Biol. Macromol. **121**, 905–912 (2019)
14. M. Sarkar, P. Majumdar, Application of response surface methodology for optimization of heavy metal biosorption using surfactant modified chitosan bead. Chem. Eng. J. **175**, 376–387 (2011)
15. P.G. Ferreira, V.F. Ferreira, F.D.C. da Silva, C.S. Freitas, P.R. Pereira, V.M.F. Paschoalin, Chitosans and nanochitosans: recent advances in skin protection, regeneration, and repair. Pharmaceutics **14**(6), 1307 (2022)
16. F. Croisier, C. Jerome, Chitosan-based biomaterials for tissue engineering. Eur. Polym. J. **49**(4), 780–792 (2013)
17. B. Focher, P.L. Beltrame, A. Naggi, G. Torri, Alkaline N-deacetylation of chitin enhanced by flash treatments. Reaction kinetics and structure modifications. Carbohydr. Polym. **12**(4):405–418 (1990)
18. A. Pelletier, I. Lemire, J. Sygusch, E. Chornet, R. Overend, Chitin/chitosan transformation by thermo-mechano-chemical treatment including characterization by enzymatic depolymerization. Biotechnol. Bioeng. **36**(3), 310–315 (1990)
19. T. Lertwattanaseri, N. Ichikawa, T. Mizoguchi, Y. Tanaka, S. Chirachanchai, Microwave technique for efficient deacetylation of chitin nanowhiskers to a chitosan nanoscaffold. Carbohydr. Res. **344**(3), 331–335 (2009)
20. I. Aranaz, M. Mengíbar, R. Harris, I. Paños, B. Miralles, N. Acosta, G. Galed, A. Heras, Functional characterization of chitin and chitosan. Curr. Chem. Biol. **3**(2), 203–230 (2009)
21. J. Cai, Y. Du YangJ, L. Fan, Y. Qiu, J. Li, J.F. Kennedy, Enzymatic preparation of chitosan from the waste Aspergillus niger mycelium of citric acid production plant. Carbohydr. Polym. **64**(2), 151–157 (2006)
22. M.M. Rahman, S. Kabir, T.U. Rashid, B. Nesa, R. Nasrin, P. Haque, M.A. Khan, Effect of γ-irradiation on the thermomechanical and morphological properties of chitosan obtained from prawn shell: evaluation of potential for irradiated chitosan as plant growth stimulator for Malabar spinach. Radiat. Phys. Chem. **82**, 112–118 (2013)
23. A. Madni, R. Kousar, N. Naeem, F. Wahid, Recent advancements in applications of chitosan-based biomaterials for skin tissue engineering. J. Bioresour. Bioprod. **6**(1), 11–25 (2021)
24. R.C. Chien, M.T. Yen, J.L. Mau, Antimicrobial and antitumor activities of chitosan from shiitake stipes, compared to commercial chitosan from crab shells. Carbohydr. Polym. **138**, 259–264 (2016)
25. C.Y. Yan, J.W. Gu, D.P. Hou, H.Y. Jing, J. Wang, Y.Z. Guo, H. Katsumi, T. Sakane, A. Yamamoto, Synthesis of TAT tagged and folate modified N-succinyl-chitosan self-assembly nanoparticles as a novel gene vector. Int. J. Biol. Macromol. **72**, 751–756 (2015)
26. M. Kong, X.G. Chen, K. Xing, H.J. Park, Antimicrobial properties of chitosan and mode of action: a state-of-the-art review. Int. J. Food Microbiol. **144**(1), 51–63 (2010)
27. A.C. Marín, M. Culcasi, M. Cassien, P. Stocker, S. Thetiot-Laurent, B. Robillard, P.S. ChinniciF, Chitosan as an antioxidant alternative to sulphites in oenology: EPR investigation of inhibitory mechanisms. Food Chem. **285**, 67–76 (2019)
28. K. Azuma, T. Osaki, S. Minami, Y. Okamoto, Anticancer and anti-inflammatory properties of chitin and chitosan oligosaccharides. J. Funct. Biomater. **6**(1), 33–49 (2015)

29. V.N. Davydova, A.A. Kalitnik, P.A. Markov, A.V. Volod'Ko, S.V. Popov, I.M. Ermak, Cytokine-inducing and anti-inflammatory activity of chitosan and its low-molecular derivative. Appl. Biochem. Microbiol. **52**(5), 476–482 (2016)
30. A. Verlee, S. Mincke, C.V. Stevens, Recent developments in antibacterial and antifungal chitosan and its derivatives. Carbohydr. Polym. **164**, 268–283 (2017)
31. D. Yu, J. Feng, H. You, S. Zhou, Y. Bai, J. He, H. Cao, Q. Che, J. Guo, Z. Su, The microstructure, antibacterial and antitumor activities of chitosan oligosaccharides and derivatives. Mar. Drugs **20**(1), 69 (2022)
32. P.S. Bakshi, D. Selvakumar, K. Kadirvelu, N.S. Kumar, Chitosan as an environment friendly biomaterial–A review on recent modifications and applications. Int. J. Biol. Macromol. **150**, 1072–1083 (2020)
33. A. Ali, S. Ahmed, A review on chitosan and its nanocomposites in drug delivery. Int. J. Biol. Macromol. **109**, 273–286 (2018)
34. A.O. Silva, R.S. Cunha, D. Hotza, R.A.F. Machado, Chitosan as a matrix of nanocomposites: a review on nanostructures, processes, properties, and applications. Carbohydr. Polym. **272**, 118472 (2021)
35. S.C. Fernandes, C.S. Freire, A.J. Silvestre, C.P. Neto, A. Gandini, L.A. Berglund, L. Salmen, Transparent chitosan films reinforced with a high content of nanofibrillated cellulose. Carbohydr. Polym. **81**(2), 394–401 (2010)
36. V. Vivcharenko, A. Benko, K. Palka, M. Wojcik, A. Przekora, Elastic and biodegradable chitosan/agarose film revealing slightly acidic pH for potential applications in regenerative medicine as artificial skin graft. Int. J. Biol. Macromol. **164**, 172–183 (2020)
37. V. Chimisso, V. Maffeis, D. Hürlimann, C.G. Palivan, W. Meier, Self-assembled polymeric membranes and nanoassemblies on surfaces: preparation, characterization, and current applications. Macromol. Biosci. **20**(1), 1900257 (2020)
38. F.J. Pavinatto, L. Caseli, O.N. Oliveira Jr., Chitosan in nanostructured thin films. Biomacromolecules **11**(8), 1897–1908 (2010)
39. E. Zhang, R. Xing, S. Liu, Y. Qin, K. Li, P. Li, Advances in chitosan-based nanoparticles for oncotherapy. Carbohydr. Polym. **222**, 115004 (2019)
40. U. Garg, S. Chauhan, U. Nagaich, N. Jain, Current advances in chitosan nanoparticles based drug delivery and targeting. Adv. Pharm. Bull. **9**(2), 195 (2019)
41. J.P. Chen, G.Y. Chang, J.K. Chen, Electrospun collagen/chitosan nanofibrous membrane as wound dressing. Colloids Surf. A Physicochem. Eng. Asp. **313**, 183–188 (2008)
42. M. Mahdavi, N. Mahmoudi, F. RezaieAnaran, A. Simchi, Electrospinning of nanodiamond-modified polysaccharide nanofibers with physico-mechanical properties close to natural skins. Mar. Drugs **14**(7), 128 (2016)
43. K. Ghosal, C. Agatemor, Z. Spitalsky, S. Thomas, E. Kny, Electrospinning tissue engineering and wound dressing scaffolds from polymer-titanium dioxide nanocomposites. Chem. Eng. J. **358**, 1262–1278 (2019)
44. F. Davani, M. Alishahi, M. Sabzi, M. Khorram, A. Arastehfar, K. Zomorodian, Dual drug delivery of vancomycin and imipenem/cilastatin by coaxial nanofibers for treatment of diabetic foot ulcer infections. Mater. Sci. Eng. C **123**, 111975 (2021)
45. L. Xing, Y.T. Fan, L.J. Shen, C.X. Yang, X.Y. Liu, Y.N. Ma, L.Y. Qi, K.H. Cho, C.S. Cho, H.L. Jiang, pH-sensitive and specific ligand-conjugated chitosan nanogels for efficient drug delivery. Int. J. Biol. Macromol. **141**, 85–97 (2019)
46. M.C. Pella, M.K. Lima-Tenorio, E.T. Tenorio-Neto, M.R. Guilherme, E.C. Muniz, A.F. Rubira, Chitosan-based hydrogels: From preparation to biomedical applications. Carbohydr. Polym. **196**, 233–245 (2018)
47. D. Archana, J. Dutta, P.K. Dutta, Evaluation of chitosan nano dressing for wound healing: characterization, in vitro and in vivo studies. Int. J. Biol. Macromol. **57**, 193–203 (2013)
48. A.J. Friedman, J. Phan, D.O. Schairer, J. Champer, M. Qin, A. Pirouz, K. Blecher-Paz, A. Oren, P.T. Liu, R.L. Modlin, J. Kim, Antimicrobial and anti-inflammatory activity of chitosan–alginate nanoparticles: a targeted therapy for cutaneous pathogens. J. Invest. Dermatol. **133**(5), 1231–1239 (2013)

49. B.S. Vasile, O. Oprea, G. Voicu, A. Ficai, E. Andronescu, A. Teodorescu, A. Holban, Synthesis and characterization of a novel controlled release zinc oxide/gentamicin–chitosan composite with potential applications in wounds care. Int. J. Pharm. **463**(2), 161–169 (2014)
50. S.M. Jung, G.H. Yoon, H.C. Lee, M.H. Jung, S.I. Yu, S.J. Yeon, S.K. Min, Y.S. Kwon, J.H. Hwang, H.S. Shin, Thermodynamic insights and conceptual design of skin-sensitive chitosan coated ceramide/PLGA nanodrug for regeneration of stratum corneum on atopic dermatitis. Sci. Rep. **5**(1), 1–12 (2015)
51. G. Behl, J. Iqbal, N.J. O'Reilly, P. McLoughlin, L. Fitzhenry, Synthesis and characterization of poly (2-hydroxyethylmethacrylate) contact lenses containing chitosan nanoparticles as an ocular delivery system for dexamethasone sodium phosphate. Pharm. Res. **33**(7), 1638–1648 (2016)
52. Y.H. Lin, J.H. Lin, T.S. Li, S.H. Wang, C.H. Yao, W.Y. Chung, T.H. Ko, Dressing with epigallocatechin gallate nanoparticles for wound regeneration. Wound Repair Regen. **24**(2), 287–301 (2016)
53. V.V.S.R. Karri, G. Kuppusamy, S.V. Talluri, S.S. Mannemala, R. Kollipara, A.D. Wadhwani, S. Mulukutla, K.R.S. Raju, R. Malayandi, Curcumin loaded chitosan nanoparticles impregnated into collagen-alginate scaffolds for diabetic wound healing. Int. J. Biol. Macromol. **93**, 1519–1529 (2016)
54. A. Gharatape, M. Milani, S.H. Rasta, M. Pourhassan-Moghaddam, S. Ahmadi-Kandjani, S. Davaran, R. Salehi, A novel strategy for low level laser-induced plasmonic photothermal therapy: the efficient bactericidal effect of biocompatible AuNPs@(PNIPAAM-co-PDMAEMA, PLGA and chitosan). RSC Adv. **6**(112), 110499–110510 (2016)
55. A. Nawaz, T.W. Wong, Microwave as skin permeation enhancer for transdermal drug delivery of chitosan-5-fluorouracil nanoparticles. Carbohydr. Polym. **157**, 906–919 (2017)
56. Y.H. Lin, J.H. Lin, Y.S. Hong, Development of chitosan/poly-γ-glutamic acid/pluronic/curcumin nanoparticles in chitosan dressings for wound regeneration. J. Biomed. Mater. Res. B Appl. Biomater. **105**(1):81–90 (2017)
57. A. Oryan, E. Alemzadeh, J. Tashkhourian, S.F.N. Ana, Topical delivery of chitosan-capped silver nanoparticles speeds up healing in burn wounds: a preclinical study. Carbohydr. Polym. **200**, 82–92 (2018)
58. L. Zhou, X. Zhao, M. Li, L. Yan, Y. Lu, C. Jiang, Y. Liu, Z. Pan, J. Shi, Antibacterial and wound healing–promoting effect of sponge-like chitosan-loaded silver nanoparticles biosynthesized by iturin. Int. J. Biol. Macromol. **181**, 1183–1195 (2021)
59. J.J. Artunduaga Bonilla, L. Honorato, D.F. Cordeiro de Oliveira, R. Araújo Gonçalves, A. Guimarães, K. Miranda, L. Nimrichter, Silver chitosan nanocomposites as a potential treatment for superficial candidiasis. Med. Mycol. **59**(10), 993–1005 (2021)
60. M. Mondejar-Lopez, A.J. Lopez-Jimenez, J.C.G. Martínez, O. Ahrazem, L. Gomez-Gomez, E. Niza, Comparative evaluation of carvacrol and eugenol chitosan nanoparticles as eco-friendly preservative agents in cosmetics. Int. J. Biol. Macromol. **206**, 288–297 (2022)
61. T. Ramasamy, J.O. Kim, C.S. Yong, K. Umadevi, D. Rana, C. Jimenez, J. Campos, M. Ramalingam, Z.S. Haidar, Novel core–shell nanocapsules for the tunable delivery of bioactive rhEGF: formulation, characterization and cytocompatibility studies. J. Biomater. Tissue Eng. **5**(9), 730–743 (2015)
62. J. Shao, B. Wang, J. Li, J.A. Jansen, X.F. Walboomers, F. Yang, Antibacterial effect and wound healing ability of silver nanoparticles incorporation into chitosan-based nanofibrous membranes. Mater. Sci. Eng. **C98**, 1053–1063 (2019)
63. C.W. Li, Q. Wang, J. Li, M. Hu, S.J. Shi, Z.W. Li, G.L. Wu, H.H. Cui, Y.Y. Li, Q. Zhang, X.H. Yu, Silver nanoparticles/chitosan oligosaccharide/poly (vinyl alcohol) nanofiber promotes wound healing by activating TGFβ1/Smad signaling pathway. Int. J. Nanomed. **11**, 373–387 (2016)
64. M. Liu, Y. Shen, P. Ao, L. Dai, Z. Liu, C. Zhou, The improvement of hemostatic and wound healing property of chitosan by halloysite nanotubes. RSC Adv. **4**(45), 23540–23553 (2014)
65. B. Veleirinho, D.S. Coelho, P.F. Dias, M. Maraschin, R.M. Ribeiro-do-Valle, J.A. Lopes-da-Silva, Nanofibrous poly (3-hydroxybutyrate-co-3-hydroxyvalerate)/chitosan scaffolds for skin regeneration. Int. J. Biol. Macromol. **51**(4), 343–350 (2012)

66. R. Huang, W. Li, X. Lv, Z. Lei, Y. Bian, H. Deng, H. Wang, J. Li, X. Li, Biomimetic LBL structured nanofibrous matrices assembled by chitosan/collagen for promoting wound healing. Biomaterials **53**, 58–75 (2015)
67. M. Salehi, S. Farzamfar, F. Bastami, R. Tajerian, Fabrication and characterization of electrospun PLLA/collagen nanofibrous scaffold coated with chitosan to sustain release of aloe vera gel for skin tissue engineering. Biomed. Eng. - Appl. Basis Commun. **28**(05), 1650035 (2016)
68. S.J.S. Tabaei, M. Rahimi, M. Akbaribazm, S.A. Ziai, M. Sadri, S.R. Shahrokhi, M.S. Rezaei, Chitosan-based nano-scaffolds as antileishmanial wound dressing in BALB/c mice treatment: characterization and design of tissue regeneration. Iran. J. Basic Med. Sci. **23**(6), 788 (2020)
69. T. Demoulins, P. Milona, K.C. McCullough, Alginate-coated chitosan nanogels differentially modulate class-A and class-B CpG-ODN targeting of dendritic cells and intracellular delivery. Nanomed.: Nanotechnol. Biol. Med. **10**(8):1739–1749 (2014)
70. R. Tian, L. Xian, Y. Li, X. Zheng, Silica modified chitosan/polyethylenimine nanogel for improved stability and gene carrier ability. J. Nanosci. Nanotechnol. **16**(5), 5426–5431 (2016)
71. G.I. Harisa, M.M. Badran, S.A. AlQahtani, F.K. Alanazi, S.M. Attia, Pravastatin chitosan nanogels-loaded erythrocytes as a new delivery strategy for targeting liver cancer. Saudi Pharm. J. **24**(1), 74–81 (2016)
72. Q. Zha, X. Wang, X. Cheng, S. Fu, G. Yang, W. Yao, R. Tang, Acid–degradable carboxymethyl chitosan nanogels via an ortho ester linkage mediated improved penetration and growth inhibition of 3-D tumor spheroids in vitro. Mater. Sci. Eng. **C78**, 246–257 (2017)
73. G. Yang, X. Wang, S. Fu, R. Tang, J. Wang, pH-triggered chitosan nanogels via an ortho ester-based linkage for efficient chemotherapy. Acta Biomater. **60**, 232–243 (2017)
74. Y. Shitrit, M. Davidovich-Pinhas, H. Bianco-Peled, Shear thinning pectin hydrogels physically cross-linked with chitosan nanogels. Carbohydr. Polym. **225**, 115249 (2019)
75. Y. Ashoori, M. Mohkam, R. Heidari, S.N. Abootalebi, S.M. Mousavi, S.A. Hashemi, N. Golkar, A. Gholami, Development and in vivo characterization of probiotic lysate-treated chitosan nanogel as a novel biocompatible formulation for wound healing. Biomed Res. Int. (2020)
76. J.P. de Mesquita, C.L. Donnici, F.V. Pereira, Biobased nanocomposites from layer-by-layer assembly of cellulose nanowhiskers with chitosan. Biomacromolecules **11**(2), 473–480 (2010)
77. J.M. Silva, J.R. Garcia, R.L. Reis, A.J. Garcia, J.F. Mano, Tuning cell adhesive properties via layer-by-layer assembly of chitosan and alginate. Acta Biomater. **51**, 279–293 (2017)
78. M.P. Sousa, J.F. Mano, Cell-adhesive bioinspired and catechol-based multilayer freestanding membranes for bone tissue engineering. Biomimetics **2**(4), 19 (2017)
79. A. Shah, I. Hussain, G. Murtaza, Chemical synthesis and characterization of chitosan/silver nanocomposites films and their potential antibacterial activity. Int. J. Biol. Macromol. **116**, 520–529 (2018)
80. C.F. Sousa, C.A. Saraiva, T.R. Correia, T. Pesqueira, S.G. Patricio, M.I. Rial-Hermida, J. Borges, J.F. Mano, Bioinstructive layer-by-layer-coated customizable 3D printed perfusable microchannels embedded in photo crosslinkable hydrogels for vascular tissue engineering. Biomolecules **11**(6), 863 (2021)
81. S. Sydow, D. de Cassan, R. Hänsch, T.R. Gengenbach, C.D. Easton, H. Thissen, H. Menzel, Layer-by-layer deposition of chitosan nanoparticles as drug-release coatings for PCL nanofibers. Biomater. Sci. **7**(1), 233–246 (2019)
82. Z.A. Khan, S. Jamil, A. Akhtar, M.M. Bashir, M. Yar, Chitosan based hybrid materials used for wound healing applications-A short review. Int. J. Polym. Mater. Polym. **69**(7), 419–436 (2019)
83. M.T. Pelegrino, A.B. Seabra, Chitosan-based nanomaterials for skin regeneration. Med. Sci. **4**(3), 352–381 (2017)
84. N. Cai, C. Li, C. Han, X. Luo, L. Shen, Y. Xue, F. Yu, Tailoring mechanical and antibacterial properties of chitosan/gelatin nanofiber membranes with Fe_3O_4 nanoparticles for potential wound dressing application. Appl. Surf. Sci. **369**, 492–500 (2016)

85. J. Liu, Z. Qian, Q. Shi, S. Yang, Q. Wang, B. Liu, J. Xu, X. Guo, H. Liu, An asymmetric wettable chitosan–silk fibroin composite dressing with fixed silver nanoparticles for infected wound repair: in vitro and in vivo evaluation. RSC Adv. **7**(69), 43909–43920 (2017)
86. M. Rahimi, R. Ahmadi, H.S. Kafil, V. Shafiei-Irannejad, A novel bioactive quaternized chitosan and its silver-containing nanocomposites as a potent antimicrobial wound dressing: Structural and biological properties. Mater. Sci. Eng. **C101**, 360–369 (2019)
87. A. Shah, M.A. Buabeid, E.S.A. Arafa, I. Hussain, L. Li, G. Murtaza, The wound healing and antibacterial potential of triple-component nanocomposite (chitosan-silver-sericin) films loaded with moxifloxacin. Int. J. Pharm. **564**, 22–38 (2019)
88. N. Masood, R. Ahmed, M. Tariq, Z. Ahmed, M.S. Masoud, I. Ali, R. Asghar, H.A. AndleebA, Silver nanoparticle impregnated chitosan-PEG hydrogel enhances wound healing in diabetes induced rabbits. Int. J. Pharm. **559**, 23–36 (2019)
89. Z. Lu, J. Gao, Q. He, J. Wu, D. Liang, H. Yang, R. Chen, Enhanced antibacterial and wound healing activities of microporous chitosan-Ag/ZnO composite dressing. Carbohydr. Polym. **156**, 460–469 (2017)
90. L. Shahzadi, A.A. Chaudhry, A.R. Aleem, M.H. Malik, K. Ijaz, H. Akhtar, F. Alvi, A.F. Khan, I.U. Rehman, M. Yar, Development of K-doped ZnO nanoparticles encapsulated crosslinked chitosan based new membranes to stimulate angiogenesis in tissue engineered skin grafts. Int. J. Biol. Macromol. **120**, 721–728 (2018)
91. L. Ding, X. Shan, X. Zhao, H. Zha, X. Chen, J. Wang, C. Cai, X. Wang, G. Li, J. Hao, G. Yu, Spongy bilayer dressing composed of chitosan–Ag nanoparticles and chitosan–Bletilla striata polysaccharide for wound healing applications. Carbohydr. Polym. **157**, 1538–1547 (2017)
92. R.A.B. Sanad, H.M. Abdel-Bar, Chitosan–hyaluronic acid composite sponge scaffold enriched with Andrographolide-loaded lipid nanoparticles for enhanced wound healing. Carbohydr. Polym. **173**, 441–450 (2017)
93. B. Lu, F. Lu, Y. Zou, J. Liu, B. Rong, Z. Li, F. Dai, D. Wu, G. Lan, In situ reduction of silver nanoparticles by chitosan-l-glutamic acid/hyaluronic acid: Enhancing antimicrobial and wound-healing activity. Carbohydr. Polym. **173**, 556–565 (2017)
94. A. Bal-Ozturk, O. Karal-Yilmaz, Z.P. Akguner, S. Aksu, A. Tas, H. Olmez, Sponge-like chitosan-based nanostructured antibacterial material as a topical hemostat. J. Appl. Polym. Sci. **136**(19), 47522 (2019)
95. S. Kumari, B.N. Singh, P. Srivastava, Effect of copper nanoparticles on physico-chemical properties of chitosan and gelatin-based scaffold developed for skin tissue engineering application. 3 Biotech **9**(102), 1–14 (2019)
96. B. Hou, M. Qi, J. Sun, M. Ai, X. Ma, W. Cai, Y. Zhou, L. Ni, J. Hu, F. Xu, L. Qiu, Preparation, characterization and wound healing effect of vaccarin-chitosan nanoparticles. Int. J. Biol. Macromol. **165**, 3169–3179 (2020)
97. J. Radwan-Pragłowska, Ł Janus, M. Piątkowski, D. Bogdał, D. Matysek, Hybrid bilayer PLA/chitosan nanofibrous scaffolds doped with ZnO, Fe_3O_4, and Au nanoparticles with bioactive properties for skin tissue engineering. Polymers **12**(1), 159 (2020)
98. M. Zahiri, M. Khanmohammadi, A. Goodarzi, S. Ababzadeh, M.S. Farahani, S. Mohandesnezhad, N. Bahrami, I. Nabipour, J. Ai, Encapsulation of curcumin loaded chitosan nanoparticle within poly (ε-caprolactone) and gelatin fiber mat for wound healing and layered dermal reconstitution. Int. J. Biol. Macromol. **153**, 1241–1250 (2020)
99. U. Kini, B.N. Nandeesh, Physiology of bone formation, remodeling, and metabolism, in *Radionuclide and Hybrid Bone Imaging*. (Springer, Berlin, Heidelberg, 2012), pp.29–57
100. E. Mackie, Y.A. Ahmed, L. Tatarczuch, K.S. Chen, M. Mirams, Endochondral ossification: how cartilage is converted into bone in the developing skeleton. Int. J. Biochem. Cell Biol. **40**(1), 46–62 (2008)
101. K. Balagangadharan, S. Dhivya, N. Selvamurugan, Chitosan based nanofibers in bone tissue engineering. Int. J. Biol. Macromol. **104**, 1372–1382 (2017)
102. V. Campana, G. Milano, E. Pagano, M. Barba, C. Cicione, G. Salonna, W. Lattanzi, G. Logroscino, Bone substitutes in orthopaedic surgery: from basic science to clinical practice. J. Mater. Sci.: Mater. Med. **25**(10):2445–2461 (2014)

103. N. Shibuya, D.C. Jupiter, Bone graft substitute: allograft and xenograft. Clin. Podiatr. Med. Surg. **32**(1), 21–34 (2015)
104. S. Murugesan, T. Scheibel, Chitosan-based nanocomposites for medical applications. J. Polym. Sci. **59**(15), 1610–1642 (2021)
105. T. Gong, J. Xie, J. Liao, T. Zhang, S. Lin, Y. Lin, Nanomaterials and bone regeneration. Bone Res. **3**, 15029 (2015)
106. D.F. Williams, There is no such thing as a biocompatible material. Biomaterials **35**(38), 10009–10014 (2014)
107. A. Bhowmick, N. Pramanik, T. Mitra, A. Gnanamani, M. Das, P.P. Kundu, Mechanical and biological investigations of chitosan–polyvinyl alcohol based ZrO_2 doped porous hybrid composites for bone tissue engineering applications. New J. Chem. **41**(15), 7524–7530 (2017)
108. J.W. Ai, W. Liao, Z.L. Ren, Enhanced anticancer effect of copper-loaded chitosan nanoparticles against osteosarcoma. RSC Adv. **7**(26), 15971–15977 (2017)
109. D. Zhou, C. Qi, Y.X. Chen, Y.J. Zhu, T.W. Sun, F. Chen, C.Q. Zhang, Comparative study of porous hydroxyapatite/chitosan and whitlockite/chitosan scaffolds for bone regeneration in calvarial defects. Int. J. Nanomed. **12**, 2673 (2017)
110. S. Banerjee, B. Bagchi, S. Bhandary, A. Kool, N.A. Hoque, P. Biswas, K. Pal, P. Thakur, K. Das, P. Karmakar, S. Das, Antimicrobial and biocompatible fluorescent hydroxyapatite-chitosan nanocomposite films for biomedical applications. Colloids Surf. B Biointerf. **171**, 300–307 (2018)
111. M. Shakir, S. Mirza, R. Jolly, A. Rauf, M. Owais, Synthesis, characterization and in vitro screening of a nano-hydroxyapatite/chitosan/Euryale ferox nanoensemble–an inimitable approach for bone tissue engineering. New J. Chem. **42**(1), 363–371 (2018)
112. S. Maji, T. Agarwal, J. Das, T.K. Maiti, Development of gelatin/carboxymethyl chitosan/nano-hydroxyapatite composite 3D macroporous scaffold for bone tissue engineering applications. Carbohydr. Polym. **189**, 115–125 (2018)
113. W. Zhou, Y. Li, J. Yan, P. Xiong, Q. Li, Y. Cheng, Y. Zheng, Construction of self-defensive antibacterial and osteogenic AgNPs/gentamicin coatings with chitosan as nanovalves for controlled release. Sci. Rep. **8**(1), 1–12 (2018)
114. B. Li, X. Xia, M. Guo, Y. Jiang, Y. Li, Z. Zhang, S. Liu, H. Li, C. Liang, H. Wang, Biological and antibacterial properties of the micro-nanostructured hydroxyapatite/chitosan coating on titanium. Sci. Rep. **9**(1), 1–10 (2019)
115. P.P. Zhao, H.R. Hu, J.Y. Liu, Q.F. Ke, X.Y. Peng, H. Ding, Y.P. Guo, Gadolinium phosphate/chitosan scaffolds promote new bone regeneration via Smad/Runx2 pathway. Chem. Eng. J. **359**, 1120–1129 (2019)
116. K. Jahan, G. Manickam, M. Tabrizian, M. Murshed, In vitro and in vivo investigation of osteogenic properties of self-contained phosphate-releasing injectable purine-crosslinked chitosan-hydroxyapatite constructs. Sci. Rep. **10**(1), 1–17 (2020)
117. M. Ahmad, K. Manzoor, S. Ikram, Chitosan nanocomposites for bone and cartilage regeneration, in *Applications of Nanocomposite Materials in Dentistry*, Woodhead Publishing (2019), pp. 307–317
118. N. Asadi, E. Alizadeh, R. Salehi, B. Khalandi, S. Davaran, A. Akbarzadeh, Nanocomposite hydrogels for cartilage tissue engineering: a review. Artif. Cells Nanomed. Biotechnol. **46**(3), 465–471 (2018)
119. E.B. Hunziker, Articular cartilage repair: basic science and clinical progress. A review of the current status and prospects. Osteoarthr. Cartil. **10**(6):432–463 (2002)
120. S.Q. Liu, B. Qiu, L.Y. Chen, H. Peng, Y.M. Du, The effects of carboxymethylated chitosan on metalloproteinase-1, − 3 and tissue inhibitor of metalloproteinase-1 gene expression in cartilage of experimental osteoarthritis. Rheumatol. Int. **26**(1), 52–57 (2005)
121. Y. Zhou, K. Liang, S. Zhao, C. Zhang, J. Li, H. Yang, X. Liu, X. Yin, D. Chen, W. Xu, P. Xiao, Photopolymerized maleilated chitosan/methacrylated silk fibroin micro/nanocomposite hydrogels as potential scaffolds for cartilage tissue engineering. Int. J. Biol. Macromol. **108**, 383–390 (2018)

122. E.B. Toloue, S. Karbasi, H. Salehi, M. Rafienia, Evaluation of mechanical properties and cell viability of poly (3-Hydroxybutyrate)-Chitosan/Al_2O_3 nanocomposite scaffold for cartilage tissue engineering. J. Med. Signals Sens. **9**(2), 111 (2019)
123. M.A. Shamekhi, H. Mirzadeh, H. Mahdavi, A. Rabiee, D. Mohebbi-Kalhori, M.B. Eslaminejad, Graphene oxide containing chitosan scaffolds for cartilage tissue engineering. Int. J. Biol. Macromol. **127**, 396–405 (2019)
124. P. BahramiMiyanji, D. Semnani, A. Hossein Ravandi, S. Karbasi, A. Fakhrali, S. Mohammadi, Fabrication and characterization of chitosan-gelatin/single-walled carbon nanotubes electrospun composite scaffolds for cartilage tissue engineering applications. Polym. Adv. Technol. **33**(1), 81–95 (2022)
125. M.H. Mirmusavi, M. Ahmadian, S. Karbasi, Polycaprolactone-chitosan/multi-walled carbon nanotube: a highly strengthened electrospun nanocomposite scaffold for cartilage tissue engineering. Int. J. Biol. Macromol. **209**, 1801–1814 (2022)
126. S.M. Nasr, N. Rabiee, S. Hajebi, S. Ahmadi, Y. Fatahi, M. Hosseini, M. Bagherzadeh, A.M. Ghadiri, M. Rabiee, V. Jajarmi, T.J. Webster, Biodegradable nanopolymers in cardiac tissue engineering: From concept towards nanomedicine. Int. J. Nanomed. **15**, 4205–4224 (2020)
127. T. Limongi, L. Tirinato, F. Pagliari, A. Giugni, M. Allione, G. Perozziello, P. Candeloro, E. Di Fabrizio, Fabrication and applications of micro/nanostructured devices for tissue engineering. Nanomicro Lett. **9**(1), 1–13 (2017)
128. K. Kalishwaralal, S. Jeyabharathi, K. Sundar, S. Selvamani, M. Prasanna, A. Muthukumaran, A novel biocompatible chitosan–Selenium nanoparticles (SeNPs) film with electrical conductivity for cardiac tissue engineering application. Mater. Sci. Eng. **C92**, 151–160 (2018)
129. A. Abedi, M. Hasanzadeh, L. Tayebi, Conductive nanofibrous Chitosan/PEDOT: PSS tissue engineering scaffolds. Mater. Chem. Phys. **237**, 121882 (2019)
130. N. Liu, J. Chen, J. Zhuang, P. Zhu, Fabrication of engineered nanoparticles on biological macromolecular (PEGylated chitosan) composite for bio-active hydrogel system in cardiac repair applications. Int. J. Biol. Macromol. **117**, 553–558 (2018)
131. P. Baei, S. Jalili-Firoozinezhad, S. Rajabi-Zeleti, M. Tafazzoli-Shadpour, H. Baharvand, N. Aghdami, Electrically conductive gold nanoparticle-chitosan thermosensitive hydrogels for cardiac tissue engineering. Mater. Sci. Eng. C **63**, 131–141 (2016)
132. S. Soltani, M. Ebrahimian-Hosseinabadi, A. Zargar Kharazi, Chitosan/graphene and poly (D, L-lactic-co-glycolic acid)/graphene nano-composites for nerve tissue engineering. Tissue Eng. Regen. Med. **13**(6), 684–690 (2016)
133. M. Habibizadeh, S. Nadri, A. Fattahi, K. Rostamizadeh, P. Mohammadi, S. Andalib, M. Hamidi, N. Forouzideh, Surface modification of neurotrophin-3 loaded PCL/chitosan nanofiber/net by alginate hydrogel microlayer for enhanced biocompatibility in neural tissue engineering. J. Biomed. Mater. Res. A **109**(11), 2237–2254 (2021)
134. H.S. Hung, Y.C. Yang, C.H. Chang, K.B. Chang, C.C. Shen, C.L. Tang, S.Y. Liu, C.H. Lee, C.M. Yen, M.Y. Yang, Neural differentiation potential of mesenchymal stem cells enhanced by biocompatible Chitosan-gold nanocomposites. Cells **11**(12), 1861 (2022)
135. H. Baniasadi, S.A.A.R, S. Mashayekhan, Fabrication and characterization of conductive chitosan/gelatin-based scaffolds for nerve tissue engineering. Int. J. Biol. Macromol. **74**, 360–366 (2015)
136. S. Saravani, M. Ebrahimian-Hosseinabadi, D. Mohebbi-Kalhori, Polyglycerol sebacate/chitosan/gelatin nano-composite scaffolds for engineering neural construct. Mater. Chem. Phys. **222**, 147–151 (2019)
137. M. Jafarkhani, Z. Salehi, T. Nematian, Preparation and characterization of chitosan/graphene oxide composite hydrogels for nerve tissue Engineering. Mater. Today: Proc. **5**(7), 15620–15628 (2018)
138. T. Tayebi, A. Baradaran-Rafii, A. Hajifathali, A. Rahimpour, H. Zali, A. Shaabani, H. Niknejad, Biofabrication of chitosan/chitosan nanoparticles/polycaprolactone transparent membrane for corneal endothelial tissue engineering. Sci. Rep. **11**(1), 1–12 (2021)
139. A.O.M. Salehi, S.H. Keshel, F. Sefat, L. Tayebi, Use of polycaprolactone in corneal tissue engineering: a review. Mater. Today Commun. **27**, 102402 (2021)

140. S. Wang, W. Liu, B. Han, L. Yang, Study on a hydroxypropyl chitosan–gelatin based scaffold for corneal stroma tissue engineering. Appl. Surf. Sci. **255**(20), 8701–8705 (2009)
141. B. Ozcelik, K.D. Brown, A. Blencowe, M. Daniell, G.W. Stevens, G.G. Qiao, Ultrathin chitosan–poly (ethylene glycol) hydrogel films for corneal tissue engineering. Acta Biomater. **9**(5), 6594–6605 (2013)
142. Y.H. Wang, T.H. Young, T.J. Wang, Investigating the effect of chitosan/polycaprolactone blends in differentiation of corneal endothelial cells and extracellular matrix compositions. Exp. Eye Res. **185**, 107679 (2019)

Chapter 11
Chitosan Nanocomposites for Biosensing Applications

Atilla Eren Mamuk, Ceyhun Işık, Sema Aslan, and Derya Bal Altuntaş

Abstract Electrochemical biosensors have a wide range of application areas from the pharmaceutical industry to the environment in terms of being simple, portable, low cost, and having short response time, high sensitivity, specificity, and selectivity, when compared with thermal, optical, and piezoelectric biosensors. Various biocomponents such as enzymes, nucleic acids, antibodies, tissue and organelles, nucleic acids, and microorganisms are used in the design of biosensors. Investigations on liquid crystal-chitosan nanocomposite sensors have been increasing recently. Liquid crystals (LCs) are very soft materials with high sensitivity towards external effects. This chapter reveals an overall idea regarding the introduction to electrochemical biosensors, nanocomposites, and liquid crystal-chitosan biosensors applications.

11.1 Introduction

Biosensors have attracted great interest in recent years due to their potential applications in many fields, especially in environmental monitoring, medical diagnosis and treatment, national security, and the food industry [1–4]. The biosensors are integrated miniaturized devices that detect signals using biological material such as an enzyme, nucleic acid, antibody, and receptor protein as a transducer-dependent sensing element [5, 6]. It has been reported that the performance of biosensors is related to the intrinsic properties of the biological materials used, such as their size, biological activity, and binding ability for target analytes, as well as the properties of the substrate on which these biological materials are immobilized [7, 8]. It is known

A. E. Mamuk
Faculty of Science, Department of Physics, Mugla Sitki Kocman University, Mugla, Turkey

C. Işık · S. Aslan
Faculty of Science, Department of Chemistry, Mugla Sitki Kocman University, Mugla, Turkey

D. B. Altuntaş (✉)
Faculty of Engineering and Architecture, Department of Bioengineering, Recep Tayyip Erdogan University, Rize, Turkey
e-mail: derya.balaltuntas@erdogan.edu.tr

S. K. Swain and A. Biswal (eds.), *Chitosan Nanocomposites*,
Biological and Medical Physics, Biomedical Engineering,
https://doi.org/10.1007/978-981-19-9646-7_11

that the performance of biosensors is affected by the intrinsic properties of the biological materials used, such as their size, biological activity, and binding abilities for target analytes, as well as the properties of the substrate on which these biological materials are immobilized [9]. Surface interactions between substrate and biological material cause immobilized materials to acquire additional chemical, physical, and kinetic properties, which can greatly increase the sensitivity and specificity of surface-based biosensors, which is crucial for sensor performance [10, 11]. For these reasons, the selection of a suitable substrate material can contribute significantly to the properties of the immobilized system.

The use of nanocomposites composed of various synthetic and natural polymers as an intermediate substrate material between electrode surfaces and biological materials has been extensively investigated to develop selective and sensitive surface-based biosensors [11–13]. A natural polymer such as chitosan has strong gel-forming ability, low cost, and environmental friendliness, so it is highly suitable for use in biosensor preparation [14]. Chitosan (CS), a natural polymer, is widely used in biosensor preparation due to its biodegradability, low cost, biocompatible, non-toxicity, and film-forming nature [15]. CS is a biopolymer derived from the only natural cationic polysaccharide derived by deacetylation of chitin [16]. Due to the abundance of amino and hydroxyl groups in the CS polymeric chain, it is easy to modify CS through various interactions such as electrostatic interactions, covalent bonding, and hydrogen bonding [17]. Therefore, the use of CS-based nanocomposites as substrate material has gradually increased instead of other polymers in the preparation of biosensors in the last decade [18].

This chapter focuses on the biosensor applications of chitosan-based nanocomposites. It also aims to provide a guide to researchers for the study and development of chitosan nanocomposite biosensors. Firstly, the operation principles and the basic concepts of biosensors, nanocomposites, and chitosan are reviewed. Then, examples of chitosan-based biosensor applications and liquid crystal-chitosan nanocomposite biosensors are presented.

11.2 Biosensors

Biosensors are devices assembled with a bio-recognizer and a transducer to obtain a meaningful response related to the concentration of analytes (Fig. 11.1). Biological material interacts with the component to be determined. Significant signals are generated and the amount of analyte is determined by using electrochemical, optical, and calorimetric properties [19]. After the development of the first glucose electrodes by Clark and Lyons in 1962, biosensors were studied extensively and were started to be used in a wide range of areas from medicine to environment [20]. Various biological materials such as cofactors, enzymes, antigen-antibodies, microorganisms, organelles, tissues, and cells are used in the designing of biosensors [21]. Out of these biomaterials, enzymes are the biomaterials most widely used due to their high sensitivity and specificity [22].

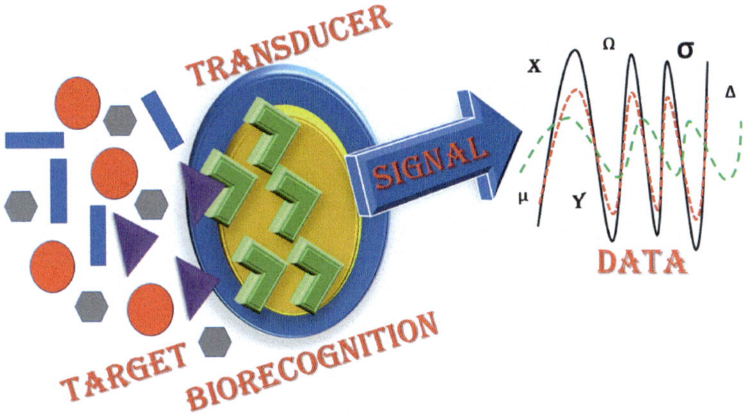

Fig. 11.1 A Schematic representation of a biosensor

11.2.1 Classification of Biosensors: Classification of Biosensors by Biocomponent

Various biocomponent such as enzymes, nucleic acids, antibodies, tissue and organelles, nucleic acids, and microorganisms are used in the design of biosensors [23]. It is possible to examine biosensors in four categories according to their biocomponents.

11.2.2 Enzyme Biosensors

They are biosensors in which enzymes are used as bio compounds in biosensor design. It is seen that the first studies in biosensor technology started with enzyme biosensors (Fig. 11.2) [24, 25]. Enzymes are biomolecules, usually with a protein structure and catalytic function. During the catalytic reaction, enzyme–substrate complex is formed and the product that is formed at the end of the reaction enables the formation of signals that can be measured directly or indirectly by the biosensor.

The high specificity of enzymes, the occurrence of reactions with the interaction of enzyme–substrate, and the rapid occurrence of biocatalysis with 100% efficiency expand the area of application of enzyme biosensors [26].

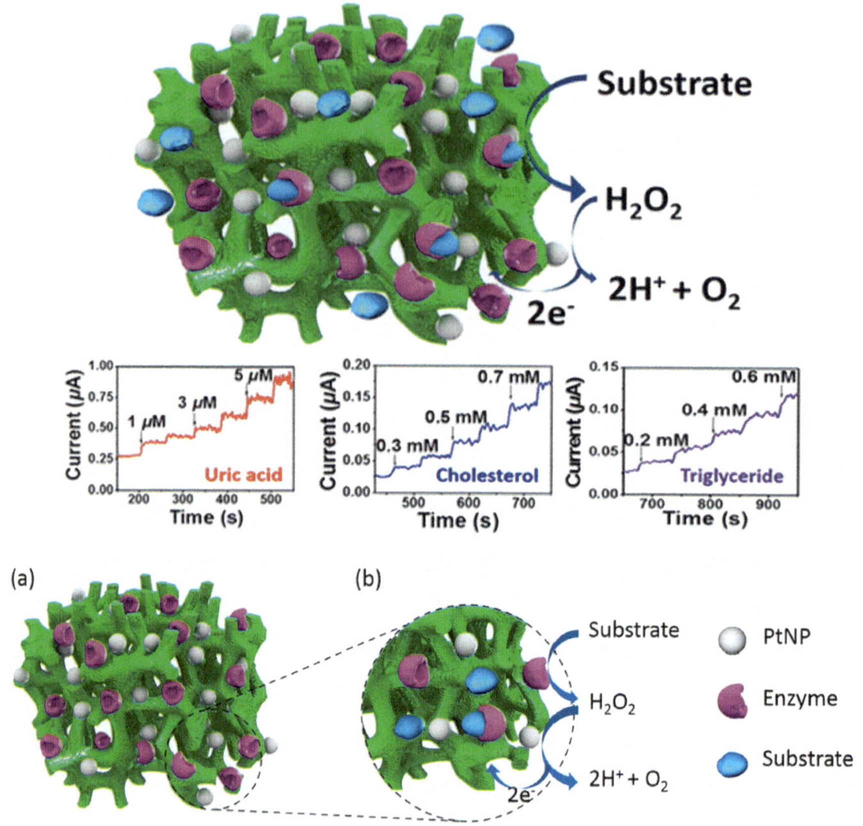

Fig. 11.2 Schematic illustration of the general sensing mechanism of CPH-based electrode platform. **a** PtNPs and enzymes loaded onto hierarchically three-dimensional (3D)-porous PAni hydrogel matrices to form PAni hydrogel/PtNPs hybrid electrodes. **b** The PtNP-catalyzed sensing process of the biosensor based on PAni/PtNPs/enzyme hybrid films. Reproduced with permission from American Chemical Society [25]

11.2.3 Microbial Biosensor

They are biosensors in which microorganisms are used as biocompounds in biosensor design. Microorganisms which are used as biological sensors have extensive advantages in detecting chemicals. Enzymes are the most used structures in biosensor production, with very high biological sensitivity. Although purified enzymes are specific for their substrates or inhibitors, the process of purifying is tiring, time consuming, and costly. Microorganisms offer an ideal alternative to these above mentioned barriers (Fig. 11.3) [27]. Since enzymes will be in natural environments, they are more resistant to external influences. Enzymes operating with coenzyme do not require external coenzyme addition; in addition, regeneration of

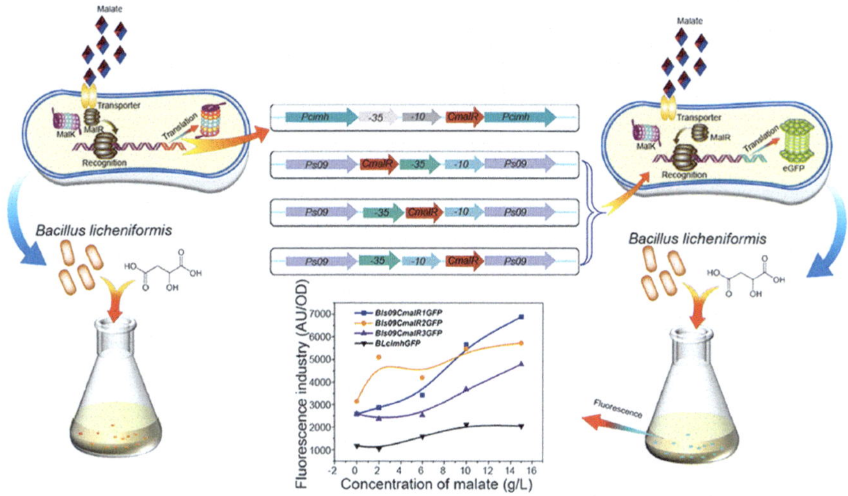

Fig. 11.3 Construction of a genetically encoded biosensor in response to malate in B. licheniformis by Zhang et al. [27]. Reproduced with permission from American Chemical Society [27]

coenzymes takes place within the cell. They usually last longer than enzyme electrodes. In addition, if a reaction sequence in which many enzymes are involved will be examined instead of a single enzyme in case of intracellular enzymes, it is again an important advantage to use cells instead of enzymes. There are also disadvantages of using microorganism.

Since the cell membrane forms a diffusion barrier, suitable biosensors cannot be prepared for molecules that cannot pass through macromolecules and membrane. Microbial biosensors have a longer response time and return to baseline signal level after use when compared with enzyme sensors. Since the cell contains many enzymes, it is possible for the targeted assay reaction to be influenced by other enzymes. A decrease in activity that may be experienced during immobilization and contamination, is also among the most common problems faced in fabrication of biosensors [28].

11.2.4 Nucleic Acid Biosensors

They are biosensors in which nucleic acids are used as bio compound in biosensor design. When compared with conventional hybridization experiments, DNA biosensors are promising for faster, simpler, and cheaper acquisition of sequence-specific information in human, viral and bacterial nucleic acid. The development of a DNA hybridization biosensor has an important place in large-scale pathogen detection and molecular diagnosis. DNA recognition surfaces are used to monitor hybridization events with a specific sequence or in the detection of substances to be analyzed (drugs, etc.) which interact with this surface. DNA biosensors are based on the hybridization of synthetic single-stranded DNA (ssDNA) oligomer with the target sequence, which has a short base sequence of 20–40 bases, corresponding to the base sequence of the target sought [29].

11.2.5 Immunosensors

They are biosensors in which antibodies are used as biocompounds. Antibodies are glycoproteins produced by the immune system. This type of biosensor is prepared based on antibody-antigen interaction [30].

11.2.6 Electrochemical Biosensors

Electrochemical biosensors have a wide range of application areas ranging from the pharmaceutical industry to the environment in terms of being simple, portable, low cost, and having short response time, high sensitivity, specificity, and selectivity, when compared with thermal, optical, and piezoelectric biosensors. In the design of electrochemical biosensors, which are obtained upon a chemical change and electron production and obtaining electrical responses proportional to concentration, electrodes obtained from gold, silver, carbon, or platinum are generally used [31]. Amperometric, potentiometric, conductometric biosensors are examples of electrochemical biosensors.

11.2.7 Amperometric Biosensors

The analytes must create an electrochemical current at a constant potential. This current value sometimes increases according to the reaction that takes place on the specific surface of an active working electrode. The response is compared with the reference electrode. The calibration line is composed of those increasing current values according to the increasing concentrations. The most important indicating point is the consumption of the measured species in this method. Controversly, in the potentiometrical measurement process is that the concentration of the analytes doesn't change [32].

11.2.8 Potentiometric Biosensors

Potentiometry usually deals with the two-electrode dipped electrolyte solution in which an active working electrode and an ion-selective special reference electrode are used. It is highly dependent on the measurement of the potential difference between working electrode and reference electrode. In this process, the potential of the electrode change is measured with respect to the changing species or concentrations of the analyte. The electrodes are classified according to their target ions or gas sensing probes such as ammonia [33].

11.2.9 Conductometric Biosensors

Conductometric biosensors are generally used in the determination of air humidity and concentration of certain gases. Although conductometric biosensors have less application areas than amperometric and potentiometric biosensors, the fact that they do not use reference electrode during analysis, they operate at low voltage and they are not sensitive to light are their disadvantages. There is a migration of ions in biosensors based on the conductivity of liquids. When a potential is applied in the solution, an electric field is created and ion migration starts [34].

11.2.10 Optical Biosensors

These are the type of transducers developed by making use of the phenomena like absorption, refraction, scattering, or brightness of the light. The reference electrode is not used in studies. Since the use of optical fibers is possible in long distance uses, there is no need for close contact with the bioreceptor and it can easily be minimized in in vivo studies. Optical methods are among the most applied transducers [35].

11.2.11 Piezoelectric Biosensors

Piezoelectric effect is voltage generation by a mechanically stretched surface. This effect is sensed with an oscillator induced by a voltage difference on a piezo-electrical sensor. This change is proportional to the mass. It is widely used in designing analytical sensors [36].

11.2.12 Calorimetric Biosensors

The energy changes during the biosensing activity sometimes reveals as a thermal change. In this type of sensor, the response is obtained through the combination of biomaterial with a physical transducer such as thermometer. It is possible to use thermal (calorimetric) biosensors in enzyme activity measurements, clinical monitoring, process control, anhydrous medium measurements, and environmental monitoring by monitoring the changes that occur in temperature. Biosensors are simple and inexpensive devices with high sensitivity and fast response time. Therefore, they are used in many areas from health to environment. With the developing technology, wireless biosensors are designed and they are used especially in the diagnosis and monitoring of patients in the field of health. For this reason, the design and use of different biosensors is important. Wireless biosensors suitable for in vivo use have been designed in recent years. In 1950s, Clark and Lyons determined glucose in vitro and in the following year's personalized commercial glucose meters were designed based on this determination. Today, intrabody sensors are designed which can be run by the energy supplied from the biochemical actions that occurs in the body. Also, the microchips can create an outer signal which can be received as an electrical data of the analyte concentration by an automation system. Wireless biosensors are fast developing, permanent, and wearable biosensors connected to networks. For these reasons, wireless biosensors facilitate monitoring disease markers in the hospital environment or daily life [37].

11.2.13 Chitosan Nanocomposite Biosensors

Polysaccharides are one of the most widely used natural polymeric biomaterials in biotechnological applications. Due to their properties such as biodegradability, biocompatibility, and high usability, the use of biopolymers, especially in life science applications is increasing day by day [38]. Chitosan is one of the most widely used biomaterials due to its unique properties in the fields of medicine, agriculture, food, environmental protection, materials science, biocatalysis, nutrition, and biotechnology [39, 40].

Chitosan nanocomposites are used extensively in recent years in drug release, biomedical, catalysis, desalination, environmental problems, and electrochemical fields due to their properties such as biomimetic, biocompatibility, and biodegradability [41]. With the developments in nanotechnology, chitosan-based nanocomposites that offer improved barrier, chemical, thermal, and mechanical properties have begun to be produced by using metal, inorganic, and organic materials [42]. These chitosan-based nanocomposites have attracted the attention of researchers due to their unique properties such as high stability, biodegradability, non-toxicity, and biocompatibility [43, 44]. For these reasons, chitosan nanocomposites are widely used as biosensors in environmental, biomedical, pharmaceutical, and catalysis research applications [45].

11.3 Nanocomposites

Solid materials consisting of at least two phases and one of the phases smaller than 100 nm are defined as nanocomposites. The sizes, structures, and physical and chemical properties of the different phases in nanocomposites contribute to obtaining composites with fascinating and unique electrochemical, thermal, mechanical, optical, and catalytic properties [46]. Composite materials are used in different applications according to their nano dimensions. Nanocomposites below the sizes of 100 nm, 50 nm, 20 nm, and 5 nm are used in mechanical reinforcement and superparamagnetism, determination of refractive index changes, changing the hardness of a hard-magnetic material and catalytic activity studies, respectively [47, 48].

The filler phase present in nanocomposites ensures that the composite has a high surface-to-volume ratio. The filler phase material consists of fibers (e.g., electrospun fiber or carbon nanotubes (CNT) s), particles (e.g., metal oxides, minerals), or sheets (e.g., organoclays, exfoliated clay stacks). The interface between the filler and the matrix affects the matrix material properties such as chemical structure, polymer chain conformation, crystallinity, degree of thermoset cure, polymer chain ordering degree, or polymer chain mobility, and makes composites become a better conventional material [49]. The nano-sized filler, which is present in small amounts in the structure of the composite, has a significant effect on the macro-scale properties of the composite. Adding fiber as a filler can improve the thermal properties of composite materials and their resistance to chemical and physical damage. In the case of using other types of nanofillers, some other properties such as electrical conductivity, optics, hardness, strength, dielectric, and stiffness are improved [50]. The nano-sized fillers are divided into three categories, namely, metal, polymer, and ceramic matrix composites, depending on the matrix they are dispersed during the composite formation process [51].

11.3.1 Polymer Matrix Nanocomposite

Nanocomposites containing nano-sized fillers in various forms (fibers, sheets, tubes, whiskers, particles, etc.) and in which a polymer acts as a matrix are defined as polymer matrix nanocomposites [46]. However, the addition of nanofillers to a polymer matrix during nanocomposite production makes it possible to obtain a high-performance composite by benefiting from the properties and nature of the nanoscale filler which directly contributes to the overall performance of the material [52]. The improvement in the properties of nanocomposites is due to the large surface area or high aspect ratio of the fillers [53].

11.3.2 Ceramic Matrix Nanocomposites

Composites formed by the dispersion of fillers such as metal, carbon nanotubes, particles, fibers, graphene, whiskers, and second polymer in a ceramic matrix are called ceramic matrix nanocomposites [54, 55]. These nanocomposites show important properties such as wear resistance, high temperature stability, good fracture toughness, resistance to thermal shocks, and superior strength [56, 57]. Due to these unique physical and mechanical properties, ceramic matrix nanocomposites are widely used in many fields, especially in the chemistry, electricity and electronics, biomedical, aerospace and defense, energy and automotive industries [58–61].

11.3.3 Metal Matrix Nanocomposites

Metal matrix nanocomposites are formed by embedding hard and stiff fillers consisting of fibers or particles into a more ductile and softer metallic matrix [62]. Although metals are more ductile and softer than ceramics, their hardness and strength are lower. In addition, due to the lower melting point of most metals and alloys used in structural applications, their properties at high temperatures are lower than ceramics [63]. Therefore, the incorporation of ceramics into alloys and metals during the production of metallic matrix nanocomposites composites allows to develop materials with better creep resistance, lower density, better stiffness and strength than metals, and better ductility and toughness than ceramics [64]. As a result of this situation, these nanocomposite systems are used in various fields such as electronic packaging, aerospace, thermal management, automotive, building and construction, and maritime and rail transport [65].

11.4 Chitosan & its Structure

Chitosan, the hydrolysis product of chitin, is a natural biopolymer and is obtained by removing the acetyl group from chitin under alkaline conditions [66]. Chitin is a natural polysaccharide, and it is usually obtained from cell walls of fungi, beaks of cephalopods exoskeletons of crustaceans and shellfish [67]. Chitosan is a biocompatible, biodegradable, and ecofriendly polymer. Chitosan, which is formed by connecting two repeating units with a ß-(1, 4)-glycosidic bond, is the N-deacetylated form of chitin and has a linear structure (poly (1, 4)-ß-linked 2-amino-2-deoxy-D-glucose). Chitosan has different degrees of deacetylation. The degree of deacetylation has an important influence on the properties, quality, and reactivity of chitosan

[41, 68]. The ratio of 2 acetamino-2-deoxy-D-glucopyranose to 2-amino-2-deoxy-D-glucopyranose in the structure of chitosan is defined as the degree of deacetylation. Therefore, chitosan dissolves in dilute acidic solutions, since the degree of deacetylation is over 55% [18].

Each glycosidic unit of chitosan contains hydrophilic functional groups such as two hydroxyl (–OH) groups and an amino ($-NH_2$) group. The presence of these functional groups in the structure of chitosan causes it to display excellent biological and chemical properties. In addition, these functional groups allow the modification of chitosan, thereby improving its physical and biological properties [69, 70].

11.5 Chitosan Nanocomposite Biosensor Applications

In general, biosensors are divided into classes as described above. The main purpose of the development of biosensors is to achieve a fast and practical application by using highly specific bioreactions of analytes at low concentrations [71, 72]. In this sense, different types of biosensors have been used to identify numerous molecules [73–75]. Some examples of these are given below for electrochemical, optical, and surface plasmon resonance techniques. The most common type of biosensor known is the glucose biosensor, and many studies in this field have been reported involving chitosan and nanocomposites. Additionally, other types of the analytes are determined in this wide range of sensing areas, as given in the Table 11.1.

One of the most effective nanomaterials is multi-walled carbon nanotubes, which were used to modify a novel chitosan-based matrix and used as an electrochemical biosensor for the determination of ifosfamide (IFO), acetaminophen (ACOP), domperidone (DOM), and sumatriptan (SUM) species. Here the sensitivity is enhanced by the addition of Ti3C2–MXene structures. MXenes are very useful two-dimensional structures for the selective and sensitive measurements. Lm-based electrochemical sensor for ifosfamide (IFO), acetaminophen (ACOP), domperidone (DOM), and sumatriptan (SUM). With the LOD values of 0.00031, 0.00028, 0.00034, and 0.00042 μM with the linear ranges of 0.0011 to 1.0 μM 0.0042–7.1, 0.0046–7.3, and 0.0033–61 μM have been developed. (Kalambate 2020).

In a study reported by Peng et al. [76] it is stated that graphene oxide addition enhanced the positively charge medium of the composite chitosan sol–gel which facilitates the biological reaction. The proposed electrochemical glucose biosensor showed a linear range from 0.02 to 5.39 mM with a low detection limit of 6.5 μM [76].

Another useful combination for carbon electrode surface coating is the gold nanoparticle and chitosan composite. A study for Monosodium L-glutamate sensing by DPV and EIS methods achieved a linear range of 100 pM to 1 μM with a detection sensitivity in the order of ~102 μA/nM [77]. Other similar electrochemical sensors are given in Table 11.1.

Optical biosensors are one of the most widely used biosensors. Recently, Basu et al., reported a chitosan-based silica nanocomposite as a label-free glucose

Table 11.1 Some of the chitosan based biosensors reported in the literature

Chitosan-nanocomposite type	Analyte	Method	LOD and Linear range	Refs.
Chitosan/reduced graphene oxide aerogel with dispersed ZrO2 nanoparticles modified glassy carbon electrode	Luteolin	Cyclic voltammetry (CV) and differential pulse voltammetry (DPV)	1 nM and 5 nM to 1000 nM	[81]
MXene/Chitosan on screen-printed electrode	Zearalenone	CV and electrochemical impedance spectroscopy (EIS)	0.4 pg ml^{-1} and 1 fg ml^{-1} to 1 ng ml^{-1}	[82]
MXene/CTS/Cu_2O	Glucose and cholesterol	CV	60.295 μA·L/(mmol·cm^2) with LOD being 52.4 μmol/L for glucose 215.71 μA·L/(mmol·cm^2) and LOD low to 49.8 μmol/L for cholesterol	[83]
Reduced graphene oxide-chitosan-ferrocene deposition of Pt–Pd bimetallic nanoparticles (RGO-CS-Fc/Pt–Pd BNPs)	GPC3 aptamers	CV	3.67 ng/mL and 0.001 to 10.0 μg/mL	[84]
Films of chitosan containing nanocomposites of graphene and gold nanoparticles (AuNPs) at a gold electrode	Glucose	CV, chronoamperometry	180 μM and 2.5 to 7.5 mM	[85]
$Ti_3C_2T_X$MXene/Chitosan	Potential prostate cancer biomarker in urine	Amperometry	18 nM and a linear range up to7.8 μM	[86]

(continued)

Table 11.1 (continued)

Chitosan-nanocomposite type	Analyte	Method	LOD and Linear range	Refs.
Riboflavin/AuPt-PPy/graphene-chitosan-modified GCE	sarcosine	CV, DPV, and EIS	0.68 μM 2.5–600 μM	[87]
Au nanoparticles-graphene-chitosan on cryogel electrode	Carcinoembryonic antigen immunosensor	CV	2.0×10^{-7} ng mL^{-1} and 1.0×10^{-6} to 1.0 ng mL^{-1} with a detection limit of	[88]
Tyrosinase/CdS quantum dots/chitosan	Catechol	Amperometric	0.3 nM and 1.0×10^{-9} to 2.0×10^{-5} M	[89]
Titanium (IV) oxide microparticles/gold nanoparticles/chitosan/CPE	CEA	CV and EIS	0.01 ng/mL and 0.01 to 20 ng/mL	[90]

biosensor with a 0,76 mM LOD value and they stated that chitosan is very adequate for the optical measurements [78]. Another optical phenylalanine determination method is reported by Fazial and Tan including graphene oxide-chitosan. In this study, phenylalanine, a marker for the fish freshness underwent an enzymatic reaction and the final reduced form of fluorogenic NADH is produced. Therefore, a linear range of 0.03 mM and 0.24 mM was achieved and the LOD value is reported as 0.02 mM for this optical biosensing approach [79]. In a surface plasmonic resonance-based biosensor developed by Mukhtar et al., different compositions of the chitosan composites as Ag-chitosan and Cu-chitosan were reported for the glucose detection [80].

11.6 Chitosan Nanocomposites Used for Sensing Different Biomolecules

It has been widely explained in the literature that chitosan is used in different forms and applied with different biomolecules, as shown in Fig. 11.4. In this sense, the gel-forming feature of chitosan comes to light. Commonly used molecules in certain types of biosensors previously described are mentioned. More specifically, it is worth mentioning the different molecules with which chitosan is used. Xia et al. reported a Chitosan used graphene-based molecular imprinting study that was previously used to determine bovine serum albumin (BSA) [10]. In this study, a linear working range of 1×10^{-10} to 1×10^{-4} g/L (R = 0.996). The LOD value has been reported as 2×10^{-11} g/L for BSA.

In the studies on the determination of antibiotic nature of chitosan, a successful study was also demonstrated in an application using norfloxacin. With the electrochemical sensor powered by nY2O3 nanoparticle loaded chitosan, a LOD value of 3.87 pM was reached in the range of 1 pM–10 μM [92]. In another interesting study, the composite structure of bacteriorhodopsin and chitosan was presented. In this

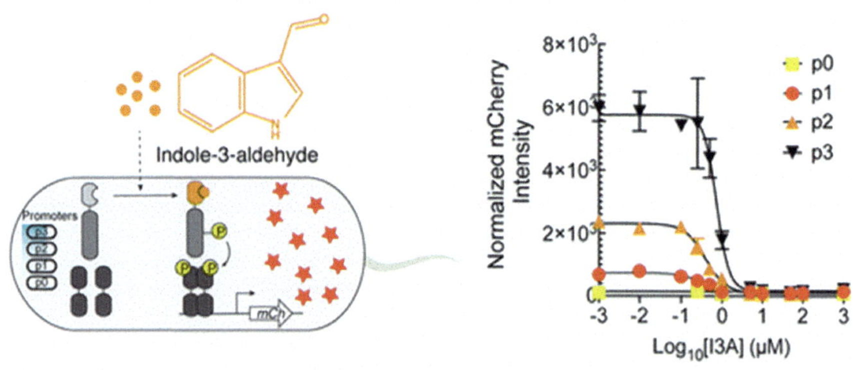

Fig. 11.4 A schematic diagram of indole-3-aldehyde detecting sensor reported by Wang et al. [91]. Reproduced with permission from American Chemical Society [91]

study, an optically active structure is presented as chitosan increases cell activation by forming a self-healing hydrogel structure [93].

Anthocyanins are one of the leading compounds in the food industry. These compounds were modified with chitosan to increase their stability. As a result, it has been reported that a more stable structure is obtained by the chitosan addition at different pHs and temperatures and also in the presence of L-ascorbic acid [94]. This is a very inspiring composition for the electrode modifications. In another study, it was observed that the use of Sage (Salvia officinalis L.) as a chitosan additive significantly increased the antioxidant activity. It has been suggested that the pinene, thujone, camphor, and limonene species found in the plant extract have this effect [95]. This is a very valuable finding for the sensor applications. In another amperometric biosensor study, it was reported that phenol was determined using laponite-doped chitosan. It is sensitive that the modifications made to increase the response sensitivity of the sensors and biosensors are biocompatible. In this sense, the use of biological molecules is more preferred. Within the scope of this study, it was emphasized that the long-term stability was increased by using this composite [96]. Another important compound is 4-nitrophenol. This compound, which is generally used in tissue staining, is an environmentally harmful dye that is important to determine. For this purpose, an enzymatic biosensor was developed by combining *Agrocybe aegerita peroxygenase* (AaP) and chitosan containing Au NP. The results obtained showed that the electrochemical biosensor detects 4-nitrophenol with a LOD value of 0.2 μM (based on the S/N = 3) in the operating range of 10 and 30 μM [97].

Vitamins are also important biological agents that increase the signal used with chitosan. In this regard, a study with tocopherols has been reported [98]. Finally, it has been reported that an impedimetric aptasensor has been developed for the detection of tetracycline antibiotics. In this study, chitosan and graphenoxide and 1-Pyrenebutyric acid–N-hydroxysuccinimide ester (Pry) were used together. As a result, a working range of 1.0×10^{-15}–3.2×10^{-7} and a LOD of 3.2×10^{-16} M have been reported [99].

11.7 Liquid Crystal-Chitosan Biosensors

Investigations on liquid crystal-chitosan nanocomposite sensors have been reported increasing recently. Liquid crystals (LCs) are soft materials with high sensitivity to external effects. The physical behavior of LC can be manipulated by a change in external state [100–106]. Rheological properties of isotropic liquids as well as crystalline order properties of solid crystals are simultaneously coming to sight by LCs in a certain temperature range, i.e., in a mesophase [100–106]. Moreover, since LCs show high biocompatibility, chemical stability, and reusability, LC-based composites are favorable photonic materials for sensor applications [107–113]. The working principle of LC-based composites is based on the sudden change in molecular orientation of LCs at the LC/doped-material boundary by the existence of biological and chemical structures. These existences can shift the surface anchoring of LC which is one of the fundamental parameters to describe the state of the molecular orientation of LC. Thus, the optical view of LC is changed and the change can be observed with naked eye when the LC-based nanocomposite is positioned between a couple of crossed polarizers [114–116]. LC-based nanocomposite sensors have various advantages, such as fast responsiveness, simple generating, high sensitivity, storability, and usability in various environments, in order for them to be used as an attractive application for monitoring biological and chemical variance or reactions [114–120].

Recently, a new trend in biosensor research has been emerging with the assembling of LCs and chitosan as a nanocomposite. LC-Chitosan nanocomposites are quite interesting materials for biosensing because chitosan shows superior bio-based properties such as being biocompatible and biodegradable. Moreover, while hydrophobic LCs are generally used especially in polymer dispersed liquid crystals (PDLCs), chitosan exhibits hydrophilic and hydrophobic behavior to the weakly acidic solutions and in solid state, respectively. However, chitosan is immiscible with a LC in solid state [121, 122].

Chitosan is a good candidate as a polymer matrix for PDLC [121, 122] and also PDLCs are suitable structures for biosensors [123, 124]. Using chitosan as a polymer matrix was done for the first time by Marin et al. In the work, 4-cyano-4'-pentylbiphenyl (5CB) which shows nematic mesophase was used as a LC. 5CB was homeotropically aligned inside the chitosan matrix and formed as a globule with a singularity defect. Since the LC dispersed in chitosan matrix resulting increase in surface, the iridescence feature of LC raised. It was noticed that a very less amount of LC was enough to prepare a PDLC with chitosan matrix [121]. In another work, Ailincai and Marin reported that chitosan and a cholesteric mesophase LC which was cholesteryl acetate—very common commercial LC—were assembled as an eco-friendly PDLC [122]. The reason why these materials were chosen was explained as both LC and chitosan contained compounds that comply with international environmental protection standards. The PDLC which was produced had reasonable wettability and surface-free energy which is the sign of being biocompatible materials.

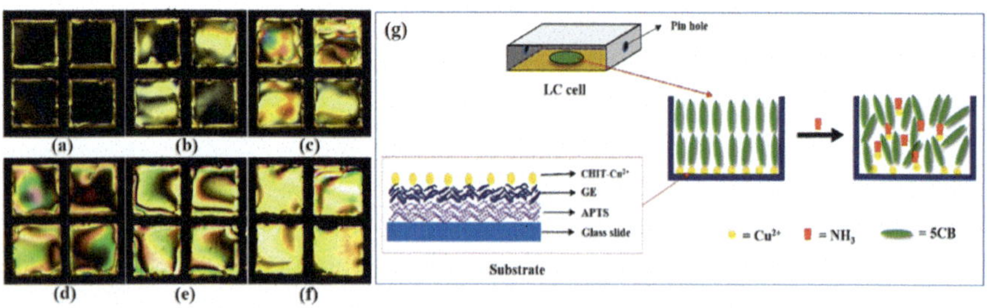

Fig. 11.5 Optical textures of LC-based optical sensor taken by polarized optical microscopy. Depending on exposing time in ammonia for **a** 0 s, **b** 10 s, **c** 25 s, **d** 50 s, **e** 75 s, and **f** 100 s, brightness of each texture changed. Also, **g** a cartoon representation of fabrication process of LC-based optical sensor for ammonia detection is given. Reproduced with permission from The Optical Society [128]

Besides, it was reported that the eco-friendly PDLC were quite interesting materials in order for them to being a promising candidate to form biosensors [122].

Apart from PDLC, various LC/chitosan nanocomposites were reported so as to investigate their biosensing activity. Niu et al. reported that a LC-based optical sensor was developed for detecting ammonia (Fig. 11.5g). 5CB was used because it was preferred several times for detecting some biomolecules in the literature [115, 125–127]. The optical images of LC-based nanocomposite appeared dark among a pair of crossed polarizers since the LC was homeotropically aligned due to existence of the state of complex formed from copper perchlorate in composite (Fig. 11.5a). It was obvious that depending on the increase in exposing time in ammonia, optical textures of LC-based optical textures appeared brighter (Fig. 11.5b–f). The working principle of the sensor based on linking between the cyano group of LCs, which existed on chitosan-Cu^{2+} composite placed on a glass, and ammonia. By means of the sensor, ammonia could detect in a wide range of 500 ppm to 1250 ppm. Besides, it was reported that the method was cost-effective, high selective, and rather sensitive, and also the method could be applied to sense biomolecules and ions of hazardous materials by the work [128].

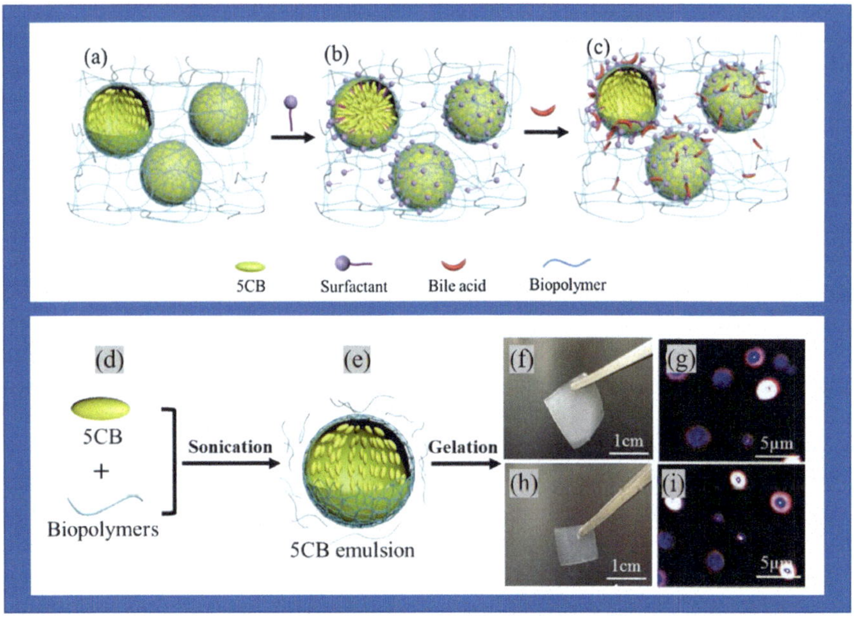

Fig. 11.6 **a–c** The cartoon representation of molecular orientation of PHDLC which is affected the existence of surfactant and bile acid. **d, e** The cartoon representation of formation of polymer-stabilized 5CB LC emulsion. **f** and **e** were the images of chitosan and agarose hydrogel dispersed 5CB LC droplet films, respectively. **g** and **i** were the textures of 5CB LC droplets which was dispersed in chitosan and agarose hydrogel films, respectively. Reproduced with permission from Elsevier [114]

In further research, it emerged as a induced polymer hydrogel dispersed LC (PHDLC) globule fragments by gelation of chitosan-5CB LC composite, whose liquid crystalline features emerge at room temperature, emulsion film [114]. In the study, viscoelastic features of the PHDLC and change in volume of PHDLC were investigated. It was shown that ionic surfactants and bile acids could achieve the boundary of dispersed LC droplets, and leach in the hydrogel layers from exterior aqueous ambient. The molecular alignment of PHDLC was induced by surfactant/LC and bile acid/LC interactions at the boundary of the LC globules and this case followed (Fig. 11.6a–c). In (Fig. 11.6d–i), the process of polymer-stabilized 5CB LC emulsion, hydrogel films, and the related textures were also presented. Thus, how the response time and diversity of the LC globules in the detection of the molecules were affected by polymer matrix could be determined. According to scientists, the emerging composite opened a door to a novel approach towards soft material-based sensing system. Despite the fact that more investigations are necessary to find out how the polymer matrix affects the sensor features of hydrogel-dispersed LC droplets [114].

At an aqueous/LC boundary, chitosan and phospholipid membrane interactions were investigated for the first time by Liu et al. [116]. The molecular alignment of LC was affected by the interactions of chitosan and lipid membranes. Thus, the optical appearance of LC could be shifted in a pair of crossed polarizers. In the study alignment of LC was shifted from homeotropic to planar, namely, the LC appeared first dark, after shifting the alignment it was observed bright [116]. By this method, a great deal of biomolecules can be sensed optically. Monitoring the antibacterial events is performed by using LC/chitosan nanocomposites as well. Wei and Jang reported a new approach on LC biosensors in order to follow the interaction between the phospholipid membrane and chitosan-modified graphene oxide, and so as to monitor antibacterial events of chitosan-modified graphene oxide structure [118]. In the study, as sensing systems (Fig. 11.7) there were two different LC structures which were a LC film confined by electron microscope grids, and sessile LC droplets. As a thermotropic LC, 5CB was used. After some treatments were applied to grids and producing the LC film over the grids, the LC film was covered with self-assembled phospholipid monolayer. Also, sessile LC droplets were generated in several steps: first, the thermotropic nematic 5CB was put in n-heptane with a low weight percentage. Then, a small amount of the mixture was dripped on an OTS-treated slide and was anchored on it by evaporating the solvent. By using polarized optical microscopy, the optical variances of LC film due to shifting molecular ordering were determined. These variances represented that graphene oxide was able to penetrate into the phospholipid membranes. Also, the variances pointed the chitosan-graphene oxide structure was more influential for demolish the phospholipid membranes. Moreover, the molecular alignment of LC at the aqueous/LC boundary was found to be homeotropic. The reason of this case was specified by the scientists as lipids on the bacteria cell membrane might be replaced to surface of LC droplets. It was reported that holding the bacteria with chitosan-graphene oxide composite diminished unity of the bacterial cells. Hereby, the chitosan-graphene oxide-abiotic bacteria could be specified by sessile LC droplets [118].

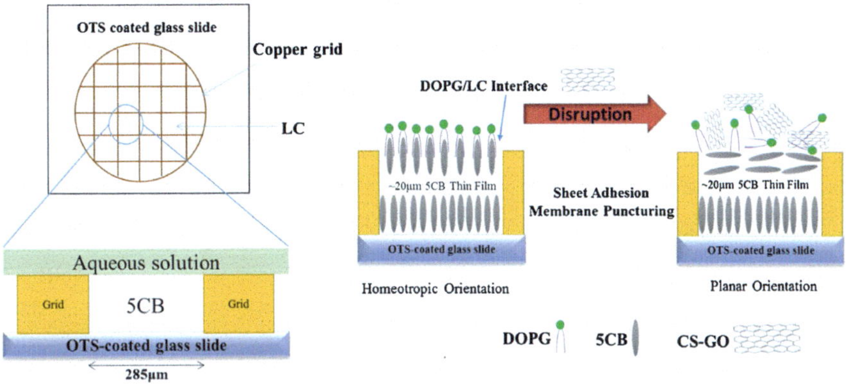

Fig. 11.7 Cartoon representation of LC thin film sensing system and display of the molecular alignment of 5CB before and after demolishment of the dioleoyl-sn-glycero-3-phospho-rac-(1-glycerol) sodium salt (DOPG) membrane by chitosan-graphene oxide (CS-GO). Reproduced with permission from Elsevier [118]

Fang et al. reported a work on biosensor applications with LC droplets dispersed in aqueous medium [129]. In the work, 5CB was dispersed in an aqueous solution. Thereafter, surface of aqueous/5CB was tailored with the adsorption of chitosan at aqueous/5CB interface and penetration of a surfactant. It was stated that cholic acid which existed in biostructures included uric acid, ascorbic acid, and urea was detected by the 5CB droplets which were coated with chitosan/surfactant as an optical detector. The detection was performed, likewise being the other most of works, by monitoring the change in the molecular orientation with the naked eye using crossed polarizers. The scientists reported the detection limit of the LC for cholic acid was diminished due to chitosan as a coating. At the same time, the detection limit of 5CB droplets which was coated with chitosan/surfactant could be shifted in a wide range, which depended on the amount of the droplets (Fig. 11.8) [129]. A novel approach on a chitosan/surfactant coated LC structures was indicated as a promising nanostructure for detection of bile acids which is used as a sign for identification of serious health problems [115, 129]. Fang and colleagues have reported another study about compositing chitosan hydrogels and LC droplets as a sensing platform for detecting bile acid [115]. Hydrogels, which are a type of matrix of hydrophilic polymers, are favorable materials for bioapplications because of showing unique features such as the ability to adsorb water, being permeable for biomolecules, and easily processible [114, 115, 119, 120, 130, 131]. It was given that composite material made up of LC droplet-stabilized chitosan hydrogel structures were considered as a mobile sensor for the detection of bile acid. The cartoon representation of formation steps of the 5CB droplet-embedded chitosan hydrogel was presented in (Fig. 11.9). Also, in this study, the working principle of biosensor based on following the change in alignment of the LC, which was affected due to adsorption of the bile acid diffused into chitosan. The scientists specified a consequence that was different from chitosan/surfactant coated 5CB droplets, 5CB droplet-embedded chitosan hydrogel films could be simply generated and by this case, bile acid could be selectively sensed. In addition, it is stated that the hydrophobicity of bile acid was responsible for value of the response time and detection limit of LC-included hydrogels [115].

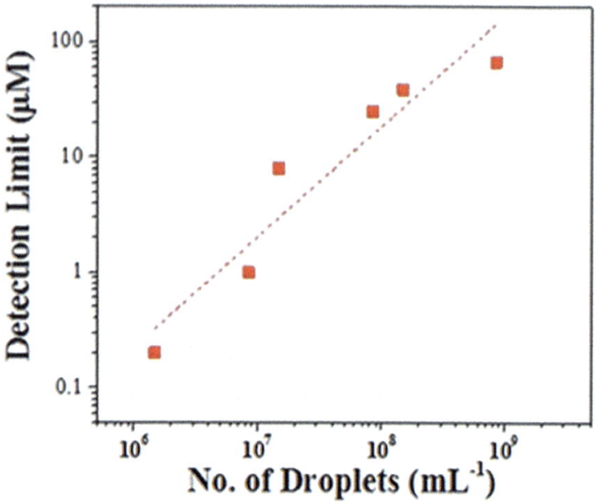

Fig. 11.8 The detection limit of chitosan/sodium tetradecyl sulfate complex-coated 5CB droplets as a function of a number of the droplets. Reproduced with permission from Royal Society of Chemistry [129]

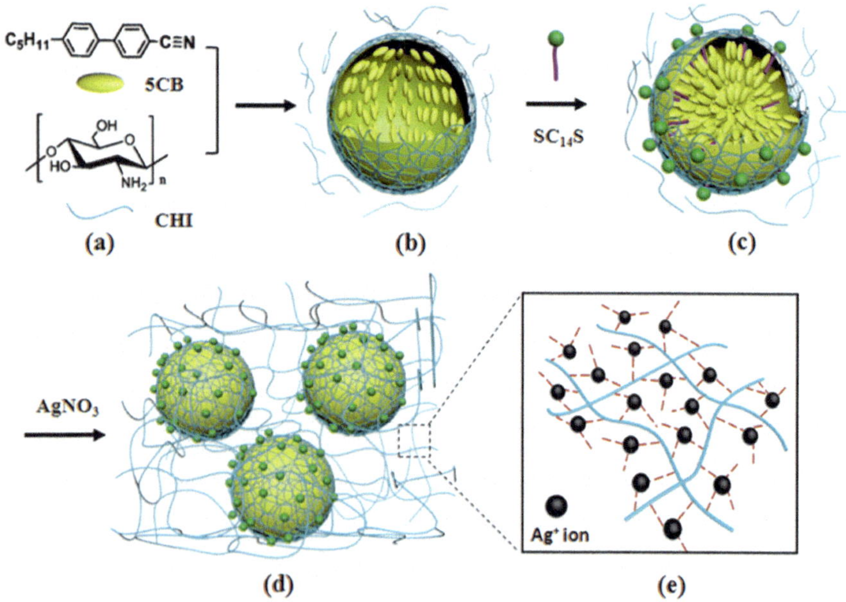

Fig. 11.9 **a** Chemical formula of 5CB and chitosan. **b–e** The cartoon representation of the formation steps of the 5CB droplet-embedded chitosan hydrogel. Reproduced with permission from the American Chemical Society [115]

LC-embedded chitosan matrix was also developed for detecting methanol ratio in a methanol-ethanol mixture. Two different chiral nematic (cholesteric) LCs mixture, which reflected different wavelengths, and chitosan matrix were used as sensing materials and a LC holder, respectively [119]. Molecular orientation of a chiral nematic LC is identified with twisted-nematic layers. The distance between the layers at which planar angle of director of the nematic layers completes one rotation is described as a pitch. Chiral nematics can selectively reflect an incident light and wavelength of the reflected light corresponds to the length of pitch [100, 101, 103]. The working principle of sensing system relied on different changes of methanol and ethanol on the pitch of chiral nematics. Thus, methanol and ethanol ratio in a mixture could be determined by following the shifting of reflection spectrum. It was reported that the new type of sensor showed good linearity, stability, and also repeatability [119].

11.8 Conclusion

Chitosan is one of the most widely used biomaterials. Chitosan-based nanocomposites have attracted the attention of researchers due to their unique properties such as high stability, biodegradability, non-toxicity, and biocompatibility. For these reasons, chitosan nanocomposites are widely used as biosensors in environmental, biomedical, pharmaceutical, and catalysis research applications. A new trend in biosensor research has been emerging with the assembling of LCs and chitosan as a nanocomposite. LC-Chitosan nanocomposites are quite interesting materials for biosensing because chitosan shows superior bio-properties. Based on this issue, the present chapter gives a concluding remark toward chitosan nanocomposites for biosensing applications, classifications, and possible synthesis of new age materials.

References

1. S. Hameed, L. Xie, Y. Ying, Conventional and emerging detection techniques for pathogenic bacteria in food science: a review. Trends Food Sci. Technol. **81**, 61 (2018)
2. M.I. Rocha-Gaso, L.J. Villarreal-Gómez, D. Beyssen, F. Sarry, M.A. Reyna, C.N. Ibarra-Cerdeña, Biosensors to diagnose chagas disease: a brief review. Sensors (Switzerland) **17**, 1 (2017)
3. V.V. Singh, Recent advances in electrochemical sensors for detecting weapons of mass destruction. A review. Electroanalysis **28**, 920 (2016)
4. S. Uniyal, R.K. Sharma, Technological advancement in electrochemical biosensor based detection of organophosphate pesticide chlorpyrifos in the environment: a review of status and prospects. Biosens. Bioelectron. **116**, 37 (2018)
5. P. Gomathi, M.K. Kim, J.J. Park, D. Ragupathy, A. Rajendran, S.C. Lee, J.C. Kim, S.H. Lee, H. Do Ghim, Multiwalled carbon nanotubes grafted chitosan nanobiocomposite: a prosperous functional nanomaterials for glucose biosensor application. Sens. Actuators, B Chem. **155**, 897 (2011)

6. E. Muthusankar, D. Ragupathy, Chitosan based nanocomposite biosensors: a recent review. Sens. Lett. **16**, 81 (2018)
7. J. Lei, H. Ju, Signal amplification using functional nanomaterials for biosensing. Chem. Soc. Rev. **41**, 2122 (2012)
8. X. Bin, E.H. Sargent, S.O. Kelley, Nanostructuring of sensors determines the efficiency of biomolecular capture. Anal. Chem. **82**, 5928 (2010)
9. C. Yang, M.E. Denno, P. Pyakurel, B.J. Venton, Recent trends in carbon nanomaterial-based electrochemical sensors for biomolecules: a review. Anal. Chim. Acta **887**, 17 (2015)
10. Y. Jiang, J. Wu, Recent development in chitosan nanocomposites for surface-based biosensor applications. Electrophoresis **40**, 2084 (2019)
11. C. Zhu, G. Yang, H. Li, D. Du, Y. Lin, Electrochemical sensors and biosensors based on nanomaterials and nanostructures. Anal. Chem. **87**, 230 (2015)
12. Z. Wang, Z. Dai, Carbon nanomaterial-based electrochemical biosensors: an overview. Nanoscale **7**, 6420 (2015)
13. B.B. Prasad, A. Prasad, M.P. Tiwari, R. Madhuri, Multiwalled carbon nanotubes bearing "Terminal Monomeric Unit" for the fabrication of epinephrine imprinted polymer-based electrochemical sensor. Biosens. Bioelectron. **45**, 114 (2013)
14. F.R.R. Teles, L.P. Fonseca, Applications of polymers for biomolecule immobilization in electrochemical biosensors. Mater. Sci. Eng. C **28**, 1530 (2008)
15. J. Ahmed, M. Mulla, M. Maniruzzaman, Rheological and dielectric behavior of 3d-printable chitosan/graphene oxide hydrogels. ACS Biomater. Sci. Eng. **6**, 88 (2020)
16. H. Yan, L. Jiang, X. Xu, Y. Li, Y. Shen, S. Zhu, Ultrastrong composite film of chitosan and silica-coated graphene oxide sheets. Int. J. Biol. Macromol. **104**, 936 (2017)
17. F. Croisier, C. Jérôme, Chitosan-based biomaterials for tissue engineering. Eur. Polym. J. **49**, 780 (2013)
18. S.K. Shukla, A.K. Mishra, O.A. Arotiba, B.B. Mamba, Chitosan-based nanomaterials: a state-of-the-art review. Int. J. Biol. Macromol. **59**, 46 (2013)
19. P. Tetyana, P. Morgan Shumbula, Z. Njengele-Tetyana, Biosensors: design, development and applications, in *Nanopores* (IntechOpen, 2021)
20. L.C. Clark, C. Lyons, Electrode systems for continuous monitoring in cardiovascular surgery. Ann. N. Y. Acad. Sci. **102**, 29 (1962)
21. D. Grieshaber, R. MacKenzie, J. Vörös, E. Reimhult, Electrochemical biosensors—Sensor principles and architectures. Sensors **8**, 1400 (2008)
22. C.J. Watson, M.T. Ledwidge, D. Phelan, P. Collier, J.C. Byrne, M.J. Dunn, K.M. McDonald, J.A. Baugh, Proteomic analysis of coronary sinus serum reveals leucine-rich A2-Glycoprotein as a novel biomarker of ventricular dysfunction and heart failure. Circ. Hear. Fail. **4**, 188 (2011)
23. N. Dewan, P. Ahmed, G. Chowdhury, S. Pandit, D. Dasgupta, Nanotechnology based biosensors and its application. Pharma Innov. J. **18**, 18 (2016)
24. D. Bal Altuntaş, F. Kuralay, MoS2/Chitosan/GOx-Gelatin modified graphite surface: preparation, characterization and its use for glucose determination. Mater. Sci. Eng. B Solid-State Mater. Adv. Technol. **270** (2021)
25. L. Li, Y. Wang, L. Pan, Y. Shi, W. Cheng, Y. Shi, G. Yu, A Nanostructured conductive hydrogels-based biosensor platform for human metabolite detection. Nano Lett. **15**, 1146 (2015)
26. V.K. Nigam, P. Shukla, Enzyme based biosensors for detection of environmental pollutants—a review. J. Microbiol. Biotechnol. **25**, 1773 (2015)
27. Y. Zhang, Y. Li, F. Xiao, H. Wang, L. Zhang, Z. Ding, S. Xu, Z. Gu, G. Shi, Engineering of a biosensor in response to malate in Bacillus Licheniformis. ACS Synth. Biol. **10**, 1775 (2021)
28. S. Singh, V. Kumar, D.S. Dhanjal, S. Datta, R. Prasad, J. Singh, Biological biosensors for monitoring and diagnosis, in *Microbial Biotechnology: Basic Research and Applications, Environmental and Microbial Biotechnology* (2020), pp. 317–335
29. Y. Du, S. Dong, Nucleic acid biosensors: recent advances and perspectives. Anal. Chem. **89**, 189 (2017)

30. D. Bal Altuntaş, H.S. Nalkıran, S. Aslan, Z. Yolcu, Development of MoS2 and Au nanoparticle including disposable CEA-based immuno-cytosensor platforms. Chem. Pap. **76**, 5217 (2022)
31. I.H. Cho, D.H. Kim, S. Park, Electrochemical biosensors: perspective on functional nanomaterials for on-site analysis. Biomater. Res. **24**, 1 (2020)
32. F. Schachinger, H. Chang, S. Scheiblbrandner, R. Ludwig, Amperometric biosensors based on direct electron transfer enzymes. Molecules **26**, 4525 (2021)
33. J. Ding, W. Qin, Recent advances in potentiometric biosensors. TrAC Trends Anal. Chem. **124**, 115803 (2020)
34. K. Berketa, O. Saiapina, L. Fayura, A. Sibirny, S. Dzyadevych, O. Soldatkin, Novel highly sensitive conductometric biosensor based on arginine deiminase from mycoplasma hominis for determination of arginine. Sens. Actuators B Chem. **367**, 132023 (2022)
35. P. Damborský, J. Švitel, J. Katrlík, Optical biosensors. Essays Biochem. **60**, 91 (2016)
36. M. Pohanka, The piezoelectric biosensors: principles and applications, a review. Int. J. Electrochem. Sci. **12**, 496 (2017)
37. N. Sivarajasekar, Ra. Dharshni, Rd. Dharshini, Thermal biosensors and their applications. Am. Int. J. Res. Sci. **50** (2019)
38. A.O. Silva, R.S. Cunha, D. Hotza, R.A.F. Machado, Chitosan as a matrix of nanocomposites: a review on nanostructures, processes, properties, and applications. Carbohydr. Polym. **272**, 118472 (2021)
39. K.V. Harish Prashanth, R.N. Tharanathan, Chitin/Chitosan: modifications and their unlimited application potential-an overview. Trends Food Sci. Technol. **18**, 117 (2007)
40. P.S. Bakshi, D. Selvakumar, K. Kadirvelu, N.S. Kumar, Chitosan as an environment friendly biomaterial—a review on recent modifications and applications. Int. J. Biol. Macromol. **150**, 1072 (2020)
41. B.R. Rizeq, N.N. Younes, K. Rasool, G.K. Nasrallah, Synthesis, bioapplications, and toxicity evaluation of chitosan-based nanoparticles. Int. J. Mol. Sci. **20** (2019)
42. A.R. Del Bakhshayesh, N. Asadi, A. Alihemmati, H. Tayefi Nasrabadi, A. Montaseri, S. Davaran, S. Saghati, A. Akbarzadeh, A. Abedelahi, An overview of advanced biocompatible and biomimetic materials for creation of replacement structures in the musculoskeletal systems: focusing on cartilage tissue engineering. J. Biol. Eng. **13**, 1 (2019)
43. M.S. Riaz Rajoka, L. Zhao, H.M. Mehwish, Y. Wu, S. Mahmood, Chitosan and its derivatives: synthesis, biotechnological applications, and future challenges. Appl. Microbiol. Biotechnol. **103**, 1557 (2019)
44. H. Jafari, M.K. Pirouzifard, M.A. Khaledabad, H. Almasi, Effect of chitin nanofiber on the morphological and physical properties of chitosan/silver nanoparticle bionanocomposite films. Int. J. Biol. Macromol. **92**, 461 (2016)
45. C. Rodrigues, J.M.M. de Mello, F. Dalcanton, D.L.P. Macuvele, N. Padoin, M.A. Fiori, C. Soares, H.G. Riella, Mechanical, thermal and antimicrobial properties of chitosan-based-nanocomposite with potential applications for food packaging. J. Polym. Environ. **28**, 1216 (2020)
46. A.K. Mishra, S.B. Mishra, A. Tiwari, Nanocomposites and their biosensor applications, in *Biosensor Nanomaterials* (2011), pp. 255–268
47. M. Jose-Yacaman, L. Rendon, J. Arenas, M. Carmen Serra Puche, An ancient nanostructured. Science (80-.) **273**, 223 (1996)
48. P.V. Braun, Natural nanobiocomposites, biomimetic nanocomposites, and biologically inspired nanocomposites, in *Nanocomposite Science and Technology* (2003), pp. 155–214
49. P.M. Ajayan, Bulk metal and ceramics nanocomposites, in *Nanocomposite Science and Technology* (2003), pp. 1–75
50. C. Picu, P. Keblinski, Modeling of nanocomposites, in *Nanocomposite Science and Technology* (2021), pp. 215–222
51. L.S. Schadler, Polymer-based and polymer-filled nanocomposites, in *Nanocomposite Science and Technology* (2003), pp. 77–153.
52. A. Usuki, M. Kawasumi, Y. Kojima, A. Okada, T. Kurauchi, O. Kamigaito, Swelling behavior of montmorillonite cation exchanged for ω -amino acids by e-Caprolactam. J. Mater. Res. **8**, 1174 (1993)

53. E. Manias, Stiffer by design. Nat. Mater. **6**, 9 (2007)
54. J. Sun, D. Ye, J. Zou, X. Chen, Y. Wang, J. Yuan, H. Liang, H. Qu, J. Binner, J. Bai, A review on additive manufacturing of ceramic matrix composites. J. Mater. Sci. Technol. **13** (2022)
55. F. Bouville, E. Maire, S. Meille, B. Van De Moortèle, A.J. Stevenson, S. Deville, Strong, tough and stiff bioinspired ceramics from brittle constituents. Nat. Mater. **13**, 508 (2014)
56. E. Feilden, C. Ferraro, Q. Zhang, E. García-Tuñón, E. D'Elia, F. Giuliani, L. Vandeperre, E. Saiz, 3D printing bioinspired ceramic composites. Sci. Rep. **7**, 1 (2017)
57. H. Bian, Y. Yang, Y. Wang, W. Tian, H. Jiang, Z. Hu, W. Yu, Effect of microstructure of composite powders on microstructure and properties of microwave sintered alumina matrix ceramics. J. Mater. Sci. Technol. **29**, 429 (2013)
58. C.E. Smith, G.N. Morscher, Z.H. Xia, Monitoring damage accumulation in ceramic matrix composites using electrical resistivity. Scr. Mater. **59**, 463 (2008)
59. E. Pietrzykowska, M. Małysa, A. Chodara, B. Romelczyk-Baishya, K. Szlązak, W. Święszkowski, Z. Pakieła, W. Lojkowski, Biodegradable ceramic matrix composites made from nanocrystalline hydroxyapatite and silk fibers via crymilling and uniaxial pressing. Mater. Lett. **293** (2021)
60. S. Tang, C. Hu, Design, preparation and properties of carbon fiber reinforced ultra-high temperature ceramic composites for aerospace applications: a review. J. Mater. Sci. Technol. **33**, 117 (2017)
61. G. Zheng, X. Yin, J. Wang, M. Guo, X. Wang, Complex permittivity and microwave absorbing property of Si 3N 4-SiC composite ceramic. J. Mater. Sci. Technol. **28**, 745 (2012)
62. A. Mortensen, J. Llorca, Metal matrix composites. Annu. Rev. Mater. Res. **40**, 243 (2010)
63. B. Basu, K. Balani, Fundamentals of nature and characteristics of ceramics, in *Advanced Structural Ceramics* (2011), pp. 3–10
64. F. Nturanabo, L. Masu, J. Baptist Kirabira, Novel applications of aluminium metal matrix composites, in *Aluminium Alloys and Composites* (2020), pp. 1–24
65. P. Maurya, N. Kota, J. Gibmeier, A. Wanner, S. Roy, Review on study of internal load transfer in metal matrix composites using diffraction techniques. Mater. Sci. Eng. A **840**, 142973 (2022)
66. C. Escudero-Oñate, E. Martínez-Francés, A review of chitosan-based materials for the removal of organic pollution from water and bioaugmentation, in *Chitin-Chitosan—Myriad Functionalities in Science and Technology* (InTech, 2018)
67. S. Kumar, F. Ye, S. Dobretsov, J. Dutta, Chitosan nanocomposite coatings for food, paints, and water treatment applications. Appl. Sci. **9** (2019)
68. S. Mallakpour, L. Khodadadzadeh, Biocompatible and biodegradable chitosan nanocomposites loaded with carbon nanotubes, in Biodegradable and biocompatible polymer composites: processing, properties and applications (Elsevier Ltd, 2017), pp. 187–221
69. J. Wang, C. Chen, Chitosan-based biosorbents: modification and application for biosorption of heavy metals and radionuclides. Bioresour. Technol. **160**, 129 (2014)
70. A. Leudjo Taka, M.J. Klink, X. Yangkou Mbianda, E.B. Naidoo, Chitosan nanocomposites for water treatment by fixed-bed continuous flow column adsorption: a review. Carbohydr. Polym. **255**, 117398 (2021)
71. G. Kravanja, M. Primožič, Ž Knez, M. Leitgeb, Chitosan-based (Nano)materials for novel biomedical applications. Molecules **24**, 1960 (2019)
72. A.K. Nayak, S. Alkahtani, M.S. Hasnain, Biomedical nanocomposites, in *Biomedical Composites: Perspectives and Applications*. ed. by A.K. Nayak, M.S. Hasnain (Springer Singapore, Singapore, 2021), pp.35–69
73. S. Amirthalingam, J. Rangasamy, Chitosan-based biosensor fabrication and biosensing applications, in *Chitosan for Biomaterials III: Structure-Property Relationships*. ed. by R. Jayakumar, M. Prabaharan (Springer International Publishing, Cham, 2021), pp.233–255
74. T. Thirugnanasambandan, Bio-Nanocomposites, in *Biomedical Application, in Polymer Based Bio-Nanocomposites: Properties, Durability and Applications*. ed. by C. Muthukumar, S.M.K. Thiagamani, S. Krishnasamy, R. Nagarajan, S. Siengchin (Springer Singapore, Singapore, 2022), pp.275–291

75. S. Sharma, B. Sharma, P. Jain, Graphene based biopolymer nanocomposites, in *Sensors, in Graphene Based Biopolymer Nanocomposites*. ed. by B. Sharma, P. Jain (Springer Singapore, Singapore, 2021), pp.273–286
76. H. Peng, Z. Huang, Y. Zheng, W. Chen, A. Liu, X. Lin, A novel nanocomposite matrix based on graphene oxide and ferrocene-branched organically modified sol-gel/chitosan for biosensor application. J. Solid State Electrochem. **18**, 1941 (2014)
77. D. Sharma, R. Devi, J. Jaiswal, H.S. Dutta, R. Khan, M. Dhayal, A Highly sensitive immunosensor based on in situ reduced gold-chitosan nanocomposite for detection of monosodium L-Glutamate. J. Biosyst. Eng. **47**, 28 (2022)
78. D. Basu, S.M. Hossain, J. Das, Surface modified chitosan-silica nanocomposite porous thin film based multi-parametric optical glucose sensor. Appl. Phys. A **128**, 688 (2022)
79. F.F. Fazial, L.L. Tan, Phenylalanine-responsive fluorescent biosensor based on graphene oxide-chitosan nanocomposites catalytic film for non-destructive fish freshness grading. Food Control **125**, 107995 (2021)
80. W.M. Mukhtar, N.A.M. Taib, A.R.A. Rashid, Sensitivity analyses of Cu/Chitosan and Ag/Chitosan based SPR biosensor for glucose detection. J. Phys. Conf. Ser. **1892** (2021)
81. X. Hou, W. Wu, F. Zhao, W. Xie, Q. Yang, Construction of an electrochemical sensor with graphene aerogel doped with ZrO2 nanoparticles and chitosan for the selective detection of luteolin. Microchim. Acta **188** (2021)
82. S. Siva Sangu, S. Chandra Bose Gopinath, M.F. Abdul Shukur, M.S. Mohamed Saheed, An electrochemical approach for ultrasensitive detection of zearalenone in commodity using disposable screen-printed electrode coated with MXene/Chitosan Film. Bionanoscience (2022)
83. T. Hu, M. Zhang, H. Dong, T. Li, X. Zang, X. Li, Z. Ni, Free-standing MXene/Chitosan/Cu2O electrode: an enzyme-free and efficient biosensor for simultaneous determination of glucose and cholesterol. J. Zhejiang Univ. A **23**, 579 (2022)
84. M. Chen, H. Li, X. Su, R. Wu, H. Feng, X. Shi, J. Liang, J. Chen, G. Li, Label-free electrochemical aptasensor based on reduced graphene Oxide–Hemin–Chitosan nanocomposite for the determination of Glypican-3. New J. Chem. **45**, 8608 (2021)
85. C. Shan, H. Yang, D. Han, Q. Zhang, A. Ivaska, L. Niu, Graphene/AuNPs/Chitosan nanocomposites film for glucose biosensing. Biosens. Bioelectron. **25**, 1070 (2010)
86. S. Hroncekova, T. Bertok, M. Hires, E. Jane, L. Lorencova, A. Vikartovska, A. Tanvir, P. Kasak, J. Tkac, Ultrasensitive Ti3C2TX MXene/Chitosan nanocomposite-based amperometric biosensor for detection of potential prostate cancer marker in urine samples. Processes **8**, 580 (2020)
87. T. Liu, B. Fu, J. Chen, K. Li, An electrochemical sarcosine sensor based on biomimetic recognition, Microchim. Acta **186** (2019)
88. S. Samanman, A. Numnuam, W. Limbut, P. Kanatharana, P. Thavarungkul, Highly-sensitive label-free electrochemical carcinoembryonic antigen immunosensor based on a novel Au nanoparticles-Graphene-Chitosan nanocomposite cryogel electrode. Anal. Chim. Acta **853**, 521 (2015)
89. E. Han, Y. Yang, Z. He, J. Cai, X. Zhang, X. Dong, Development of tyrosinase biosensor based on quantum dots/chitosan nanocomposite for detection of phenolic compounds. Anal. Biochem. **486**, 102 (2015)
90. S. Aslan, Ü. Anik, Development of TiO2 and Au nanocomposite electrode as CEA immunosensor transducer. Electroanalysis **26**, 1373 (2014)
91. J. Wang, C. Zhang, W.S. Childers, A biosensor for detection of indole metabolites. ACS Synth. Biol. **10**, 1605 (2021)
92. F. Mohammad, H.A. Al-Lohedan, H.N. Al-Haque, Chitosan-Mediated fabrication of metal nanocomposites for enhanced biomedical applications. Adv. Mater. Lett. **8**, 89 (2017)
93. F.Y. Hsieh, H.W. Han, X.R. Chen, C.S. Yang, Y. Wei, S. Hui Hsu, Non-Viral delivery of an optogenetic tool into cells with self-healing hydrogel. Biomaterials **174**, 31 (2018)
94. J. Ge, P. Yue, J. Chi, J. Liang, X. Gao, Formation and stability of anthocyanins-loaded nanocomplexes prepared with chitosan hydrochloride and carboxymethyl chitosan. Food Hydrocoll. **74**, 23 (2018)

95. N. Vosoughi, M. Gomarian, A. Ghasemi Pirbalouti, S. Khaghani, F. Malekpoor, Essential oil composition and total phenolic, flavonoid contents, and antioxidant activity of sage (Salvia Officinalis L.) extract under chitosan application and irrigation frequencies. Ind. Crops Prod. **117**, 366 (2018)
96. Q. Fan, D. Shan, H. Xue, Y. He, S. Cosnier, Amperometric phenol biosensor based on laponite clay-chitosan nanocomposite matrix. Biosens. Bioelectron. **22**, 816 (2007)
97. L. Peng, U. Wollenberger, M. Hofrichter, R. Ullrich, K. Scheibner, F.W. Scheller, Bioelectrocatalytic properties of agrocybe aegerita peroxygenase. Electrochim. Acta **55**, 7809 (2010)
98. S. Kadyan, A.K. Singh, N.R. Panjagari, Electrospinning: technology for encapsulation of bioactive compounds. Indian Food Ind. Mag. **1** (2020)
99. A. Benvidi, M. Nikmanesh, M.D. Tezerjani, Graphene oxide/chitosan based impedimetric aptasensor along with an ester linker for the detection of tetracycline. J. Nanostructures **12**, 204 (2022)
100. L. M. Blinov, *Structure and Properties of Liquid Crystals* (2011)
101. I. Dierking, *Textures of Liquid Crystals* (2003)
102. P. Özden, A.E. Mamuk, N. Avcı, Investigation of the viscoelastic properties of 4-Propoxy-Biphenyl-4-Carbonitrile. Liq. Cryst. **46**, 2190 (2019)
103. M. Kleman, O.D. Lavrentovich, *Soft Matter Physics: An Introduction* (Springer, New York, NY, 2004)
104. N. Avci, V. Borshch, D.D. Sarkar, R. Deb, G. Venkatesh, T. Turiv, S.V. Shiyanovskii, N.V.S. Rao, O.D. Lavrentovich, Viscoelasticity, dielectric anisotropy, and birefringence in the nematic phase of three four-ring Bent-Core liquid crystals with an L-Shaped molecular frame. Soft Matter **9**, 1066 (2013)
105. B. Zappone, A.E. Mamuk, I. Gryn, V. Arima, A. Zizzari, R. Bartolino, E. Lacaze, R. Petschek, Analogy between periodic patterns in thin smectic liquid crystal films and the intermediate state of superconductors. Proc. Natl. Acad. Sci. **117**, 17643 (2020)
106. Ç. Koçak, A.E. Mamuk, E.Ç. Demirci Dönmez, M. Poyraz, Electrospun fibers of liquid crystal mixtures. J. Mater. Sci. Mater. Electron. **33**, 15209 (2022)
107. C.G. Reyes, A. Sharma, J.P.F. Lagerwall, Non-electronic gas sensors from electrospun mats of liquid crystal core fibres for detecting volatile organic compounds at room temperature. Liq. Cryst. **43**, 1986 (2016)
108. K. Schelski, C.G. Reyes, L. Pschyklenk, P.-M. Kaul, J. Lagerwall, Quantitative Volatile Organic Compound (VOC) sensing with liquid crystal core fibers. SSRN Electron. J. 100661 (2021)
109. J. Hu, D. Zhou, Y. Su, S. Liu, P. Miao, Y. Shi, W. Sun, Y. Liu, Fiber micro-tip temperature sensor based on cholesteric liquid crystal. Opt. Lett. **45**, 5209 (2020)
110. D.H. Yeo, S.Y. Park, Liquid-crystal-based biosensor for detecting Ca 2+ in human Saliva. J. Ind. Eng. Chem. **74**, 193 (2019)
111. N. Ul Amin, H.M. Siddiqi, Y.K. Lin, Z. Hussain, N. Majeed, Bovine serum albumin protein-based liquid crystal biosensors for optical detection of toxic heavy metals in water. Sensors (Switzerland) **20** (2020)
112. P. Popov, L.W. Honaker, E.E. Kooijman, E.K. Mann, A.I. Jákli, A Liquid crystal biosensor for specific detection of antigens. Sens. Bio-Sens. Res. **8**, 31 (2016)
113. J. Wang, A. Jákli, J.L. West, Liquid crystal/polymer fiber mats as sensitive chemical sensors. J. Mol. Liq. **267**, 490 (2018)
114. J. Deng, W. Liang, S. Rhodes, J. Fang, Influence of polymer networks on the sensor properties of hydrogel dispersed liquid crystal droplets. Colloids Surf. A Physicochem. Eng. Asp. **570**, 438 (2019)
115. J. Deng, W. Liang, J. Fang, Liquid crystal droplet-embedded biopolymer hydrogel sheets for biosensor applications. ACS Appl. Mater. Interfaces **8**, 3928 (2016)
116. D. Liu, Q.Z. Hu, C.H. Jang, Orientational behaviors of liquid crystals coupled to chitosan-disrupted phospholipid membranes at the aqueous-liquid crystal interface. Colloids Surf. B Biointerf. **108**, 142 (2013)

117. D.K. Nguyen, C.H. Jang, An acetylcholinesterase-based biosensor for the detection of pesticides using liquid crystals confined in microcapillaries. Colloids Surf. B Biointerf. **200**, 111587 (2021)
118. Y. Wei, C.H. Jang, Liquid crystal as sensing platforms for determining the effect of graphene oxide-based materials on phospholipid membranes and monitoring antibacterial activity. Sens. Actuators B Chem. **254**, 72 (2018)
119. Y. Su, Z. Lan, J. Wang, L. Zeng, D. Zhou, Z. Peng, W. Sun, Y. Liu, Optical fiber sensor for determination of methanol ratio in methanol-doped ethanol based on two cholesteric liquid crystal droplets embedded in chitosan. J. Light. Technol. **39**, 5170 (2021)
120. Y. Zheng, L. Zhang, B. Duan, Anisotropic chitosan/tunicate cellulose nanocrystals hydrogel with tunable interference color and acid-responsiveness. Carbohydr. Polym. **295**, 119866 (2022)
121. L. Marin, M.C. Popescu, A. Zabulica, H. Uji-I, E. Fron, Chitosan as matrix for bio-polymer dispersed liquid crystal systems. Carbohydr. Polym. **95**, 16 (2013)
122. D. Ailincai, L. Marin, Eco-friendly PDLC composites based on chitosan and cholesteryl acetate. J. Mol. Liq. **321**, 114466 (2021)
123. D. Ailincai, D. Pamfil, L. Marin, Multiple bio-responsive polymer dispersed liquid crystal composites for sensing applications. J. Mol. Liq. **272**, 572 (2018)
124. P.C. Wu, A. Karn, M.J. Lee, W. Lee, C.Y. Chen, Dye-liquid-crystal-based biosensing for quantitative protein assay. Dye. Pigment. **150**, 73 (2018)
125. R.J. Carlton, J.T. Hunter, D.S. Miller, R. Abbasi, P.C. Mushenheim, L.N. Tan, N.L. Abbott, Chemical and biological sensing using liquid crystals. Liq. Cryst. Rev. **1**, 29 (2013)
126. Q.Z. Hu, C.H. Jang, Using liquid crystals to report molecular interactions between cationic antimicrobial peptides and lipid membranes. Analyst **137**, 567 (2012)
127. X. Niu, D. Luo, R. Chen, F. Wang, X. Sun, H. Dai, Optical biosensor based on liquid crystal droplets for detection of cholic acid. Opt. Commun. **381**, 286 (2016)
128. X. Niu, Y. Zhong, R. Chen, F. Wang, D. Luo, Highly sensitive and selective liquid crystal optical sensor for detection of ammonia. Opt. Express **25**, 13549 (2017)
129. T. Bera, J. Deng, J. Fang, Tailoring the surface of liquid crystal droplets with chitosan/surfactant complexes for the selective detection of bile acids in biological fluids. RSC Adv. **5**, 70094 (2015)
130. N.A. Peppas, *Hydrogels in Biology and Medicine: From Molecular Principles to Bionanotechnology*, Advanced materials.
131. N. Bhattarai, J. Gunn, M. Zhang, Chitosan-based hydrogels for controlled, localized drug delivery. Adv. Drug Deliv. Rev. **62**, 83 (2010)

Chapter 12
Chitosan Based Hybrid Polymeric Systems in Diabetes Monitoring

Lovepreet Kaur, Neelam, Younis Ahmad Hajam, Rajesh Kumar, Mohd Salim Reshi, and Seema Rai

Abstract Diabetes is an endocrinometabolic disorder caused due to the irregular glucose metabolism. Insulin is secreted by pancreatic islets of β-cells. Hence, the pancreas is known as a "glucose regulator" that facilitates in regulating the synthesis and secretion of insulin which in turn facilitates in keeping blood glucose in balance. The complications related to diabetes range from hyperglycemia, metabolic dysfunction, reproductive abnormalities, neuronal and renal damage. Various drugs and remedies are available for the treatment of diabetes coming from allopathy, homeopathy and ayurveda and are available since the past few decades. Moreover, in current era, nanomedicines prepared from various different plant parts and other polymeric compounds are are under consideration for the treatment of polygenic disease and associated complications. However, they are also having some demerits. Therefore, chitosan based nanopolymeric might be a better option for the treatment of diabetes and associated complication. In this chapter chitosan based nanoformualtions has been discussed along with their role in treating of diabetes.

L. Kaur · Neelam · Y. A. Hajam (✉)
Department of Life Sciences and Allied Health Sciences, Sant Baba Bhag Singh University, Jalandhar, Punjab 144030, India
e-mail: younismajeed64@gmail.com

R. Kumar
Department of Biosciences, Himachal Pradesh University, Summerhill, Shimla 171005, India

M. S. Reshi
Department of Zoology, School of Biosciences and Biotechnology, Baba Ghulam Shah Badshah University, Rajouri, Jammu and Kashmir 185234, India

S. Rai
Department of Zoology, Guru Ghasidas Vishwavidayalaya (A Central University), Bilaspur, Chhattisgarh 495009, India

S. K. Swain and A. Biswal (eds.), *Chitosan Nanocomposites*,
Biological and Medical Physics, Biomedical Engineering,
https://doi.org/10.1007/978-981-19-9646-7_12

12.1 Diabetes

Diabetes is the most common and difficult endocrinal-metabolic disease that ends in excessive blood glucose level (hyperglycemic condition) due to the inadequate secretion or motion of insulin. Insulin is secreted by β-cells in the pancreas. Hence, pancreas is known as a "glucose regulator" which facilitates synthesis and secretion of insulin which in turn keeps the blood glucose in regulated stage. Diabetes mellitus, being a complicated metabolic ailment progresses by puzzling interactions among biology and environmental factors. High blood glucose levels result in multi-organ failure coupled with urinary tract infections, adult-onset blindness, coronary heart ailments and stroke, immoderate blood stress and nerve damage and lower limb amputations [1]. In today's era immense alteration in physical activity patterns, dietary patterns, sedentary habits, and energy imbalances due to increasing consumption of high-caloric food stuffs results in increasing body mass index (BMI). Simultaneously, fast urbanization is greatly prevailing that leads to an increase in metabolic disorders, further in urban areas highest dip in metabolic functions had been pronounced due to changes in dietary patterns as well as bodily inactivity [2].

All these variations directly influence the metabolic processes such as carbohydrates, proteins and lipid metabolism. Besides, all these metabolic disorders and other biochemical processes, further symptoms include obesity, cardiovascular diseases (CVD's), neurodegeneration, renal failure, hormonal imbalances as well as reproductive complications [3]. The prevalence of obesity further leads to kidney and liver disorders, cardiovascular disorders and reproductive disorders [4]. Moreover, unhealthy lifestyles and physical inactivity provides great hold over the obesogenic environment. Globally, over-weight is the leading cause of severe health problems. In past decades (2002–2012) obesity has been increased adversely from 22.8% to 30.1%. Recent studies evidences that more than 1.9 billion adults are suffering from over-weight moreover; 650 million are completely obese widely. Unfortunately, around 2.8 million deaths occured because of the consequences of being over-weight or obese [5].

Diabetes is likewise named as a "silent killer" because of its moderate agnostic responses due to its negligible glycemic load. Hence, this circumstance is called pre-diabetes, characterized via means of obese patients. Further, if out of control and untreated, pre-diabetic circumstance, can also additionally cause everlasting diabetes. In 2019, a study dictated that 463 adults have been affected by diabetes out of 1000 people and is expected that during 2045 it might attain to around seven hundred million [6]. Raman et al. [7] proposed the type of diabetes mellitus into 4 types: particularly Type 1 diabetes mellitus, Type 2 diabetes mellitus, Type-three diabetes and gestational diabetes mellitus (GDM) (American Diabetes Association 2014) [7]. Figure 12.1 represents the types of diabetes.

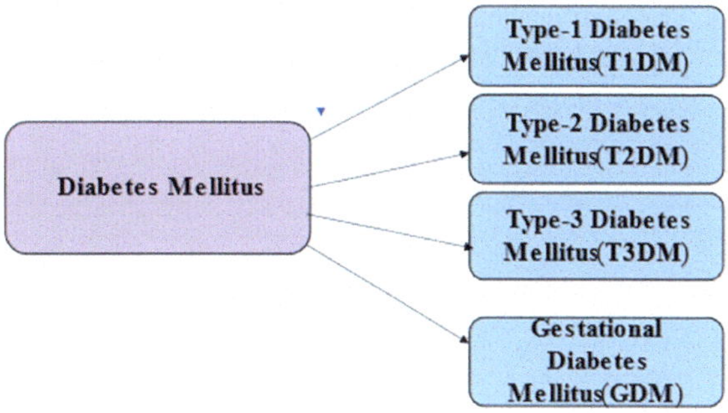

Fig. 12.1 Illustrative scheme showing types of diabetes

12.1.1 Type-One Diabetes Mellitus

Type 1 diabetes mellitus condition often arises, while pancreatic islets of β-cells exhibit insufficient secretion of insulin for metabolizing glucose to control high glucose level (hyperglycemia). T1DM nearly affects children, hence named as "juvenile diabetes" or "insulin-dependent diabetes". Besides, it only affects 5–10% of the population [8]. Variety of antibodies produced during T1DM are referred to as islet cell antibodies (ICA). The three primary antibodies produced in opposition to pancreas in the course of T1DM are Anti-insulin (IAA), Glutamic acid enzyme (GADA) and Insulinoma-affiliation macromolecule antibodies (IA2A). Moreover, ICA is the primary auto-antibody usual in T1DM patients. Macromolecule antibodies are produced in opposition to insulin and pro-insulin and had been detected earlier than supplementing exogenous insulin in the course of T1DM. GADA and IA2A were detected around 75–80% in populace produced via way of means of β-cells in the course of T1DM. GADA plays a role in synthesizing amino-butyrate in pancreatic cells. The human leucocyte complex (HLA) exactly the DR… and DQ genes, plays an essential role in pathogenesis of T1DM [9]. Genes belonging to HLA family are HLA-DQA1, HLA-DQB1 and HLA-DRB1. The HLA system plays an efficient role in immune system by differentiating native proteins and proteins produced by foreign viruses and bacteria. T-lymphocyte (proteins), CeCmotif, MHC (Class II DQα1), MHC (Class II DQβ1), IL-2a, IL-6, insulin, 2′-5′- oligoadenylate synthetase 1, PTP (Protein tyrosine phosphate) and the small ubiquitin-like modifier 4 are a variety of genetic loci relevant with T1DM. Modulation of insulin synthesis is caused due to variation in genetic loci present in altered chromosomes. Etiology (The cause or origin of disease) particularly of T1DM includes various environmental conditions, such as stress and viral infections [10]. Figure 12.2 illustrates the cause of type 1 diabetes.

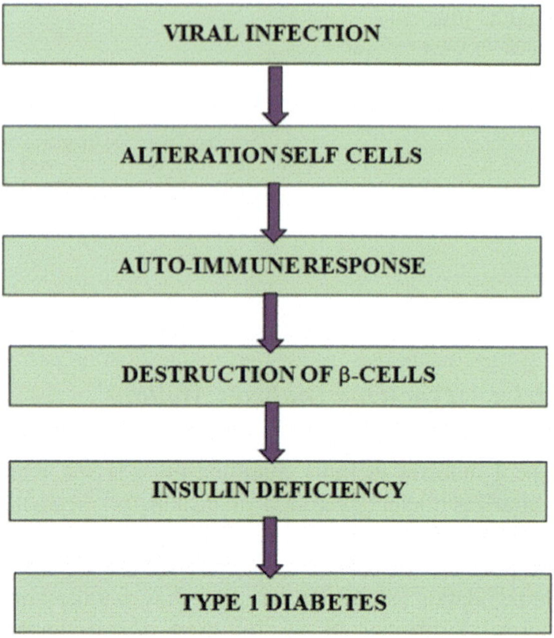

Fig. 12.2 Showing the cause of type 1 diabetes

12.1.2 Type-Two Diabetes Mellitus

Type 2 diabetes mellitus circumstance is not unusual in vintage age and therefore named as "Adult-onset diabetes" and about 95% of populace instances are evidenced globally [11]. T2DM is characterized via the means of reducing the manufacturing of insulin or failure of pancreatic β-cells. Decreased shipping of glucose into the liver, adipocytes and muscle tissues are the consequences of T2DM [12]. Besides, the prognosis of T2DM stays doubtful for plenty years, which lead persistent effects due to persisting hyperglycemic circumstance. Moreover, T2DM may be a genetic defect that is the end result of interactions among numerous genes and environmental aspects. Generally, T2DM is accompanied with ageing and way of lifestyle along with bodily inactivity, sedentary habits, and extended consumption of alcohol in addition to cigarette smoking that extensively ends in T2DM severity. Surprisingly, T1DM and T2DM possesses, an identical genetic historical past while; the genetic foundation of T1DM is extra big in assessment to T2DM [13]. Triggle et al. [14] have evidenced that threatened hazard elements aren't handiest constricted to a place or an organ in fact, it develops due to more than one gene interactions localized on complete genome [14]. TCF7L2 (Transcription aspect like 7-like 2), PPARG (Peroxisome Proliferator Activated Receptor Gamma), CDKAL1 (Cdk5 regulatory related protein 1-like 1), JAZF1 (Juxtaposed with every other zinc finger gene 1), HHEX (Hematopoietically Expressed Homebox), SL30 A8 (Solute provider own

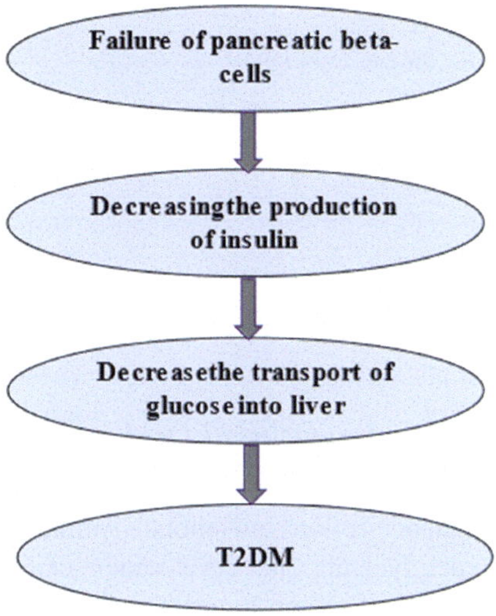

Fig. 12.3 Illustration listing the causes of type -2 diabetes

circle of relatives 30 member 8) and IGF2BP2 (Insulin-like boom aspect 2 mRNA-binding protein 2) are the numerous genes. Figure 12.3 predicts the causes of type 2 diabetes.

12.1.3 Type-Three Diabetes

Type-3 diabetes (T3DM) may be a neuro-endocrinal disorder that immediately affects the development of T2DM to Alzheimer's disorder (AD) [15]. Moreover, T1DM leads to disorder of diverse signaling pathways which includes inflammatory responses, insulin-increase issue signaling, ApoE4A allele, acetylcholine esterase interest in addition to vascular alteration of mind capillaries. Recent research has verified a hyperlink among systematic alterations, which includes dementia, obesity, diabetes, mind-insulin resistance and metabolic syndrome [16]. Generally, Alzheimer's disorder is marked through down-regulation of hormone insulin and neuronal insulin receptors compared to age-matched controls. These deformity changes, results in sluggish irregularity in entire insulin signaling cascade, which in turn leads to improvement in insulin resistance [17]. A link among deassociation of proteins, mind-insulin resistance in addition to cognitive impairment become evidenced through Sultana and their co-workers [18]. Figure 12.4 represents the causes of type-3 diabetes.

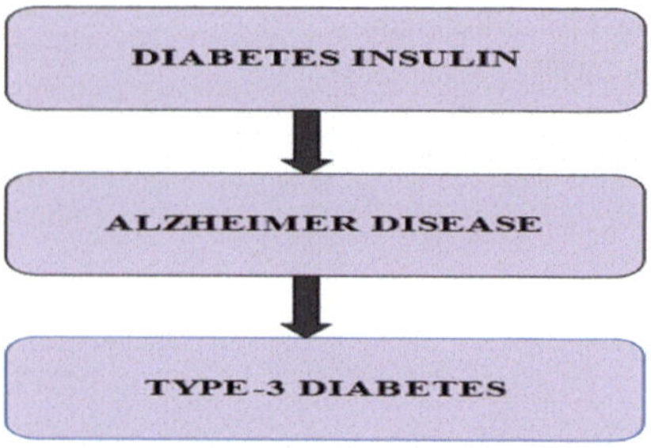

Fig. 12.4 Illustration listing the causes of type-3 diabetes

12.1.4 Gestational Diabetes Mellitus

Pregnancy-related metabolic dysfunction, markedly elevated blood sugar levels (hyperglycemia), the development of diabetic-like symptoms in the mother, and the negative effects on the growing foetus are what give rise to the term "gestational diabetes mellitus." Pregnant women who have high blood sugar have a higher incidence of harmful maternal, foetal, and neonatal diseases. Pregnant women who are carbohydrate intolerant tend to develop gestational diabetes [19]. Women diagnosed with gestational diabetes mellitus typically possess a higher chance of developing frank diabetes (pre-diabetes). Furthermore, gestational diabetes mellitus (GDM)

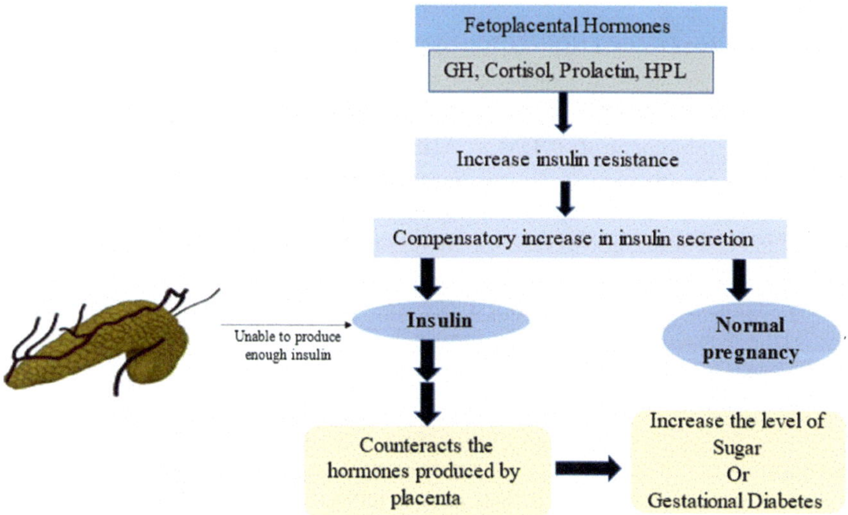

Fig. 12.5 Representation of the cause of gestational diabetes mellitus

significantly increases the chance of neonatal mortality as it reasons maternal hyperglycemia, which triggers the forced release of more and more insulin, which in turn stimulates foetal growth abnormally and increases birth weight. After delivery, this hyperglycemia in women may be reversed, according to [15]. Figure 12.5 suggests the causes of Gestational diabetes mellitus.

12.2 Treatments for Diabetes

12.2.1 Allopathic Drugs

Various drugs are used in the treatment of diabetes mellitus. However, exclusively prescribed anti-diabetic drugs are Glibenclamide and Lipoic acid and due to its efficiency as well as compliancy, the development of new formulation including both the substances which came into existence, therefore acting as a multi-goal drug system. Primary motive of this chapter is to focus on the formulation including physicochemical characterization of recent polymeric progresses primarily based on chitosan (CS) in which matrix are encapsulated with glibenclamide (Gly) and lipoic acid (LA). Micro-particles (CS-Gly, CS-LA, and CS-Gly-LA) victimisation ionic gelatin process together with pentasodium tripolyphosphate (TPP), a cross-linking agent can be preferred relevantly for the preparation of polymeric systems. Principally, other allopathic drugs are "sulfonylureas", "biguanides", "glinides", "thiazolidinediones", "alpha-gluconidase inhibitors", "SGLT-2 (Sodium-glucose Cotransporter-2) inhibitors" and "DPP-4 inhibitors".

The main aim of anti-diabetic remedy is to decrease the elevated glucose level that in most of the cases, fails to attain. Likewise, mainly anti-diabetic drugs results in hypoglycemia, liver injury as well as kidney failure, diabesity, hypersensitivity reactions and gastrointestinal complications.

Moreover, glibenclamide being a second-technology, sulfonylurea is more regularly recommended in case of type 2 DM (T2DM); while remedy with biguanides or first-technology fails to control hyperglycemia. It mainly acts by control endocrine secretion in duct gland islets of β-cells and subsequently leads to decrease in blood glucose level. Although it's intense as well as long-term duration, yet Gly, possesses poor water-solubility that results, in deduce and variable oral bioavailability (Table 12.1).

12.2.2 Plant Extracts

Plant-derived polyphenols also plays a significant role in curing diabetes. It comprises of a polyphenolic compound that helps in regulating blood glucose level. Polyphenols acts as notable antioxidant molecules. Moreover, to some extent, they are successful

Table 12.1 Showing allopathic drugs and their side effects

S.No	Allopathic drugs	Side effects
1	Biguanides	GI side effects (diarrhea) Kidney damage, Anxiety, Nerve damage, Vitamin B12 deficiency, Lactic acidosis (rare)
2	Sulfonylureas	↑Cardiovascular risk (CV) risk, Hypoglycemia (Due to excessive dosage), Obesity, Hepatic failure
3	Thiazolidisnediones	↑Weight, Bone fractures, ↑LDL-C (rosiglitazone), Anemia, Swelling (edema) from fluid retention, Mascular edema (in eyes), Heart attack (Myocardial infarction)
4	GLP-1 analogs	GI side effects (diarrhea), ↑Heart rate, Acute pancreatitis (Inflammation), C-cell hyperplasia/medullary thyroid tumors in animals, Injectable, Upper respiratory infection, Weight loss
5	DPP-4 inhibitors	Angioedema/utricaria and other immune-mediated dermatological effects, Acute pancreatitis, ↑Heart failure, Urinary tract infection
6	α-gluconidase inhibitors	Gastrointestinal issues (Diarrhea), Flatulence
7	SGLT-2 inhibitors	Urinary tract infection, Urination
8	Meglitinide	↑Weight, Hypoglycemia
9	Bromocriptine	Constipation, Heartburn, Loss of apetite, Stomach cramps

in treating diabetes, but in most of the cases, it fails to do so. Besides, they also lead to some side effects (Table 12.2).

Table 12.2 Enlisted below depicts some plants and their parts, which prefers to treat diabetes

S.No	Plant name	Plant part	References
1	*Urticadioica* (Stinging nettle)	Leaves	[20]
2	*Carthamustinctorius* (Safflower)	Flower	[21]
3	*Bauhiniaforficate* (Brazilian orchid tree)	Leaves	[22]
4	*Salvia nemorosa* (Sage)	Aerial part	[23]
5	*Ginseng sp.* (Asian ginseng)	Roots, stalk, leaves, and berries	[24]
6	*Cinnamomum verum* (Cinnamon)	Whole plant	[25]
7	*Dendrobium chrysotoxum (Golden-bow Dendrobium)*	Aerial parts	[26]
8	*Zingiber zerumbet* (Bitter ginger)	Roots	[27]
9	*Kaempferia parviflora*	Roots	[28]
10	*Opuntia monacantha* (Culinary)	Leaves	[29]
11	*Aloe vera*	Leaf	[30]
12	*Trigonella foenumgraecum*	Seed	[31]
13	*Bauhinia forficate*	Leaf	[32]
14	*Gymnemasylvestre*	Leaf	[33]
15	*Swertia punicea*	Whole plant	[34]
16	*Combretum micranthum*	Leaves	[35]
17	*Sarcopoterium spinosum*	Roots	[36]
18	*Liriope spicata*	Leaves	[37]
19	*Caesalpinia bonducella*	Seeds	[38]
20	*Terminalia chebula*	Seeds	[39]
21	*Achillea santolia*	Aerial parts	[40]
22	*Cynomorium coccineum*	Flower, root	[41]
23	*Avicennia marina*	Branches	[42]
24	*Moringa oleifera*	Fruit	[43]
25	*Ocimumforskolei*	Leaf	[44]
26	*Azadirachta indica*	Leaf, bark	[45]

12.2.3 Nanoparticles

Nanoparticles can be defined as a nanodimensional solid entity that is used to deliver RNA and proteins in order to analyze the diseases as well as to detect the extent or progression of disease. Polysaccharides, lipids, metals and polymers are the building blocks of nanoparticles. Nanoparticles are smaller than 100 nm in size. Nanoparticles used for treatment of diabetes are of two types: nanocapsules and nanospheres. Generally, nanocapsules encapsulate drugs in outermost empty shell while in nanospheres, drugs are dispersed in the matrix of a particle. Actually, nanocapsules and nanospheres together constitute, the term nanocarriers hence, known as nanoparticles. Above all, these nanoparticles are specialized in conjugating to compounds including cell-penetrating peptides (CPPs) that results in facilitating the transduction of proteins likewise, insulin into cells by penetrating the plasma membrane.

On the basis of insulin delivery in diabetes treatment, nanoparticles can be categorized into following:

Polymeric biodegradable nanoparticles

Polymeric micelles

Ceramic nanoparticles

Liposomes

Dendrimer

Qiao et al. [46] reported that nanotechnology may be characterized because the monitoring, construction, repairing as well as control of human organic structures at cell degree via means of production substances and systems at molecular degree [46]. Hence, nanomedicines are available for the treatment of diabetes by the integration of medicines and nanotechnology. Nanoparticles, are to be exploited efficiently for nanomedicines [47]. Nanomedicines can be an applicant for the detection of molecules including DNA or proteins, imaging enhancers and focused on precise tissues in order to supply healing agents [48].

There are significant disadvantages for nanotechnology in the field of its users. The surface area of nanoparticles is greater than their volume. Poor drug delivery is caused by friction and the inevitable clumping of nanoparticles into bigger structures. As a result, because of their tiny size, these drug carriers are eliminated by the body's excretory systems. Initially, it was shown that nanoparticles cause medication carriers to build up in the skin and eyes. Therefore, nanotechnology does not successfully aid in the treatment of diabetes, TB, or cardiovascular disorders. For the treatment of diabetes, none of these therapeutic choices are adequate.

12.3 Chitosan

Chitin is the second abundant sustainable biopolymer that exists in nature, cellulose being first [49]. This polysaccharide with marvellous properties is extracted from the residual matter formed chemically or enzymatically treatment of shells of crabs and shrimps [50]. Over the past few years, chitosan has been recognized for its excellent physical and biological properties like low viscosity, biocompatibility, maximum water-solubility and perishable in nature as compared to chitin, which enables

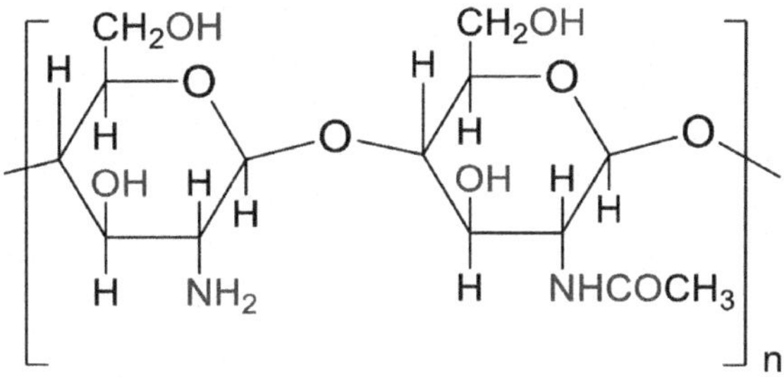

Fig. 12.6 The 3-dimensional shape of chitosan

wide scale applications [51]. Figure 12.6 represents the 3-dimensional structure of chitosan.

Over the last 3 years, various studies have been conducted on chitosan due to its various applications that specially emphasized on pharmacology, chemistry and polymeric sciences including life sciences [52].

Additionally, chitosan is produced chemically (through acid hydrolysis and oxidative degradation) or biologically (via microwave degradation and UV radiation, including an enzymatic degradation process) [53]. Although industrial scale chitosan manufacturing relies heavily on chemical decomposition methods such as acid hydrolysis due to their efficient operating procedures and affordable output.

For the degradation of chitin, a few of the acidic reagents most frequently utilized are hydrochloric acid, nitric acid, and hydrofluoric acid. The majority of the time, hydrochloric acid is used in catalytic settings at 80 °C for 1–2 h [54]. However, one of the main downsides of chemical degradation is the convolution of the resultant chemicals that must be removed or purified. Additionally, these hazardous chemical complexes frequently contaminate the environment by discharging a sizable quantity of hazardous waste in response to bodily deterioration [55].

During the past few years, enzymatic degradation has been known profoundly because of its excessive stage of analyzation of chemical reaction sites. Chitinases is one among the more often used enzymes that are generally preferred by latest studies [56]. The enzymatic degradation being non-poisonous and controllable, hence not needing the elimination of salt. Nevertheless, enzymatic methods possess a mild impact over chitosan products, likewise chemical shape and organic activities [57]. However, this method provides certain drawbacks which includes excessive manufacturing costs, terrible practicality and tough response circumstances that leads to difficulties in industrial manufacturing. Recently, merging of above strategies is excessively utilized in manufacturing of commercial chitosan. After purification process, in order to analyze the degree of deacetylation and polymerization, chitosan is in addition distinguished by nuclear resonance or mass spectrometry [58]. Figure 12.7 illustrates the biomedical applications of chitosan.

Chitosan provides an amazing chemical as well as pharmacokinetic attributes i.e., (bioavailability, biocompatibility and low toxicity) [59]. Chitosan, because of the presence of glucosamine possesses a positive net surface charge at an acidic

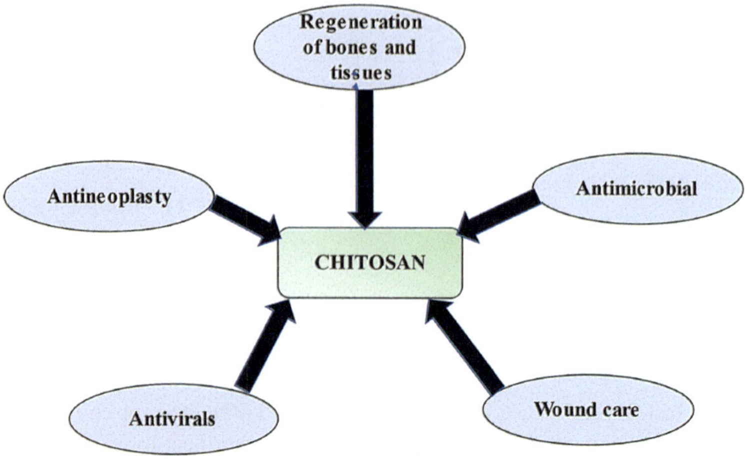

Fig. 12.7 Biomedical applications of Chitosan

pH scale that facilitates chitosan in linking with anion macromolecule in the secretion layer. Mucoadhesion attributes of chitosan had been achieved electrostatically, hydrophobic interaction as well as hydrogen bonding [60]. Therefore, the hydroxyl and amino groups presence within the spine of chitosan provides a wide range of chemical modifications which increases solubility, progresses mobiliary absorption and facilitates future launch of medication [61]. Hence, this attribute favors chitosan absorption of intestinal and nasal cells and leads to an amazing passage for mobilization of medication [62].

Though the majority of biomaterials for medication administration don't seem to be allowed by the Food and Drug Administration (FDA), chitosan is surprisingly permitted for tissue engineering [63]. Because of this, chitosan-based drug transporters are designated as another type of biomaterial due to their distinctive property. Numerous formulations, such as capsules, tablets, and micro/nanocomposite, can be made from chitosan to generate a wide range of medical and pharmaceutical devices [64]. Amazingly, different types of chitosan have a considerable impact on improving diabetes related fat and glucose metabolism. More recently, chitosan offers certain cutting-edge carriers that are useful in delivering anti-diabetic medications to specific areas [65].

Chitosan is considered as a promising and excellent carbohydrate polymer due to its significant chemical as well as pharmacokinetic properties in biomedical applications. Chitosan plays an essential role in diminishing hyperglycemia, and depending on diabetes mellitus that is associated with imbalanced lipid metabolism and proliferating pancreatic β-cells have been reported recently. In order to decrease glucose level, chitosan has been used in various nanocarriers for delivering variety of anti-diabetic drugs. Additionally, chitosan-based nanocarriers are utilized as an anti-diabetic drug in diminishing hyperglycemia. Hence, chitosan is recognized as the first anti-diabetic drug preferred for oral administration against diabetes mellitus.

Therefore, chitosan also plays an imperative role such as in delivering and operating stem cells and further distinguishing them into pancreatic like β-cells for curing type 1 diabetes mellitus.

Diabetes is acknowledged as the most prevalent and difficult health issue in the twenty-first century. High glucose levels (hyperglycemia), a collection of endocrine-metabolic disorders brought on by inadequate insulin production, are a defining feature of diabetes. Hepato-renal damage, adult-onset blindness, lower limb amputations, heart disorders, stroke, nerve damage, and high blood pressure were some of the various organ failures it caused. Additionally, diabetics are more likely to develop cardiovascular issues such as atherosclerosis, hypertension, aberrant lipoproteins, and cerebrovascular illness.

12.3.1 Chitosan Nanoparticles

Numerous insulin-loaded nanoparticles, including those made of lipids, metals, natural polymers, proteins, and synthetic polymers, have been shown to be effective for oral administration [66]. Due to their small size and higher surface area to volume ratio, nanoparticles are thought to be effective because their retention time before reaching intestinal absorption is too long, enhancing penetration and bioavailability. Consequently, decreasing encapsulant frequency and doses and increasing patient compliance [67].

Different natural nanoparticles have been used over the years to alleviate the problems associated with oral insulin absorption while utilizing cutting-edge nanotechnology. Due to its incredible characteristics, chitosan and its derivatives have been widely used in formulations for oral insulin delivery. As a result, by acting as a mucoadhesive and having the ability to open tight connections of epithelial cells, chitosan serves as an effective insulin transporter. These characteristics, which provide positive feedback, are present in the chitosan surface amine groups [68].

Typically, nanoparticles made from chitosan or its derivatives have a positively charged surface. Surprisingly, chitosan's distinctive functional groups have shown that it can be exploited to satisfy specific aims by creating a polymer with a wide range of prospective uses [69]. Therefore, a variety of techniques, including ionic gelation, emulsion solvent diffusion, polyelectrolyte complexion, emulsion-droplet coalescence, reverse micelle creation, complex coacervation, or solvent evaporation procedures, can be used to create chitosan nanoparticles.

12.3.1.1 Role of Chitosan Nanoparticles

Diabetes mellitus, a more complex form of metabolic syndrome that causes elevated blood glucose levels, is comparatively more common worldwide. It has been established throughout the last few decades that diabetes mellitus has a major negative impact on health. As of now, diabetes mellitus requires sufficient care and supervision while using multifactorial hazard reduction strategies [70]. Notably, the main goal is to maintain normal plasma glucose levels because failure to control persistent hyperglycemia can cause serious health issues such as cardiovascular disease, neuropathic conditions, nephropathy, and retinopathy. In order to maintain blood glucose levels in people with type 1 diabetes and recently in people with diabetes 2, subcutaneous insulin infusion is necessary [71].

12.3.1.2 Role of Chitosan in Enhancing the Pathophysiology of Diabetes

Chitosan and its derivatives work wonders at reducing high glucose levels in diabetic mice [72]. The liver and muscles are two primary glucose routes that are both up and down regulated by chitosan's antihyperglycemic actions. For studies to lessen the effects of diabetes mellitus, chitosan of various sizes and molecular weights (MW) are needed. Chito-oligosaccharide biguanide (COSG), a type of chitosan that is commercially accessible and structurally similar to metformin, inhibits pancreatic beta-mobile cell death and increases insulin production in diabetic rats [73]. Additionally, consumption of chitosan accelerates the production of hepatic glycogen in diabetic rats and increases the expression of the glucose transporter gene 4 (GLUT4) in adipose tissues [74].

Chitosan increases glucose uptake through skeletal muscle, which may be a result of increased GLUT4 and Akt (Ak strain transforming) phosphorylation in rats with diabetes that was induced by streptozotocin (STZ). Additionally, chitosan aids in lowering phosphoenolpyruvate carboxykinase expression as well as the degree of phosphorylation of the p38 protein in the liver. Amazing enzyme phosphoenolpyruvate carboxykinase (PEPCK), advances the irreversible gluconeogenesis process and plays a critical role in maintaining glucose homeostasis [75].

Adenosine monophosphate protein kinase (AMPK) is a crucial protein that regulates the energy levels of cells; it was shown that diabetic rats treated with chitosan had higher levels of AMPK phosphorylation. Subsequently, law of AMPK and PEPCK (Phosphoenolpyruvate carboxykinase) may be a superb method in retrieving diabetes with chitosan remedy [76]. Advancingly, while chitosan oligosaccharide anti-diabetic properties was utilized to examine rats with induced STZ-caused diabetes the results implied that plasma glucose tiers had been lowering more or less from 19% with 0.3% chitosan oligosaccharide supplementation. Likewise, some others have a look at demonstrated the LDL cholesterol-decreasing impact of excessive and lower molecular weight chitosan. Besides, the large molecular mass of chitosan laid a key impact in decreasing general plasma LDL cholesterol with STZ-caused diabetes in cases of rats is considered as the highest one on deducting hyperglycemia with various pathways. Therefore, the advanced biochemical serum parameters together with serum glucose in addition to triglyceride levels, have been identified to have reduced the molecular weight of chitosan. Furthermore, long-time period intake of low-molecular chitosan oligosaccharides owing to excessive calorie weight loss program in diabetes mouse version become relevant with significant discounts in speedy glucose, glycosylated hemoglobin, and inhibition in the small gut. Additionally, it is observed that chitosan ingestion in rats fed with fructose-wealthy weight loss program; Adiponectin, a protein secreted with the aid of using adipose tissue, elevated; inhibition of gluconeogenesis, resulting in declining glucose level.

12.3.1.3 Chitosan for Diabetes Mellitus as a Drug Shipping Provider

Every medication provides a special property to hit the goal at a foremost dosage. Moreover, due to an amazing development in therapeutic research on chitosan; it has become a multi-faceted anti-diabetic agent in addition to a green provider of medications. Chitosan possesses the ability to move and launch biomolecules including genes, hormones in addition to vitamins. Especially drug delivery and drug routes can be provided by chitosan-based transporters who prefer each of oral as well as subcutaneous injection. Till now, diversity of drug shipping research has been completed through the usage of chitosan nanocarriers for one-of -a-kind diseases, likewise cancer, cardiovascular disease, etc. Among them, in vitro and in vivo experiments, plays a vital role in anti-diabetic drug assessments.

12.3.1.4 Glucagon-Like Peptide-1

GLP-1 could be a well-known peptide that provides an anti-diabetic impact with the aid of using evoked internal secretion in exocrine gland β-cells. Hence, Chitosan based nano and micro formulations were utilized for the stimulus-sensitive and site-goal delivery of GLP-1. The researchers formulated a site-unique multi-functional oral drug transport version for the treatment of type 2 Diabetes mellitus. Porous nanoparticles obtained from semiconducting material (PSi) became implemented to the pH-sensitive nano matrix polymer.

Previous studies reported that co-cultures of Caco-2 and HT29-MTX, and floor functionalization with chitosan PSi nanoparticles resulted in significant permeability. A positive penetration of PSi nanoparticles covered with chitosan within the intestinal obstructor, those difunctionalized nanoparticles resulted in delivery of anti-diabetic drug impact of GLP-1(Glucagon-like peptide 1) and dipeptidyl peptidase-four (DPP-4) protein inhibitors. Moreover, it has been studied that the formulations are fabricated in such an excellent way that they could launch GLP-1 in addition to DPP-4 at both pH (1.2 and 6.8). Hence, coadministration of GLP-1 and DPP-4 inhibitors with chitosan-working PLGA (Poly D, L-lactic-co-glycolic acid: most effective biodegradable polymer) nanoparticles conjugated to cell-penetrating peptides might be used to combat the diabetic ailments. While, in case of non-over-weight rats, twin dosage induces hypoglycaemic impact consequently, and blood sugar tiers have been decreased.

12.3.1.5 Small Peptide Exendin-4 Delivery

It has been studied that in vivo and in vitro conditions, chitosan possesses the ability to combine and release the tiny peptides including exendin-four.

The exendin-4 anti-diabetic peptide and lower molecular weight chitosan (LMCS) may be combined into a novel form for oral administration that has disulphide linkages that may dissolve in a physiological setting. Therefore, by combining positively charged LMCS and negatively charged exendin-4, the peptide will be protected against GI hydrolytic enzymes. This integration can preserve the form of the exendin-4 to a 90% degree while having no impact on its biological characteristics. After being administered orally to diabetic mice, these nanocomposites caused insulin secretion and had a wonderful hypoglycaemic effect.

Additionally, chitosan and poly (-glutamic acid) nanoparticles (chitosan/-PGA NP) united with exendin-4 could also be created and utilized in combination with insulin to lessen the discomfort of type 2 diabetes. The authors hypothesised that because of the chitosan fragments' lifestyles, chitosan/-PGA NP improves the shipping of insulin and exendin-4 over the enteral barrier. As a result, oral administration of nanoparticles dramatically decreased plasma glucose levels in type 2 diabetic rats. Additionally, an examination of the Materia medica of insulin and exendin-4 signifies their stability.

12.3.1.6 Role of Chitosan and Its Derivatives by Suppressing Supermolecule-Hydrolysing Enzymes

Carbohydrate-hydrolysing enzymes disintegrates carbohydrates into monosaccharides thereby, decreasing the diversion of internal organ carbohydrate-hydrolysing enzymes which in turn restricts the absorption of glucose in gut succeeding in decreasing glucose levels. Consequently, inhibition of carbohydrate-hydrolysing enzymes can be useful in ameliorating diabetes. Carbohydrate organic process enzymes that include α-amylases and α-glucosidases found in mammals play an essential role in remodelling of monosaccharides such as glucose as well as levulose from digestible carbohydrates to be quickly absorbed via., bloodstream. Therefore, obstructing those enzymes sports as much as a positive degree will normalizes the elevation in blood glucose levels. Additionally, water soluble chitosan derivatives result in undertaking α-amylase and α-glucosidase.

Supplementation of long-time period COS (Chito-oligosaccharide) can drastically harmonize hyperglycaemia while, glycated hemoglobin A1c(HbA1c) degrees in db/db in case of mice suggests that of decreasing intestinal sucrase as well as glucoamylase diversion and managing the expression of mRNA of sucrase-isomaltase (SI) compound. Moreover, MW of COS with thousand Da establishes a great capacity in governing postprandial degrees of blood glucose in Sprague–Dawley (SD) rats as it becomes more frequently absorbed into move rather instead of COS with better MW.

12.3.1.7 Role of Chitosan and Its Derivatives in Diminishing Hypoglycaemic Agent Resistance Thereby Progressing Aldohexose Uptake

Hypoglycaemic agent resistance is one among the first indicators of diabetes, distinguishes with the aid of using a decreased amount of insulin in case of goal tissues along with adipose tissue, liver and musculus. In fact, GLUT4 is the primary glucose transporter in case of fat tissue and skeletal muscle while GLUT2 is the primary glucose transporter in case of duct gland islets. COS supplementation ameliorates insulin sensitivity index in addition to glucose tolerance in STZ-triggered diabetic mice accompanied by high-calorie diet and additionally in a look at utilizing db/db mice, similarly diminished insulin resistance.

An associated degree offered specific in vivo anti-diabetic impact of COS with 1.5×103 Da greatly leads to the expression of GLUT4. COS prevents apoptosis of duct gland β-cells in vivo and provides protection against rupturing of pancreas. However, COS plays a significant role in differentiation of INS-1 cells in vitro and increases GLUT2 with the aid of using complete release of insulin. Further, these consequences verifies that COS provides a great capacity in defending apoptosis of pancreatic β-cells triggering with the aid of using STZ.

12.3.2 Chitosan Polymers

In order to enhance the biodistribution of flavonoid naringenin (flavonoid) associated with diabetic in vivo model, Smruthi et al. [77] created a polymeric NP [77]. Chitosan and alginate, two natural carbohydrate polymers, was used to organize naringenin NPs. With 50 mg/kg of naringenin NPs and orally ingested naringenin for 19 days, male Wistar rats that had developed diabetes due to streptozotocin were shown to have an anti-diabetic effect. Compared to oral loose compound, oral naringenin NPs play a critical role in effectively lowering blood sugar, LDL cholesterol, and triglycerides. Additionally, it is believed that naringenin NPs work far more effectively than loose oral naringenin at regulating pancreatic abnormalities seen during diabetes management. Additionally, scientists showed that the creation of intermediate shell NPs with naringenin loaded chitosan and alginate promotes an outstanding. Additionally, researchers showed that producing middle shell chitosan-alginate nanoparticles with naringenin loaded on them causes great anti-diabetic responses in rats with diabetes mellitus brought on by streptozotocin.

Kaushal et al. [78] demonstrated polymeric NPs in order to assess the pharmaceutical efficiency in anti-diabetic research favring oral quercetin (flavonoid) delivery [78]. Quercetin NPs consumption was analyzed in vitro through utilizing human colonic epithelial cellular line HT29 and observed that cellular ingestion is predicted on the amount of quercetin NPs. Diabetic male Wistar rats, had been brought about with unfastened quercetin and quercetin NPs toanalyze the in vivo anti-diabetic effects after oral remedy of 100 mg/kg frame weight for 28 days. Quercetin NPs exhibits maximum reduction in blood glucose tiers, without displaying any hypoglycaemic effect. Additionally, free oral quercetin furtherance LDL cholesterol and triglycerides levels, kidney and coronary heart cells, including regeneration of pancreatic β-cells. Hence, authors evidenced significant potential of quercetin as an oral delivery anti-diabetic drug.

Taghipour et al. [79] fabricated amphiphilic chitosan derivatives loaded with scutellarin (flavonoid) distinguished as intestinal target NP carrier [79]. Scutellarin-loaded chitosan-derivatives (nutrition B12) NPs were induced in normal male Sprague–Dawley rats, with a dose of 40 mg/kg for 8 weeks through single oral administration. It was observed that overall plasma concentrations of scutellarin had been better in animals while dealing with scutellarin NPs as compared to unfastened scutellarin. Scutellarin NPs showed positive results in curing retinal damages in diabetic rats due to the excellent efficiency of scutellarin NPs. Admittedly, authors stated that scutellarin NPs proves to be more efficient in improving anti-diabetic consequences than unfastened scutellarin through oral administration.

Samadder et al. [80] formulated PLGS NPs laden with pelargonidin (flavonoid) with an average size of 12 nm approximately [80]. Pelargonidin NPs synthesis is a time dependent method so the rate of distribution varies from cell living substance to nucleus that can be attained within two hours. Generally, pelargonidin NPs prefers thirteen minutes only whereas, free pelargonidin demands one hour nearly in case of cellular entry. Consequently, it was concluded, that pelargonidin NPs possesses higher maintenance in cellular aldohexose uptake in comparison to free pelargonidin. Pelargonidin NPs exhibits harmonizing effects in adjusting glucose homeostasis by assessing GLUT4 expression interpretation. Generally, Pelargonidin NPs prefers thirteen minutes only whereas, free pelargonidin demands one hour nearly in case of cellular entry. Pelargonidin NPs reveals 10 times more defensive effects over free pelargonidin in triggering glucose stability. Moreover, in vitro studies evidenced that pelargonidin NPs suppresses the circumstances of DM.

Carbonare et al. [81] formulated polymeric NPs containing fisetin (flavonoid) for suitable oral administration [81]. Fisetin NPs can be prepared by comprising poly (ε-caprolactone) PCL and PLGA-PEG carboxylic acid (PLGA-PEG-COOH). Fisetin NPs possesses a diameter that ranges from 140 to 200 nm, consisting of fisetin of about 70–82%. In the scope of oral administration, it was evidenced that these polymeric NPs provide protection as well as preservation of fisetin in gastric restoring conditions. In non-cellular assay (in vitro) experiments fisetin NPs also showed α-glucosidase inhibition activity. Fisetin NPs as well as free fisetin provides excellent inhibition capability of α-glucosidase in comparison with acarbose (commercial inhibitor). At last, it is concluded that these in vitro experiments evidenced the preserved antioxidant and anti-diabetic properties through nanoencapsulation method.

References

1. A.A. Abejew, A.Z. Belay, M.W. Kerie, Diabetic complications among adult diabetic patients of a tertiary hospital in Northeast Ethiopia. Adv. Public Health (2015)
2. M. Bos, C. Agyemang, Prevalence and complications of diabetes mellitus in Northern Africa, a systematic review. BMC Public Health **13**, 1–7 (2013)
3. V. Ormazabal, S. Nair, O. Elfeky, C. Aguayo, C. Salomon, F.A. Zuniga, Association between insulin resistance and the development of cardiovascular disease. Cardiovasc. Diabetol. **17**, 1–14 (2018)
4. K.C. McPherson, C.A. Shields, B. Poudel, B. Fizer, A. Pennington, A. Szabo-Johnson, W.L. Thompson, D.C. Cornelius, J.M. Williams, Impact of obesity as an independent risk factor for the development of renal injury: implications from rat models of obesity. Am. J. Physiol. Renal. Physiol. **316**, 316–327 (2019)
5. T. Tan, C.W. Leung, Associations between perceived stress and BMI and waist circumference in Chinese adults: data from the 2015 China health and nutrition survey. Public Health Nutr. 1–10 (2020)
6. R.L. Thomas, S. Halim, S. Gurudas, S. Sivaprasad, D.R. Owens, IDF Diabetes Atlas: a review of studies utilising retinal photography on the global prevalence of diabetes related retinopathy between 2015 and 2018. Diabetes Res. Clin. Pract. **157**, 107840 (2019)
7. S. Raman, M.P. Rajakumar, J. Ramya, Medical information system for classification of diabetes mellitus using layered neural network. J. Contr. Eng. Appl. Inf. **23**(4), 95–104 (2021)
8. M. Badpa et al., Association of long-term environmental exposures in pregnancy and early life with islet autoimmunity development in children in Bavaria, Germany. Environ. Res. 113503 (2022)
9. B. Liu, S. Yuanyuan, R. Fu, Current research status of HLA in immune-related diseases. Immun. Inflamm. Dis. **9**(2), 340–350 (2021)
10. K. Zorena et al., Environmental factors and the risk of developing type 1 diabetes—Old disease and new data. Biology **11**(4), 608 (2022)
11. J. Burahmah, D. Zheng, R.D. Leslie, Adult-onset type 1 diabetes: a changing perspective. Eur. J.Internal Med. (2022)
12. T.M. Batista, N. Haider, C. Ronald Kahn, Defining the underlying defect in insulin action in type 2 diabetes. Diabetologia **64**(5), 994–1006 (2021)
13. T. Swapna et al., Burden and disease pathogenesis of influenza and other respiratory viruses in diabetic patients (2022)
14. C.R. Triggle et al., A comprehensive review of viral characteristics, transmission, pathophysiology, immune response, and management of SARS-CoV-2 and COVID-19 as a basis for controlling the pandemic. Front. Immunol. **12**, 631139 (2021)
15. P. Sharma et al., Complementary and alternative medicine for the treatment of diabetes and associated complications: a review on therapeutic role of polyphenols. Phytomed. Plus **2**(1), 100188 (2022)
16. E. Barone et al.,The interplay among oxidative stress, brain insulin resistance and AMPK dysfunction contribute to neurodegeneration in type 2 diabetes and Alzheimer disease. Free Radic. Biol. Med. **176**, 16–33 (2021)
17. V. Latina et al.,Tau cleavage contributes to cognitive dysfunction in strepto-zotocin-induced sporadic alzheimer's disease (sAD) mouse model. Int. J. Mol. Sci. **22**(22), 12158 (2021)
18. M. Sultana, S. JadoonandArif Malik,The development of dementia in diabetic and alzheimer patients
19. P. Ustianowski et al.,PPARG, TMEM163, UBE2E2, and WFS1 gene polymorphisms are not significant risk factors for gestational diabetes in the polish population. J. Personal. Med. **12**(2), 243 (2022)
20. A. Gohari, A. Noorafshan, M. Akmali, F. Zamani-Garmsiri, A. Seghatoleslam, Urtica dioica distillate regenerates pancreatic beta cells in streptozotocin-induced diabetic rats. Iran. J. Med. Sci. **43**(2), 174 (2018)

21. S.S. Gomashe, K.P. Ingle, Y.A. Sarap, D. Chand, S. Rajkumar, Safflower (Carthamus tinctorius L.): an underutilized crop with potential medicinal values. Ann. Phytomed. **10**(1), 242–248 (2021)
22. C.A. Tonelli, S.Q. de Oliveira, A.A. da Silva Vieira, M.W. Biavatti, C. Ritter, F.H. Reginatto, A.M. de Campos, F. Dal-Pizzol, Clinical efficacy of capsules containing standardized extract of Bauhinia forficata Link (pata-de-vaca) as adjuvant treatment in type 2 diabetes patients: a ran (2022)
23. M. Ivanov et al.,Bioactivities of Salvia nemorosa L. inflorescences are influenced by the extraction solvents. Ind. Crops Products **175** (2022): 114260.domized, double blind clinical trial. Journal of Ethnopharmacology, 282, p.114616
24. V. Todorova, K. Ivanov, S. Ivanova, Comparison between the biological active compounds in plants with adaptogenic properties (Rhaponticumcarthamoides, Lepidium meyenii, Eleutherococcussenticosus and Panax ginseng). Plants **11**(1), 64 (2021)
25. N. Singh et al., Phytochemical and pharmacological review of Cinnamomum verum J. Presl-a versatile spice used in food and nutrition. Food Chem. **338**, 127773 (2021)
26. B. Moradi, S. Abbaszadeh, S. Shahsavari, M. Alizadeh, F. Beyranvand, The most useful medicinal herbs to treat diabetes. Biomed. Res. Ther. **5**(8), 2538–2551 (2018)
27. A.P. Raina, R.C. Misra, Zingiber zerumbet (L.) Roscoe ex Smith: a potential source of zerumbone rich essential oil. Med. Plants-Int. J. Phytomed. Related Ind. **14**(1), 162–168 (2022)
28. T. Begum, R. Gogoi, N. Sarma, S.K. Pandey, M. Lal, Direct sunlight and partial shading alter the quality, quantity, biochemical activities of Kaempferia parviflora Wall., ex Baker rhizome essential oil: a high industrially important species. Ind. Crops Products **180**, 114765 (2022)
29. R. Tajpuriya et al.,various pathogenesis involved in the development of Type-II Diabetes Mellitus and its management processes. Sci. Lett. 1(8) (2021)
30. F. Haghani, M.R. Arabnezhad, S. Mohammadi, A. Ghaffarian-Bahraman, Aloe vera and streptozotocin-induced diabetes mellitus. RevistaBrasileira de Farmacognosia, 1–14 (2022)
31. S. Jiang, Q. Cai, Y. Gao, H. Dong, Y. Zhao, F. Wu,... F. Lu, Trigonella foenum-graecum L. and Psoralea corylifolia L. Improve erectile dysfunction in streptozotocin-induced diabetic rats through suppression of oxidative stress, in *Evidence-Based Complementary and Alternative Medicine* (2022)
32. J.N. de Sousa et al., Modulation of the resistance to norfloxacin in Staphylococcus aureus by Bauhinia forficata link. Nat. Product Res. **35**(4), 681–685 (2021)
33. A. Chavan, K. Daniel, A.M. Patel, In-silico exploration of phytoconstituents of gymnemasylvestre as potential glucokinase activators and DPP-IV inhibitors for the future synthesis of silver nanoparticles for the treatment of type 2 diabetes mellitus. Curr. Enzym. Inhib. **18**(1), 47–60 (2022)
34. H. Xie, Q. Tong, Z. Xiang, C. Zhou, L.S. Wan, J. Chen, Demethylbellidifolin, a potential aldose reductase inhibitor ameliorates diabetic nephropathy by regulating the polyol pathway. Phytomed. Plus **2**(1), 100152 (2022)
35. O. Zannou et al.,Green and highly extraction of phenolic compounds and antioxidant capacity from kinkeliba (Combretum micranthum G. Don) by natural deep eutectic solvents (NADESs) using maceration, ultrasound-assisted extraction and homogenate-assisted extraction. Arab. J. Chem. **15**(5), 103752 (2022)
36. M. Oz, T. Aouizerat, M. Salmon-Divon, T. Rosenzweig, J. Gorelick, E. Drori, M.Y. Sherman, B.C.R. Lubin,Effects on mammalian cell transcriptome reveal a plant extract's medicinal properties: potential anti-COVID-19 and anti-cancer properties of Sarcopoterium spinosum (2022)
37. S. Pattanayak, Prevention and control of diabetes by intake of succulent biomedicines and following of designed lifestyle: a ready plan for execution (2022)
38. S. Kumar et al.Herbal medicines for diabetes management and its secondary complications. Curr. Diabetes Rev. **17**(4), 437–456 (2021)
39. A. Das et al.Low-glycemic foods with wheat, barley and herbs (Terminalia chebula, Terminalia bellerica and Emblica officinalis) inhibit α-amylase, α-glucosidase and DPP-IV activity in high fat and low dose streptozotocin-induced diabetic rat. J. Food Sci. Technol. **59**(6), 2177–2188 (2022)

40. M.S. Faisal et al. Screening of achillea santolina for anti-diabetic activity and its comparison with carallumatuberculata. Prof. Med. J. **27**(7), 1414–1419 (2020)
41. S.S Huang, C.-Y. Ko, M.-H. Chi, Y.-H. Lee, K.-H. Su, M.-M. Lee, J. Chao,Botanical species utilized in Taiwan traditional chinese medicine for diabetes mellitus treatment
42. M. Sohaib, et al. Comparative study among Avicennia marina, Phragmites australis, and Moringa oleifera based ethanolic-extracts for their antimicrobial, antioxidant, and cytotoxic activities. Saudi J. Biol. Sci. **29**(1), 111–122 (2022)
43. F. Wang, Y. Bao, C. Zhang, L. Zhan, W. Khan, S. Siddiqua,... J. Xiao, Bioactive components and anti-diabetic properties of Moringa oleifera Lam. Crit. Rev. Food Sci. Nutr. **62**(14), 3873–3897 (2022)
44. H.F. Khalil et al. Ameliorative effect of OcimumforskoleiBenth on diabetic, apoptotic, and adipogenic biomarkers of diabetic rats and 3T3-L1 fibroblasts assisted by in silico approach. Molecules **27**(9), 2800 (2022)
45. N.T.L. Chi et al., Fabrication, characterization, anti-inflammatory, and anti-diabetic activity of silver nanoparticles synthesized from Azadirachta indica kernel aqueous extract. Environ. Res. **208,** 112684 (2022)
46. K. Qiao et al. The advances in nanomedicine for bone and cartilage repair. J. Nanobiotechnol. **20**(1), 1–42 (2022)
47. C. Liu, et al. Pnictogens in medicinal chemistry: evolution from erstwhile drugs to emerging layered photonic nanomedicine. Chem. Soc. Rev. **50**(4), 2260–2279 (2021)
48. S. Sun et al. Design and application of DNA nanostructures for organelle-targeted delivery of anticancer drugs. Expert Opinion on Drug Delivery just-accepted (2022)
49. S. Hong, Y. Yuan, Q. Yang, P. Zhu, H. Lian, Versatile acid base sustainable solvent for fast extraction of various molecular weight chitin from lobster shell. Carbohydr. Polym. **201**, 211–217 (2018)
50. V.P. Santos, N.S. Marques, P.C. Maia, L.d.O. Franco, M.A.B.d. Lima, G.M.d. Campos-Takaki, Seafood waste as attractive source of chitin and chitosan production and their applications, Int. J. Mol. Sci. **21**(12), 4290 (2020)
51. B.A. Al Jahdaly, N.S. Al-Radadi, G.M. Eldin, A. Almahri, M. Ahmed, K. Shoueir, I. Janowska, Selenium nanoparticles synthesized using an eco-friendly method: dye decolorization from aqueous solutions, cell viability, antioxidant, and antibacterial effectiveness, J. Mater. Res. Technol. **11**, 85–97 (2021)
52. N.Z. Shaban, S.A. Yehia, K.R. Shoueir, S.R. Saleh, D. Awad, S.Y. Shaban, Design, DNA binding and kinetic studies, antibacterial and cytotoxic activities of stable dithiophenolato titanium (IV)-chitosan nanocomposite. J. Mol. Liq. **287**, 111002 (2019)
53. S. Kou, L. Peters, M. Mucalo, Chitosan: a review of sources and preparation methods. Int. J. Biol. Macromol. **169**, 85–94 (2021)
54. W. Wang, C. Xue, X. Mao, Chitosan: structural modification, biological activity and application. Int. J. Biol. Macromol. **164**, 4532–4546 (2020)
55. H. El Knidri, R. Belaabed, A. Addaou, A. Laajeb, A. Lahsini, Extraction, chemical modification and characterization of chitin and chitosan. Int. J. Biol. Macromol. **120**, 1181–1189 (2018)
56. M. Bonin, S. Sreekumar, S. Cord-Landwehr, B.M. Moerschbacher, Preparation of defined chitosan oligosaccharides using chitin deacetylases. Int. J. Mol. Sci. **21**(21), 7835 (2020)
57. X. Yuan, J. Zheng, S. Jiao, G. Cheng, C. Feng, Y. Du, H. Liu, A review on the preparation of chitosan oligosaccharides and application to human health, animal husbandry and agricultural production. Carbohydr. Polym. **220**, 60–70 (2019)
58. M. Kaisler, L.A. van den Broek, C.G. Boeriu, in Chitin and Chitosan as Sources of Bio-based Building Blocks and Chemicals. Chitin and Chitosan: Properties and Applications, pp 229–244 (2019)
59. M.R. Abukhadra, N.M. Refay, A. Nadeem, A.M. El-Sherbeeny, K.E. Ibrahim, Insight into the role of integrated carbohydrate polymers (starch, chitosan, and ßcyclodextrin) with mesoporous silica as carriers for ibuprofen drug; equilibrium and pharmacokinetic properties. Int. J. Biol. Macromol. **156**, 537–547 (2020)

60. M. Ways, W.M. Lau T.M., V.V. Khutoryanskiy, Chitosan and its derivatives for application in mucoadhesive drug delivery systems, Polymers **10**(3), 267 (2018)
61. M.H. Teaima, M.K. Elasaly, S.A. Omar, M.A. El-Nabarawi, K.R. Shoueir, Ecofriendly synthesis of functionalized chitosan-based nanoantibiotic system for potential delivery of linezolid as antimicrobial agents. Saudi Pharm. J. **28**(7), 859–868 (2020)
62. S. Shawky, N.M. El-Shafa, I.M. El-Mehasseb, K.R. Shoueir, M.A. El-Kemary, Spectroscopic study of self-assembly of anti-hepatitis C virus Sofosbuvir drug with bio-polymeric nanoparticles for improving the drug release effect. Spectrochim. Acta A Mol. Biomol. **261**, 120008 (2021)
63. S. Asadpour, S. Kargozar, L. Moradi, A. Ai, H. Nosrati, J. Ai, Natural biomacromolecule based composite scaffolds from silk fibroin, gelatin and chitosan toward tissue engineering applications. Int. J. Biol. Macromol. **154**, 1285–1294 (2020)
64. S. Islam, M.R. Bhuiyan, M. Islam, Chitin and chitosan: structure, properties and applications in biomedical engineering, J. Environ. Polym. Degrad. 25 (3) (2017) 854–866. S.I. Othman et al. International Journal of Biological Macromolecules 190 (2021) 417–432 430
65. P. Sutthasupha, A.J.F. Lungkaphin, Function. The Potential Roles of Chitosan Oligosaccharide in Prevention of Kidney Injury in Obese and Diabetic Conditions **11**(9), 7371–7388 (2020)
66. S.L. Crossen, T. Goswami, Nanoparticulate carriers for drug delivery. J. Pharm. Biopharm. Res. **4**(1), 237–247 (2022)
67. R. Tiwari, K. Pathak, Nanostructured lipid carrier versus solid lipid nanoparticles of simvastatin: comparative analysis of characteristics, pharmacokinetics and tissue uptake. Int. J. Pharm. **415**(1–2), 232–243 (2011)
68. Y.Y. Luo, X.Y. Xiong, Y. Tian, Z.L. Li, Y.C. Gong, Y.P. Li, A review of biodegradable polymeric systems crossen, Samantha Lokelani, and Tarun Goswami.Nanoparticulate carriers for drug for Oral Insulin Delivery. Drug Deliv. **23**, 1882–1891 (2015)
69. M. Ussia, M. Pumera, Towards micromachine intelligence: potential of polymers. Chem. Soc. Rev. (2022)
70. X. Guo et al. Association between exposure to organophosphorus pesticides and the risk of diabetes among US adults: cross-sectional findings from the National Health and Nutrition Examination Survey. Chemosphere **301,** 134471 (2022)
71. S. Menini, C. Iacobini, M. Vitale, G. Pugliese, The inflammasome in chronic complications of diabetes and related metabolic disorders. Cells **9**(8), 1812 (2020)
72. J. Moreno-Fernandez, M. Pazos-Couselo, M. Gonz´ alez-Rodriguez, P. Rozas, M. Delgado, M. Aguirre, J.M. Garcia-Lopez, Clinical value of flash glucose monitoring in patients with type 1 diabetes treated with continuous subcutaneous insulin infusion. Endocrinol. Diabetes Nutr. (Engl. Ed.) **65**(10), 556–563 (2018)
73. S.M. Constantin, F.G. Lupascu, M. Apotrosoaei, A.V. Focsa, I.M. Vasincu, L.G. Confederat, G. Dimitriu, C.E. Lupusoru, S. Routier, F. Buron, Antidiabetic effects and safety profile of chitosan delivery systems loaded with new xanthinethiazolidine-4-one derivatives: in vivo studies. J. Drug Deliv. Sci. Technol. **60**, 102091 (2020)
74. Y. Zou, Y. Wang, S. Zhang, Y. Wu, X. Liu, Chitooligosaccharide biguanide repairs islet ß-cell dysfunction by activating the IRS-2/PI3K/Akt signaling pathway in type 2 diabetic rats. Adv. Ther. **2**(5), 1800136 (2019)
75. M. Naveed, L. Phil, M. Sohail, M. Hasnat, M.M.F.A. Baig, A.U. Ihsan, M. Shumzaid, M.U. Kakar, T.M. Khan, M. Akabar, Chitosan oligosaccharide (COS): an overview. Int. J. Biol. Macromol. **129**, 827–843 (2019)
76. M. Venkatesan, V. Arumugam, R. Ayyasamy, K. Ramachadran, S. Namasivayam, U. Sundaresan, A. Govindaraju, R. Saravanan, Role of marine polysaccharides in treatment of metabolic disorders, in *Enzymatic Technologies for Marine Polysaccharides*. CRC Press (2019), pp. 335–354
77. G.V. Morgunova, A.A. Klebanov, Age-related AMP-activated protein kinase alterations: from cellular energetics to longevity. Cell Biochem. Funct. **37**(3), 169–176 (2019)
78. M.R. Smruthi, I. Nallamuthu, T. Anand, A comparative study of optimized naringenin nanoformulations using nano-carriers (PLA/PVA and zein/pectin) for improvement of bioavailability. Food Chem. **369,** 130950 (2022)

79. N. Kaushal, M Singh, R.S. Sangwan, Flavonoids: food associations, therapeutic mechanisms, metabolism and nanoformulations. Food Res. Int., 111442 (2022)
80. A. Samadder, D. Tarafdar, S.K. Abraham, K. Ghosh, A.R. Khuda-Bukhsh, Nanopelargonidin protects hyperglycemic-induced L6 cells against mitochondrial dysfunction. Planta Med. **83**(5), 468–475 (2017)
81. L. DalleCarbonare et al.,Fisetin: an integrated approach to identify a strategy promoting osteogenesis. Front. Pharmacol. 13, 890693 (2022). https://doi.org/10.3389/fphar.2022.890693. www. frontiersin.org 13

Chapter 13
Application of Chitosan Nanostructures Embedded Composite Materials in Cancer Therapy

Sora Yasri and Viroj Wiwanitkit

Abstract Chitosan is a naturally occurring polymer formed from the chitin found in the shells of many different animals, including shrimp, crabs, and squid. Weak organic acids dissolve their amino groups. It is capable of binding to both physiologically active chemicals and metal ions. Chitosan and its derivative biomaterials have recently piqued the interest of biomedical researchers due to their unusual biological features. Non-toxicity, biodegradability, biocompatibility, immunostimulating, anticancer, antibacterial, and antimicrobial activity are the most important qualities of chitosan in the biomedical field. Encapsulating and delivering various substances improves water solubility, stability, and permeability into cells and helps manage the release of these molecules, particularly in medical and pharmaceutical applications. It can be used to treat and prevent a variety of illnesses more efficiently. It is also used in textile, food, agriculture, cosmetics, and other industries. Chitosan nanostructures integrated into composite materials are among the most recent types of nanomaterials being researched. Although using chitosan alone may be insufficient to result in an adequate antibacterial effect for a variety of purposes, combining chitosan with other active substances such as metals, drugs and natural compounds in nanosystems is a common strategy for increasing its application potential. These specific nanoparticles are actively being researched and developed. The authors will dwell deeper and look into one particular use of composite materials with embedded chitosan nanostructures in clinical cancer in this chapter. Chitosan has a number of

S. Yasri (✉)
KM Center, Bangkok 1014400, Thailand
e-mail: sorayari@outlook.co.th

V. Wiwanitkit
Dr. D. Y. Patil Medical College, Hospital and Research Centre, D. Y. Patil Vidyapeeth, Pune, India

Department of Eastern Medicine, Government College University Faisalabad, Faisalabad, Pakistan

Hainan Medical University, Haikou, China

Faculty of Medicine, University of Nis, Nis, Serbia

Joseph Ayobalola University, Ikeji-Arakeji, Nigeria

S. K. Swain and A. Biswal (eds.), *Chitosan Nanocomposites*,
Biological and Medical Physics, Biomedical Engineering,
https://doi.org/10.1007/978-981-19-9646-7_13

roles in the management of malignancy. Chitosan can be used in both diagnostic and treatment. Many cancers, including lung, brain, and colon tumors, can be managed using composite materials with incorporated chitosan nanostructures.

13.1 Introduction

Chitosan (CAS 9012-76-4) is a natural polymer derived from chitin, which can be found in shrimp shells, crab shells, and squid cores, among other places, having amino groups that are soluble in dilute organic acids (Fig. 13.1). It has a strong affinity for metal ions and physiologically active compounds. Chitosan has numerous applications at the moment. Because of chitosan's many unique features, including its non-toxicity, biocompatible (biocompatibility), biological degradability (biodegradability), and does not pose any environmental risks [1] Chitosan has been known as a natural material for over a century, but its application has not been extensively investigated. In 1975, a global gathering of scientists to research the features of this chemical revealed the fundamental qualities of chitosan. Scientists have researched chitosan and found that it is a positively charged material that may trap certain lipids that are negatively charged [1]. Japan conducted the first chitosan study in biological wastewater treatment. Chitosan has since been used in a variety of industrial applications. In 1859, Professor C. Rouget developed chitosan by boiling chitin in a strong solution of potassium hydroxide, resulting in a material having qualities distinct from chitin. It is soluble in dilute organic solutions, and Hoppe-Seiler termed it "chitosan" in 1894 [1–3].

Chitin, one of the most common natural polysaccharides on the planet, is largely used in the deacetylation process to generate chitosan [3]. Chitosan is a bioactive polymer with a wide range of uses due to its beneficial properties such as antibacterial activity, non-toxicity, ease of modification, and biodegradability [3]. Chitosan is an extract derived from shrimp or crab shells, as previously stated. As a result, it is regarded as a significant natural product. When you separate proteins and minerals, you generate chitin, which is a natural dietary fiber [4, 5]. It makes people feel full and aids in the burning of extra fat. Furthermore, it possesses a mesh-like structure that may absorb 8–10 times its weight in fat or fat in the digestive tract. The fat cannot be absorbed by the body. However, it will be eliminated with feces [4, 5].

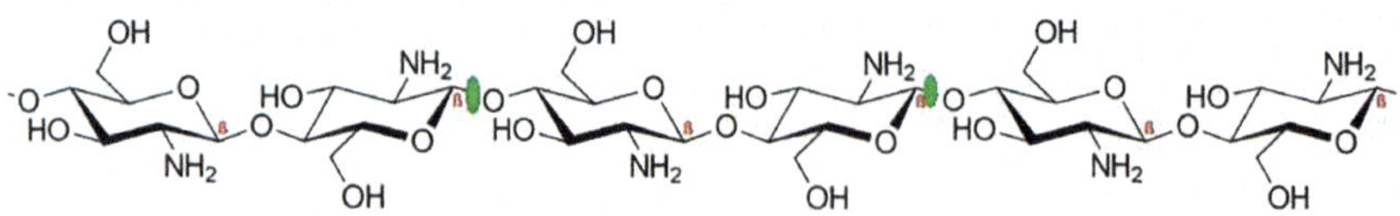

Fig. 13.1 Chitosan. In the picture, three monomers of the molecule are shown, the green dot is the intermolecular bond

Chitosan has been advocated as a dietary supplement that can aid in weight management and blood cholesterol reduction. Numerous chitosan products are widely available in pharmacies and health food stores worldwide [5]. Chitosan is a polycationic (+) fibrillar biopolymer with cellulose-like characteristics that form films with negatively charged surfaces. Chitosan is not highly hydrolyzed by digestive enzymes in humans, but there may be some limited digestion of chitosan due to bacterial flora and general enzymes. Chitosan's positively charged tertiary amino group ($-NH_3^+$) strongly binds negatively charged molecules in the stomach [5]. As a result of its interactions with anionic carboxyl groups of fatty and bile acids, chitosan inhibits the emulsification of neutral lipids (such as cholesterol and other sterols) and decreases fat absorption from the gastrointestinal system [5]. Chitosan is currently one of the most readily accessible dietary supplement products on the market (Fig. 13.2).

Even if employing chitosan by itself might not be sufficient to provide a suitable biomedical application, more research and development are still required. The new technique for material modification to increase the biological application is difficult. A novel biomedical application in medicine is the use of composite materials with incorporated chitosan nanostructures. It is already used in a number of medical specialties, including clinical oncology. Multiple functions of chitosan are involved in the treatment of cancer. Both diagnostic and therapeutic uses for chitosan are possible. Composite materials with chitosan nanostructures integrated can be used to treat a variety of malignancies, including lung, brain, and colon tumors. For biomedical professionals, knowledge of the use of chitosan nanostructures embedded composite materials in clinical cancer is new and not well discussed. The authors will continue to summarize, go into more detail, and address the use of composite materials with embedded chitosan nanostructures in cancer therapy in this chapter.

Fig. 13.2 Example of chitosan food supplementation product. The pictures are some famous locally available products in Southeast Asia

13.2 Chitosan Nanostructures Embedded Composite Materials

Chitosan is made by deacetylating chitin, which is plentiful in nature. Chitosan is created by alkaline deacetylating chitin, an abundant polymeric by-product of natural biosynthesis found primarily in crustaceans. It is available in both a simple, unstructured structure and a microcrystalline structure [5]. The difficulty of chitosan to dissolve in water and the majority of organic solvents greatly limits the variety of applications in which it can be employed [2]. Chitosan is a polycationic (+) fibrillar biopolymer with cellulose-like characteristics that form films with negatively charged surfaces [6]. Human digestive enzymes do not considerably hydrolyze chitosan, but there may be a small amount of limited chitosan digestion caused by bacterial flora and general enzymes [6].

Chitosan's positively charged tertiary amino group ($-NH_3^+$) strongly binds negatively charged molecules in the stomach [6]. As a result of interacting with the anionic carboxyl groups of fatty and bile acids, chitosan inhibits the emulsification of neutral lipids (such as cholesterol and other sterols) and decreases fat absorption from the gastrointestinal system [6]. Chitosan is a low-cost, natural, and biodegradable biopolymer. Chitosan is gaining popularity as a nanocomposite matrix due to its innovative properties for a wide range of applications [7]. There are numerous current alternative methods for creating embedded or coated nanoparticles, as well as shaping techniques that have been used (three-dimensional printing, electrospinning), and the burgeoning applications for nanocomposites in areas such as corrosion prevention, tissue engineering, and medicine [7].

Chitosan and its derivative biomaterials have recently attracted a lot of attention in the biomedical arena due to their unusual biological properties. Chitosan's most significant biological properties are non-toxicity, biodegradability, biocompatibility, immunostimulating, anticancer, antibacterial, and antimicrobial activity [8]. Chitosan's biodegradability was established in vitro and in vivo by dividing macromolecules into smaller monomer parts [8]. Living cells can interact with chitosan and its enzymatic degradation products without damage to their health, indicating that chitosan is biosafe. Chitosan can reduce chain oxidation by screening free radicals, lowering cholesterol absorption, and acting as an antibiotic and antibacterial agent against a variety of bacteria and fungi [8]. Only a few of the variables that are reported to affect the antibacterial activity of chitosan and its derivative biomaterials are the degree of deacetylation, molar weight, pH, the presence of metal cations, pKa, and the kind of bacterium [8].

Encapsulating and delivering various chemicals enhances water solubility, increases stability, increases permeability into cells, and can manage the release of these molecules, particularly in medical and pharmaceutical applications. It can be used to treat and prevent a variety of disorders. It's also useful in cosmetics, food, agriculture, textiles, and other industries [1, 2]. Chitosan nanostructures incorporated in composite materials are a novel type of nanomaterial that is being explored extensively [1, 2]. Incorporating this biopolymer with other active substances such as medicines, metals, and natural compounds in nanosystems is a typical technique for increasing its application potential, even if chitosan alone may not create an effective antibacterial effect to fulfill diverse goals [9, 10]. Chitosan has been altered by nitration, alkylation, sulphonation, phosphorylation, xanthation, Schiff's base generation, acylation, hydroxylation, and graft copolymerization [10].

It has been established that composite materials containing chitosan nanostructures have biomedical applications. Numerous studies have been published on the use of antibacterial drugs to treat infections. Among the suggested mechanisms for the intrinsic anti-pathogenic properties of chitosan-based biomaterials include interactions with microbial nucleic acids, chelation of microbial structures, and electrostatic interactions between cationic chitosan and negatively charged pathogenic cell surfaces [11, 12]. The nanostructured composite biopolymeric nanostructure demonstrates the required properties for delayed and regulated drug release [11, 12]. Nanoparticle-based medication formulations have been shown to be useful in the treatment of difficult ailments. A quick search of the major scientific publication databases reveals an astounding number of papers on nanoparticles designed for drug delivery purposes. Because of their inherent physicochemical diversity and configurable functionality, polymer particles and capsules are suitable candidates for modern pharmacology. This allows for more effective loading and encapsulation of various medicines (including genetic components, proteins, growth factors, and cells), as well as triggerable or programmed release mechanisms [11, 12]. There are numerous studies and developments on those specific nanomaterials at the moment.

13.3 Application of Chitosan in Biomedicine

Among the various qualities of chitosan are its water solubility, stimulus responsiveness, biocompatibility, biodegradability, and antibacterial activities [10]. As previously stated, chitosan is an extract obtained from shrimp or crab shells. As a result, it is recognized as an important natural product. Chitin, a natural dietary fiber, is produced when proteins and minerals are separated [4, 5]. It makes people feel full and promotes fat burning. It also has a mesh-like structure that can absorb 8–10 times its own weight in fat or fat in the digestive tract. Fat cannot enter the bloodstream and must be removed by feces [4, 5]. Chitosan has been promoted as a dietary supplement that can help with weight loss and cholesterol lowering. Several chitosan preparations can be found in pharmacies [1]. Because of its biodegradability, biological compatibility, antibacterial and antioxidant activity, and high safety, chitosan has a wide range of uses [3, 4]. Fibers, gels, films, sponges, nanoparticles, or beads could all be used. The numerous biological activities of chitosan and its derivatives are being extensively explored in order to broaden the application sectors in a wide range of disciplines. Chitosan's natural properties are mainly dependent on its solubility in water and other solvents. Chitosan oligosaccharides with low polymerization degrees are gaining popularity in pharmaceutical and medical applications due to their reduced viscosity and higher water solubility than chitosan [13].

Furthermore, for a variety of applications, chitosan can be turned into scaffolds, fiber meshes, nanoparticles, nanovehicles, nanocapsules, nanoparticles, and three-dimensional printed scaffolds [10]. Because nanocomposites interact with cells faster than bulk or micron-sized substrates or scaffolds, nanoparticles contained in chitosan matrices have been shown to have higher biological activity [10]. Chitosan-based nanomaterials have sparked substantial interest in a wide range of biological applications because of their particular chemical properties, such as desired biodegradability, compatibility, and non-toxicity. Chitosan is a biomaterial that can be used to construct extracellular tissue matrixes in tissue engineering [8]. It can be used to carry drugs to the brain as well as a number of therapeutic chemicals for the eyes, such as proteins, genes, and medications used to treat visual system disorders. Another medicinal purpose is its use in hemodialysis and renal failure [8]. Chitosan is frequently used as a delivery vehicle for active compounds and drugs [8], as well as in gene and cancer therapy, biosensor monitoring, and bioimaging. The authors aim on summarizing and explaining the specific use of chitosan nanostructures embedded composite materials in clinical cancer in this chapter.

13.4 Chitosan Nanostructures Embedded Composite Materials in Cancer Therapy

Chitosan has been widely employed in a variety of biomedical applications, including wound dressing as an antibacterial agent [8], medication delivery as a nano-sized carrier to target tumor tissue while impacting normal tissue locations slightly, gene transfer, hemodialysis, and dentistry [8]. Chitosan can be utilized on its own or in combination with other polymers, active agents, and metallic nanocomposites. Chitosan-based nanoparticles have a high permeability and retention impact, as well as the potential to cause apoptosis, which limits tumor cell proliferation [8]. Larger particles must be phagocytosed before being taken up by receptor-mediated endocytosis, which can only take up particles as small as 100–200 nm [8]. Chitosan has various applications in clinical oncology. The two major goals of using chitosan nanostructures embedded composite materials in cancer therapy are diagnostic and therapeutic (Table 13.1) (Fig. 13.3). The application for diagnostic purposes entails the use of chitosan nanostructures implanted in composite materials to aid in illness investigation via clinical laboratory or clinical imaging approaches. The use of chitosan nanostructures implanted composite materials for therapeutic purposes is typically associated with anticancer drugs and the mediation process. Dual application, theranostics, is also possible in some situations. It has been confirmed that the new chitosan nanostructures implanted in composite materials are quite useful for cancer management.

As previously stated, there are numerous current studies on the use of chitosan nanostructures implanted in composite materials for cancer therapy. The applications

Table 13.1 Application of chitosan nanostructures embedded composite materials in cancer therapy

Purposes	Examples
Diagnostic purpose	Diagnostic assay Biosensor Nanoimaging diagnosis Theranostics
Therapeutic purpose	Drug formulation Drug delivery system Magic bullet and cancer targeting Tissue engineering and regenerative repair Photothermal cancer therapy Theranostics

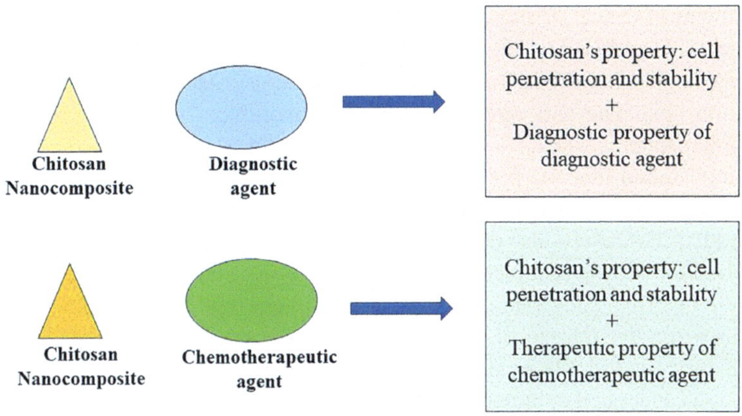

Fig. 13.3 Schematic diagram showing application of chitosan nanostructures embedded composite materials for diagnostic and therapeutic purposes in cancer therapy

of chitosan nanostructures embedded composite materials for cancer treatment will be listed below.

13.4.1 Drug Formulation

Chitosan nanoparticle composites can be made by utilizing ultrasonics to encourage consecutive precipitation; these composites can then be tested for their atomic-scale microstructure, surface charge, drug release properties, and anti-cancerous activity. For instance, Glavas-Dodov et al. [14] presented 5-Fluorouracil in topical liposome gels for anticancer treatment. A study by Glavas-Dodov et al. examined the in vitro drug release characteristics of liposome gels containing the antineoplastic agent 5-fluorouracil [14]. According to this study, hydrogels released liposomes more quickly than those implanted in chitosan-structured vehicles [14]. According to Glavas-Dodov et al. research, the liposomes served as reservoir systems for the continuous distribution of the drug that was encapsulated and the drug released from them followed the Higuchi diffusion model [14].

13.4.2 Drug Delivery System

Building multifunctional microspheres for remotely controlled drug release requires a careful selection of composite materials and production methods. Chitosan nanostructure-infused composite materials may be employed in drug delivery systems and may support the idea of remote-control release. This is helpful in the treatment of cancer. Chitosan was combined with the inorganic nanomaterials, graphene oxide and Fe_3O_4 and electro-sprayed to create homogenous microspheres with very small dimensions for anti-cancerous purposes, according to the Li et al. publication [15]. The anti cancer medication doxorubicin was then administered to the microspheres via an adsorption or embedding technique [15]. The findings demonstrated that the presence of Fe_3O_4 causes the microsphere to respond to magnetic fields, and the addition of graphene oxide increases the drug loading capacity [15]. The results also demonstrated that the quick stimuli-responsive release of doxorubicin may be easily controlled by using NIR irradiation due to the considerable photothermal conversion of Fe_3O_4 and graphene oxide [15]. Another extrinsic mechanism for the release of doxorubicin was ultrasonic stimulation [15]. In a different investigation, Tamoxifen, a crucial anticancer medication, was added to MCF-7 and cancer cell cultures and examined for antiproliferative activity by Rossi et al. [16]. Tamoxifen was also embedded in sodium alginate microparticles from Kelco and Fluka. According to Rossi et al., the formulation of the microparticles may improve selective toxicity based on the alginate used; changes in the chemical alginate composition can greatly increase anticancer drug activity and minimize anticancer drug toxicity [16].

13.4.3 Tissue Engineering

Tissue loss could be an issue in some malignancies. An intriguing medical therapeutic strategy is the use of composite materials with embedded chitosan nanostructures for tissue engineering and regeneration. Using a combination of tissue flap, nanotherapeutic, silk fibroin, and chitosan scaffold, Gupta et al. reported on the healing and regeneration of a resected tumor defect [17]. The scaffold-emodin nanoparticle composites were developed and examined for drug entrapment and release, mechanical strength, and efficacy against breast cancer cells in vitro and in vivo in an animal tumor model [17]. The amount of entrapment was determined by the composition of the scaffold and the emodin loading concentration when emodin nanoparticles were implanted in silk fibroin and chitosan scaffolds [17]. Gupta et al. discovered that tumors treated with emodin-loaded silk fibroin and chitosan scaffolds had decreased presence and size as well as equal regeneration of new tissue as compared to untreated tumors [17]. However, there was no discernible difference in tumor size across the in vivo test groups.

A quick-prototyped tissue engineering scaffold with an embedded multicomponent matrix for regulated drug release was reported in a different study by Chen et al. [18]. By using freeze-drying manipulation, a quick prototype macroporous polycaprolactone scaffold was implanted into a porous matrix made of nanoclay, -tricalcium phosphate, and chitosan [18]. By adjusting the kind and amount of each component, Chen et al. theorized that the scaffold might be used therapeutically in reconstructive surgery after bone tumor removal and in other tissue engineering applications that need local sustained drug release [18].

13.4.4 Photothermal Therapy

In clinical oncology, photothermal therapy is a novel and intriguing treatment option for cancer. Chitosan nanostructures can be integrated into composite materials for this specific function. Due to their inherent localized and controlled release properties, hydrogels are excellent drug or nanoparticle delivery systems. In particular, the use of hydrogels for the delivery and release of photoresponsive drugs or nanoparticles is gaining popularity [19]. In a recent study by Xia et al. [19], meso-tetrakis(1-methylpyridinium-4-yl) porphyrin (TMPyP) was enclosed in an injectable hydrogel comprised of glycol chitosan and dibenzaldehyde-terminated telechelic poly (ethylene glycol).

Xia et al. found that the hydrogel's reduced self-quenching action resulted in a markedly increased fluorescence intensity of TMPyP [19]. The TMPyP-hydrogel system's extraordinary anticancer efficacy and great fluorescence emission property, according to Xia et al., advance the development of imaging-guided photodynamic treatment [19]. A novel iodinated chitosan-backboned conjugate was created and synthesized by Lim et al. in a distinct study to create self-assembled biopolymeric nanoparticles with better singlet oxygen production and biological advantages [20] According to this work, the self-assembled nanoparticles with biocompatible, positively charged, and tumor-homing properties have a high tumor targetability both in vitro and in vivo due to their glycol chitosan-surfaced exterior [20]. The intraparticle heavy-atom effect also helps them produce singlet oxygen more effectively [20]. When using composite materials with chitosan nanostructures for photothermal therapy, theranostics is occasionally integrated with a diagnostic strategy [21]. A notable example is the study by Liao et al. on gold/magnetic core–shell hybrid nanoparticles for targeting and imaging-guided photothermal treatment [21].

13.4.5 Theranostics

The words "therapeutics" and "diagnostics" are the roots of the term "theranostics." For cancer theranostics, composite materials incorporating chitosan nanostructures can be employed (Table 13.2). Several new nanocomposites have been reported for this purpose [21–25]. Table 13.2 compiles key reports on these issues.

Many intriguing applications of the chitosan nanostructures embedded composite materials for cancer therapy have been reported. The authors will provide a summary of a few significant reports here.

13.4.5.1 Lung Cancer

The cells that line the airways in the lungs are most frequently impacted by the malignant illness known as lung cancer. It is the main factor behind cancer related fatalities in both men and women. Small cell and non-small cell lung cancer are the two most common kinds. These two categories are handled and grown in quite different ways. One of the most prevalent malignancies in the world is this one. This form of cancer typically advances to an advanced stage and is highly lethal.

It has been proposed that implanted composite materials with chitosan nanostructures may have pharmacological uses for the treatment of lung cancer. Guo et al. [26] reported inhalable microspheres for 2-methoxyestradiol that had chitosan-coated PLGA nanoparticles. Guo et al. [26] used soluble excipients as the matrix for the respirable MS with poly(d, l-lactide-co-glycolide) nanoparticles for 2-methoxyestradiol. According to Guo et al. [26], microspheres matrix, respirable embedding poly(d,l-lactide-co-glycolide) nanoparticles have significant potential as an inhalation agent for targeted, extremely effective, and secure therapy of lung cancer. They also noticed that 2-methoxyestradiol markedly boosted cytotoxicity in experimental animals' lungs without obviously causing inflammation. Even if there is still no proof of the successful use of composite materials with embedded chitosan nanostructures for cancer therapy, the in vitro examination confirms the viability. The study by Sachdev et al. [25] provides the finest illustration of the advantages of using composite materials containing chitosan nanostructures as theranostics.

13.4.5.2 Brain Cancer

A lethal tumor, brain cancer has a high fatality rate. Brain cancer treatment is still not effective. The new hope for brain cancer is anotherapy application. Implanted composite materials with chitosan nanostructures have been theorized to have pharmacological effects on the therapy of brain cancer, just like they do for lung cancer. Kutlu et al. reported on an in vitro experiment. Double the effectiveness of brain tumor therapy with chitosan scaffold-PLGA nanoparticles [27].

Table 13.2 Some reports on applied chitosan nanostructures embedded composite materials as theranostics

Authors	Details
Liao et al. [21]	The hybrid particles are composed of gold shells on chitosan (CTS) and magnetic nanoparticles implanted in poly (methyl methacrylate) (PMMA) cores (-Fe_2O_3 @PMMA/CTS@Au), according to a study by Liao et al. [21]. The -Fe_2O_3@PMMA/CTS@Au particles, according to Liao et al.'s analysis can act as a multifunctional anti-tumor theranostic nanoplatform [21]. This new platform can be the model for magnetically targeted thermophototherapy [21]
Fernández-Álvarez et al. [22]	The development of stealth (maghemite/PLGA)/chitosan (core/shell)/shell nanocomposites with potential uses for combined MRI and hyperthermia against cancer was described by Fernández-lvarez et al. in their report [22]. In vivo magnetic resonance imaging investigations and ex vivo histological analyses of iron deposits, according to Fernández-lvarez et al., suggested the usefulness of chitosan to give the nanocomposites long-circulating capabilities, delaying nanoparticle detection by the mononuclear phagocyte system [22]
Ghoshal et al. [23]	Ghoshal e al. reported on targeting Wnt canonical signaling by recombinant sFRP1 bound luminescent Au-nanocluster embedded nanoparticles in cancer theranostics [23]
Dutta e al. [24]	Dutta et al. [24] described bimetallic silver nanoparticle-gold nanocluster embedded composite nanoparticles for cancer theranostics. According to Dutta et al. [24], by using chitosan as a template, the bimetallic system has been turned into composite nanoparticles that can be easily transported into malignant cells. According to the result reached by Dutta et al. [24], the synthetic composite nanoparticles benefit from the combinatorial properties of the metals present at two separate nanoscale levels, enabling both the killing and bioimaging of cancer cells
Sachdev et al. [25]	The easy development of a hydrogel formulation based on chitosan that contains highly fluorescent carbon dots and is loaded with the model anticancer medication 5-Fluorouracil was the focus of Sachdev et al. [25]. This study found that the developed particle treated cells' induction of apoptosis was demonstrated by modifications in cell cycle patterns and the identification of distinctive apoptotic bodies using field emission scanning electron microscope [25]. Studies on specific apoptotic gene expression shed more light on the molecular process that causes apoptosis [25]. According to Sachdev et al. analysis, hydrogels' ability to combine fluorescent carbon dots with chemotherapeutic drugs opens up new possibilities for their possible usage in cancer theranostics [25]

In this study, poly-lactic-co-glycolic acid (50:50) nanoparticles containing the anticancer drug 5-fluorouracil were made using the emulsion-solvent evaporation method [27]. Then, during fabrication, the scaffold was infused with the antivascularization medication bevacizumab and the nanoparticles by freeze-drying and embedding, respectively [27]. According to Kutlu et al. [27], the chitosan scaffold containing 100 g of 5-fluorouracil and bevacizumab has the ability to prevent the growth of malignancies in in vitro conditions. In a different study, Shamsipour et al. successfully created core/shell chitosan-poly ethylene oxide–carbon quantum dots/carboxymethyl cellulose-polyvinyl alcohol nanofibers utilizing coaxial electrospinning as a biodegradable polymeric implant for the local administration of temozolomide [28]. When temozolomide is combined with quantum dots embedded in chitosan, it is shown that the cytotoxicity of cancer cells is boosted [28].

13.4.5.3 Colon Cancer

Colon cancer is another lethal cancer that has a high mortality rate. Treatment for colon cancer is still ineffective. The use of anotherapy is the new treatment option for brain cancer. Implanted composite materials with chitosan nanostructures may have pharmacological effects on the treatment of colon cancer, similar to what has happened with lung and brain malignancies. As shown in Table 13.3 [29–32], there are a number of intriguing findings on the use of composite materials with chitosan nanostructures for colon cancer therapy (Table 13.3).

13.4.5.4 Ovarian Cancer

A serious malignancy in females is ovarian cancer. The aberrant abdominal mass is how this cancer typically manifests. Surgery is the mainstay of ovarian cancer treatment, which is followed by chemotherapy. The fifth leading cause of cancer-related mortality and the ninth most prevalent cancer in women, ovarian cancer kills more people than any other malignancy of the female reproductive system. The use of composite materials with embedded chitosan nanostructures for the management of ovarian cancer is recorded in the literature, just like it is for the treatment of other cancers. For the treatment of ovarian cancer, Chandran et al. investigated the possible use of drug-loaded nanocomposite pectin scaffolds [33]. To extend the half-life of the drug and provide localized therapy for ovarian cancer, Chandran et al. proposed the use of a highly porous, biomimetic, and implantable pectin scaffold embedded with gemcitabine-loaded fibrin nanoconstructs [33]. The controlled and sustained release of the chemokine from the scaffold structure was thoroughly examined in vitro under various pH conditions in this study [33]. When examined with mammalian cell lines, Chandran et al. discovered that the composite scaffolds were highly biocompatible

Table 13.3 Reports on using composite materials with incorporated chitosan nanostructures applied in colon cancer therapy

Authors	Details
Ullah et al. [29]	The delivery of calcium alginate beads with specified targeting for colon cancer was described by Ullah et al. in their paper on folate-modified chitosan 5-flourouraci nanoparticles [29]
Vijayakumar et al. [30]	Marine originated polysaccharide laminarin embedded zinc oxide (ZnO) nanoparticles and their based chitosan-capped ZnO nanocomposites were the subject of a report by Vijayakumar et al. [30]. The green synthesis of chitosan-capped ZnO nanocomposites and laminarin-embedded ZnO nanoparticles is discussed in the current work [30]. Additionally, the assessment of their in vivo (Zebrafish embryo) toxicity, phytotoxicity, and in vitro cytotoxicity was described [30]
Hosseinzadeh et al. [31]	Chitosan-pluronic nanoparticles were used to deliver the anticancer drug gemcitabine orally, according to Hosseinzadeh et al. [31]. Based on an in vitro examination of a colon cancer cell line, a cytotoxicity assay of gemcitabine-loaded nanoparticles revealed an increase in the cytotoxicity of the drug if it was compared to the naïve drug alone [31]
Cao et al. [32]	With the aid of oral nanomotor-enabled mucus traversal and tumor penetration, Cao et al. reported on targeted sono-immuno-chemotherapy for colon cancer [32]. This study employed silk fibroin and chondroitin sulfate to construct mesoporous manganese oxide-based nanomotors that systematically dual-functionalize their surfaces and allow passing of mitochondrial sonosensitizers into the physical mesopores [32]

[33]. According to Chandran et al., implanted "therapeutic wafers" made of nanocomposite Pectin scaffolds loaded with a particular chemokine can be utilized to divert metastatic cancer cells and increase the survival rate of ovarian cancer patients [33].

13.4.5.5 Breast Cancer

Breast cancer is a dangerous female malignancy. This malignancy often presents as an abnormal breast lump. Chemotherapy is the second line of defense in the fight against breast cancer after surgery. Just like it is for the treatment of other malignancies, the use of composite materials with embedded chitosan nanostructures for the management of breast cancer is documented in the literature. Tamoxifen, a common breast cancer treatment drug, was added to MCF-7 and Vero cultures in a study by Rossi et al., and its antiproliferative effects were assessed both alone and when embedded in sodium alginate microparticles made with sodium alginate from Kelco and from Fluka [16]. According to the alginate used, Rossi et al. determined that microparticle formulation may improve selective toxicity: variations in the chemical composition of the alginate can significantly alter both drug activity and toxicity [16].

13.4.5.6 Gastric Cancer

A significant gastrointestinal malignancy is gastric cancer. When this cancer is initially diagnosed, it is typically advanced. The use of composite materials with embedded chitosan nanostructures for the management of stomach cancer is recorded in the literature, just like it is for the treatment of other malignancies. The fabrication process used in the study by Wang et al. to create magnetic cores involved the coprecipitation of Fe^{2+} and Fe^{3+}. The magnetic nanoparticles were then covered in layers of chitosan. Next, using sodium citrate as a natural reducing agent, magnetite-gold composite nanoparticles with spherical forms and diameters between 20 and 30 nm were created [34]. Fe_3O_4@CS/AuNPs showed the best cytotoxicity effects when applied to the HCT 116 cell line, according to Wang et al. According to Wang et al., the current nanoparticles may one day be employed to treat various gastro-duodenal tumors, particularly gastric cancer [34]. Chang et al. reported on a novel exceptionally sensitive fluorescence sensor for the magnetic separation-based detection of microRNA that is based on composite nanoparticles of lucigenin, chitosan, and silicon [35]. A good linear relationship between the fluorescence intensity sensitizing value and the target miRNA concentrations was achieved when the difference in fluorescence intensity increased along with the target DNA/miRNA concentration [35]. In order to directly detect let-7a in human gastric cancer cell samples without the use of an enzyme, a label, or any other immobilization, Chang et al. came to the conclusion that a new fluorescence analysis method had been developed using lucigenin/chitosan/SiO_2 composite nanoparticles as a DNA hybrid indicator [35].

Another study by Rizwan et al. reported the creation of a new organosoluble and pro-angiogenic chitosan derivative through the reaction of chitosan with triethylorthoformate and 1,3-dimethylbarbituric acid [36]. In a study by Rizwan et al., it was discovered that the novel derivative had no adverse effects on gastric cancer cells when tested for cytotoxicity and apoptosis [36].

13.4.5.7 Liver Cancer

Another common cancer is liver cancer. Alcoholism and chronic viral hepatitis are linked to it. The fight against liver cancer is still ineffective. The hope is that a novel anticancer medicine will be developed for the treatment of liver cancer. Similar to how it is used to treat other cancers, the use of composite materials containing embedded chitosan nanostructures for the management of liver cancer is documented in the literature. The core–shell type Fe_3O_4 nanoparticle (Ag/CS-Agar@Fe_3O_4) was described by Cai et al. in a unique Ag NPs-fabricated chitosan-agarose composite [37]. Ag/CS-Agar@Fe_3O_4 nanocomposite was shown by Cai et al. to be an effective chemotherapeutic agent against the lung and hepatocellular cancer cells [37]. A targeted drug delivery system was developed by functionalizing graphene oxide (GO) with carboxymethyl chitosan (CMC), fluorescein isothiocyanate, and lactobionic acid (LA) in a different study by Pan et al. [38]. According to Pan et al., the composite containing LA demonstrated the ability to specifically promote cell death in malignant cells, but the LA-free homologue was inactive in this situation [38]. According to Pan et al. analysis, the modified GO materials provide excellent prospects for use in drug new delivery systems to target malignancy [38]. In a different study by Huang et al., the positively charged HTCC-OREC nanocomposites and the negatively charged sodium alginate were immobilized on cellulose nanofibrous mats using the layer-by-layer (LBL) technique [39]. The intercalated nanocomposites were made using organic rectorite and N-(2-hydroxyl) propyl-3-trimethyl ammonium chitosan chloride, respectively [39]. According to Huang et al.'s research, the produced nanofibrous mats significantly inhibited the growth of human hepatoma cells [39]. Based on the numerous clinical examples provided, it is possible to infer that composite materials with embedded chitosan nanostructures might be suitable for the treatment of liver cancer.

13.4.5.8 Leukemia

Instead of a solid tumor, leukemia is a cancer of the white blood cell. The primary therapeutic approach for treating leukemia is chemotherapy. The use of composite materials with embedded chitosan nanostructures for the management of liver cancer is recorded in the literature, much like how it is used to treat other malignancies. ZnO-TiO_2-chitosan-farnesol nanocomposites (NCs) were studied for their synergistic antitumor potential against leukemia MOLT-4 cells by Elderdery et al. [40]. According to Elderdery et al., this is a novel method for enhancing the properties of ZnO-TiO_2-chitosan-farnesol composites and synergistically exhibiting anticancer activities in human leukemic cancer cells [40].

13.5 Conclusion

Chitosan is derived from chitin, a naturally occurring polymer present in several kinds of animal shells, including those of shrimp, crabs, and squid. Its amino groups are soluble in weak organic acids. Both chemical compounds having physiological activity and metal ions can be successfully bound by it. Due to their distinctive biological properties, Chitosan and its derivative biomaterials have recently attracted a lot of interest in the biomedical field. Some of chitosan's most crucial properties in the biomedical sector include its non-toxicity, biodegradability, biocompatibility, immunostimulating, anticancer, antibacterial, and antimicrobial activity. Encapsulating and delivering different chemicals enhances their water solubility, stability, permeability into cells, and capacity to control the release of these molecules, particularly in medical and pharmaceutical applications. It aids in the better treatment and prevention of numerous diseases. It can also be used in textile, food, agricultural, cosmetic, and other industries. An in-depth study is currently being carried on the newest types of nanomaterials, such as the chitosan nanostructures found in composite materials. Chitosan is often combined with other active chemicals, including medicines, metals, and natural compounds, in nanosystems to increase its potential for application, even if employing chitosan alone might not be sufficient to generate an acceptable antibacterial effect to fulfill a range of purposes. These particular nanoparticles are the subject of current research and development. The authors have tried to include details about one particular use of composite materials with integrated chitosan nanostructures in clinical cancer. The development of numerous new chitosan-based nanomaterials has shown promise in the treatment of various malignancies. Future effective cancer treatments may be made possible by the implanted composite materials utilized in cancer therapy that incorporate chitosan nanostructures.

References

1. S.G. Kou, L.M. Peters, M.R. Mucalo, Chitosan: a review of sources and preparation methods. Int. J. Biol. Macromol. **1**(169), 85–94 (2021)
2. W. Wang, Q. Meng, Q. Li, J. Liu, M. Zhou, Z. Jin, K. Zhao, Chitosan derivatives and their application in biomedicine. Int. J. Mol. Sci. **21**(2), 487 (2020)
3. A. Muxika, A. Etxabide, J. Uranga, P. Guerrero, K. de la Caba, Chitosan as a bioactive polymer: processing, properties and applications. Int. J. Biol. Macromol. **105**(Pt 2), 1358–1368 (2017)
4. D. Heber, Herbal preparations for obesity: are they useful? Prim. Care **30**(2), 441–463 (2003)
5. R. Ylitalo, S. Lehtinen, E. Wuolijoki, P. Ylitalo, T. Lehtimäki, Cholesterol-lowering properties and safety of chitosan. Arzneimittelforschung. **52**(1), 1–7 (2002)
6. S. Chu, J. Wang, F. Gao, The application of Chitosan nanostructures in stomatology. Molecules **26**(20), 6315 (2021)
7. A.O. Silva, R.S. Cunha, D. Hotza, R.A.F. Machado, Chitosan as a matrix of nanocomposites: a review on nanostructures, processes, properties, and applications. Carbohydr. Polym. **15**(272), 118472 (2021)
8. G. Kravanja, M. Primožič, Ž Knez, M. Leitgeb, Chitosan-based (Nano)materials for novel biomedical applications. Molecules **24**(10), 1960 (2019)
9. D.R. Perinelli, L. Fagioli, R. Campana, J.K.W. Lam, W. Baffone, G.F. Palmieri, L. Casettari, G. Bonacucina, Chitosan-based nanosystems and their exploited antimicrobial activity. Eur. J. Pharm. Sci. **30**(117), 8–20 (2018)
10. S. Murugesan, T. Scheibel, Chitosan-based nanocomposites for medical applications. J. Polymer Sci. **59**(15), 1610–1642 (2021)
11. V. Grumezescu, O. Gherasim, I. Negut, S. Banita, A.M. Holban, P. Florian, M. Icriverzi, G. Socol, Nanomagnetite-embedded PLGA spheres for multipurpose medical applications. Materials (Basel) **12**(16), 2521 (2019)
12. L. Midya, A.S. Patra, C. Banerjee, A.B. Panda, S. Pal, Novel nanocomposite derived from ZnO/CdS QDs embedded crosslinked chitosan: an efficient photocatalyst and effective antibacterial agent. J. Hazard Mater. **5**(369), 398–407 (2019)
13. M.E. Abd El-Hack, M.T. El-Saadony, M.E. Shafi, N.M. Zabermawi, M. Arif, G.E. Batiha, A.F. Khafaga, Y.M. Abd El-Hakim, A.A. Al-Sagheer, Antimicrobial and antioxidant properties of chitosan and its derivatives and their applications: a review. Int. J. Biol. Macromol. **1**(164), 2726–2744 (2020)
14. M. Glavas-Dodov, E. Fredro-Kumbaradzi, K. Goracinova, S. Calis, M. Simonoska, A.A. Hincal, 5-Fluorouracil in topical liposome gels for anticancer treatment–formulation and evaluation. Acta Pharm. **53**(4), 241–250 (2003)

15. S. Li, L. Xiao, H. Deng, X. Shi, Q. Cao, Remote controlled drug release from multi-functional Fe3O4/GO/Chitosan microspheres fabricated by an electrospray method. Colloids Surf B Biointerf. **1**(151), 354–362 (2017)
16. T. Rossi, V. Iannuccelli, G. Coppi, E. Bruni, G. Baggio, Role of the pharmaceutical excipients in the tamoxifen activity on MCF-7 and vero cell cultures. Anticancer Res. **29**(11), 4529–4533 (2009)
17. V. Gupta, G.H. Mun, B. Choi, A. Aseh, L. Mildred, A. Patel, Q. Zhang, J.E. Price, D. Chang, G. Robb, A.B. Mathur, Repair and reconstruction of a resected tumor defect using a composite of tissue flap-nanotherapeutic-silk fibroin and chitosan scaffold. Ann. Biomed. Eng. **39**(9), 2374–2387 (2011)
18. M. Chen, D.Q. Le, S. Hein, P. Li, J.V. Nygaard, M. Kassem, J. Kjems, F. Besenbacher, C. Bünger, Fabrication and characterization of a rapid prototyped tissue engineering scaffold with embedded multicomponent matrix for controlled drug release. Int. J. Nanomed. **7**, 4285–4297 (2012)
19. L.Y. Xia, X. Zhang, M. Cao, Z. Chen, F.G. Wu, Enhanced fluorescence emission and singlet oxygen generation of photosensitizers embedded in injectable hydrogels for imaging-guided photodynamic cancer therapy. Biomacromolecules **18**(10), 3073–3081 (2017)
20. C.K. Lim, J. Shin, I.C. Kwon, S.Y. Jeong, S. Kim, Iodinated photosensitizing chitosan: self-assembly into tumor-homing nanoparticles with enhanced singlet oxygen generation. Bioconjug. Chem. **23**(5), 1022–1028 (2012)
21. J. Liao, Y. Jia, L. Chen, L. Zhou, Q. Li, Z. Qian, D. Niu, Y. Li, P. Li, Magnetic/gold core-shell hybrid particles for targeting and imaging-guided photothermal cancer therapy. J. Biomed. Nanotechnol. **15**(10), 2072–2089 (2019)
22. F. Fernández-Álvarez, C. Caro, G. García-García, M.L. García-Martín, J.L. Arias, Engineering of stealth (maghemite/PLGA)/chitosan (core/shell)/shell nanocomposites with potential applications for combined MRI and hyperthermia against cancer. J. Mater. Chem. B. **9**(24), 4963–4980 (2021)
23. A. Ghoshal, U. Goswami, A.K. Sahoo, A. Chattopadhyay, S.S. Ghosh, Targeting Wnt canonical signaling by recombinant sFRP1 bound luminescent au-nanocluster embedded nanoparticles in cancer theranostics. ACS Biomater. Sci. Eng. **1**(12), 1256–1266 (2015)
24. D. Dutta, A.K. Sahoo, A. Chattopadhyay, S.S. Ghosh, Bimetallic silver nanoparticle-gold nanocluster embedded composite nanoparticles for cancer theranostics. J. Mater. Chem. B **4**(4), 793–800 (2016)
25. A. Sachdev, I. Matai, P. Gopinath, Carbon dots incorporated polymeric hydrogels as multifunctional platform for imaging and induction of apoptosis in lung cancer cells. Colloids Surf B Biointerf. **1**(141), 242–252 (2016)
26. X. Guo, X. Zhang, L. Ye, Y. Zhang, R. Ding, Y. Hao, Y. Zhao, Z. Zhang, Y. Zhang, Inhalable microspheres embedding chitosan-coated PLGA nanoparticles for 2-methoxyestradiol. J. Drug Target **22**(5), 421–427 (2014)
27. C. Kutlu, A.S. Çakmak, M. Gümüşderelioğlu, Double-effective chitosan scaffold-PLGA nanoparticle system for brain tumour therapy: in vitro study. J. Microencapsul. **31**(7), 700–707 (2014). https://doi.org/10.3109/02652048.2014.913727
28. M. Shamsipour, A.M. Mansouri, P. Moradipour, Temozolomide conjugated carbon quantum dots embedded in core/shell nanofibers prepared by coaxial electrospinning as an implantable delivery system for cell imaging and sustained drug release. AAPS PharmSciTech **20**(7), 259 (2019)
29. S. Ullah, A. Nawaz, A. Farid, M.S. Latif, M. Fareed, S. Ghazanfar, C.M. Galanakis, A.S. Alamri, M. Alhomrani, S.M.B. Asdaq, Folate-modified chitosan 5-flourouraci nanoparticles-embedded calcium alginate beads for colon targeted delivery. Pharmaceutics **14**(7), 1366 (2022)
30. S. Vijayakumar, J. Chen, V. Kalaiselvi, K. Tungare, M. Bhori, Z.I. González-Sánchez, E.F. Durán-Lara, Marine polysaccharide laminarin embedded ZnO nanoparticles and their based chitosan capped ZnO nanocomposites: synthesis, characterization and in vitro and in vivo toxicity assessment. Environ. Res. **213**, 113655 (2022)

31. H. Hosseinzadeh, F. Atyabi, R. Dinarvand, S.N. Ostad, Chitosan-Pluronic nanoparticles as oral delivery of anticancer gemcitabine: preparation and in vitro study. Int. J. Nanomed. **7**, 1851–1863 (2012)
32. Y. Cao, S. Liu, Y. Ma, L. Ma, M. Zu, J. Sun, F. Dai, L. Duan, B. Xiao, Oral nanomotor-enabled mucus traverse and tumor penetration for targeted chemo-sono-immunotherapy against colon cancer. Small. e2203466 (2022). https://doi.org/10.1002/smll.202203466. Online ahead of print
33. S. Chandran, G. Praveen, K.S. Snima, S.V. Nair, K. Pavithran, K. Chennazhi, V.K. Lakshmanan, Potential use of drug loaded nano composite pectin scaffolds for the treatment of ovarian cancer. Curr Drug Deliv. **10**(3), 326–335 (2013)
34. X. Wang, H.S. Almoallim, Q. Cui, S.A. Alharbi, H. Yang, In situ decorated Au NPs on chitosan-encapsulated Fe3O4-NH2 NPs as magnetic nanocomposite: investigation of its anti-colon carcinoma, anti-gastric cancer and anti-pancreatic cancer. Int. J. Biol. Macromol. **28**(171), 198–207 (2021)
35. Z. Chang, J. Feng, X. Zheng, A highly sensitive fluorescence sensor based on lucigenin/chitosan/SiO2 composite nanoparticles for microRNA detection using magnetic separation. Luminescence **35**(6), 835–844 (2020)
36. M. Rizwan, R. Yahya, A. Hassan, M. Yar, A.A. Abd Halim, A. Rageh Al-Maleki, L. Shahzadi, W. Zubairi, Novel chitosan derivative based composite scaffolds with enhanced angiogenesis; potential candidates for healing chronic non-healing wounds. J. Mater. Sci. Mater. Med. **30**(6), 72 (2019)
37. Y. Cai, B. Karmakar, M.A. Salem, A.Y. Alzahrani, M.Z. Bani-Fwaz, A.A.A. Oyouni, O. Al-Amer, G.E. Batiha, Ag NPs supported chitosan-agarose modified Fe3O4 nanocomposite catalyzed synthesis of indazolo[2,1-b]phthalazines and anticancer studies against liver and lung cancer cells. Int. J. Biol. Macromol. **31**(208), 20–28 (2022)
38. Q. Pan, Y. Lv, G.R. Williams, L. Tao, H. Yang, H. Li, L. Zhu, Lactobionic acid and carboxymethyl chitosan functionalized graphene oxide nanocomposites as targeted anticancer drug delivery systems. Carbohydr. Polym. **20**(151), 812–820 (2016)
39. R. Huang, X. Zhou, X. Liu, Q. Zhang, H. Jin, X. Shi, W. Luo, H. Deng, Nanofibrous mats layer-by-layer assembled by HTCC/layered silicate composites with in vitro antitumor activity against SMMC-7721 cells. J. Biomed Nanotechnol. **10**(3), 485–499 (2014)
40. A.Y. Elderdery, B. Alzahrani, S.M.A. Hamza, G. Mostafa-Hedeab, P.L. Mok, S.K. Subbiah, Synthesis of Zinc Oxide (ZnO)-Titanium Dioxide (TiO2)-chitosan-farnesol nanocomposites and assessment of their anticancer potential in human leukemic MOLT-4 Cell Line. Bioinorg. Chem. Appl. **28**(2022), 5949086 (2022)

Chapter 14
Chitosan-Based Nano Biomaterials and Their Applications in Dentistry

Biswajit Parhi, Debasrita Bharatiya, Shuvendu S. Purohit, and Sarat Kumar Swain

Abstract Herein, a detailed study of the development of chitosan (CS) based nanobiomaterials (NBMs) and their applications in focusing on dentistry purposes have been discussed elaborately. CS-based NBMs are multipurpose components with assured catalytic, physicochemical, biological and smart properties. These nanomaterials (NMs) have captivated their value, especially in the biomedical domain including bone tissue engineering, wound healing, biosensors, and gene delivery. The measurement of their physicochemical properties at nanoscale has been done by extensive methods such as XRD, FTIR, ZETA, FESEM, and TEM. The stimuli-responsive activity, multifunctional behaviour, morphological variance, oxygen-abundant, stable, easy synthetic methods, cost-effectiveness, spatial and temporal control, and flexible functionalization of CS offer such NBMs extensive applicabilities. The other precursors especially metal oxide and amino acid-based polymers when combined with CS NBMs develop other properties such as limited toxicity level, high mechanical strength, low breakdown strength, and high stability and durable properties which supports such materials to be used in the field of dentistry, such as to prevent dental caries, carcinogenic effect in teeth, and reduction of tooth decay, root infection and gum diseases. Further, the extensive application of such materials has been witnessed in different fields with appropriate case studies.

14.1 Introduction

The service of nanotechnology is touching every aspect of the current world due to its surface and quantum size effect [1]. The extremely small sizes of NMs and NBMs cause the positioning of a high percentage of surface atoms to yield better properties and reactivity to bulk materials [2]. The benefits of these materials are significantly seen in its usage in medical equipment, energy, storage devices, drug development,

B. Parhi · D. Bharatiya · S. S. Purohit · S. K. Swain (✉)
Department of Chemistry, Veer Surendra Sai University of Technology, Burla, Sambalpur 768018, Odisha, India
e-mail: skswain_chem@vssut.ac.in

S. K. Swain and A. Biswal (eds.), *Chitosan Nanocomposites*, Biological and Medical Physics, Biomedical Engineering,
https://doi.org/10.1007/978-981-19-9646-7_14

information technology, and sensors [3]. The primary dominance of such materials is governed by the presence of graphene oxide (GO), carbon nanotubes (CNT), metal, metal oxides (MOs), hydroxides nanoparticles ((OH)NPs), quantum dots, inorganic and small organic molecules, polymers, biopolymers including CS. The condensation, sonication, chemical precipitation, pyrolysis, ion implantation, hydrothermal method, and sol-gel methods are mainly responsible for the development of NBMs [4]. The size dependency effect modified the chemical, optical, mechanical, electrical and other physical-chemical properties to show better performance of NBMs compared to previously reported bulk materials [5, 6]. The superior cytocompatibility property exhibited by NMs is one of the important reason behind their successful in pharmacological activity [7].

The nonviable materials targeting their interaction with biological medium to evaluate, treat, and replace any tissue, organ or function of the body are called biomaterials. These are very promising to the living world and are evolving in nature. The requirement and uses of such materials are permanent and highly recommended. There are many biomaterials which are not clinically approved and the detailed research on their properties is going on. Biomaterials such as wood, and bacterial biofilms are living cells with the ability to be procured from different sources with modification in the original structure, composition and performance at several environmental factors such as pH, mechanical stress, salinity, and temperature. Natural synthesized materials such as wood, silk, bone, and cellulose have participated in the development of such biomaterials as they possess biological properties. Contrary to natural synthesized material, synthetic materials are also available as biomaterials such as bacteria, fungi, algae and animal cells incorporated into polymer or hydrogels, which are otherwise called engineered living material [8]. The close correlation between biomaterials and their biological responses is evident from cell adhesion, cell shape, proliferation, elongation, spreading, direction, migration, surface modification and stiffness [9]. Biomaterials inferred with metallic NPs and inorganic biomaterials are utilized in biomedical devices due to their corrosion-resistant, durable, biocompatible and multifunctional properties [10]. Controlled drug delivery, immunological responses, cell modulation, RNA delivery, gene modifications, and regenerative medicines are the main target study of current researchers using biomaterials for the advancement of biological study [11, 12].

In the literature, the involvement of many NBMs such as metal-based, piezoelectric metals/polymers, and small organic molecule based have been established with their efficacy in biomedical applications including dentistry purposes. Ultra-small size, high surface area of nano-sized materials is the main reason behind their high efficacy towards several biomedical applications. Superparamagnetic iron oxide (Fe_3O_4), Ag, Au NPs have shown intrinsic cytotoxicity, great therapeutic and diagnostic properties, proving themselves as a promising candidate as NBMs [13]. Inorganic-originated piezoelectric materials such as ZnO, boron nitride (BN), hydroxyapatite (HA), lithium niobate (LN), lithium sodium potassium niobate (LNKN), lead zirconate titanate (PZT) and organically modified piezopolymers such as poly(l-lactic acid) (PLLA), poly(vinylidene fluoride) (PVDF) possess multifunctionality, high piezoelectric coefficient and improved mechanical properties, which enables these NBMs for pharmaceutical, stem cell, biomedicine, tissue regeneration research and developments [14–16]. The charge capacity, permeability control, fouling resistance and microbial resistance of metal NPs are the important factors to act against microbial growth [17]. Crosslinking is another fascinating property observed by nano biopolymers with a convenient path to study the mechanical response and their use in implants [18]. The advancement of NMs based on the compositions also played an important role in regulating complex biological responses in different environmental conditions focusing few parameters such as tailored surface solubility, and high volume/surface ratio [19, 20]. NBMs are popular in current days for dental purposes owing to their wear-resistant, increased hardness, enhanced polished resin, inhibited shrinkage and stain and superior flexural strength.

There are several NBMs available in the literature discussing their biological responses to different stimuli conditions. Among them, polymer-based NBMs are one of the most promising candidates for biomedical applications including dentistry. Meanwhile, cost-effective, solid polysaccharide CS is a type of polymer which shows great efficacy in the biomedical and pharmaceutical industry with its bio-renewable, stable, biocompatibility, multifunctional, low toxic environment, controlled drug release kinetics, self-assembled and micellar association behaviour [5]. This is one of the most bountiful precursors extracted from different sources such as crabs, shrimps, prawns, and shellfish. The solubility of CS occurs in water and other organic solvents depending on the functional groups present. The acidic condition promotes the dissolution of CS in water following the amino protonation causing a gelation medium. The presence of active and reactive -NH_2 groups it made convenient way for pharmaceutical use by promoting side group attachment. The physical and chemical modification processes such as blending, radiation, grafting, photochemical, chemical etc. also enhanced the structural, chemical, mechanical and biological strengths of CS-modified materials [21]. The versatility of CS deals with the antimicrobial effect involving the ion-surface interaction following cell-wall breakage, entry of CS NPs to inside microorganism nuclei, inhibiting protein and mRNA synthesis via the penetration of CS into the nuclei of the microorganisms and stopping the entry of essential nutrients to microbial growth. The versatility developed mainly due to the presence of the acetyl group in it and the molecular weight (MW) of the CS. The low degree of acetylation (DA) and MW helps in the reduction of microorganism growth and multiplication [22]. Being a basic polysaccharide, this can attribute to the formation of chelates, biofilms and polyoxysalts improving its ability for stable interaction with metal-metal oxide NPs, and other natural polymers. Even the structural modification of CS and its derivatives played an important role in biological activity with the presence of -OH, -NH_2, quaternary ammoniumyl, carboxyalkyl, hydroxyalkyl, thiol and hydrophobic groups, such as long alkyl chains, and substituted aromatic rings. Being a hydrophobic polymer, the solubility of CS and its derivatives are limited to water solutions. To avoid such instances, the requirement of surface modification came into the picture [23]. CS and its derivatives have shown tremendous use in medical aids, wound healing, textile, membranes, metal ion collection, molecular sieves, catalysts and photography [24]. To promote sustainable agriculture and the protection of plants from microorganism attacks, the CS-based material played a vital role in the defence mechanism caused by the functional groups [25]. Even the mucoadhesive properties of such materials also enhance the gene delivery, drug loading and releasing mechanism in the solution medium and other pharmacological tasks [26]. The advanced properties of CS and its derivates give a clear impression of its usage in different areas of the biomedical field. The extraordinary and unique properties such as viscosity, solubility in various media, mucoadhesive, polyoxysalt production, and film formation of CS also motivated us to discuss its related materials for application in dentistry.

There has been a lot of work going on with the development of inorganic@CS, small organic molecule@CS, bioplolymer@CS, and inorganic polymer@CS-based nanocomposites in the last decade. These works have given a clear direction regarding the application of such NBMs in different biomedical applications. Inorganic metal oxides such as ZnO, CuO, Fe_2O_3, Ag_2O, and MnO_2 have been used in broader paths due to their stability, inertness, and high durability [27]. Even the MOs in the nano range develops superior property as compared to the bulk materials. The primary properties such as biocompatibility, stability, solubility, biodegradability, and durability are the fascinating key points of CS-based NBMs which made the scientists explore their application in various areas. The inorganic/MO@CS NBMs are quite impressive due to their interesting inherited properties such as morphology, nano size, and composition [28]. The introduction of the organic molecule to inorganic nanostructure has tremendously affected the well-defined structure and muti-defined functions. The soft organic filler inhibits the aggregation properties and allows the creation of interesting morphologies, which allows the nanocomposite to show better properties [29]. The DNA-templated polymer/inorganic nanocomposite has actively participated in biomedical applications [30]. The convenient way of the utilization of such nanocomposites has influenced its superior behaviour when conferred with CS due to the smart functional attitude. The high surface area, more porosity, better mechanical and tensile strength and multifunctional properties have increased the use of such nano-biomaterials. Bandara et al. have reported on many factors responsible for CS's biomedical and therapeutic response [20, 31]. These materials have shown great efficacy in agricultural and biomedical fields by managing abiotic stress in plants, inhibiting pathogens, and water availability for crops. The team also discussed the influence of morphological-biased NBMs against cancer cell lines. They observed that morphological variable CS@Graphite, CS@Ag nanotriangles, and CS@Au nanorods NBMs have shown activity against CNE human nasopharyngeal, NCI-H460 human non-small lung cancer cells, MDA-MB-231 human breast cancer cells respectively. Even CS-based NMs seem to be used for drug delivery therapy influenced by features such as allowing the formation of new reaction sites, biocompatibility, good absorption of drugs at specific positions and protecting unstable drugs against strong acids like hydrochloric acid and blood flow in living cell line [32, 33]. It is well understood from previous reports that the organic molecule and biobased polymers seem to represent excellent biological responses to different conditions like CS-based materials [34–37].

Perfect oral health is a necessity for each human being for the maintenance of a healthy life. The lack of oral hygiene, unhealthy diet, polluted water caused by

Fig. 14.1 Illustrates several diseases caused by dental infection

industries, and excess fluorides in water are the important causes of dental problems. Common dental problems are root infections, tooth decay, gum diseases, bad breath, sensitive teeth, cracked bone, receding gums, teeth grinding and dry mouth. Figure 14.1 depicts several problems caused by infected teeth. Root infection, tooth decay cavity and gum diseases are mainly caused by bacteria growth in the form of a film on the teeth surface. The acid production from food sugar is caused by them and damages the enamel or tooth's outer surface, this is the most common dental problem that dentists see in patients. Sensitive teeth can be caused by gum infection, enamel corrosion, cracked bone, and crown loss. Cracked bone is occurred by injury after accidents, hard-food chewing, or teeth grinding during sleep. Receding gums, mainly found in women due to hormonal imbalance, and it depends on other factors such as smoking, high blood pressure, and poor oral hygiene. A dry mouth is generally observed in old people due to less intake of the required amount of water which is also caused by nerve damage, salivary gland disease, medication intake, and cancer treatments. The current nanotechnology made an easier path to solve these issues. The path of resolving dental problems via the use of nanotechnology is called 'Nanodentistry' [38]. These therapeutic and diagnostic tools are beneficiated by the addition of biopolymer, natural polymer, small organic molecule, and inorganic metal. Enamel, dentin pulp, cementum, and periodontal ligament are the major parts of the tooth. The devices designed using CS-based NBMs pose as effective solution to many dentistry problems. As the implants are in contact with both hard and soft tissues, surface chemistry plays an important role to solve the issue. CS-based NBMs are rich in surface functionalization which plays an extraordinary role to resolve such issues by the virtue of antimicrobial, mucoadhesive and bioactive properties of CS particles [39–41].

The size related toxicity is the primary challenge faced by the CS-based NBMs. But this can be resolved by surface and size modification which may prevent accumulation of nanoparticles in the GI tract and intestine. Another important challenge is tooth decay and bad odour. In order to overcome this, oral drug delivery is the most convenient way. But the oral drug uptake is much unstable and gets accumulated in the various body parts and very less permeable. But CS-based NBMs help to open the intercellular tight junctions of epithelium and has excellent mucoadhesive property, allowing the unwanted drug molecule to pass through excreta. The CS/polylactide-coglycolide acid (CS/PLGA) NPs showed better properties as compared to PLGA NPs within the range of 25.0–100.0 μg/ml [42]. CS/biopolymer-based NBMs challenged many drawbacks in dentistry by improving the release kinetics of drugs in cell lines, and increasing drug uptake making these materials more available for treating oral defects.

To overcome a few challenges in CS-based nanomaterial, scientific progress takes a diversion towards the development of CS-based NBMs to fulfil the need of dentistry in an economical and easy pathway. The major components of such NBMs are CS/polymer, CS/metal, CS/metal oxides, CS/small organic molecule, CS/inorganic adducts and CS/ceramics-glass. The CS/Polymer mostly constituted of biopolymer, amino acid-based polymers and organic-inorganic polymers. CS/metal-based NMs are CS/Cu, CS/Zn CS/Au NPs etc. show effective oral drug delivery kinetics. CS/metal oxide involves ZnO and Ag_2O for inhibition of microbial growth in the mouth which lightens oral decay. CS/small organic molecule resembles the structure of CS/quarternary ammonium polyethyleneimine (QAS-PEI) incorporated in restorative material. This induces the formation of covalent bonds between the polymer and the resins of antibacterial agent. CS/inorganic-based NMs are the most promising category of NBMs. The inertness, durability, stability, low toxicity, and possession of essential dental nutrients attracted many scientists to use it for dentistry applications. Sharan et al. have discussed CS/ceramics-glass and CS/inorganic adducts for application in oral and maxillofacial surgery and restorative dentistry predicting hard tissue generation [43]. CS/heparin showed a significant oral drug delivery capacity for dental infection [44]. The hydrogel of CS/nano ZnO composite inhibits bacterial growth by promoting blood clotting also has applications in dental surgery [45]. Murali et al. have discussed the integration of (±)-2-aminopimelic acid, 5-aminovaleric acid amino acid and DL-2-aminocaprylic acid into CS/polyglycolic acid scaffold, which showed enhanced osteoblast biocompatibility indicating its use in dentistry [46]. A detailed study on types of CS-based NBMs has been discussed in Sect. 14.3. The CS/$CaPO_4$, CS/CaI_2, CS/HA, CS/floruohydroxy apeteite are dominant in the dentistry field.

The multifunctional properties, mechanically strengthened attitude, high longevity, good sensitivity, low toxicity and inertness are the important factors possessed by CS-based NBMs which play a big role in dentistry applications. Thus, the present work discusses the development of versatile chitosan-based NBMs, how to overcome these challenges and their efficacy in various dentistry issues. The current progress in the prevention of oral diseases, diagnostic, and therapeutic methods that could hold significant results in dentistry is the main focus of this work. Even the detailed problem with preventive measures for several oral health issues is discussed and resolved by different CS-based NBMs. We tried to understand the requirement of these materials for each problem of dentistry with many view angles depending on chemical interaction, adhesive property, stability, toxicity range, durability, cost-effectiveness and inert behaviour.

14.2 Characterization

To determine the physicochemical behaviour of CS-based NBMs for application in various fields, extensive techniques such as XRD, FTIR, HRTEM, TGA, UV, and mechanical actions are studied at different scientific platforms. Scientific experiments are required to analyse the versatility of CS-based NBMs and their importance towards dentistry.

X-ray diffraction (XRD) is a non-destructive technique to analyse the crystallinity of CS-based NBMs and to check d-spacing and planes involved during the experiment. It is reported that CS NPs shows amorphous property due to the dense network formation inside the polymer whereas the insertion of metal NPs in it increases its crystallinity by breaking the network [47]. Choudhary et al. have designed CS/Cu NBMs and studied its diffraction pattern at $2\theta = 10°–60°$ with a Cu $K\alpha = 1.5418$ Å. The CS NPs are amorphous in nature which is caused by the polymeric substitution among themselves. But the CS/Cu nanomaterial shows prominent peak at (2θ = 19.5 and 21.0) confirming the attachment of Cu to the functional groups of CS NPs invading microbe growth [48].

Fourier Transform Infrared (FTIR) is the most fundamental technique to know the chemical composition and interaction available in the designed material. CS-based NBMs are multifunctional and studies need to know the chemical interactions occurring in the material and how those interactions are responsible for the development of dental implants and oral drugs. The network-like porous CS NPs show prominent peaks at 3422 cm^{-1}, 3310 cm^{-1}, 2923.56 cm^{-1} and 2859.92 cm^{-1} corresponding to –O-H, $–NH_2$, CH_3 and CH_2 in both chitin and chitosan. Other peaks identified at 2221 cm^{-1}, 1650 cm^{-1} and 1040 cm^{-1} are attributed to the presence of the C–N group of $C–NH_2$ and the stretching band of glycosidic linkage of C–O [47]. Anitha et al.

have reported CS/poly acrylonitrile (CS/PAN) showing the successful incorporation of polymer into CS NPs with the presence of C = N at 2220 cm^{-1} [49]. The team also suggested the roughness, good porosity and high surface area of CS/PAN polymer which helps in the enhancement of the adsorption of metal NPs and biocompatibility.

High-resolution transmission electron microscopy (HRTEM) and field emission scanning electron microscopy (FESEM) are important morphological analytical tools to know the detailed structure of CS-based nanomaterials. The surface morphology gives a clear idea about the surface porosity, texture and roughness of the materials for designing of effective dental implants, oral drugs and other dentistry aids. The small size and interesting shapes of CS-based NBMs can attain specific biomedical targets such as suitable drug carrier medium, target and controlled release medium. The CS NPs are non-uniform and spherical nanomaterials with diameters ranging from 20 to 100 nm [50]. But the incorporation of any other material into it leads to the formation of different morphology. This modified morphology and fabrication of CS-based NBMs are also responsible for different biomedical response [51, 52]. Iswanti et al. have discussed uniform spherical type CS/cytosine-phosphate guanine oligodeoxynucleotides (CS/CpG ODNs) which is confirmed by TEM analysis. These NBMs seem to show compatibility with murine macrophage cells and induce the secretion of inflammatory cytokine IL-6 and Th-1 type cytokine IFN-c in human PBMCs [53].

Zeta potential technique provides a clear interpretation of the surface potential and electrokinetic potential of the designed CS NBMs. It is essential to know the stability of CS-based NBMs at the colloidal stage. Iswanti et al. have also discussed the zeta potential of CS/CpG ODNs -11.3 ± 1.5 mV which is more compared to bare CS NPs of 3.3 ± 0.4 mV. The increment of zeta potential is due to the interaction between negatively charged CpG ODNs with positively charged CS NPs. The electrokinetic potential measurement of CS-based NBMs seems to possess good stability which can be beneficial for several biomedical purposes including dentistry.

14.3 Application of Different Types of CS-Based Nano Biomaterials for Dentistry

To fulfil the aim of this discussion, various CS-based NBMs and their impact on different areas of dentistry have been discussed. Figure 14.2 represents the classification of CS-based NBMs and their implications on the treatment of dental issues. The formation of different types of CS-based NMs involves the attachment of polymer, metal, metal oxide, organic molecules, inorganic adducts, ceramics-glass and mixed materials. These different CS-based NBMs have been incredibly used for tissue engineering, drug delivery, and targeting wound healing purposes depending on their adhesiveness, stability, rich abundance, nontoxicity and biodegradable properties.

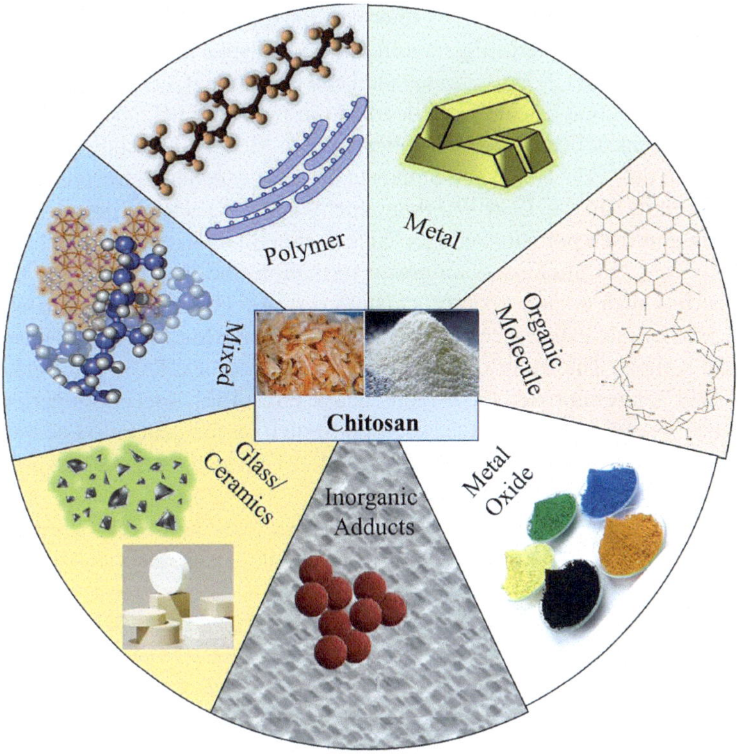

Fig. 14.2 Different types of chitosan-based nano-biomaterials based on composition

14.3.1 Chitosan/Polymer Nano Biomaterials

There are two types of polymers associated with the development of CS-based NBMs i.e., natural and synthetic polymers. Natural polymers such as collagens, and curcumin and synthetic polymers such as polyvinyl alcohol (PVA), polycaprolactone (PCL), polyethene oxide (PEO), PLA (poly lactic acid), and sodium alginate (SA) are the main candidates for formation of CS/polymer NBMs. The smart, stimuli-responsive, multifunctional, network, low toxicity, cost-effectiveness, and covalent and hydrogen bonding interaction, have promoted the importance of chitosan over other biopolymers derived from amino acids. The biocompatibility, repairability and aesthetic processability made these polymers more beneficial for the development of CS-based nanomaterial. These polymers are also limited by low fracture resistance, low ductility, and poor strength making it a little difficult for these polymers to be used in a dental implant. But the incorporation of CS NPs into such polymers improves its suitability in oral drug delivery and microbial cleansing activity. Bakopoulou et al. have developed biomimetic chitosan/gelatin (CS/Gel) scaffolds for tissue engineering of several tissues. It is also reported that the combination of CS/Gel scaffolds with dental pulp stem cells (DPSCs) improves the targeted alveolar/orofacial bone reconstruction [54]. An interesting nanofibrous electrospun nanocomposite made up of biomaterials such as CS, polyvinylalcohol and curcumin (Cur) with high porosity, hydrophilic distribution causing safe degradation in human pH, encapsulation efficiency, multi-drug delivery and controlled release kinetics has shown fascinating results in drug delivery. This nanocomposite also has excellent adsorption capacity for proteins on its surface. CS/PVA/glutaraldehyde (GTA) acts as a successful scaffold with an improved porosity and mechanical strength. It is reported that bare GTA has cytotoxicity, but after nanocomposite formation, the material shows remarkable properties such as non-toxicity andanti-corrosion effect [55]. Zhu et al. have reported another interesting electrospun fibre CS/PCL NBMs and studied its in vitro osteogenic activity. The study suggests that the addition of GTA to the metformin-loaded CS/PCL has improved the osteoinductive effect [56]. Successful periodontal treatment is observed and the efficacy is contributed by the generation of thermos-responsive CS/gelatin/β-glycerolphosphate [57]. An injectable periodontal treatment is preferred over conventional drug delivery.

Xu et al. have developed thermos-responsive hydrogel CS/gelatin/β-glycerolphosphate to treat inflammatory periodontitis. The fabricated hydrogel has exhibited controlled and faster release of aspirin and good regeneration function of erythropoietin (EPO) drugs. Both these drugs showed a remarkable releasing profile with no toxicity in both in vivo and in vitro for 3 weeks [58]. CS-based collagen [59] and metronidazole benzoate (MET) [60] are the main components which are used for dental reconstruction. These materials are also implemented to treat periodontal diseases in patients because of their mucoadhesive properties, effective drug delivery mechanism and low toxicity.

The effect of other factors such as concentration, volume, and MW play an important role in treating dental infections. Kilicarslan et al. have prepared and discussed clindamycin phosphate-loaded CS/alginate polyelectrolyte complex film. This targeted drug clindamycin phosphate is a suitable agent to treat periodontal infections depends on its mucoadhesiveness. The study elaborated the importance of drug-loaded percentage, swelling, morphology, the texture of designed material, focusing on the concentration, MW and the volume of the polymer. It is observed that good release kinetics depend on the high concentration and volume of alginate and low MW of CS particles [61]. The treatment of gingivitis requires targeted application by drug concentration raising, improved pharmacodynamic effects and prevention of side effects contradicting the conventional drug formulation [62]. To resolve these issues, Perugini et al. have designed a novel periodontal packet made up of ipriflavone loaded poly (d,l-lactide-co-glycolide) (PLGA)/CS nanocomposite for release of ipriflavone dosage into the periodontal pocket [63]. The development of microsphere and porosity have induced dentin formation in TGF-β1 loaded CS NPs and became suitable for dog models for the treatment of odontoblast cells [64]. The formulated microsphere having average size of 74–461 nm is used for encapsulation ofCS/calcium alginate (CS-Ca-SA) to enhance the particle entrapment efficiency and drug release [65].

The insertion of NPs in CS/polymer NBM inspired the desired material to resolve many dental issues. The CS/PLGA/Ag NPs have been extensively used for the development of materials for periodontal tissue engineering which allows the regeneration of new tissues in the teeth area. This material also helped in decrease in the recurrence of periodontal diseases. The biocompatibility, inertness, antibacterial property, non-reactiveness of metal NPs incited their proliferation and mineralization of periodontal membranes [66]. Saboktakin et al. have designed MET NPs incorporated in thiolated chitosan (TCS)/poly (methacrylic acid) (PMAA). This material ensures the sustained and target drug release at absorption sites [67]. Extensive work was performed by Li et al. with the development of PLGA/CS with a minimum inhibitory concentration (MIC) of 0.0156 mg/ml to treat peri-implantitis and other oral diseases related to implants [68]. The fabricated CS-based NBM has been used as tissue engineering scaffolds, dental implant purposes for the alveolar bone to periodontal complex healing. It also shows long pulp-friendly restorative properties, pulp regeneration, high surface area for drug reactivity antibacterial activities, and controlled drug release path [69]. To break the current tooth carcinoma trend, Farias et al. designed a smart, safe, natural mouthwash based on CS/peppermint (Mentha Piperita) essential oil (PEO). The designed NBM was studied against biosurfactants produced by Pseudomonas aeruginosa UCP 0992 (PB), Bacillus cereus UCP 1615 (BB) and Candida bombicola URM 3718 (CB) in carcinoma cell lines mouse fibroblast L929 and mouse macrophage RAW264.7. The toxicity effect of the designed mouthwash showed comparatively lower value than the commercially available mouthwash in the market [70]. The loading of protein drugs such as bone morphogenetic protein-7 (BMP-7), and ornidazole (ORN) have shown great efficacy in the treatment of gingivitis, dental caries and periodontal diseases [71]. The biodegradable, natural, safe, cost-effective, controlled drug release kinetics, biocompatible, less toxic, mucoadhesive properties of CS/polymer nano biomaterial instigated their importance for prolonged drug release and the formation of periodontal pockets to manage the infectious disease such as dental caries, gingivitis, oral infections.

14.3.2 Chitosan/Metal Nano Biomaterials

Metals NPs are in high demand for dental implantation accessories due to their improved mechanical action, structural reformative, small size, high surface area, anti-corrosive attitude, high specific strength, elastic modulus and biocompatibility. These nano-sized particles have shown great efficacy for antimicrobial, and anti-inflammatory properties by enhancing their reactiveness in intracellular processes. The preparation of dental implants is mostly dominated by metallic alloys such as titanium (Ti), cobalt-chromium (Co-Cr), cobalt-chromium-molybdenum (Co-Cr-Mo), cobalt-chromium-tungsten (Co-Cr-W), stainless steel and gold (Au). Though this category is bountiful for utilization in designing of dental implants but the implementation in real life application is limited due to allergies, and skin-mucous membrane inflammation. To overcome these issues, the involvement of CSNPs along with metallic NPs has improved biological integration, surrounding the site of oral infections. Ag and Au NPs are specially used for prosthetic, restorative, endodontic, orthodontic and osteogenic treatments with the incorporation of polymers. The production of reactive oxygen species (ROS), increment of cell permeability and hindrance to the DNA replications are the primary processes of the generation of metal ions during dental treatments [72].

Carvahlo et al. have designed fluoride-based stannous and CS (F/Sn/CS) toothpaste for the treatment of enamel erosion and abrasion at their initial stage. The incorporation of Sn NPs caused the formation of amorphous deposits on the enamel surface and eroded enamel with the instigation of protection activity. The layered-like structure of Sn/CS seems to possess acid-resistant property which helps in protection against dental enamel deterioration under harsh conditions. The chances of increment in the protection capability of this nanomaterial increases with increase in the interaction of Sn with the salivary proteins, affording a better preventive effect against erosion–abrasion. The antibacterial effect, induced inactivation of certain physiological functions, high affinity towards the negatively charged groups of biological molecules, structural modification in the cell membrane and high surface area of Ag NPs became made this NPs popular in consumer products. A promising biofilm agent of lactose-substituted CS/Ag nanoparticles has shown great efficacy towards antibacterial activity against *Streptococcus mitis*, *Streptococcus mutans*, and *Streptococcus oralis* planktonic phase and biofilm growth mode as well as on saliva samples with the hindrance to the plaque formation [73]. The strong bonding formation between Ag and CS and the high inflammatory property of Ag NPs are responsible for the production of its related materials for the treatment of dental caries. Zhu et al. have designed bioactive glass (BG) based Ag/CS NBM. Ag-BG/CS has inhibited the growth of odontoblastic cells in the dental area with the suppression of phosphorylation and translocation mechanisms [74]. CS/Au and CS/Ag nanocomposites are also mainly used for wound healing, nano vehicles, and adhesive coatings purposes.

The chelating effect of CS spread its direction to metal, metal ions and metal oxides to strengthen the materials without disturbing their shear bond strength. Also, the CS addition amount left a strong impression of being a suitable material nanocomposite for better material against bacterial infections as it acts as a reducing and capping agent in presence of metal NPs. Sharifianjazi et al. have discussed the preparation of Co NPs/CS nanomaterial and its activity on a bare 316L stainless steel alloy (316L SS) by electrodeposition technique and concluded that the developed material exhited improved mechanical property. This acts as a very good dental implant [75]. The protein-based biomimetic mineralization is also another important strategy for the treatment of dental diseases by the metal NP incorporated CS nanomaterials due to the facile, environmentally benign and bio-inspired processes [76]. The green, biologically inspired, facile, environment-friendly, cheap, presence of binding sites, reducing and capping sites of CS/metal nanomaterials dragged its attention towards the inhibition of dental infections.

14.3.3 Chitosan/Metal Oxides Nano Biomaterials

The multifunctional, stable, morphologically variant, inert and mechanically strengthened metal oxides have shown multiple applicabilities in different medical fields including dentistry. Metal oxides are the most stable material in harsh conditions. Dipolar oxygen plays an important role in the easy binding mechanism with other precursors for the development of innovative materials. It has a strong effect on healing of tooth defects with the reduction of plaque formation inside the oral cavity. ROS is one of the important mechanism for healing of dental infections. Metal oxides such as TiO_2, Ag_2O and ZnO-embedded CS show selective toxicity to bacteria with minimal effects on human cells. The metal ions penetrate into the cell medium and reduces the chances of biofilm formation [72].

Host-guest interaction is another reason why the CS-based metal oxide NMs show promising results in dental applications. Here, CS acts as the cavity-like host where the metal oxides are invited as a guest to the cavity. At a proximate and optimized condition, the metal oxide and metal ions are released into the solution medium to exhibit their original biological activity. Being a multifunctional polymer CS has more affinity towards interaction with the MOs. The host-guest mechanism is the pillar of this binding interaction [76]. The chemical interactions of metal oxides with CS enhance the possible toxicological effects. It is found that the toxicological effects are driven by the content of metal NPs in the nanomaterial, aggregation, and crystallinity but are independent of size and shape [77, 78]. This theory has a synergistic effect on the MW of the nanomaterial which develops the biological response to solve dental issues. The aggregation and crystalline CS/metal oxide nanomaterials have shown less impact on biological behaviour as compared to free/amorphous nanomaterials. It is reported that CuO/CS NBMs reduced toxicity and ROS production in A549 cells and have shown prolonged, sustained and controlled drug release kinetics [79]. The advantage of such an extensive result lies in the covalent-coordinate bond formation by the metal of variant morphology-based ZnO, ZrO_2 and functional groups of CS NPs [80, 81].

Factors such as multifunctionality, cavity-like host, MW, ability to accept metal and metal oxide NPs such as ZnO, Ag_2O, TiO_2 etc. make CS derivated biomaterials reactive antimicrobial species following ROS and host-guest mechanism. But many limitations lie with the development and application of CS-based metal oxides NBMs such as aggregation [82], solubility, low antimicrobial-antibacterial activity and less bonding interaction, hence the choosing of suitable metal oxides becomes important. To avoid these instances, the of the addition of inorganic metal oxides such as CaO, Na_2O, K_2O, MgO and d block metal oxides such as ZnO, CuO and AgO can incite interesting biological responses [83].

14.3.4 Chitosan/Organic Molecules Nano Biomaterials

The biocompatiblity, easily solublity, arranged structure, rich in functional groups are the few essential features of the organic biomolecules. These, when combined with CS, impart great attention towards applicability in dentistry. Polyphenolic curcumin is one of the best examples of organic molecule-embedded CS NBMs developed to reduce dental infections. Mustafa et al. have designed curcuminoid/CS mouthwash to treat oral mucosa under denture or denture stomatitis [84]. Zhao et al. have developed an injectable antibacterial conductive material made up of carbon nanotube (CNT)/glycidyl methacrylate/CS-based cryogel for haemorrhage and wound healing applications. Such types of NBMs could act as good medicinal gels for root canal infection [85]. The ease of delivery of the encapsulated drugs, binding site retention and controlled release of the drugs ensures a longer stay in the oral cavity to impact activity against the periodontal pathogen. The incorporation of the organic molecule chlorhexidine gluconate (Chx) in CS NPs has enhanced the bioadhesive property and antimicrobial activity of the overall material [86]. Thomas et al. designed an injectable hydrogel prepared from oxidized hyaluronic acid (HDA) embedded CS NPs and studied the elasticity or stiffness of the NBMs. With the increment of HAD/CS nanomaterials, the stiffness increased from lower to the highest value of 130.78 ± 19.83 kPa to 181.47 ± 19.77 kPa respectively with a decrement in gelling time which maintained the phenotypic integrity [87]. Sukpaita et al. have designed a novel CS-based organic molecule NBMs of CS/dicarboxylic acid (CS/DA) and its usefulness for bone tissue engineering material. This material can be used as a bone regenerative material with good osteoinductive and osteoconductive properties [88].

14.3.5 Chitosan/Inorganic Adducts Nano Biomaterials

The cell proliferation, reparative approach, self-assembling hierarchical structure, and mechanically stable, bio-mimicked inorganic adducts such as hydroxyapatite, NaF, and $SnCl_2$ made the CS/metformin-based nano biomaterials. Calcium phosphate cement (CPC) and alkaline phosphatase (ALP) embedded CS/metformin nanomaterial are developed and used for scaffold preparation, osteo/odontogenic composite is a highly promising scaffold with the potential for tissue engineering and dentin regeneration [89]. Liao et al. have designed an interesting mesoporous mHA/CS nanocomposite scaffold to treat periodontal issues loaded with recombinant human amelogenin (rhAm). The *in-vivo*, in vitro biological effects of the 7 nm sized mHA/CS/rhAm possessed a surface area of 33.95 m^2/g and 20 $\mu g/mL$ rhAm loaded nanocomposites inhibited the formation of periodontal pathogens and promote the formation of bone and cementum-like tissue [90]. The development of electrospun fibre like meloxicam (MX) immobilized biodegradable CS/PVA/HA) via solvent casting method. The sustainable and controlled drug release kinetics proven this material is suitable for soft tissue-engineered implants and has anti-inflammatory properties [91]. The host-guest interaction also played an important role in the reduction of periodontitis. This mechanism induced the interactions between the host immune system and plaque microorganisms. Xu et al. have synthesized a hydrogel of CS/b-sodium glycerophosphate (b-GP)/gelatin with the loading of drugs such as aspirin and erythropoietin (EPO) to check pharmacological effects of anti-inflammation and tissue regeneration [92]. Natural and cost-effective dental varnishes prepared from propolis, mask, and CS/NaF have acted as a very good antimicrobial effect against in vitro bacterial-induced enamel demineralization [93]. Graphene oxide (GO) seems to possess excellent mechanical strength and its reach on biocompatibility is successful. The development of GO/CS/HA nanocomposite was preferred for the improvement of osteointegration between the implant and the bone [94]. Nanorod-structured CS/HA nanomaterials enhance the biomineralization of enamel and act as reinforcement materials for the prevention of dental caries. The enamel remineralization experiment was performed by using CS/$CaPO_4$ NPs oriented the biomimetic replication. The orientated assembly of ACP guided by amelogenin in the biomineralization of enamel. This sophisticated approach improved the formation of enamel-like crystals with an advanced mechanical property performance [95].

14.3.6 Chitosan/Ceramics/Glass Nano Biomaterials

CS-based ceramics NBMs are quite an emerging material for dentistry purposes. The low thermal and electrical conductivity, good corrosion resistivity, good fracture toughness, translucency and good mechanical strength of ceramics-glass nanomaterial dragged attention to them. Similar to polymer, few limitations like brittleness and ductility exists, but have been resolved after the interaction with CS NPs. CS/zinc incorporated HA has shown good dental applications due to the participation of metal and inorganic adducts in the CS cavity. This intra-pocket like nanocomposite is suitable for the treatment of alveolar bone loss in the rat model of periodontitis [96].

14.3.7 Chitosan Mixed Nano Biomaterials

CS mixed NBMs always comprise of two or more precursors based on their different physical and chemical properties. The superiority and advanced properties enhanced the utilization of such material for dental treatments [97, 98]. Table 14.1 represents a list of different structured CS-based NBMs, and their application in dentistry with effective treatment and results.

Table 14.1 Details of types of nanomaterials and their treatment, target application and methods

Structural CS-based NBMs	Application	Treatments and results	References
Network and fibrous CS/PVA/PCL	Scaffold	Successful bone tissue regeneration with bone morphogenetic protein 2 and vascular endothelial growth factor	[77]
Micron-sized CS/liposome	Oral medicine, bioadhesive	Treatment for periodontal disease, oral candidiasis with the reduction in plaque formation with the enhancement in viscosity, permeabilizer, antimicrobial, anti-adhesive and anti-dental caries	[79]
CS/curcuminoid	Mouthwash	Used for denture stomatitis with the effective therapeutic result	[84]
Spherical CS/HDA	Cartilage production	Growth and functionality of encapsulated chondrocytes	[87]
Chitosan/DA	Bone tissue generation	Good osteoinductive/osteoconductive properties by tissue generation of human periodontal ligament cells	[88]
CS/calcium phosphate	Scaffold	Dentin regeneration stimulating the odontogenic differentiation of human induced pluripotent stem cells	[89]
A fibrous film like meloxicam/CS/PVA/HA	Soft tissue-engineered implants	Periodontal disease treatment with anti-inflammatory and nontoxicity effects	[91]
Rutin-loaded CS	A drug in the saliva fluid	Prevents dental caries with anticarcinogenic effect	[92]
CS-NaF	Dental varnishes	Acted a significant role against in vitro bacterial-induced enamel demineralization with an antibacterial effect against S. mutans	[93]
CS/HA nanorods	Prevents periodontitis	Restorative dentistry osteogenic and cell differentiation with the enhancement of the enamel biomineralization	[94]
Carboxymethyl CS/$CaPO_4$	Teeth	Biometric replication	[95]
Oriented and ordered CMC/alendronate (ALN)	Tooth enamel	Remineralization and biomineralization of the enamel via protease decomposing amelogenin	[96]

(continued)

Table 14.1 (continued)

Structural CS-based NBMs	Application	Treatments and results	References
CS/biopolymer	Scaffold	Osteoconductive purpose with bone tissue regeneration	[99]
Microporous CS/gelatin/bioactive glass-ceramics	Scaffold	A potential candidate for alveolar bone regeneration applications	[100]
Nano-sized CS/gelatin/HA	Dentistry	Good swelling character and biological response to MG 63 cells	[97]
CS/rose-bengal (RB)	Dentine collagen matrix	Dental caries and bad breath treatment by the production of antibiofilm effect with no dark toxicity	[98]

14.4 Current and Future Prospects of Chitosan-Based Nano Biomaterials in Dentistry

Practical Importance and Limitations

Climate change, environmental pollution, hormonal imbalance, stress and lack of oral hygiene cause numerous dental problems. Despite these problems, significant dental implants and dental care products have been designed with smart nanotechnology. Conventional materials are costly and less durable. Thus, to narrow down the gap between conventional materials and modern nanotechnology and to take care of the need for dental care products, nanotechnology developed an efficient and eco-friendly technology to satisfy these needs. The cost of conventional dental implant alloys is one of the important limitations that need to be eradicated. To solve many dentistry related issues, the requirement of smart materials came into the picture. We discussed CS-based NBMs for medical and dentistry exposures such as varnishes, dental implants, drug ailments and oral hygiene toothpaste. In the present study, we discussed many CS-based NBMs for dentistry purposes depending on the factors such as composition, concentration, MW, and other physical properties. Figure 14.3 represents the chitosan-based nanomaterials and their utilization to solve dental problems.

14.5 Concluding Remark

This chapter discusses chitosan-based nano-biomaterials and their significant uses for dentistry purposes. These engineered nano-sized materials are biocompatible, biodegradable, cost-effective, stable, and durable in behaviour. The multifunctional,

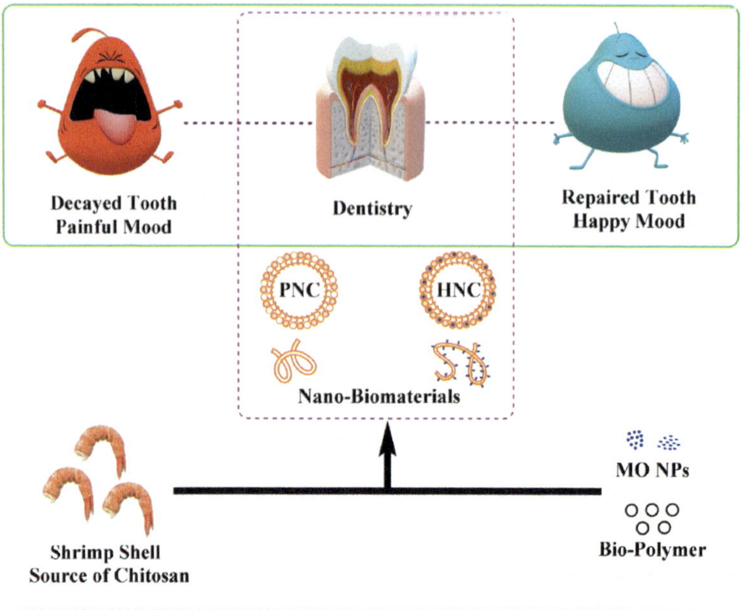

Fig. 14.3 Illustrates the chitosan-based nanomaterials and their utilization to solve the dental problems

macromolecular, variant morphological attributes, oxygen-rich groups, easy preparation, and rich abundance of these NBMs made scientists work on them. These materials have induced significant cytotoxicity and antimicrobial properties through oxidative stress and ROS mechanisms. This work also suggests that morphological texture and surface properties of CS-based nanomaterials determined by chemical composition possibly play a critical role in ROS production. Besides ROS, host-guest interactions also played an important role in the development of CS-based nanomaterials and promote space-carrying capacity. This smart material and its derivatives improved their applicability with the help of organized nanotechnology into several directions like electronics, energy industries, pharmaceuticals, and medical and biological purposes.

Acknowledgements The authors also acknowledge the research support grants awarded by University Grant Commission (UGC) India, DS Kothari Post-Doctoral fellowship to Dr. Biswajit Parhi. Dr. Debasrita Bharatiya acknowledges Veer Surendra Sai University of Technology (VSSUT), Burla, India for the support in research facilities as a Research Associate (RA).

Conflict of Interest The authors declare there is no conflict of interest in publishing this article.

References

1. T.A. Saleh, Nanomaterials: Classification, properties, and environmental toxicities. Environ. Technol. Innov. **20**, 101067 (2020). https://doi.org/10.1016/j.eti.2020.101067
2. E. Roduner, Size matters: Why nanomaterials are different. Chem. Soc. Rev. **35**, 583–592 (2006). https://doi.org/10.1039/b502142c
3. J.R.L. Gregory V Lowry, Kelvin B, Simon C. Apte, Transformations of nanomaterials in the environment. Environ. Sci. Technol. **46**, 6893–9899 (2012). https://doi.org/10.1016/B978-0-08-099408-6.00002-5
4. J.R. Peralta-Videa, L. Zhao, M.L. Lopez-Moreno, G. de la Rosa, J. Hong, J.L. Gardea-Torresdey, Nanomaterials and the environment: A review for the biennium 2008–2010. J. Hazard. Mater. **186**, 1–15 (2011). https://doi.org/10.1016/j.jhazmat.2010.11.020
5. D. Rohilla, S. Chaudhary, A. Umar, An Overview of Advanced Nanomaterials for Sensor Applications, Eng. Sci. 16 (2021) 47–70. https://doi.org/10.30919/es8d552.
6. L.A. Kolahalam, I. V. Kasi Viswanath, B.S. Diwakar, B. Govindh, V. Reddy, Y.L.N. Murthy, Review on nanomaterials: Synthesis and applications. Mater. Today Proc. **18**, 2182–2190 (2019). https://doi.org/10.1016/j.matpr.2019.07.371
7. L. Zhang, T.J. Webster, Nanotechnology and nanomaterials: Promises for improved tissue regeneration. Nano Today **4**, 66–80 (2009). https://doi.org/10.1016/j.nantod.2008.10.014
8. A. Rodrigo-Navarro, S. Sankaran, M.J. Dalby, A. del Campo, M. Salmeron-Sanchez, Engineered living biomaterials. Nat. Rev. Mater. **6**, 1175–1190 (2021). https://doi.org/10.1038/s41578-021-00350-8
9. L. Yang, S. Pijuan-Galito, H.S. Rho, A.S. Vasilevich, A.D. Eren, L. Ge, P. Habibović, M.R. Alexander, J. De Boer, A. Carlier, P. Van Rijn, Q. Zhou, High-throughput methods in the discovery and study of biomaterials and materiobiology. Chem. Rev. **121**, 4561–4677 (2021). https://doi.org/10.1021/acs.chemrev.0c00752
10. N. Eliaz, Corrosion of metallic biomaterials: A review. Materials (Basel) **12**, 1–91 (2019). https://doi.org/10.3390/ma12030407
11. O.S. Fenton, K.N. Olafson, P.S. Pillai, M.J. Mitchell, R. Langer, Advances in biomaterials for drug delivery. Adv. Mater. **30**, 1–29 (2018). https://doi.org/10.1002/adma.201705328
12. A.M. Brokesh, A.K. Gaharwar, Inorganic biomaterials for regenerative medicine. ACS Appl. Mater. Interfaces **12**, 5319–5344 (2020). https://doi.org/10.1021/acsami.9b17801
13. Y. Patil-Sen, Advances in nano-biomaterials and their applications in biomedicine. Emerg. Top Life Sci. **5**, 169–176 (2021). https://doi.org/10.1042/ETLS20200333
14. K. Kapat, Q.T.H. Shubhra, M. Zhou, S. Leeuwenburgh, Piezoelectric nano-biomaterials for biomedicine and tissue regeneration. Adv. Funct. Mater. **30**, 1909045 (2020). https://doi.org/10.1002/adfm.201909045
15. R. Sharma, D. Sharma, L.D. Hazlett, N.K. Singh, Nano-biomaterials for retinal regeneration. Nanomaterials **11**, 1–25 (2021). https://doi.org/10.3390/nano11081880
16. N.S. Rejinold, R. Jayakumar, Y.C. Kim, Radio frequency responsive nano-biomaterials for cancer therapy. J. Control. Release **204**, 85–97 (2015). https://doi.org/10.1016/j.jconrel.2015.02.036
17. M.T. Amin, A.A. Alazba, U. Manzoor, A review of removal of pollutants from water/wastewater using different types of nanomaterials. Adv. Mater. Sci. Eng. **2014**, 1–24 (2014). https://doi.org/10.1155/2014/825910
18. S.T. Chang, L.C. Chen, S. Bin Lin, H.H. Chen, Nano-biomaterials application: Morphology and physical properties of bacterial cellulose/gelatin composites via crosslinking. Food Hydrocoll. **27**, 137–144 (2012). https://doi.org/10.1016/j.foodhyd.2011.08.004
19. J. Gao, B. Xu, Applications of nanomaterials inside cells. Nano Today **4**, 37–51 (2009). https://doi.org/10.1016/j.nantod.2008.10.009
20. S. Bandara, H. Du, L. Carson, D. Bradford, R. Kommalapati, Agricultural and biomedical applications of chitosan-based nanomaterials. Nanomaterials **10**, 1–31 (2020). https://doi.org/10.3390/nano10101903

21. V. Arias, F.I. Khan, S. Ali, Chitosan-based nanomaterials: A state-of-the-art review. Int. J. Biol. Macromol. **59**, 46–58 (2013)
22. R.C. Goy, D. De Britto, O.B.G. Assis, A review of the antimicrobial activity of chitosan. Polimeros **19**, 241–247 (2009). https://doi.org/10.1590/S0104-14282009000300013
23. P. Sahariah, M. Másson, Antimicrobial chitosan and chitosan derivatives: A review of the structure-activity relationship. Biomacromol **18**, 3846–3868 (2017). https://doi.org/10.1021/acs.biomac.7b01058
24. M.N.V.R. Kumar, A review of chitin and chitosan applications. Raective Funct. Polym. **46**, 1–27 (2000)
25. A. El Hadrami, L.R. Adam, I. El Hadrami, F. Daayf, Chitosan in plant protection. Mar. Drugs **8**, 968–987 (2010). https://doi.org/10.3390/md8040968
26. Ž.K., M.L. Gregor Kravanja, M. Primožiˇc, Chitosan-based (nano)materials for novel biomedical applications. Molecules **24**, 1960 (2019). https://doi.org/10.1016/B978-0-12-816911-7.00023-2
27. M.A. Malik, M.Y. Wani, M.A. Hashim, F. Nabi, Nanotoxicity: Dimensional and morphological concerns. Adv. Phys. Chem. **2011**, 1–15 (2011). https://doi.org/10.1155/2011/450912
28. H. Yang, C. Liu, D. Yang, H. Zhang, Z. Xi, Comparative study of cytotoxicity, oxidative stress and genotoxicity induced by four typical nanomaterials: The role of particle size, shape and composition. J. Appl. Toxicol. **29**, 69–78 (2009). https://doi.org/10.1002/jat.1385
29. J. Yuan, A.H.E. Müller, One-dimensional organic-inorganic hybrid nanomaterials. Polymer (Guildf) **51**, 4015–4036 (2010). https://doi.org/10.1016/j.polymer.2010.06.064
30. X. Xu, P. Winterwerber, D. Ng, Y. Wu, DNA-programmed chemical synthesis of polymers and inorganic nanomaterials. Top. Curr. Chem. **378**, 57–81 (2020). https://doi.org/10.1007/s41061-020-0292-x
31. D. Lauritano, L. Limongelli, G. Moreo, G. Favia, F. Carinci, Nanomaterials for periodontal tissue engineering: Chitosan-based scaffolds. A systematic review. Nanomaterials **10**, 1–16 (2020). https://doi.org/10.3390/nano10040605
32. J. Li, C. Cai, J. Li, J. Li, J. Li, T. Sun, L. Wang, H. Wu, G. Yu, Chitosan-based nanomaterials for drug delivery. Molecules **23**, 1–26 (2018). https://doi.org/10.3390/molecules23102661
33. R. Jayakumar, D. Menon, K. Manzoor, S.V. Nair, H. Tamura, Biomedical applications of chitin and chitosan based nanomaterials—A short review. Carbohydr. Polym. **82**, 227–232 (2010). https://doi.org/10.1016/j.carbpol.2010.04.074
34. B. Parhi, D. Bharatiya, S.K. Swain, Effect of polycaprolactone on physicochemical, biological, and mechanical properties of polyethylene oxide and polyamino acids nano block copolymers. J. Appl. Polym. Sci. 1–13 (2022). https://doi.org/10.1002/app.52116
35. P.K. Sethy, P. Mohapatra, S. Patra, D. Bharatiya, S.K. Swain, Antimicrobial and barrier properties of polyacrylic acid/GO hybrid nanocomposites for packaging application. Nano-Struct. Nano-Objects **26**, 100747 (2021). https://doi.org/10.1016/j.nanoso.2021.100747
36. D. Bharatiya, S. Patra, B. Parhi, S.K. Swain, A materials science approach towards bioinspired polymeric nanocomposites: A comprehensive review. Int. J. Polym. Mater. Polym. Biomater. **1**, 1–16 (2021). https://doi.org/10.1080/00914037.2021.1990057
37. B. Parhi, D. Bharatiya, S.K. Swain, Application of quercetin flavonoid based hybrid nanocomposites: A review. Saudi Pharm. J. **28**, 1719–1732 (2020). https://doi.org/10.1016/j.jsps.2020.10.017
38. P.K.P. Sreenivasalu, C.P. Dora, R. Swami, V.C. Jasthi, P.N. Shiroorkar, S. Nagaraja, S.M.B. Asdaq, M.K. Anwer, Nanomaterials in dentistry: Current applications and future scope. Nanomaterials **12**, 1676 (2022). https://doi.org/10.3390/nano12101676
39. E.A.A. Neel, L. Bozec, R.A. Perez, H.W. Kim, J.C. Knowles, Nanotechnology in dentistry: Prevention, diagnosis, and therapy. Int. J. Nanomed. **10**, 6371–6394 (2015). https://doi.org/10.2147/IJN.S86033
40. A. Bhattacharyya, G. Janarthanan, I. Noh, Nano-biomaterials for designing functional bioinks towards complex tissue and organ regeneration in 3D bioprinting. Addit. Manuf. **37**, 101639 (2021). https://doi.org/10.1016/j.addma.2020.101639

41. R.V. Kumaraswamy, S. Kumari, R.C. Choudhary, A. Pal, R. Raliya, P. Biswas, V. Saharan, Engineered chitosan based nanomaterials: Bioactivities, mechanisms and perspectives in plant protection and growth. Int. J. Biol. Macromol. **113**, 494–506 (2018). https://doi.org/10.1016/j.ijbiomac.2018.02.130
42. X. Lang, T. Wang, M. Sun, X. Chen, Y. Liu, Advances and applications of chitosan-based nanomaterials as oral delivery carriers: A review. Int. J. Biol. Macromol. **154**, 433–445 (2020). https://doi.org/10.1016/j.ijbiomac.2020.03.148
43. J. Sharan, S. Singh, S.V. Lale, M. Mishra, V. Koul, O.P. Kharbanda, Applications of nanomaterials in dental science: A review. J. Nanosci. Nanotechnol. **17**, 2235–2255 (2017). https://doi.org/10.1166/jnn.2017.13885
44. S. Das, S. Mitra, S.M.P. Khurana, N. Debnath, Nanomaterials for biomedical applications. Front. Life Sci. **7**, 90–98 (2013). https://doi.org/10.1080/21553769.2013.869510
45. C.-T. Ng, G.-H. Baeg, L.E. Yu, C.-N. Ong, B.-H. Bay, Biomedical applications of nanomaterials as therapeutics. Curr. Med. Chem. **25**, 1409–1419 (2017). https://doi.org/10.2174/0929867324666170331120328
46. A. Murali, G. Lokhande, K.A. Deo, A. Brokesh, A.K. Gaharwar, Emerging 2D nanomaterials for biomedical applications. Mater. Today **50**, 276–302 (2021). https://doi.org/10.1016/j.mattod.2021.04.020
47. M.E.A. Ali, M.M.S. Aboelfadl, A.M. Selim, H.F. Khalil, G.M. Elkady, Chitosan nanoparticles extracted from shrimp shells, application for removal of Fe(II) and Mn(II) from aqueous phases. Sep. Sci. Technol. **53**, 2870–2881 (2018). https://doi.org/10.1080/01496395.2018.1489845
48. R.C. Choudhary, S. Kumari, R. V Kumaraswamy, A. Pal, Chapter 5 Characterization methods for chitosan—based nanomaterials. Genome Data Anal. **1**, 103–116 (2019). https://doi.org/10.1007/978-3-030-12496-0
49. T. Anitha, P. Senthil Kumar, K. Sathish Kumar, B. Ramkumar, S. Ramalingam, Adsorptive removal of Pb(II) ions from polluted water by newly synthesized chitosan-polyacrylonitrile blend: Equilibrium, kinetic, mechanism and thermodynamic approach. Process Saf. Environ. Prot. **98**, 187–197 (2015). https://doi.org/10.1016/j.psep.2015.07.012
50. V. Thamilarasan, V. Sethuraman, K. Gopinath, C. Balalakshmi, M. Govindarajan, R.A. Mothana, N.A. Siddiqui, J.M. Khaled, G. Benelli, Single step fabrication of chitosan nanocrystals using penaeus semisulcatus: Potential as new insecticides, antimicrobials and plant growth promoters. J. Clust. Sci. **29**, 375–384 (2018). https://doi.org/10.1007/s10876-018-1342-1
51. M. Morsy, K.M. Mostafa, H.A.M. Amyn, A.A. hameed El-Ebissy, A.M. Salah, M.A. Youssef, Synthesis and characterization of freeze dryer chitosan nanoparticles as multifunctional eco-friendly finish for fabricating easy care and antibacterial cotton textiles. Egypt. J. Chem. **62**, 1277–1293 (2019). https://doi.org/10.21608/EJCHEM.2019.6995.1583
52. N. Schlueter, J. Klimek, C. Ganss, Randomised in situ study on the efficacy of a tin/chitosan toothpaste on erosive-abrasive enamel loss. Caries Res. **47**, 574–581 (2013). https://doi.org/10.1159/000351654
53. F.C. Iswanti, I. Nurulita, S. Djauzi, M. Sadikin, A.B. Witarto, T. Yamazaki, Preparation, characterization, and evaluation of chitosan-based nanoparticles as CpG ODN carriers. Biotechnol. Biotechnol. Equip. **33**, 390–396 (2019). https://doi.org/10.1080/13102818.2019.1578690
54. A. Bakopoulou, A Georgopoulou, I. Grivas, C. Bekiari, O. Prymak, K Loza, M. Epple, G.C. Papadopoulos, P. Koidis, M Chatzinikolaidou, Dental pulp stem cells in chitosan/gelatin scaffolds for enhanced orofacial bone regeneration. Dent. Mater. **35**, 310–327 (2019). https://doi.org/10.1016/j.dental.2018.11.025
55. A. Golchin, S. Hosseinzadeh, M. Staji, M. Soleimani, A. Ardeshirylajimi, A. Khojasteh, Biological behavior of the curcumin incorporated chitosan/poly(vinyl alcohol) nanofibers for biomedical applications. J. Cell. Biochem. **120**, 15410–15421 (2019). https://doi.org/10.1002/jcb.28808
56. J. Zhu, H. Ye, D. Deng, J. Li, Y. Wu, Electrospun metformin-loaded polycaprolactone/chitosan nanofibrous membranes as promoting guided bone regeneration membranes: Preparation and characterization of fibers, drug release, and osteogenic activity in vitro. J. Biomater. Appl. **34**, 1282–1293 (2020). https://doi.org/10.1177/0885328220901807

57. Y. Pakzad, F. Ganji, Thermosensitive hydrogel for periodontal application: In vitro drug release, antibacterial activity and toxicity evaluation. J. Biomater. Appl. **30**, 919–929 (2016). https://doi.org/10.1177/0885328215614191
58. X. Xu, Z. Gu, X. Chen, C. Shi, C. Liu, M. Liu, L. Wang, M. Sun, K. Zhang, Q. Liu, Y. Shen, C. Lin, B. Yang, H. Sun, An injectable and thermosensitive hydrogel: Promoting periodontal regeneration by controlled-release of aspirin and erythropoietin. Acta Biomater. **86**, 235–246 (2019). https://doi.org/10.1016/j.actbio.2019.01.001
59. A. Di Martino, A. Drannikov, N.S. Surgutskaia, K. Ozaltin, P.S. Postnikov, T.E. Marina, V. Sedlarik, Chitosan-collagen based film for controlled delivery of a combination of short life anesthetics. Int. J. Biol. Macromol. **140**, 1183–1193 (2019). https://doi.org/10.1016/j.ijbiomac.2019.08.228
60. A.H. El-Kamel, L.Y. Ashri, I.A. Alsarra, Micromatricial metronidazole benzoate film as a local mucoadhesive delivery system for treatment of periodontal diseases. AAPS PharmSciTech **8**, E184–E194 (2007). https://doi.org/10.1208/pt0803075
61. M. Kilicarslan, M. Ilhan, O. Inal, K. Orhan, Preparation and evaluation of clindamycin phosphate loaded chitosan/alginate polyelectrolyte complex film as mucoadhesive drug delivery system for periodontal therapy. Eur. J. Pharm. Sci. **123**, 441–451 (2018). https://doi.org/10.1016/j.ejps.2018.08.007
62. E.Y. and M.G.Ö. A. Alper Öztürk, (PLGA) Nanoparticles for oral administration : Effect of polymer molecular weight and surface modification with chitosan on formulation. Polymers (Basel) **11**, 1632 (2019)
63. P. Perugini, I. Genta, B. Conti, T. Modena, F. Pavanetto, Periodontal delivery of ipriflavone: New chitosan/PLGA film delivery system for a lipophilic drug. Int. J. Pharm. **252**, 1–9 (2003). https://doi.org/10.1016/S0378-5173(02)00602-6
64. F. Li, X. Liu, S. Zhao, H. Wu, H.H.K. Xu, Porous chitosan bilayer membrane containing TGF-β1 loaded microspheres for pulp capping and reparative dentin formation in a dog model. Dent. Mater. **30**, 172–181 (2014). https://doi.org/10.1016/j.dental.2013.11.005
65. S.K. Yadav, G. Khan, G.V. Bonde, M. Bansal, B. Mishra, Design, optimization and characterizations of chitosan fortified calcium alginate microspheres for the controlled delivery of dual drugs. Artif. Cells Nanomed. Biotechnol. **46**, 1180–1193 (2018). https://doi.org/10.1080/21691401.2017.1366331
66. Y. Xue, X. Hong, J. Gao, R. Shen, Z. Ye, Preparation and biological characterization of the mixture of poly(lactic-co-glycolic acid)/chitosan/Ag nanoparticles for periodontal tissue engineering. Int. J. Nanomed. **14**, 483–498 (2019). https://doi.org/10.2147/IJN.S184396
67. M.R. Saboktakin, R.M. Tabatabaie, A. Maharramov, M.A. Ramazanov, Development and in vitro evaluation of thiolated chitosan-Poly(methacrylic acid) nanoparticles as a local mucoadhesive delivery system. Int. J. Biol. Macromol. **48**, 403–407 (2011). https://doi.org/10.1016/j.ijbiomac.2010.12.014
68. Y. Li, R. Na, X. Wang, H. Liu, L. Zhao, X. Sun, G. Ma, F. Cui, Fabrication of antimicrobial peptide-loaded PLGA/Chitosan composite microspheres for long-acting bacterial resistance. Molecules **22**, 1637 (2017). https://doi.org/10.3390/molecules22101637
69. S. Husain, K.H. Al-Samadani, S. Najeeb, M.S. Zafar, Z. Khurshid, S. Zohaib, S.B. Qasim, Chitosan biomaterials for current and potential dental applications. Materials (Basel) **10**, 1–20 (2017). https://doi.org/10.3390/ma10060602
70. J.M. Farias, T.C.M. Stamford, A.H.M. Resende, J.S. Aguiar, R.D. Rufino, J.M. Luna, L.A. Sarubbo, Mouthwash containing a biosurfactant and chitosan: An eco-sustainable option for the control of cariogenic microorganisms. Int. J. Biol. Macromol. **129**, 853–860 (2019). https://doi.org/10.1016/j.ijbiomac.2019.02.090
71. S. Zang, R. Mu, F. Chen, X. Wei, L. Zhu, B. Han, H. Yu, B. Bi, B. Chen, Q. Wang, L. Jin, Injectable chitosan/β-glycerophosphate hydrogels with sustained release of BMP-7 and ornidazole in periodontal wound healing of class III furcation defects. Mater. Sci. Eng. C. **99**, 919–928 (2019). https://doi.org/10.1016/j.msec.2019.02.024
72. F. Liu, T. Hong, J. Xie, X. Zhan, Y. Wang, Application of reactive oxygen species-based nanomaterials in dentistry: A review. Curr. Comput.-Aided Drug Des. **11**, 266 (2021). https://doi.org/10.3390/cryst11030266

73. M. Di Giulio, S. Di Bartolomeo, E. Di Campli, S. Sancilio, E. Marsich, A. Travan, A. Cataldi, L. Cellini, The effect of a silver nanoparticle polysaccharide system on streptococcal and saliva-derived biofilms. Int. J. Mol. Sci. **14**, 13615–13625 (2013). https://doi.org/10.3390/ijms140713615
74. N. Zhu, X. Chatzistavrou, L. Ge, M. Qin, P. Papagerakis, Y. Wang, Biological properties of modified bioactive glass on dental pulp cells. J. Dent. **83**, 18–26 (2019). https://doi.org/10.1016/j.jdent.2019.01.017
75. F. Sharifianjazi, S. Khaksar, A. Esmaeilkhanian, L. Bazli, S. Eskandarinezhad, P. Salahshour, F. Sadeghi, S. Rostamnia, S.M. Vahdat, Advancements in fabrication and application of chitosan composites in implants and dentistry: A review. Biomolecules **12**, 155 (2022). https://doi.org/10.3390/biom12020155
76. C.C. Carrion, M. Nasrollahzadeh, M. Sajjadi, B. Jaleh, G.J. Soufi, S. Iravani, Lignin, lipid, protein, hyaluronic acid, starch, cellulose, gum, pectin, alginate and chitosan-based nanomaterials for cancer nanotherapy: Challenges and opportunities. Int. J. Biol. Macromol. **178**, 193–228 (2021). https://doi.org/10.1016/j.ijbiomac.2021.02.123
77. A. Aguilar, N. Zein, E. Harmouch, B. Hafdi, F. Bornert, O. Damien, F. Clauss, F. Fioretti, O. Huck, N. Benkirane-jessel, G. Hua, Biodental engineering V. Biodental Eng. **24**, 1–17 (2019). https://doi.org/10.1201/9780429265297
78. C. Buzea, I. Pacheco, Nanomaterials and their classification (2017). https://doi.org/10.1007/978-81-322-3655-9_1
79. L.A. Frank, G.R. Onzi, A.S. Morawski, A.R. Pohlmann, S.S. Guterres, R.V. Contri, Chitosan as a coating material for nanoparticles intended for biomedical applications. React. Funct. Polym. **147**, 104459 (2020). https://doi.org/10.1016/j.reactfunctpolym.2019.104459
80. M. Dash, F. Chiellini, R.M. Ottenbrite, E. Chiellini, Chitosan - A versatile semi-synthetic polymer in biomedical applications. Prog. Polym. Sci. **36**, 981–1014 (2011). https://doi.org/10.1016/j.progpolymsci.2011.02.001
81. S. Hein, K. Wang, W.F. Stevens, J. Kjems, Chitosan composites for biomedical applications: Status, challenges and perspectives. Mater. Sci. Technol. **24**, 1053–1061 (2008). https://doi.org/10.1179/174328408X341744
82. B.R. Rizeq, N.N. Younes, K. Rasool, G.K. Nasrallah, Synthesis, bioapplications, and toxicity evaluation of chitosan-based nanoparticles. Int. J. Mol. Sci. **20**, 5776 (2019). https://doi.org/10.3390/ijms20225776
83. Saad Bin Qasim and Ihtesham Ur Rehman, Application of nanomaterials in dentistry. Micro Nanomanuf. **2**, 319–336 (2017). https://doi.org/10.1007/978-3-319-67132-1
84. M.W. Mustafa, S. Ungphaiboon, N. Phadoongsombut, K. Pangsomboon, S. Chelae, S. Mahattanadul, Effectiveness of an alcohol-free chitosan-curcuminoid mouthwash compared with chlorhexidine mouthwash in denture stomatitis treatment: A randomized trial. J. Altern. Complement. Med. **25**, 552–558 (2019). https://doi.org/10.1089/acm.2018.0459
85. X. Zhao, B. Guo, H. Wu, Y. Liang, P.X. Ma, Injectable antibacterial conductive nanocomposite cryogels with rapid shape recovery for noncompressible hemorrhage and wound healing. Nat. Commun. **9** (2018). https://doi.org/10.1038/s41467-018-04998-9
86. G. Ikinci, S. Şenel, H. Akincibay, S. Kaş, S. Erciş, C.G. Wilson, A.A. Hincal, Effect of chitosan on a periodontal pathogen Porphyromonas gingivalis. Int. J. Pharm. **235**, 121–127 (2002). https://doi.org/10.1016/S0378-5173(01)00974-7
87. L. V Thomas, R. VG, P. D Nair, Effect of stiffness of chitosan-hyaluronic acid dialdehyde hydrogels on the viability and growth of encapsulated chondrocytes. Int. J. Biol. Macromol. **104**, 1925–1935 (2017). https://doi.org/10.1016/j.ijbiomac.2017.05.116
88. V.E. Teerawat Sukpaita, S. Chirachanchai, P. Suwattanachai, A.P. and R.S. Ampornaramveth, In vivo bone regeneration induced by a Scaffold of chitosan/dicarboxylic acid seeded with human periodontal ligament cells. Int. J. Mol. Sci. **20**, 4883 (2019)
89. W. Qin, J.Y. Chen, J. Guo, T. Ma, M.D. Weir, D. Guo, Y. Shu, Z.M. Lin, A. Schneider, H.H.K. Xu, Novel calcium phosphate cement with metformin-loaded chitosan for odontogenic differentiation of human dental pulp cells. Stem Cells Int. **2018**, 1–10 (2018). https://doi.org/10.1155/2018/7173481

90. Y. Liao, H. Li, R. Shu, H. Chen, L. Zhao, Z. Song, W. Zhou, Mesoporous hydroxyapatite/chitosan loaded with recombinant-human amelogenin could enhance antibacterial effect and promote periodontal regeneration. Front. Cell. Infect. Microbiol. **10**, 1–14 (2020). https://doi.org/10.3389/fcimb.2020.00180
91. M. Yar, A. Farooq, L. Shahzadi, A.S. Khan, N. Mahmood, A. Rauf, A.A. Chaudhry, I.U. Rehman, Novel meloxicam releasing electrospun polymer/ceramic reinforced biodegradable membranes for periodontal regeneration applications. Mater. Sci. Eng. C **64**, 148–156 (2016). https://doi.org/10.1016/j.msec.2016.03.072
92. A.G. Patil, A.H. Jobanputra, Rutin-chitosan nanoparticles: Fabrication, characterization and application in dental disorders. Polym. Plast. Technol. Eng. **54**, 202–208 (2015). https://doi.org/10.1080/03602559.2014.935425
93. M.O. Wassel, M.A. Khattab, Antibacterial activity against Streptococcus mutans and inhibition of bacterial induced enamel demineralization of propolis, miswak, and chitosan nanoparticles based dental varnishes. J. Adv. Res. **8**, 387–392 (2017). https://doi.org/10.1016/j.jare.2017.05.006
94. V. Bonilla-Represa, C. Abalos-Labruzzi, M. Herrera-Martinez, M.O. Guerrero-Pérez, Nanomaterials in dentistry: State of the art and future challenges. Nanomaterials **10**, 1–27 (2020). https://doi.org/10.3390/nano10091770
95. K.D. Jandt, D.C. Watts, Nanotechnology in dentistry: Present and future perspectives on dental nanomaterials. Dent. Mater. **36**, 1365–1378 (2020). https://doi.org/10.1016/j.dental.2020.08.006
96. L.K. Foong, M.M. Foroughi, A.F. Mirhosseini, M. Safaei, S. Jahani, M. Mostafavi, N. Ebrahimpoor, M. Sharifi, R.S. Varma, M. Khatami, Applications of nano-materials in diverse dentistry regimes. RSC Adv. **10**, 15430–15460 (2020). https://doi.org/10.1039/d0ra00762e
97. M. Peter, N. Ganesh, N. Selvamurugan, S.V. Nair, T. Furuike, H. Tamura, R. Jayakumar, Preparation and characterization of chitosan-gelatin/nanohydroxyapatite composite scaffolds for tissue engineering applications. Carbohydr. Polym. **80**, 687–694 (2010). https://doi.org/10.1016/j.carbpol.2009.11.050
98. A. Besinis, T. De Peralta, C.J. Tredwin, R.D. Handy, Review of nanomaterials in dentistry: Interactions with the oral microenvironment, clinical applications, hazards, and benefits. ACS Nano **9**, 2255–2289 (2015). https://doi.org/10.1021/nn505015e
99. B.N. Singh, V. Veeresh, S.P. Mallick, Y. Jain, S. Sinha, A. Rastogi, P. Srivastava, Design and evaluation of chitosan/chondroitin sulfate/nano-bioglass based composite scaffold for bone tissue engineering. Int. J. Biol. Macromol. **133**, 817–830 (2019). https://doi.org/10.1016/j.ijbiomac.2019.04.107
100. M. Peter, N.S. Binulal, S.V. Nair, N. Selvamurugan, H. Tamura, R. Jayakumar, Novel biodegradable chitosan-gelatin/nano-bioactive glass ceramic composite scaffolds for alveolar bone tissue engineering. Chem. Eng. J. **158**, 353–361 (2010). https://doi.org/10.1016/j.cej.2010.02.003
101. J. Wojnarowicz, T. Chudoba, W. Lojkowski, A review of microwave synthesis of zinc oxide nanomaterials: Reactants, process parameters and morphoslogies. Nanomaterials **10**, 1086 (2020). https://doi.org/10.3390/nano10061086
102. T.S. Carvalho, A. Lussi, Combined effect of a fluoride-, stannous- and chitosan-containing toothpaste and stannous-containing rinse on the prevention of initial enamel erosion-abrasion. J. Dent. **42**, 450–459 (2014). https://doi.org/10.1016/j.jdent.2014.01.004

Chapter 15
Investigation of Chitosan/Metal and Metal Oxide Nanocomposites as a New Strategy for Enhanced Anti-Biofilm Efficacy with Reduced Toxicity

Govindan Rajivgandhi, Muthuchamy Maruthupandy, and Franck Quero

Abstract Worldwide, biofilm producing bacteria are the most challenging and careful concern, responsible for variety of infections in human every year. The clumping of sessile bacterial cells attached to each other in the inherent surfaces and embedded in an extracellular polymeric matrix is referred as a biofilm. Biofilm producing bacteria and their mode of dissemination are more resistance to conventional antibiotics and immune response killing. Researchers are more struggled and investigated alternative anti-biofilm drugs with novel strategies to combat antibiotic resistant biofilm producing bacteria. In recent years, metal and metal oxide nanoparticles are widely used nanomaterials with promising applications in medical field and offer novel approaches to tackle different kind of pathogens. On contrary, increased concentrations of nanomaterial with considerable level of toxicity against various infections in medical field are great challenge. To avoid this problem, nanoformulation demonstrates hopeful choice to succeed improved bioavailability and targeted drug delivery. Chitosan (Cs) is a natural polymer derived from derivatives of chitin which is extensively used in biomedical research due to the complete degradation and biocompatibility. Metal nanoparticles (M NPs) and metal oxide nanoparticles (MO NPs) have wide surface area and frequently utilized in biomedical research because of their inert biological properties. In recent years, the preparation of Cs/metal nanocomposites (Cs/M NCs)and Cs/metal oxide nanocomposites (Cs/MO NCs) are safer, chemically stable, and have excellent bioavailability and biocompatibility with minimized toxicitics against various infectious bacteria. Hence, the main objective of this work is to discuss the alternative way to increase the bioavailability of M NPs and MO NPs in the presence of Cs to treat biofilm producing bacteria with minimized toxicity.

G. Rajivgandhi · M. Maruthupandy (✉) · F. Quero (✉)
Laboratorio de Nanocelulosa y Biomateriales, Departamento de Ingeniería Química, Biotecnología y Materiales, Universidad de Chile, Avenida Beauchef-851, Santiago, Chile
e-mail: maruthupandym@hotmail.com

F. Quero
e-mail: fquero@uchile.cl

S. K. Swain and A. Biswal (eds.), *Chitosan Nanocomposites*,
Biological and Medical Physics, Biomedical Engineering,
https://doi.org/10.1007/978-981-19-9646-7_15

15.1 Introduction

In recent years, usages of biopolymer including Chitosan (Cs), alginates, cellulose and pectin in biological risk assessment has increased attention worldwide in various fields including agricultural, industrial, biological, biotechnological, pharmaceutical, medicinal and health care sciences. Cs is the naturally occurring polysaccharide rich biopolymer that can be obtained from derivatives of chitin [1, 2]. In addition, it is widely used as a cationic biocompatible polymer in various fields, and also improves the production of nano-chemicals, thereby increasing the productivity, growth and immunity. Owing to the presence of hydroxyl and amino functional groups on Cs, the surface capping of metal and metal oxide-based nanoparticles using Cs provides a novel cost effective and toxic free production method, and helps in improvement of biological efficiency of nanomaterials effectively [3]. The positive charged low pH of the water-soluble cationic Cs acts as an excellent biocompatible material and is easily degradable and non-toxic in nature. Owing to these properties, Cs is suitable for implementation in biomedical applications. The positive charges of the cationic rich Cs surface have the ability to bond with negatively charged molecules or substances such as other polymers, drugs, bioactive compounds, chemical derivatives, antibiotics, metal and metal oxide nanomaterial's, enzymes and etc. It provides significant physiological and biological benefits with potential biomedical applications in drug delivery [4] (Fig. 15.1).

15.2 Chitosan

Cs is a poly (1 $\rightarrow$ 4)-2-amino-2-deoxy-d-glucan and cationic multifunctional natural polymer. It is approved by the U.S. FDA for drug delivery and tissue engineering approach [5]. It is an important colloidal drug carrier molecule which provides great drug loading capacity to deliver drug, gene and antibiotics in specific sites and is sometimes utilized in cancer therapy. This drug delivery process may increase the drug concentration inside the cells, might overcome the shortcoming result from infections and promoted the conventional drug delivery strategies [6]. Cs nanocomposites with charged nanoparticles have shown mucoadhesive properties in drug to deliver for prolonged period of time at target site locations. Cs is a linear polyamine and it stimulated the process of cross-linkages because of more amino groups and its cationic nature, also it helped to allow the ionic-cross-linking process with multivalent anions. In addition, the process of nanoencapsulation between the nanomaterials, drugs, antibiotics and Cs are manipulated in aqueous solution without using toxic solvents [7] (Fig. 15.2).

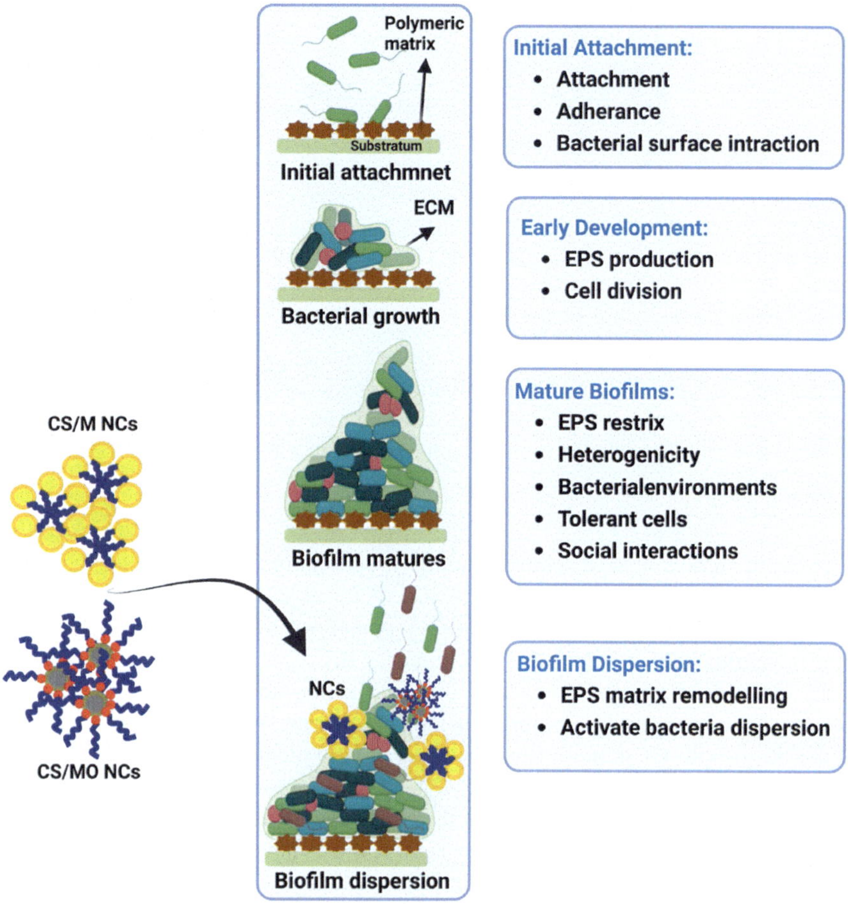

Fig. 15.1 Cs/M NCs and Cs/MO NPs as excellent biomaterials to eradicate biofilm formation with reduced toxicity

15.3 Importance of Chitosan

Cs is a second most natural biopolymer after cellulose, which can be obtained from the shells of crustaceans, fungi cell walls, algae and some bacteria. It is a promising carrier material used in various biomedical applications including nano drug delivery, vaccine delivery and inherent biological and chemical properties [8]. These properties include excellent biocompatibility, biodegradability, nonimmunogenecity, and increased bioadhesion ability, wound healing capacity and low toxicity [9]. Based on these advantages, Cs and its derivatives have gained attention in drug delivery research and will have performed as unlimited application perspective. In recent years, Cs based nanoparticles are extensively implemented in animal experiments for human related infection due to compromised toxicity. In addition, it is concentrated

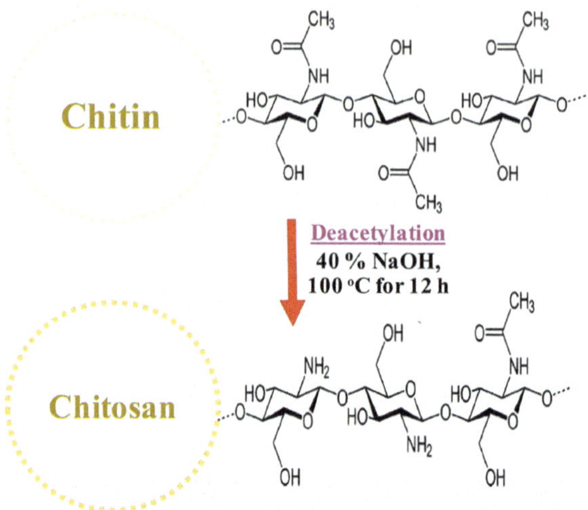

Fig. 15.2 Extraction of Cs from chitin using degree of deacetylation

in in-depth studies against wound healing, tissue engineering, scaffold materials, drug carrier molecule, vaccine, vehicle and vector as usage in a variety of forms. Interestingly, green productions in the development of Cs based products are highly used in various fields for the benefit of humans [10].

Cs has unique properties such as easy utilization and processing ability and capacity to release the drugs through long term time interval. The combination of the drugs with Cs is more economic and environmental friendliness compared to other classes of materials. Recently, Zheng et al. reported that Cs has excellent drug delivery materials, and maintains the continuous drug release capacity in long term basis. In addition, high molecular weight of the Cs composes the basic structures of microbes and plays a major role in important processes [11]. Potential natural Cs is derived from plants, animals, bacteria, fungi and some other living organisms by different synthesis methods [12]. These synthesized Cs will be used as biomaterial in the medical field, food additives in food industry, adsorbents, packaging, cosmetics, water treatment, energy conversion, biosensor, plastic degradation, fabric and cloths in wool industry. In addition, the excessive production of biodegradable biopolymers such as cellulose, starch, glycogens are used in energy conversion and various biological functions are also of considerable interest in the recent research study.

Notable functions of carbohydrate biopolymers are intercellular interaction, cell variations, arrangement of multicellular structure and malignant neoplasms growth [13]. In advance, solubility, viscosity, chemical, gelation, physical, mechanical, availability, constituents, surface nature, bond variation, degree of polymerization and other structural capacities of carbohydrate polymer are helpful to decide its potentiality in various biomedical, industrial and biopharmaceutical applications [14]. Recently, cancer targeting, tissue engineering, wound healing, tissue regeneration using biopolymers would be beneficial to focus on developing Cs based materials which are the best choice to serve as biological substitutes for body tissues at the nanoscale [15]. Therefore, biopolymers are excellent reservoirs than other materials to promote the drug carrier role and releasing efficiency in the target site of different infections.

Among various polysaccharides, Cs is the superior biopolymer with diverse functionalization, and is an easy crosslinking non-toxic, biocompatibility material. The mechanical and physiological strengths are very high due to the cross-linking nature. It has higher drug loading capacity because of its higher porous network and larger surface area and most importantly it prevents the burst release of drugs and attains controlled drug release [16]. When Cs is employed as a drug delivery carrier molecule, it can be diffuse strong electrostatic interaction with other materials like bioactive compounds, proteins, enzymes, genes through the positive charged surface containing amino groups. It can improved the bioavailability of drug and activate the drugs for sustainable delivery with a long term period [17].

More number of methods are used for drug delivery against various infections including physical, chemical and biological and the stimuli induced responses are the most favoured attributes for site specific drug delivery. This process is used for significant increase in drug delivery that trigged the drug release and increases the bioavailability and loading of drug loading in long term [18]. Further, Cs is the most suitable biopolymer for scaffolding materials due to the increased biodegradable efficiency and minimized toxicity. Recent report of Vinit et al. documented that the amino sugars of the Cs are completely absorbed by the body during the process of degradation. In addition, the promising chemical and thermal stability of the Cs structure is more helpful for the degradation process in human body. In drug delivery process, the modified structural and functional moieties of Cs scaffolds are important. Also, scaffold-based Cs is use in broad range of applications in translational medicine [19] (Table 15.1).

Cs contains nitrogen groups in its molecular structure, and its cationic nature helps to form polyelectrolyte complexes. It also exhibited its extra ordinary characteristic features like mucoadhesion, transfection, and in situ gelation. Advantages of these properties can help the primary amino acids to attach to the polymeric chains. In the scaffold/Cs preparation, initially the cationic nature of the Cs helps in the formation of carboxylate salts including lactate acetate and citrate. Cs is divided into many kinds based on molecular weight, degree of deacetylation and degree of crystallinity. Sometimes, the poor solubility nature of the Cs has limited polymeric nature. In this stage, the chemical modifications like alkylation and carboxy methylation helped to improve the properties of Cs molecules. It helps in improvement of Cs solubilization and it helps in protonation of the amino group (-NH_2) of the D-glucosamine moieties. The polyelectrolyte is formed after transferring of polysaccharides in acidic media. So, the protonation process is heightened and also solubility of Cs is automatically increased.

Table. 15.1 Important Cs, Cs/metal and metal oxide nanocomposites and scaffold material usage in biomedical application

No.	Cs	Metal, metal oxide nanocomposites/Scaffold materials	Drug delivery application for various infections	References
1	Cs/metal formulation	Cs/silver nanocomposites	Anti-cancer therapy	[20]
2	Cs/metal formulation	Cu/Cs	Tissue engineering	[8]
3	Cs/metal formulation	Cs/Cu/Ag	Environmental pathogens	[21]
4	Cs/metal formulation	Cs/graphene	Anti-cancer therapy	[22]
5	Cs/hydrogel formation	Cs/hydrogel formation	Biomedical application	[23]
6	Cs/metal oxide nanocomposites	Cs/CuS	Photothermal cancer therapy	[24]
7	Cs/metal oxide nanocomposites	Cs/rGO/Fe_3O_4	Anti-biofilm and anti-oxidant activities	[25]
8	Cs/metal oxide nanocomposites	Cs/graphene oxide	Cancer therapy	[26]
9	Cs/metal oxide nanocomposites	Graphene oxide in Cs/alginate	Anti-fungal	[27]
10	Cs/Scaffold materials	Hydroxyapatite/Cs/gelatin	Bone regeneration	[28]
11	Cs/Scaffold materials	Cs/polycaprolactone	Bone defect repairing	[29]
12	Cs/Scaffold materials	Cs/HPMC scaffold	Sore throat	[30]
13	Cs/Scaffold materials	Cs/gelatin	Bone reformation therapy	[31]
14	Cs/Scaffold materials	Cs/calcium phosphate	Cytocompatibility	[32]
15	Cs/Scaffold materials	Cs microspheres/HPCH	Osteochondral regeneration	[33]
16	Cs/Scaffold materials	Cs/collagen	Rheological and release studies	[34]

15.4 Mechanism of Chitosan in Drug Delivery Approach

Drug delivery against various infectious sites has created high demands in the biomedical applications and also it is one of the major issues in the controlled drug delivery approach [35]. The controlled drug delivery process is affected by various environmental stimuli changes such as temperature, pH, light sources, proteins,

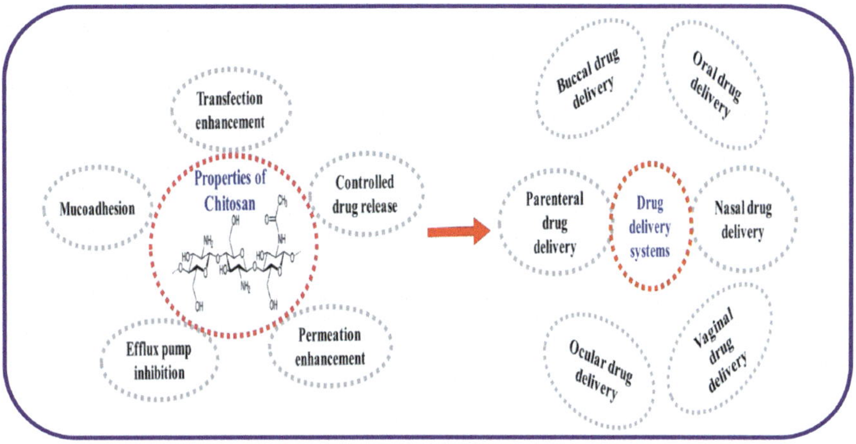

Fig. 15.3 Properties and different drug delivery systems of Cs molecules

enzymes, ultrasound, magnetic field, some organic and inorganic materials. The significant changes in environment affect the swelling and de-swelling of the stimuli responsive hydrogel, and it released the entrapped drugs at an undefined rate [11, 12]. It also depends on the physical behaviours, ionic charges or synthesis method of drug preparation (Fig. 15.3).

Usually, the structure of porous material allows all kind of drugs, antibiotics, bioactive materials, enzymes for entrapping and it accommodates by adjusting internal and outer environment in a hydrogel form. This kind of hydrogel based controlled drug delivery approach is a good way to deliver the drugs compared with other gels because of the cross linkages of injectable hydrogels in the charged surface of the objects. It is prominantly used in scaffold-based controlled drug delivery approach in biomolecule entrapment field [21, 36]. Recently, complete drug delivery processes are needed for additional support with ethical rules. All the entrapment of drugs into the hydrogels or scaffold methods should be followed the defined ethical rules. It must be followed to explain the results clearly. They are as follows: (1). Confirm the gel degradation in target sites, (2). Avoid toxic solvents during entrapment process and (3). Should not produce toxicity. These criteria are used to increase the physiological and mechanical properties within entrapped drugs that are more sufficient to inactivate the infectious pathogens after loading the drugs [23].

Based on the criteria, controlled drug delivery approach with efficient drug transfer into the target sites, the chemically as well as physically cross-linked polymers is the best choice. Compared with the physical and chemical methods of the cross-linking process, the physical method of cross-linking is the safer preparation technique than chemical preparation method. The increased mechanical properties and drug efficiency with the long-term process are achieved by the physical cross-linking method [15]. Also, the controlled drug delivery based on the covalent bond transfer is more safe and has high performance in injectable drug delivery at the target sites.

Recent reports found that Cs is the dominantly utilized in drug delivery applications. [24]. In particular, biodegradable, biocompatible, decreased immunogenicity and intrinsic antimicrobial activity are more helpful in biomedical application without causing toxicity. Notably, the bio-substitutes, anti-microbial, food packaging, wound dressing and drug carrier approaches by Cs based materials are very effective than any other chemical molecules. All the applications of Cs based drug delivery approaches using various drug molecules are reported by Neema et al. Danya et al. Wentao et al. [25, 28, 29]. The agreed result of Wei-Ting et al. [37] reported that the Cs loaded with cisplatin and demethoxycurcumin drug molecules entered into the bacterial and cancer cells causing complete destruction due to the target site drug delivery.

Particularly, the primary amine groups and hydroxyl groups of the Cs containing more active sites, helped to deliver the drugs into the cells very efficiently. The amino groups present in the Cs molecules help it to become soluble in the acidic solution but Cs is insoluble in neutral or alkaline solution. This inherent property of the Cs helps it to acts as an excellent carrier molecule in some in vivo inhibition studies [38]. The anhydride modification of carboxylated Cs in amphiphilic condition acts as an excellent nanocarrier for solubilizing drugs or achieves the controlled release of drug. In addition, the amphipathic conditioned Cs nanocarrier drugs are excellent reservoir drug candidates in the field of anti-tumour drug delivery and also cancer therapy. Based on the above said information, nanocarriers of Cs based drugs are the excellent controlled drug delivery materials and also deliver sufficient doses of drugs to eradicate the infectious pathogens in target sites and avoid the inherent disadvantages of the conventional therapeutic drugs. Interestingly, the most important advantages using the Cs are, the drug release time because the drug is released for a longer period of time and can minimize the side effect and deliver the drug to the target site position. When the drugs are released in a controlled manner, the infection will be cured completely, and also the long-time delivery of the drug may possible to avoid re-entry of infection to the particular site. In addition, when the drug loaded Cs molecule hits the target, a complex mechanism will be activated in the drug delivery process and drug will be released in a controlled level in target sites of the desired region. It is more helpful to implant transfer, catheterized biofilm formed cells, wound dressing, tissue regeneration and cancer cells drug loading process as a controlled delivery approach [2, 3].

15.5 Target Site Drug Delivery

In recent years, the drug resistance in human infections is heightened due to the poor solubility, poor degradability, low stability nature, short period of life, decreased physiochemical properties and increased toxicity of drug carriers [15]. Anti-microbial, anti-biofilm, anti-cancer and anti-immunogenic infections are threatened worldwide population due to the scarcity of administrated drugs into the cells. To improve the effectiveness of drugs within the cells, the optimization of drug delivery system is the better choice to entrap drugs including antibiotics, bioactive compounds and chemical drugs to achieve a therapeutic drug effect. In particular, this kind of approach is the excellent choice for cancer therapy [24]. Specifically, the absorption, distribution and the extent of, drug release by the drug delivery model are very important, while achieving optimize therapeutic effects, improvement of patients' condition, safety and convenience. In addition, the recent drug delivery model approach is mainly focusing on the optimal speed of drug release and enhancing the ability of drug delivery and the ability of the target site drug delivery into the cells. In this drug delivery approach, the side effect of drugs through in vitro and in-vivo experiment is an important criteria to consider for utilization in the medical field. To overcome this problem, surface modification of carrier molecules that improve the nature of their targeting ability in humans and also decrease in the toxicity nature of drugs are necessary.

Among the various carrier molecules, polymers are the best choice to deliver the drugs into the cells using as a carrier molecule. Initially, doxorubicin is the important cancer drug used within the polymers as a drug carrier to reach the target site in cancer cells and it enhances the cellular uptake and destroy the cancer cells vigorously [23]. The target delivery of selecting drugs against specific diseases without damaging the original cells is proposed by Paul Ehrlich, and the process called hypothetical drug is the magic bullet. Based on this advantage, scientists are concentrating their attention on the development of ideal drugs that specifically target the particular site of action. The progress of this kind of drug delivery is very effective and highly used in the biomedical field as a format of nanomedicine. In addition, it is composed of three different roles, (1) therapeutic agent, (2) target moiety, (3) carrier system [39]. Among these three, carrier molecules are very important because it affects the pharmacokinetics and pharmacodynamics of the drugs. In this material, synthetic or natural polymers such as biopolymers, lipids, dendrimers and surfactant are important components [40]. Out of these polymers, Cs is an excellent polysaccharide rich biopolymer and it the most utilized biopolymers in the biomedical application worldwide. The positively charged Cs molecules have excellent biocompatibility, abundant bioavailability, unique mucoadhesivity, complete biodegradability, low immunogenicity, inherent pharmacological and biomedical properties [34]. All the properties of Cs are influenced by molecular weight and degree of deacetylation, and chemically modified Cs are also a great opportunity in the medical field to act as a carrier molecule which is available in the form of trimethyl quaternized, carboxyalkyl, sugar bearing, bile acid modified, cyclodextrin linkage and thiolated Cs [41]. Notably, thiolation of Cs is composed of disulfide bonds with more cysteine rich subdomains of mucus glycoproteins and has the excellent improved capacity in mucoadhesive property. In this format, enormous amphiphilicity nature creates a suitable drug delivery vehicle for cancer therapy. In addition, the small size nanoparticles are an excellent reservoir, which can be administrated for different antibiotics, bioactive compounds, enzymes, drugs and also used intravenously for target drug delivery. This is similar to all other Cs as an excellent target site specific drug delivery vehicle in various biomedical applications [10].

Therefore, Cs has excellent strength in physical, chemical and biological properties for drug delivery approaches especially mucoadhesive nature that influenced the mucous layer and significantly heightened the drug nature in the vigorous time of infection at the mucosal surface. Based on this advantage, the Cs can increase the penetration ability into the human tissue cells by proving its ability in the drug delivery process.

15.6 Biofilm

Worldwide, millions of death evidences are reported due to the rapid spread of bacterial infections every year [42]. Till-date third generation cephalosporin group of antibiotics are the major choice to treat bacterial infections. Extensive and continuous use of inappropriate antibiotics led to the increase in drug resistance in bacteria. The drug resistance bacteria and their virulence effect in human beings are very dangerous especially carbapenem resistant and ESBL producing Enterobacteriaceae including *A. baumannii*, *P. mirabilis*, *K. pneumonia*, and also gram-negative *P. aeruginosa* [43]. In 2017, WHO announced that drug resistance bacteria listed in various classes based on the criteria, 1. critical, 2. high and 3. medium [44]. Among these carbapenem and ESBL producing bacteria are the fatal pathogens which cause severe infections and considerable death in developed countries. Particularly, antibiotic resistant bacteria are not only involved in the transformation of structure and gene mutation in bacteria, but also involved in the bacterial biofilm formation.

Biofilm is a well-organized bacterial community incorporated in a self-assembled extracellular polymeric matrix which adhered permanently to the inherent surfaces [45]. It consists of rich polysaccharides, proteins, amino acids, lipids, extracellular DNA and nucleic acids. In specific, exopolysaccharides provide not only a satisfactory environment for bacterial growth but also serve as an excellent physical barrier [46]. This exopolysaccharide protects the bacterial outer layer by providing high antibiotic resistance and antibiotics infiltration by restricting the penetration of drugs [47]. Drug resistant bacteria causing biofilm formation are very dangerous in medical and clinical field. This is because biofilm forming bacteria are 1,000 times resistance than drug resistant bacteria due to the signalling behaviour of plenty of pathogenic gram-negative bacteria [48]. Exopolysaccharide creates automatic antibiotic inactivation and enhances the suitable surface environment for foreign pathogenic particles, so as to prevent the infections caused by biofilms that is the most obstinate challenge.

In environmental media, the biofilm pathogens triggered the microbial contamination and it is the most severe threats to public health worldwide [49]. Recent researchers strongly reported that the biofilm is a special group of bacterial community which develops the strong physical and chemical barrier with the help of autogenerated extracellular polymeric substances. It protects the embedded bacteria and developed the immune defences as well as environmental stress against antibiotics, chemical compounds, drugs and any other format of inhibiting factors [50]. Sometimes, with the partial elimination of biofilm bacteria in water, food and other industrial materials, it easily develops continuous microbial contamination. All the food, agricultural and industrial materials are highly compromised by biofilm bacteria and chance for contamination. In the industry, the biofilm bacteria are key factors for high level of damage in industrial devices resulting in economic loss. In this condition, the intracellular survival of biofilm forming bacteria is more resistant to all kinds of existing antibiotics and also has a chance for treatment failure. Also, the extracellular biofilm pathogens are acting as intracellular pathogens as pportunistic. In recent years, the enterobacteriaceae contributed to 90% of biofilm formation in different surfaces including *K. pneumoniae*, *E. coli*, *A. baumannii*, *P. mirabilis*, *P. aeruginosa* and all the carbapenem and ESBL resistant pathogens. The survival of all these bacteria within the biofilm is important and their host defences nature is 1,000 times vigorous than other drug resistant pathogens [51]. The important reason of the host defence mechanism is the extracellular matrix and alteration of metabolic activity in bacteria, which adopt dormant phenotypes and low responsiveness to antibiotics.

15.7 Mechanism of Biofilm

The more viscous, 3D arrangement of bacterial communities within the outer cellular matrix layer are formatted to protect and fence the pathogenic microorganisms enveloped within [52]. The extracellular matrix of the biofilm structure is constructed with more polysaccharides, eDNA and eRNA of nucleic acid, more lipid content, high protein and amino acid and some other host substances [53]. Extracellular layers of the biofilm cells provide more viscoelasticity and these layers help to stablize the biofilm cells in the surface of the biofilm and it also influenced by mechanical stress and thermo tolerance. Biofilm formation prevents the sequestration and diffusion of the molecules, which can impact biofilm function including metabolism of cells and mobility of cells that lead to antimicrobial resistant populations. These continuous persistent cells can utilize the repopulation of the remaining biofilm materials leading to recurring infections. In addition, the cell-to-cell communication model of the biofilm formation is very dangerous and its resistant rates are unpredictable [54]. Recent years, the researchers reported the cell to cell communicating factors of biofilm cells. Recent reports of biofilm formation are mediated by the cell to cell communication process and it called as quorum sensing [55]. The cell density dependent signalling system relies on the making and identifying of small, diffusible, extracellular molecules called as an autoinducers, which act upon gene expression. Quorum sensing regulates the motility, virulence, sporulation, conjugation, in biofilm formation [56]. The known signalling factors of quorum sensing system are N-acyl homoserine lactones that act as an important signalling factor in gram negative bacteria. Hence, current time is critical and there is an emerging need to discover novel strategies to combat biofilm bacteria in bacterial infections [49].

15.8 Effect of Chitosan on Biofilm Formation

In recent years, researchers are concentrating in the biological applications of Cs in various infectious diseases. This is because Cs has unique properties, excellent biodegradability, biocompatibility and decreased toxicity than other chemical [45]. It has significant antimicrobial activities against various bacteria, fungi and cancer cells. In addition, Cs is used in all the fields including food, agriculture, flocking in water treatment, sensor, industrial enzyme preparation, waste water treatment, plant defences, food additive, supplement in food preservation, dehydrating agent in cosmetics, hydrogel film in various biopharmaceutical industry and drug delivery carriers. Among these, the antimicrobial properties of Cs against multi drug resistant bacteria, fungi and cancer cells are heightened worldwide and its antimicrobial properties depend on its physiochemical properties. In this chapter, we highlight the antimicrobial applications of Cs against biofilm producing bacteria.

In the last few decades, the antimicrobial properties of Cs against bacteria are thoroughly identified and reported in plenty of articles [46]. Mechanistically, the antimicrobial effect of Cs is highly depending on their structure, physiochemical characteristics and environmental conditions due to the activation of reactive hydroxyl groups at the c-3 and C-6 position [43]. In addition, Cs is classified in two different ways of approach against bacteria, i.e., High molecular weight Cs and low molecular weight Cs. Both the Cs molecules are very effective in bacterial inhibition and their mode of actions are also very effective against intracellular effects and extracellular effects or both based on the targeting site of the antimicrobial effects [42]. In particular, due to the high molecular weight of Cs, it cannot enter into the cell wall and cell membrane, and also the potential antimicrobial properties are involved as a chelator. It can prevent the external nutrients from being taken up by cells and also help to avoid the cell permeability of external pathogens [43]. Importantly, the low molecular weight Cs is very important in antimicrobial research, because it has the ability to inhibit the intracellular and extracellular bacterial cells. Further, it can be altering the RNA, protein synthesis, amino acid production, signalling factors, mitochondrial function, enzyme production, extension of cell cycles and excess nutrient. Then, the inhibition effect of Cs is highly dependent on the type of targeted microorganisms [48].

15.9 Mechanism of Chitosan on Bacterial Biofilm

Based on the anti-bacterial activity, the potential inhibition ability of Cs is depending on the bacterial class such as gram-positive bacteria and gram-negative bacteria, and Cs is affected by these bacteria based on their surface nature [50]. Usually, gram positive bacteria have more peptidoglycan and teichoic acid [51]. Instead, the gram-negative bacteria have more lipopolysaccharide. Based on the different cell surfaces structures of the gram-positive bacteria and gram-negative bacteria are the lead target sites for Cs and susceptibility also depends on surface nature of these bacteria. In addition, more negatively charged gram-negative bacterial surface charges are easily attached to phosphorylated groups of positive charged Cs due to the influence of lipopolysaccharides. Also, the cationic Cs with phospholipids is more susceptible to negative charged gram-negative bacterial surface.

In Fig. 15.4, the positively charged Cs molecules may interact electrostatically with negatively charged teichoic acids and lipopolysaccharides (found in Gram-positive bacteria) and lipopolysaccharides (found in Gram-negative bacteria) to block intracellular and extracellular exchanges, disrupt cell walls, and ultimately cause cytoplasmic leakage. Importantly, the environmental pH comes below 6.5, the gram-negative bacterial surface and Cs surface is linked each other and can easily enter into the bacterial cells, and then destroy the complete bacterial cells [43]. These evidences are more suitable to inhibit the gram-negative bacterial infections, and also confirmed that the gram-negative bacteria are more susceptible to Cs than gram positive bacteria. Even some researchers are reported that the gram-positive bacteria are also very sensitive to Cs, because rich teichoic acid of the gram-negative bacteria has a negative charge due to the presence of phosphate group in their structure. Only, the mutation of teichoic acid in the biosynthesis pathway is quite different to the Cs mechanism and also resistance to Cs, indicating that the mode of action of Cs is little complex than simple electrostatic interactions. Apart from this, the thick cell wall of the gram-positive bacterial surface is also preventing the Cs from binding directly to the cell membrane.

The antifungal effect of Cs against plant and human affected with fungal infection is reported by Hua et al. [57]. The anti-fungal activity between the fungi and Cs is mainly interpreted the interaction of Cs with the cell wall or cell membrane. In addition, the minimum inhibition concentration effect of the Cs against fungi differed and is highly linked with the molecular weight and degree of deacetylation of Cs, solvent pH and type of fungus and fungal target sites. Further, the fungal cell

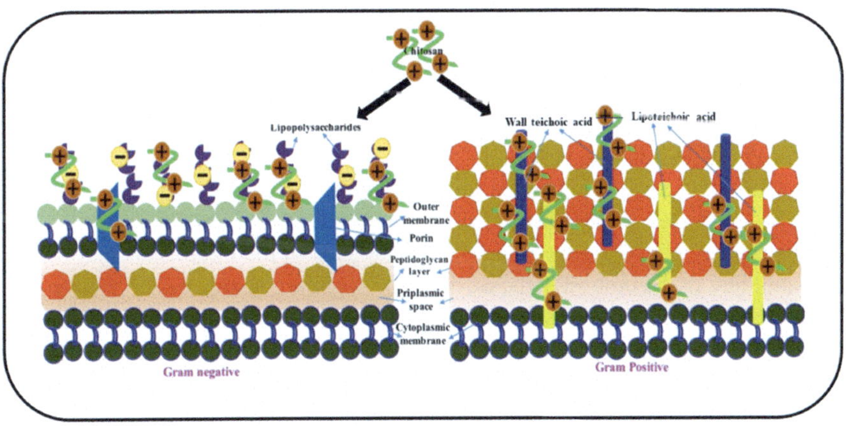

Fig. 15.4 Mechanism of Cs on biofilm producing gram-negative bacteria interactions

membrane has more unsaturated fatty acid content and it can be correlated with the Cs formed due to the interaction of negative charged surface and positive charged counter surface nature [4]. So, fungal surface membrane cells are more susceptible to Cs due to the excellent membrane fluidity of unsaturated fatty acids. Importantly, the highly dangerous *Candida albicans*, *Candida tropicalis* and other *Candida* species are more susceptible to Cs due to excess production of negative charged particles [14]. Also, some fungi including *C. tropicalis* expressed and increased more than 1,000-fold susceptibility to Cs [43]. Similarly, the low molecular weight of Cs help it to easily penetrate into the extracellular cell wall and cell surfaces, and acted as an antifungal agent. After entering into the cell wall, it can inhibit the DNA/RNA and protein synthesis. Evidently, some researchers reported that the Cs could be inhibited by the mitochondrial activity in various fungal cells.

Cs is a natural biopolymer that is composed of rich polysaccharides of monosaccharides of glucosamine and N-acetyl glucosamine. Each monosaccharide has three essential functional groups, amino group at C_2 position and hydroxyl groups at C_3 and C_6 positions [2]. Different monosaccharides of the Cs provide various physiochemical properties, particularly important structural changes including degree of deacetylation, viscosity and molecular weight. Among these three structural changes, degree of deacetylation and different molecular weights of the Cs are involved in the increased biomedical applications including anti-bacterial activity and anti-biofilm activity. In addition, it helped to determine the solubility and viscosity of Cs. Recent studies agreed that the polycationic nature of the amino groups and N-acetylglucosamine groups of the Cs have unique anti-biofilm property [4, 5]. In biofilm formation, a greater number of the virulence behaviour exhibited negative charged surfaces that are highly sensitive to cationic nature of the Cs through electrostatic interaction. Notably, exopolysaccharides, proteins, amino acids, enzymes, DNS generated more negative charged molecules, and it is compromised by positive charged Cs through electrostatic interactions. After the Cs penetrated into the biofilm bacterial cells and interacts with the cell cycle process it completely arrests the biofilm organelles [6]. Also, the biofilm forming gram negative bacterial cells and their internal components of carbohydrates, proteins, cytoplasmic granules, cell membrane leakage materials are compromised to Cs through functional group interaction. After this process, the permeability of bacteria is lost due to its internal ability to produce the virulence factors, resulting in the leakage of more cytoplasmic content, which obviously leads to cell death [7].

After this process, Cs enhanced the cell membrane components and it can be further be stimulated to the conjugation process with cationic antimicrobial peptides [9]. Then, it actively prevents the biofilm organelles in the target sites and creates internal failure in the biofilm cells, particularly in gram negative bacteria and sometimes in gram positive bacteria. Finally, the polymeric nature of the Cs residues is involved in the chelation process with the help of many important metal ions including zinc, calcium, and magnesium. All the metal ions act as a precursor molecule and are involved in the transcription and translation process of bacterial biofilm genes; thus, these processes get altered or stopped causing cell death [10].

Further, Cs is used to prevent the biofilm contamination in hospitalized infections and medical devices such as catheters, orthopaedic implants by gram positive bacteria of *S. aureus* and *S. hemolysis*, and gram-negative bacteria of *E. coli*, *P. mirabilis*, *A. baumannii* and *P. aeruginosa*. Interestingly, Cs prevents the biofilm formation in catheterized materials and implant surfaces made up of titanium, catheter tubes, titanium alloy, aluminium steels and stainless steel. Especially, the gram-positive bacteria i.e., *S. aureus* and *S. epidermis* are completely eradicated and also affects significant bacterial shrinkages. This efficient protection procedure for medical device surface using Cs coating method can be considered by reducing the high risk of device related infections in the medical and clinical field. Specifically, the concentration dependent Cs and its timing of prevention has very important advantages to prevent the biofilm cells in long term. In food industry, Cs served as packaging materials for food preservation including junk foods, vegetables, fruits, sea foods and processed meat, gel, fibre and edible film [13]. Furthermore, Cs and Cs made films acted as biodegradable materials with antimicrobial ability to avoid the nutritional behaviour of moisture, gaseous and microbial spoilage for safe health and without environmental impacts [14].

Importantly, the anti-biofilm activity of Cs is available in many structural forms and they are very effective against biofilm forming bacteria through intracellular and extracellular levels as a format of unmodified nature, hydrogel, high and low molecular weights, chemical modification, nanoparticle and conjugation. Overall, Cs, owing to its highly reactive chemical molecules provides greater efficiency in the entire research field and is applied in different life aspects in order to destroy microbial contamination and decrease health and environmental concerns.

15.9.1 Chitosan/Metal/Metal Oxide Nanocomposites on Biofilm Formation

Nanotechnology is of greatest interest worldwide due to the potency of application in diverse fields, particularly medicine, clinical, pharmaceutical and cosmetic industry [58]. It deals with the synthesis of various nanoparticles having particle size below 100 nm. Among various usage materials, nanoparticle gained attractive attention due to their unique properties in biological, chemical and physiological materials to their macro and micro counterparts. Especially, green synthesized nanoparticles are involved and produced the greatest revolution in biomedical application being cost effective, less time consuming and low toxicity [59]. In the medical field, the frequent usage of antibiotics may lead to the development of multi drug resistant behaviour in pathogens, and also emerging need of new classes of antibiotics to eradicate these multi drug resistant pathogens. The existing antibiotics have lower biocompatibility with decreased biological properties. On the other hand, nanoparticles can cross the outer cellular layer of resistant microbes, overcome the efflux barrier and decrease the colonial formation in drug resistant bacteria. Due to this defect, more number of different antibiotics are synthesized for complete eradication of drug resistant and biofilm forming pathogens. Likewise, more number of nanoparticles including silver, copper, zinc, gold, titanium dioxide, and the recently emerged 2D materials of graphene, Mxene having broad spectrum antimicrobial activity against multi drug resistant bacteria, are synthesized and applied in biological activities [60, 61]. All these nanomaterials are used to eradicate the bacteria through ROS damages. After entry of nanoparticles into the outer membrane, it simulates the ROS in higher rate and stops the bacterial mechanism.

Different nanoparticles having different properties are also used in different fields such as food, agriculture, water industry, energy, environment, catalysis, biomedical application, pharmaceutical industry, drug delivery approach, photo electrochemical sciences, electronic, and mechanics [62]. In addition, nanomaterials are very effective than other bioactive compound, antibiotics and drugs due to the unique structure and properties. It can be synthesized from various methods like physical, chemical and biological having very effective biological properties against infectious pathogens. Among these, chemical method is very effective and is used different types of chemicals for production process, and it takes lot of time and has high production rate. In addition, the usage of different chemicals may change due to the formation of toxic by products, hence the recovery and production cost is very high. Also, it cannot be used sometimes in all the applications. On the other hand, the physical method of nanoparticle synthesis is more time consuming and highly expensive due to the usage of energetic radiations. Moreover, the physical method alters the physiochemical properties in nanoparticles when using the energetic environment. Also, it produced some side effects and could change into toxic.

Instead, the biological method of nanoparticle synthesis is very effective than physical and chemical methods, using natural materials as a reducing agent such as plant materials, microorganisms, enzymes, polysaccharides, proteins, peptides and sponges. Compared with the other two methods, the biological mediated nanoparticle synthesis is safe, low cost, time consuming, and eco-friendly. It is a more reliable method having the potential role of beneficial effect and easily accessible sources with large scale production in a short period of time. Further, the production rate of plant mediated nanoparticle is very high than the other biological sources like microbes, algae, sponges, peptides and also than the chemical and physical methods. Also, to avoid the decreased inhibition of nanoparticle efficiency, new strategies are needed to deliver the nanoparticles completely to the target sites and also minimum time for drug delivery has been proposed. In recent years, polymeric nanoparticles or polymer-based metal and metal oxide nanoparticles are used as carrier molecules for non-toxic mode.

Cs is the second most important natural cationic biopolymer after cellulose that has excellent biomedical and pharmaceutical activities. It has been reported previously for its adhesiveness, non-irritability, biocompatibility and is a biodegradable material widely used in various biomedical and pharmaceutical applications. Importantly, Cs can serve as carrier molecules to be applied in the matrix to immobilize and release macromolecules in the inside of the cells. Specifically, Cs are well known for their controlled drug delivery release properties and are used both in-vitro and in-vivo applications. Non-toxic and non-immunogenic properties of the Cs are used systematically in the treatment of medical especially biofilm eradication. Previously, the nanocomposite formation of Cs and nanoparticles has been investigated, as Cs itself is known to produce antimicrobial activity, and also play a role of stabilizing the nanoparticles [63]. Hence, the result of the investigation is concluded that the Cs acts as an antimicrobial agent as well as stabilize and protect the formed nanoparticles from agglomeration and growth in the same pot. This proposed mechanism is evidently proved by Sanaz et al. [64] and the nanoparticles are synthesized using plant sources as a reducing agent, and the differentiation is confirmed by physiochemical characterization of both nanoparticle alone and Cs/metal nanoparticle combination. Moreover, further result of Cs/metal nanoparticle combination is providing the evidence that the particle size, zeta potential and morphology changed the activity against biofilm producing bacteria.

Further, Cs molecules improve the metal nanoparticle behaviour and immune responses against pathogenic microbes. In the receptor of M and dendritic cells of the Cs simulated the available nanoparticles potency into the cells, and increase the intestinal permeability, and improve humoral and cellular immunity by increasing the immune cells that helped to directly kill the bacteria. Recently, Yulong et al. reported that the Cs nanoparticles enhance the anti-biofilm activity against various biofilm bacteria. In this study, Cs and its derivatives are reported as suitable carrier molecules for drug delivery. The biofilm matrix is the target in this study, and the negatively charged surface of the biofilm formation is solubilized by Cs nanoparticles [65] (Fig. 15.5).

The biofilm matrix materials of polysaccharides, proteins, extracellular DNSA and other components are reported. Also, the detachment of biofilm structure is formed and more number of dead biofilm cells are produced. Importantly, the combination of antimicrobial drugs and DNase helps to inactivate the promoter genes in the biofilm cells and the biofilm matrix is disassembled and enhanced the drug nature to kill the biofilm cells. This new strategy of antibiotic delivery combined with Cs molecule is a better choice to eradicate biofilm formation and biofilm forming pathogens. The polymeric nanoparticle of positive charges is easily bound with negative charges of biofilm surface genes and easily enters into the biofilm cells. The drugs are diffused within the target places of biofilm cells and deactivate the biofilm signalling factors

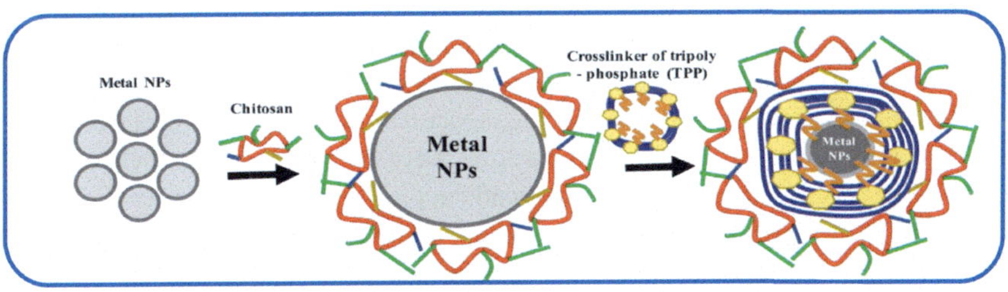

Fig. 15.5 Mechanism of the ex-situ approach for Cs-metal nanoparticle synthesis

and then delivered the drugs into the biofilm matrix. In this condition, the drug is needed only at the lowest concentration to inhibit the biofilm formation. This new strategy is proposed where the cationic charges of Cs and negative charges of microbial membranes are binding each other. But Cs alone could not inhibit the biofilm growth effectively, instead positive charged drug molecules with Cs combination easily penetrated into the negative charged biofilm surface and interferes with microbial cells inside the biofilm. Pérez-Díaz et al. documented that the biocompatible and biodegradable hydrogels of Cs exhibited structural similarities with extracellular molecules as well as the forms stable complexes with the addition with molecules having excellent antimicrobial and anti-biofilm properties. For this advantage, Food and Drug Administration has approved Cs for use in various biomedical applications especially, antimicrobial activity, anti-biofilm activity, anti-cancer activity and wound dressing [66]. Rajagopalan et al. worked with Cs/Ag/ZnO nanocomposite against biofilm producing bacteria, and achieved the remarkable antimicrobial activity. In the biofilm eradication process, potential anti-biofilm activity of Cs/ZnO nanocomposites is acheived due to the addition of controlling amount of Zn salt during its preparation through sol-gel transformation. Nanocomposites formation using Cs is a new generation and excellent biomaterials against infectious pathogens, and it requires inhibiting the intracellular components of the biofilm. Also, it creates the inhibition of phagocytosis and inflammatory responses against various biofilm forming bacteria [67].

15.9.2 Chitosan Diminishes the Toxicity Level in Nanoparticles

Recent years, the nanoparticles usage in the biomedical field has been increased in the entire world and is being done to enhance this efficiency compared with existing drugs and antibiotics. All the nanoparticles are having different biomedical properties in different concentrations. Most of the nanoparticles are owing to be aware in the environmental concern due to their toxicity. Since, more nanoparticles are used as a food additive, which exposes them to most of the living organisms and it affects significant health issues for humans. Sometimes, nanomaterials are used for anticancer agents as a form of light sources. It could possible to cause toxicity in human and there is a need to reduce the toxicity on the environment, food and medicine. In particular, the bioactive compounds and organic compound moieties of plant materials and microorganisms are the best choice to decrease the toxicity nature in metal oxide nanoparticles, especially secondary metabolites, proteins, amino acids, enzymes, fatty acids, nucleic acids, vitamins, polysaccharides and other biological active molecules as a format of reducing and capping agents.

In the modern era, the essential need to fabricate readymade ecological and biopolymer-based metal nanoparticle encapsulation to kill infectious pathogens. In particular, polymer-based nanoparticles helped the metal oxide nanoparticles for improving the metal oxide nanoparticles. In addition, the detached effect between the infectious bacteria and surface of the human surface is increased by the matrix of the biopolymer. Then it helps to deliver the drug at the target site. The natural polymer encapsulated metal oxide nanoparticles are getting many advantages and also approaching in drug delivery research by most of the researchers in the world. Based on the excellent specificity and unique mechanism of drug delivery approaches, it is very important to inhibit the bacterial infections. Recently, the synthesis of metal oxide nanoparticles using biological routes are more important due to the enhancement of organic compounds effect into the metal oxide nanoparticles as precursors. In recent years, the metal precursor methods of metal oxide nanoparticles synthesis using plant extract, leaves, bark, seeds, stems, seaweeds, microbes, sponges, animal sources are having incomparable repercussions. Biological mediated metal oxide nanoparticles, notably silver oxide nanoparticles, zinc oxide nanoparticles, copper oxide nanoparticles, graphene oxide nanoparticles possess established antimicrobial, anticancer and antimalarial activities in the biomedical applications.

15.9.3 Toxicity Potential of Nanodrug Carriers

Despite of its considerable potential, using of drug carriers based on metal and metal oxide nanoparticles has significant drawbacks. The use of nanoparticles can cause issues because of their decreased stability, propensity to aggregate, potential to release metal ions, or the ability to change composition by oxidizing their surface. The characteristics of nanoparticles, i.e., their morphology (shape, size), chemical purity, and type of solvent used for dispersion medium are connected to the selection of the preparation method, type of functionalization, and type of biofunctionalization of nanocarriers, as well as their stability and aggregation susceptibility and agglomeration. These factors can affect the level of cytotoxicity in the case of nanoparticles [68, 69]. The way metal ions are produced or how they influence healthy cells, leading to their damage and death, are just two examples of how changes in the aforementioned characteristics affect the toxicity of the substance.

Increased cytotoxicity of nanoparticles is influenced by a variety of variables, including delivery method and accumulation site. When modified nanoparticles are appropriately created, they can prolong the drug's release without harming healthy cells. The size, structure, and content of nanoparticles all affect how well they interact with living things. The following are the key categories of parameters that influence nanoparticle biocompatibility and their non-toxicity. Use of metallic nanoparticles as carriers: the metallic nanocarriers utilized in a certain application may be harmful to particular cell types. On other tissues, though, they might not have the same impact. Depending on the duration and route of drug administration, or the half-life of nanoparticles, these carriers may have cytotoxic effects. Drug-carrier particles' biocompatibility has an impact on how stable they are in a biological environment.

The environment will disintegrate nanoparticles. The pH of the surrounding environment, at low levels may completely dissolve nanoparticles, blocking their function and promoting the migration of metallic compounds throughout the body.

Although the exact mechanisms underlying nanoparticle cytotoxicity are unknown, it is thought that the ROS they produce are the main reason behind the cytotoxicity. If the amount of ROS surpasses the antioxidant capacity of the cell, biomolecules are destroyed in the processes of severe oxidation, which can become a direct cause of cell death. Cell defence mechanisms deal with tiny amounts of ROS. A three-stage model has been used to describe how ROS affect cells. Antioxidant enzyme activity is boosted in the initial stage to protect the cell. Inflammation results from the increase in pro-inflammatory cytokines in the second stage and cell death results from mitochondrial disturbance in the third stage. Based on the observation of the toxic effects of ZnO NPs on phagocytic or bronchial epithelial cells, which manifested in the release of lactate dehydrogenase, damage to DNA, proteins, and lipids, ultimately leads to death by apoptosis or necrosis [70].

The dispersion of ions throughout the body and discharge ofions into the environment are two of the key reasons through why the metallic nanoparticles are harmful. The quantity of biologically active ions produced and their accessibility for interaction with the bacterial cell wall are directly correlated with the antibacterial activity of metal nanoparticles. For instance, AgNPs react by breaking their original structure and altering their original characteristics when present with sulphides and chlorides [71]. According to Levard et al., insoluble silver sulphides can be prepared at very low levels of AgNP sulphidation. This made silver less poisonous in this form, but it also altered its shape, which could not be good for carriers [72]. After entering the bloodstream, metal ions can permeate into cells, even healthy ones. Particle size, shape, the existence of additional surface coatings, or the environment in which they are found are just a few of the variables that could lead to the release of metal ions from nanoparticles. Cell viability is impacted by free ions, which also cause oxidative stress and cytokine production [71]. The nanometallic carrier surface's chemical make-up enables the development of a biocompatible covering for human body cells. The coatings serve two purposes: first, they prevent oxidation of the metallic cores of the particles; second, they enable the attachment of particular functional molecules to the core, such as medicines, targeting molecules, and contrast agents. Additionally, the hazardous activity of metal oxide nanoparticles is intimately related to the release of individual metal ions by dissociation method when contact with cells or biological fluids. It's also crucial how nanoparticles enter the cell. When compared to nanoparticles that entered the cell through a different pathway, endocytosed nanoparticles can harm the cell even more. It is related to the lysosomes' acidic environment, which encourages the release of ions from nanoparticles. When these ions interact with the cytoplasm, they immediately produce oxygen radicals (superoxide radicals).

Another issue that impacts nanoparticle stability and restricts their usage as drug carriers is aggregation. Brownian motion particle collisions cause particle agglomeration and precipitation. Therefore, obtaining nanoparticles that are evenly disseminated and stable in the solution phase is highly desirable (especially in physiological serum). Increasing the repulsion between nanoparticles, which raises their colloidal dispersion, is one of the potential approaches. However, chemical stability in the biological environment is compromised in the case of nanoparticles used in biomedical therapy. For instance, cancer cells' acidic environment (pH = 6–7) might lead to the aggregation of many nanoparticles. However, techniques for producing stable nanoparticles at neutral or acidic pH levels are also known. In order to enhance the interfacial properties of the particles, Shakiba et al. produced Au NPs with good stability under acidic circumstances by forming a self-assembled monolayer. The development of an aqueous layer surrounding the particles leads to increased stability and has improved particle dispersion in the biological medium [73].

Particles smaller than 30 nm are utilized for aesthetic and marketing purposes to achieve transparency in tanning UV protection lotions and creams, where both ZnO and TiO_2 are extensively used. Given the extremely small particle size, it should be taken into account that nanoparticles might penetrate the skin's surface layers and enter living cells. Fortunately, investigations carried out in vitro and in vivo in this case which clearly shows that these nanoparticles did not penetrate into the deeper layers of the skin [74]. Inhalation and oral ingestion are two further ways, in addition to skin exposure, that nanoparticles can enter the human body. Once within, nanoparticles can quickly move to many organs and parts of the body. They can also easily penetrate cells via phagocytic and endocytic pathways. ZnO nanoparticles given orally to mice eventually reach the liver, heart, spleen, pancreas, and bone, according to Wang et al. They also observed that when the dose of the 20 nm nanoparticles increased, the harmful effect on the mouse body decreased. As a result, careful consideration should be given to oral exposure to tiny dosages of ZnO NPs [75]. The absence of pertinent experimental and theoretical data is an undeniable restriction when using metal nanoparticles and metal oxides. In a relatively short period of time, nanoparticles have grown in prominence as drug carriers. Given the impact of metallic nanoparticles on organisms after prolonged use, knowing the biological compatibility of nanoparticles currently requires ongoing research [69] (Table 15.2).

There are numerous ways to stop or restrict the hazardous effects of metal oxides and nanoparticles. The production of nanoparticles with the necessary qualities without hazardous effects can be achieved by altering the shape, size, and techniques to modify their surface [76]. Drug delivery nanoparticles are exposed to a physiological medium that contains a lot of salt and different proteins. These two elements have an impact on the stability of nanoparticles. While proteins are adsorbed on the surface of nanoparticles, changing their size and surface charge, and high salt concentration lower the electrostatic repulsion between the nanoparticles, causing them to aggregate [77]. For these materials to be used safely in biomedical diagnostics and therapy, it is essential to understand how nanoparticles behave in the actual systems and their potential interactions with biological systems.

Table 15.2 Cs diminishes the toxicity level of metal oxide nanoparticles in the combination, and enhances the anti-biofilm activity against various pathogens

S. no.	Metal oxide nanoparticles	Concentration	Cs/metal oxide nanocomposites	Concentration	Toxicity
1	Ag NPs	32 μg/mL	Ag NPs-Cs NCs	62.5 μg/mL	Confirmed against bacteria
2	Au NPs	1000 μg/mL	Au NPs-Cs NCs	250 μg/mL	Confirmed against bacteria
3	ZnO NPs	128 μg/mL	Cs-ZnO NCs	32 μg/mL	Confirmed against bacteria
4	Fe_2O_3	140 μg/mL	Fe_2O_3-Cs NCs	3.12 μg/mL	Confirmed against bacteria
5	Graphene NPs	70 μg/mL	G-Cs NCs	40 μg/mL	Confirmed against marine bacteria
6	CuO NPs	60 mg/ml	CuO-Cs	60 mg/mL	Confirmed against styrofoam
7	ZnO NPs	50 mg/ml	Cs-ZnO NCs	0.625 mg/mL	Confirmed against bacteria
8	Cs NPs	125 μg/mL	Cs-TiO_2	3.9 μg/mL	Confirmed against bacteria
9	Ag NPs	80 μg/mL	Cs-Ag NPs	80 μg/mL	Confirmed against marine shrimp
10	EOs	500 μg/mL	Cs-EOs	100 μg/mL	Confirmed against bacteria

15.9.4 Conclusion

Cs is an excellent biopolymers to enhance the applicability of metal and metal oxide nanoparticles for eradicating bacterial infections, especially biofilm eradication. Cs helps the metal oxide nanoparticle surface to attract the negatively charged biofilm producing bacterial surface by electrostatic interaction. In addition, Cs also helps metal and metal oxide nanoparticles to increase their efficiency and long term stability on target sites. Further, Cs acts as an excellent biocompatibility and toxic free carrier molecules due to the chemical and structure similarities with metal and metal oxide nanoparticles. The toxicity reduction effect of Cs mediated metal oxide nanocomposite materials is effectively proved against marine shrimp and buffer fish model. Altogether, the current chapter deals with the role of Cs that enhances the anti-biofilm activity with reduced toxicity.

Acknowledgements Dr. G. Rajivgandhi and Dr. Franck Quero acknowledge the financial support from ANID-FONDECYT (Chile) under the Postdoctoral Fellowship No. 3220019.

References

1. R. Swarup, R. Jong-Whan, Fabrication of chitosan-based functional nanocomposite films: Effect of quercetin-loaded chitosan nanoparticles. Food Hydrocolloids **121**, 107065 (2021)
2. S. Sandeep, S. Gurpreet, B. Niraj, A. Kanika, Characterization and preparation of Fe_3O_4 nanoparticles loaded bioglass chitosan nanocomposite coating on Mg alloy and in vitro bioactivity assessment. Int. J. Biol. Macromol. **151**, 519–528 (2020)
3. M.M. Hany, M.E. Nagi, M.A. Ahmed, M.I. Mohamed, M.E. Ibrahim, Potential of chitosan nanoparticle/fluoride nanocomposite for reducing the toxicity of fluoride an in-vivo study on the rat heart functions: Hematopoietic and immune systems. Int. J. Biol. Macromol. **216**, 251–262 (2022)
4. F.N. Ozioma, S. Sudarshan, S. Paosena, K. Vongkamjan, P.V. Supayang, Enhancement of food shelf life with polyvinyl alcohol-chitosan nanocomposite films from bioactive Eucalyptus leaf extracts. Food Biosci. **36**, 100609 (2020)
5. L. Huixing, S.M. Maryam, A. Azadeh, B. Hadi, L. Xiang, S.-S. Saeed, T. Davood, K. Amirsalar, Fabrication of tragacanthin gum-carboxymethyl chitosan bio-nanocomposite wound dressing with silver-titanium nanoparticles using freeze-drying method. Mater. Chem. Phys. **279**, 125770 (2022)
6. R. Mahdi, A. Raman, S.K. Hossein, S.-I. Vahid, A novel bioactive quaternized chitosan and its silver-containing nanocomposites as a potent antimicrobial wound dressing: Structural and biological properties. Mater. Sci. Eng. C **101**, 360–369 (2019)
7. Z. Xiaonan, Y. Hui, X. Lu, L. Zhikun, M. Chuang, X. Wei, W. Yingbo, Chitosan regulated electrochemistry for dense hydroxyapatite/MgO nanocomposite coating with antibiosis and osteogenesis on titanium alloy. Colloid Interface Sci. Commun. **48**, 100616 (2022)
8. B. Nishat, K. Asha, T. Nitika, S. Gaurav, S. Ragini Raj, S. Rahul, Phytochemically stabilized chitosan encapsulated Cu and Ag nanocomposites to remove cefuroxime axetil and pathogens from the environment. Int. J. Biol. Macromol. **212**, 451–464 (2022)
9. A.-T. Ghasem, I. Alireza, O.A. Zahra, H.T. Adel, Synthesis and characterization of chitosan encapsulated zinc oxide (ZnO) nanocomposite and its biological assessment in pepper (*Capsicum annuum*) as an elicitor for *in vitro* tissue culture applications. Int. J. Biol. Macromol. **189**, 170–182 (2021)
10. D.S. Vidya, A. Atif, A. Oyouni, M.H. Yousef, S.A. Saad Ali, M.Al.A. Osama, Al.Z. Waseem, H.M. Ahmad, Chemopreventive role of tin oxide-chitosan-polyethylene glycol-crocin nanocomposites against lung cancer: An in vitro and in vivo approach. Process Biochem. **120**, 186–194 (2022)
11. J. Zheng, H. Gaowei, Z. Kai, Mannose-anchored quaternized chitosan/thiolated carboxymethyl chitosan composite NPs as mucoadhesive carrier for drug delivery. Carbohyd. Polym. **283**, 119174 (2022)
12. W. Jun, W. Xinyu, L. Haiwen, Q. Mingyue, S. Wenbin, Y. Xu, Z. Zheng, L. Binbin, A hollow chitosan-coated PLGA microsphere to enhance drug delivery and anticancer efficiency. J. Drug Deliv. Sci. Technol. **73**, 103482 (2022)
13. G. Rajivgandhi, C. Gnanasekaran, R. Govindan, C. Chenthis Kanisha, M. Maruthupandy, Q. Franck, A.M. Ramzi, M.N. Omar, A.S. Nasir, L. Wen-Jun, Effective removal of biofilm formation in Acinetobacter baumannii using chitosan nanoparticles loaded plant essential oils. J. King Saud Univ. Sci. **34**, 101845 (2022)
14. M. Su, M. Doris, H. Feng, L. Matthias, S.S. Berit, T. Yulong, Preparation and antibiofilm studies of curcumin loaded chitosan nanoparticles against polymicrobial biofilms of Candida albicans and Staphylococcus aureus. Carbohyd. Polym. **241**, 116254 (2020)
15. E. Dounia, F.E. Walid, S.El-S. Gharieb, Promising advances in nanobiotic-based formulations for drug specific targeting against multidrug-resistant microbes and biofilm-associated infections. Microb. Pathogen. **170**, 105721 (2022)
16. P. Krishna Kumar, T. Muktanand, P. Nidhi, A. Ashish Kumar, G. Shilpkala, A. Meraj, T. Ragini, S. Sanjay, Alginate lyase immobilized chitosan nanoparticles of ciprofloxacin for the improved

antimicrobial activity against the biofilm associated mucoid *P. aeruginosa* infection in cystic fibrosis. Int. J. Pharmaceut. **563**, 30–42 (2019)
17. P. Andi Dian, A. Qonita Kurni, E.U. Sartini, V.-Z. Fabiana, J.P. Alejandro, M.E. Yayu, A.M. Sandra, P. Muh. Rezky, N.T. Irma, F.D. Ryan, Selective delivery of silver nanoparticles for improved treatment of biofilm skin infection using bacteria-responsive microparticles loaded into dissolving microneedles. Mater. Sci. Eng. C **120**, 111786 (2021)
18. T. Nuttaya, P. Rattiyaphorn, P. Jakkapat, C. Pisit, K. Somdej, K. Kwanjai, C. Sorujsiri, Chitosan biological molecule improves bactericidal competence of ceftazidime against *Burkholderia pseudomallei* biofilms. Int. J. Biol. Macromol. **201**, 676–685 (2022)
19. R. Vinit, Y.-G. Yeseul Kim, J.H. Kim, J.L. Lee, Chitosan-gum Arabic embedded alizarin nanocarriers inhibit biofilm formation of multispecies microorganisms. Carbohydr. Polym. **284**, 118959 (2022)
20. A.A. Zeinab, S. Zahra, S. Reza, H. Elham, Esmaeil, Darabpour Cu(II) metal-organic framework@Polydimethylsiloxane nanocomposite sponges coated by chitosan for antibacterial and tissue engineering applications. Microporous Mesoporous Mater. **336**, 111866 (2022)
21. S. Zahra, M.-J. Azin, Molecular dynamics simulations on chitosan/graphene nanocomposites as anticancer drug delivery using systems. Chin. J. Phys. **66**, 362–382 (2020)
22. S. Ishita, B. Aalok, Chitosan based injectable hydrogels for smart drug delivery applications. Sens. Int. **3**, 100168 (2022)
23. A.M. Sheril, P. Arul Prakash, M.S. Mohamed Jaabir, S. Dhanavel, R. Manikandan, A. Stephen, Dopamine-conjugated CuS/chitosan nanocomposite for targeted photothermal drug delivery: *In vitro* cytotoxicity study to establish bio-compatibility. J. Drug Deliv. Sci. Technol. **61**, 102193 (2021)
24. K. Viswanathan, S.A. Mohamad, D. Sandhanasamy, G. Kasi, A. Ayyakannu, G. Marimuthu, Chitosan overlaid Fe3O4/Rgo nanocomposite for targeted drug delivery, imaging, and biomedical applications. Sci. Report. **10**, 18912 (2020)
25. P. Neema, B. Bhashkar Singh, T. Himani, P. Mintu, N. Pushpa Bhakuni, D. Anirban, S.P.S. Mehta, S. Nanda Gopal, Development of biodegradable chitosan/graphene oxide nanocomposite via spray drying method for drug loading and delivery application. J. Drug Deliv. Sci. Technol. **74**, 103555 (2022)
26. J. Huang, J. Jacobsen, N. Genina, S.W. Larsen, H.M. Nielsen, A. Müllertz, H. Mu, Investigating the effect of graphene oxide in chitosan/alginate-based foams on the release and antifungal activity of clotrimazole in vitro. Eur. J. Pharm. Sci. **174**, 106204 (2022)
27. Y.E. Noura, A.M. Nadia, A.A. Nahed, Evaluation of the antimicrobial and anti-biofilm activity of novel salicylhydrazido chitosan derivatives impregnated with titanium dioxide nanoparticles. Int. J. Biol. Macromol. **205**, 719–730 (2022)
28. C. Danya, L. Pei, L. Mengna, Z. Changqing, G. Youshui, G. Yaping, Nacre-mimetic hydroxyapatite/chitosan/gelatin layered scaffolds modifying substance P for subchondral bone regeneration. Carbohyd. Polym. **291**, 119575 (2022)
29. S. Wentao, Z. Xuan, B. Lu, D. Yao, W. Zhe, Z. Yanjun, Y. Shuang, Z. Zhijian, Z. Peng, T. Hong, W. Qing, L. Xiaojie, Alendronate crosslinked chitosan/polycaprolactone scaffold for bone defects repairing. Int. J. Biol. Macromol. **204**, 441–456 (2022)
30. H. Sana, M.S. Rai, A.S. Muhammad, M. Asif, H. Zahid, Smart mucoadhesive buccal chitosan/HPMC scaffold for sore throat: *In vitro, ex vivo* and pharmacokinetic profiling in humans. J. Drug Deliv. Sci. Technol. **71**, 103271 (2022)
31. L. Roshini Yadav, K. Balagangadharan, K. Lavanya, N. Selvamurugan, Orsellinic acid-loaded chitosan nanoparticles in gelatin/nanohydroxyapatite scaffolds for bone formation in vitro. Life Sci. **299**, 120559 (2022)
32. X. Ji, S. Hongwei, L. Xiaohong, W.U. Muhammad, L. Guowen, X. Ziyang, M. Limin, H. Xuecheng, L. Zehua, L. Qian, J. Xulin, Y. Guang, Z. Yu, Injectable immunomodulation-based porous chitosan microspheres/HPCH hydrogel composites as a controlled drug delivery system for osteochondral regeneration. Biomaterials **285**, 121530 (2022)
33. P.M. Eduardo, C.A.M. Virginia, M.H. Marilia, M.G.P. Ana, Influence of blend ratio and mangosteen extract in chitosan/collagen gels and scaffolds: Rheological and release studies. Carbohyd. Polym. **292**, 119647 (2022)

34. P. Narmatha Christy, S. Khaleel Basha, V. Sugantha Kumari, Nano zinc oxide and nano bioactive glass reinforced chitosan/poly(vinyl alcohol) scaffolds for bone tissue engineering application. Mater. Today Commun. **31**, 103429 (2022)
35. G. Huang, Y. Liu, L. Chen, Chitosan and its derivatives as vehicles for drug delivery. Drug Deliv. **24**, 108–113 (2017)
36. A. Azin, H. Shadi, K. Sepideh, Green synthesized-silver nanoparticles coated with targeted chitosan nanoparticles for smart drug delivery. J. Drug Deliv. Sci. Technol. **74**, 103554 (2022)
37. H. Wei-Ting, L. Mikael, L. Yi-Chi, L. Dean-Mo, C. Guang-Yuh, Dual drug-loaded biofunctionalized amphiphilic chitosan nanoparticles: Enhanced synergy between cisplatin and demethoxycurcumin against multidrug-resistant stem-like lung cancer cells. Eur. J. Pharm. Biopharm. **109**, 165–173 (2016)
38. W. Chenqu, Z. Yanyan, L. Fan, B. Songhua, P. Ming, F. Li, Precise engineering of cholesterol-loaded chitosan micelles as a promising nanocarrier system for co-delivery drug-siRNA for the treatment of gastric cancer therapy. Process Biochem. **120**, 265–274 (2022)
39. S. Kariminia, A. Shamsipur, M. Shamsipur, Analytical characteristics and application of novel chitosan coatedmagnetic nanoparticles as an efficient drug delivery system forciprofloxacin Enhanced drug release kinetics by low-frequencyultrasounds. J. Pharmaceut. Biomed. Anal. **129**, 450–457 (2016)
40. Y. Ding, Z.S. Shirley, S. Huadong, S. Kangning, L. Futian, Q. Yushi, Y. Jun, Design and construction of polymerized-chitosan coated Fe_3O_4 magnetic nanoparticles and its application for hydrophobic drug delivery. Mater. Sci. Eng. C **48**, 487–498 (2015)
41. M. Ikram, K. Chaudhary, A. Shahzadi, A. Haider, I. Shahzadi, A. Ul-Hamid, N. Abid, J. Haider, W. Nabgan, A.R. Butt, Mentha pulegium extract for packaging applications. Surf. Interfaces **34**, 102349 (2022)
42. K. Sanket, S.P. Balwant, N.S. Brahma, Superior inhibition of virulence and biofilm formation of *Pseudomonas aeruginosa* PAO1 by phyto-synthesized silver nanoparticles through anti-quorum sensing activity. Microb. Pathog. **170**, 105678 (2022)
43. M.J. Kanchan, S. Amruta, K. Umesh, K.N. Latesh, A.P. Ramdas, S. Jaiprakash, B.K. Bharat, V.S. Ajay, P. Rajendra, G.C. Manohar, Biofilm inhibition in Candida albicans with biogenic hierarchical zinc-oxide nanoparticles. Biomater. Adv. **134**, 112592 (2022)
44. G. Rajivgandhi, M. Maruthupandy, T. Muneeswaran, M. Anand, F. Quero, N. Manoharan, L. Wen-Jun, Biosynthesized silver nanoparticles for inhibition of antibacterial resistance and biofilm formation of methicillin-resistant coagulase negative *Staphylococci*. Bioorganic Chem. **89**, 103008 (2019)
45. L. Jin-Hyung, K. Yong-uy, H.C. Moo, L. Jintae, ZnO nanoparticles inhibit Pseudomonas aeruginosa biofilm formationand virulence factor production. Microbiol. Res. **169**, 888–896 (2014)
46. M.-C. Elizabeth, G.-R. Myriam De la, I.R. C´esar, T.-G. Patricia, A.G.-N. Marco, A.G-N., U.-B. Victor, A.M.R. María, G.-F. Ricardo, Antimicrobial activity and inhibition of biofilm formation *in vitro* and on human dentine by silver nanoparticles/carboxymethyl-cellulose composites. Arch. Oral Biol. **120**, 104943 (2020)
47. M. Maruthupandy, G. Rajivgandhi, S. Kadaikunnan, T. Veeramani, N.S. Alharbi, T. Muneeswaran, J.M. Khaled, Wen Jun, L., K.F. Alanzi, Anti-biofilm investigation of graphene/chitosan nanocomposites against biofilm producing P. aeruginosa and K. pneumonia. Carbohydr. Polym. **230**, 115646 (2020)
48. Y.K. Mohanta, K. Biswas, S.K. Jena, A. Hashem, E.F. Abd_Allah, T.K. Mohanta, Anti-biofilm and antibacterial activities of silver nanoparticles synthesized by the reducing activity of phytoconstituents present in the Indian medicinal plants. Front. Microbiol. **11**, 1143 (2020)
49. G. Rajivgandhi, G. Ramachandran, M. Maruthupandy, N. Manoharan, N.S. Alharbi, S. Kadaikunnan, J.M. Khaled, T.N. Almanaa, Wen-Jun, L., Anti-oxidant, anti-bacterial and anti-biofilm activity of biosynthesized silver nanoparticles using Gracilaria corticata against biofilm producing K. pneumonia. Colloids Surf. A **600**, 124830 (2020)
50. Q. Faizan Abul, A. Iqbal, A. Mohammad, M. Salim, F.Al.-R. Basel, S.A.A. Mohammed, A.A. Yaser, Biofabricated silver nanoparticles exhibit broadspectrum antibiofilm and antiquorum sensing activity against Gram-negative bacteria. RSC Adv. **11**, 13700 (2021)

51. J. Tang, W. Wang, W. Chu, Antimicrobial and anti-quorum sensing activities of phlorotannins from seaweed (Hizikia fusiforme). Front. Cell. Infect. Microbiol. **10**, 586750 (2020)
52. A.Q. Faizan, S.K. Mohammad, A. Iqbal, M.H. Fohad, A.K. Rais, H. Iftekhar, A.S. Syed, A. Walaa, Coumarin exhibits broad-spectrum antibiofilm and antiquorum sensing activity against gram-negative bacteria: In Vitro and In Silico Investigation. ACS Omega **6**, 18823–18835 (2021)
53. S.D. Murugesan, T. Ramar, S. Periasamy, A. Sridhar, S. Srinivasan, T. Ramasamy, Combined effect of a natural flavonoid rutin from Citrus sinensis and conventional antibiotic gentamicin on Pseudomonas aeruginosa biofilm formation. Food Control **90**, 282–294 (2018)
54. S. Xueying, W. Lei, L. Liu, L. Jianke, W. Xiaoxia, Impact of tea tree essential oil and citric acid/choline chloride on physical, structural and antibacterial properties of chitosan-based films. Food Control **141**, 109186 (2022)
55. M. Maruthupandy, G. Rajivgandhi, F. Quero, L. Wen-Jun, Anti-quorum sensing and anti-biofilm activity of nickel oxide nanoparticles against *Pseudomonas aeruginosa*. J. Environ. Chem. Eng. **8**, 104533 (2020)
56. K. Suganya, S. Gowtham, M. Murugan, Decrease of growth, biofilm and secreted virulence in opportunistic nosocomial *Pseudomonas aeruginosa* ATCC 25619 by glycyrrhetinic acid. Microb. Pathog. **126**, 332–342 (2019)
57. L. Hua, C. Beibei, L. Jiangyang, L. Wenzhe, W. Dianhui, W. Jian, C. Xingguang, Effect of hydroxypr1opylation on physical properties, antifungal and mycotoxin inhibitory activities of clove oil emulsions coated with chitosan. Food Biosci. **50**, Part B, 102159 (2022)
58. M.M. Zirar, A.O. Akeem, Y. Elvan, Chitosan/metal oxide nanocomposites: synthesis, characterization, and antibacterial activity. Int. J. Polym. Mater. Polym. Biomater. **70**, 383–391 (2021)
59. A.M.F. Mohammad, Y.F.T. Farha, S.I.A. Sundos, K. Haliza, Roles of chitosan in synthesis, antibacterial and anti-biofilm properties of bionano silver and gold. RSC Adv. **12**, 19297 (2022)
60. K. Achyut, K. Sanjeeb, K. Jibon, C. Devasish, Chitosan−iron oxide coated graphene oxide nanocomposite hydrogel: A robust and soft antimicrobial biofilm. ACS Appl. Mater. Interfaces **8**, 20625–20634 (2016)
61. M. Muthuchamy, R. Govindan, K. Shine, V. Thangasamy, S.A. Naiyf, M. Thillaichidambaram, M.K. Jamal, L. Wen Jun, F.A. Khalid, Anti-biofilm investigation of graphene/chitosan nanocomposites against biofilm producing P. aeruginosa and K. pneumonia. Carbohydr. Polym. **230**, 115646 (2020)
62. P. Abdul Rasheed, A.J. Khadeeja, R. Kashif, P.P. Ravi, H.S. Mostafa, H. Mohamed, S. Ayman, M.A. Aboubakr, A.M. Khaled, Controlling the biocorrosion of sulfate-reducing bacteria (SRB) on carbon steel using ZnO/chitosan nanocomposite as an eco-friendly biocide. Corros. Sci. **148**, 397–406 (2019)
63. E.M.B. Mona Shaban, M.R. Omnia Karem, F.A. Taher, A.Z. Samar, Chitosan and chitosan-zinc oxide nanocomposite inhibit expression of *Las*I and *Rhl*I genes and quorum sensing dependent virulence factors of *Pseudomonas aeruginosa*. Int. J. Biol. Macromol. **149**, 1109–1117 (2020)
64. A. Sanaz, M. Omid, J. Fatemeh Nasiri, T. Majid Jafar, G. Morteza Sasani, S. Somayeh Salmani, M.A. Mohammad Hossein, L. Kyu-Yeon, P. Hyung-Ho, Green synthesis of multifunctional ZnO/chitosan nanocomposite film using wild Mentha pulegium extract for packaging applications. Surfaces Interfaces **34**, 102349 (2022)
65. T. Yulong, M. Su, L. Matthias, M. Doris, M.H. Greta, W. Jia, E. Dominik, S.S. Berit, Enhancing antibiofilm activity with functional chitosan nanoparticles targeting biofilm cells and biofilm matrix. Carbohyd. Polym. **200**, 35–42 (2018)
66. M. Pérez-Díaz, E. Alvarado-Gomez, M. Magaña-Aquino, R. Sánchez Sánchez, C. Velasquillo, C. Gonzalez, A. Ganem-Rondero, G. Martínez-Castañon, N. Zavala-Alonso, F. Martinez-Gutierrez, Anti-biofilm activity of chitosan gels formulated with silver nanoparticles and their cytotoxic effect on human fibroblasts. Mater. Sci. Eng. C **60**, 317–323 (2016)
67. T. Rajagopalan, M. Balasubramanian, V. Sekar, S. Jeyachandran, J. Raja, S. Sathappan, V. Baskaralingam, R. Palaniappan, S. Avinash, Chitosan coated Ag/ZnO nanocomposite and their antibiofilm, antifungal and cytotoxic effects on murine macrophages. Microb. Pathog. **100**, 124–132 (2016)

68. J. Hua, M.G. Vijver, M.K. Richardson, F. Ahmad, W.J.G.M. Peijnenburg, Particle-specific toxic effects of differently shaped zinc oxide nanoparticles to zebrafish embryos (Danio rerio) Environ. Toxicol. Chem. **33**, 2859–2868 (2014). https://doi.org/10.1002/etc.2758
69. G. Rudramurthy, M. Swamy, U. Sinniah, A. Ghasemzadeh, Nanoparticles: Alternatives against drug-resistant pathogenic microbes. Molecules **21**, 836 (2016). https://doi.org/10.3390/molecules21070836
70. J.W. Rasmussen, E. Martinez, P. Louka, D.G. Wingett, Zinc oxide nanoparticles for selective destruction of tumor cells and potential for drug delivery applications. Expert Opin. Drug Deliv. **7**, 1063–1077 (2010). https://doi.org/10.1517/17425247.2010.502560
71. L. Wei, J. Lu, H. Xu, A. Patel, Z.-S. Chen, G. Chen, Silver nanoparticles: Synthesis, properties, and therapeutic applications. Drug Discov. Today. **20**, 595–601 (2015). https://doi.org/10.1016/j.drudis.2014.11.014
72. C. Levard, E.M. Hotze, B.P. Colman, A.L. Dale, L. Truong, X.Y. Yang, A.J. Bone, G.E. Brown, R.L. Tanguay, R.T. Di Giulio et al., Sulfidation of silver nanoparticles: Natural antidote to their toxicity. Environ. Sci. Technol. **47**, 13440–13448 (2013). https://doi.org/10.1021/es403527n
73. A. Shakiba, O. Zenasni, M.D. Marquez, L.T. Randall, Advanced drug delivery via self-assembled monolayer-coated nanoparticles. AIMS Bioeng. **4**, 275–299 (2017). https://doi.org/10.3934/bioeng.2017.2.275
74. A.V. Zvyagin, X. Zhao, A. Gierden, W. Sanchez, J.A. Ross, M.S. Roberts, Imaging of zinc oxide nanoparticle penetration in human skin in vitro and in vivo. J. Biomed. Opt. **13**, 064031 (2008). https://doi.org/10.1117/1.3041492
75. B. Wang, W. Feng, M. Wang, T. Wang, Y. Gu, M. Zhu, H. Ouyang, J. Shi, F. Zhang, Y. Zhao et al., Acute toxicological impact of nano- and submicro-scaled zinc oxide powder on healthy adult mice. J. Nanopart. Res. **10**, 263–276 (2008). https://doi.org/10.1007/s11051-007-9245-3
76. D. Ungor, I. Dékány, E. Csapó, Reduction of tetrachloroaurate(Iii) ions with bioligands: Role of the thiol and amine functional groups on the structure and optical features of gold nanohybrid systems. Nanomaterials **9**, 1229 (2019). https://doi.org/10.3390/nano9091229
77. A. Sironmani, K. Daniel, Drug discovery and development present and future (IntechOpen, London, UK, 2011). Silver nanoparticles—Universal multifunctional nanoparticles for bio sensing, imaging for diagnostics and targeted drug delivery for therapeutic applications

Chapter 16
Chitosan Nanocomposites as Scaffolds for Bone Tissue Regeneration

Fouad Damiri, B. H. Jaswanth Gowda, Swetha Andra, Satheeshkumar Balu, Satish Rojekar, and Mohammed Berrada

Abstract The concept of bone tissue engineering for the treatment of significant bone abnormalities becomes essential when tissue damage exceeds the human body's intrinsic regenerating potential. Over the course of the last decade, an extensive range of biodegradable polymers and scaffolding methods have been produced. Chitosan (CS) is an excellent material to be used as a scaffold in bone tissue engineering because it can promote cell adhesion, cell growth, and bone matrix mineralization. The presence of primary amino groups in CS are responsible for its unique characteristics, including mucoadhesion, transfection, in situ gelation and regulated drug release. Scaffolds made from CS that are sensitive to their surroundings have been shown to be effective in delivering drugs locally and over the long term to massive bone defects.

F. Damiri (✉) · M. Berrada
Laboratory of Biomolecules and Organic Synthesis (BIOSYNTHO), Department of Chemistry, Faculty of Sciences Ben M'Sick, University Hassan II of Casablanca, Casablanca 20000, Morocco
e-mail: fouad.damiri@outlook.fr

B. H. J. Gowda
Department of Pharmaceutics, Yenepoya Pharmacy College & Research Centre, Yenepoya (Deemed to Be University), Mangalore, Karnataka 575018, India

S. Andra
Department of Chemistry, Rajalakshmi Institute of Technology, Chennai, Tamil Nadu 600124, India

S. Balu
Department of Oral Pathology, Saveetha Dental College and Hospitals, Saveetha Institute of Medical and Technical Sciences, Chennai, Tamil Nadu 600077, India

S. Rojekar
Department of Pharmacological Sciences, Icahn School of Medicine at Mount Sinai, New York, NY 10029, USA

S. K. Swain and A. Biswal (eds.), *Chitosan Nanocomposites*,
Biological and Medical Physics, Biomedical Engineering,
https://doi.org/10.1007/978-981-19-9646-7_16

16.1 Introduction

Bone tissue has a complex hierarchical structure as a dynamic living tissue. Bone serves several key tasks in the body, which includes structural integrity, shields key organs from impact, and allows for steady blood flow to nourish the bone marrow [1–3]. Bone has the innate ability to repair itself over the course of time, if there is little trauma or deficiency [4, 5]. However, there are cases when the critical-size bone lesion is formed by accident or trauma, and it is difficult to perform adequate healing of bone tissue defects spontaneously. It is a severe physiological worry to repair this kind of injury [6, 7]. Furthermore, external transplantable implant-based drugs are employed in clinical settings to treat greater or severe bone damage or abnormalities. In general, most individuals accept this kind of implant well, and there have been no severe adverse effects or immunogenic responses. However, only a small number of individuals are immunogenic, and the contamination from a second operation makes it hard to use them widely [8]. Moreover, bone tissue engineering (BTE) comprises major parts that include cells, scaffolds, and growth factors that attain excellent bone simulations and overcome the problem with the existing problem with transplantations mentioned earlier. However, conventional tissue engineering methods are unsuitable for treating this kind of defect because they are time-consuming and inconvenient. Tissue harvesting, cell separation, ex-vivo co-culture on a scaffold, and two invasive surgical procedures are all part of the process [8–10]. It is always preferable to the host's endogenous healing ability since it regenerates and cures bone defects without any adverse response or immunogenicity; hence, facilitated endogenous bone tissue engineering (FEBTE) has therefore been recommended as the most practicable technique for direct bone formation [11, 12]. Most crucially, unlike BTE, it requires no ex-vivo cell culture of autologous cells, hence avoiding the risky invasive medical operation [13]. However, if the same reparative impact as BTE is sought, the FEBTE approach often employs bioactive scaffold, which in-situ stimulates the inherent regeneration capacity of native bone tissue and promotes bone formation and tissue healing. Therefore, the material and composition of the scaffold are of utmost importance, and the bone-like morphology is essential for loading endogenous stem cells and growth medium into the injured region of the host [3, 14].

Scientists have prepared several metals and polymer based implants (titanium and polymethyl acrylate) with the help of implant-based techniques. However, host-material interactions and biosafety of this method are challenging; another challenge is the surgical insertion and related infections [15, 16]. Therefore, scientists further focused on natural and biodegradable materials, which would help to reduce the side effects and limitations of the existing technologies. Natural polymers include hyaluronic acid, bacterial cellulose, and silk firoin etc. These materials are biodegradable, biocompatible, nonimmunogenic and fulfil the requirements of in-vivo transplantation [17, 18]. Still, this material is not that efficient for this kind of application and has some limitations. Hyaluronic acid (HA) faces the issue of low mechanical strength and difficult purification during manufacturing; cellulose has poor biodegradability, and fibroin is not easily accessible. Therefore, this limits the applicability of these materials in BTE. Thus, it is of the utmost importance to identify the most appropriate biocompatible and easily accessible material for the advancement of BTE [19, 20].

Bone defects are a serious concern and require grafts to promote healing; this is a costly problem in healthcare management. CS is a naturally occurring mucoadhesive biodegradable biopolymer [21] and is primarily produced from the exoskeleton of marine crustaceans. It has considerably attracted attention in recent years as a scaffolding biodegradable material for bone regeneration/regenerative medicine and tissue engineering because it has a molecular structure and biological activity like the bone extracellular matrix. When it comes to bone scaffolds, CS stands out as a magnificent biopolymer that facilitates adhesion, promotes osteoblast cell growth, and mineralization of bone matrix [22–28].

CS has excellent biocompatibility, biodegradability, and nontoxicity, which is broadly used in bone repair and regeneration [3, 29, 30]. The hydrophilic properties of CS, on the other hand, could promote better cellular attachment and proliferation on the scaffold interfaces and surfaces [31, 32]. Moreover, CS chemistry is fascinating because the abundant hydroxyl and amino groups on its surface which are amenable to chemical modification and would result in the efficient generation of polymeric materials [33, 34]. Crosslinking CS with collagen would increase the scaffolds'mechanical strength and porosity, making them ideal for supporting the proliferation of MC3T3-E1 cells [35]. By raising the concentration of HA nanoparticles, CS/hydroxyapatite-based smart material composite may be able to encourage the outstanding osteogenic differentiation of hMSC [36]. Osteogenesis might be aided by a CS-based scaffold, which serves as a carrier to efficiently regulate the delivery of osteo inductive substances such proteins, peptides, and medications [37–40]. Biomineralization is very essential for bone health and development; CS is highly used and combined with the other molecules and could help in bone mineralization [41]. CS mineralization template could induce in-situ crystallization of the inorganic functional bioactive materials like magnetic ferric tetroxide and hydroxyapatite [3, 40]. Also, chitosan could be easily modified and used for developing several types of thermos-sensitive-gels/hydrogels and microneedle-based drug delivery in bone repairs and therapeutics [42–45]. Therefore, CS-based scaffolds are a viable choice in endogenous bone defect repairs and regenerative medicine and therapies due to their high processability and designability.

16.2 BTE Scaffolds

Designing scaffolds that may be implanted into the body and aid in bone regeneration is an important part of bone tissue engineering. BTE scaffolds let osteogenic stem cells to adhere and offer a favourable conditions for cells to develop and cause regeneration of new bone; this is essential since cells alone cannot grow in 3D. Ceramic, metal and polymer scaffolds are the three major categories of scaffolds based on their

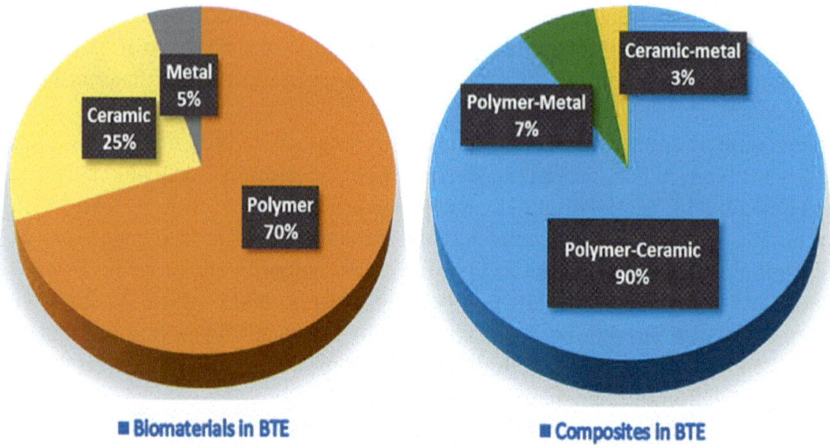

Fig. 16.1 The percentage of different types of pure biomaterials as well as their composite materials that are used in BTE applications. Reproduced with permission from MDPI [47]

primary constituent materials [8, 46]. The Web of Science® database for the years 2017 to 2021 has 7915 articles. These articles demonstrate that polymer scaffolds continue to be the most advanced, following ceramic scaffolds that are combined with polymer scaffolds to produce hybrid scaffolds (Fig. 16.1).

16.3 Bone Repair Strategies

16.3.1 Traditional Bone Tissue Engineering

Despite the fact that auto/allogeneic transplants, the gold standard in the healthcare setting, have been shown to be effective in repairing critical sized bone lesions and their broad adoption is hampered by a lack of donors, immunological concerns, and the danger of infection associated with second surgical surgery [48, 49]. In the last 30 years, BTE has been widely embraced as a viable solution to the problems inherent in auto/allogeneic transplants [50]. The BTE approach may be broken down into its three primary components: a scaffold, cells, and growth regulators. Figure 16.2 depicts the first step of the process, the separation and collection of autologous target tissues, followed by the revival and multiplication of stem cells in specialised culture apparatus. The cells are grown to a suitable quantity, and then planted onto a prefabricated scaffold for in vitro co-culture. Furthermore, favourable circumstances are maintained by the constant addition of growth stimulants and nutrients. This process culminates in the implantation of the newly developed tissues into the patient's lesions, where they may interact with the host tissue and stimulate in vivo repair of the damaged tissue [51–53].

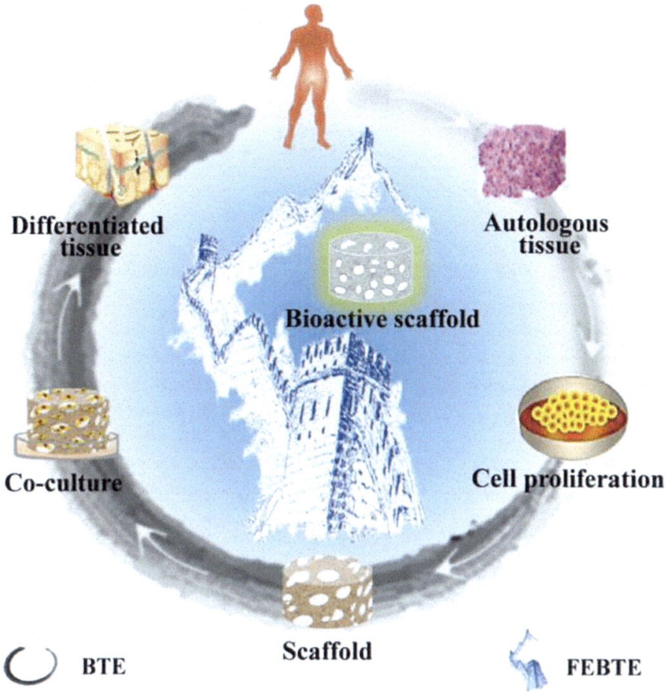

Fig. 16.2 A comparison between two approaches towards the regeneration of bone. Traditional bone tissue engineering (BTE) requires tissue harvesting, cell separation, and co-culturing with a scaffold that is grown outside of the patient's body, but facilitated endogenous bone tissue engineering (FEBTE) eliminates these time-consuming and potentially dangerous steps by employing a biologically active scaffold. Reproduced with permission from MDPI [3]

16.3.2 Facilitated Endogenous Bone Tissue Engineering (FEBTE)

In order to prevent BTE from becoming a costly treatment and reserved primarily for the rich, the challenges posed by its complexities and high price should be resolved. To be more specific, in order to replace labor-intensive manufacturing processes, researchers need to design massive production platforms that are automated and can be replicated. Due to the organism's inherent capacity for repair, the physiological surroundings are used as a bioreactor in order to streamline the laborious BTE procedure and, eventually, accomplish in situ bone restoration. As a result, the methodology known as FEBTE came into existence [11].

In contrast to BTE strategy, FEBTE strategy is a new practical method that makes an effort to get rid of procedures that are laborious, time-consuming, and costly. These processes include tissue harvest, cell separation, and ex-vivo co-culture with a scaffold. Intriguingly, all that is required to implement this method is the placement of a pharmacologically active scaffold into the bone defect areas in order to begin

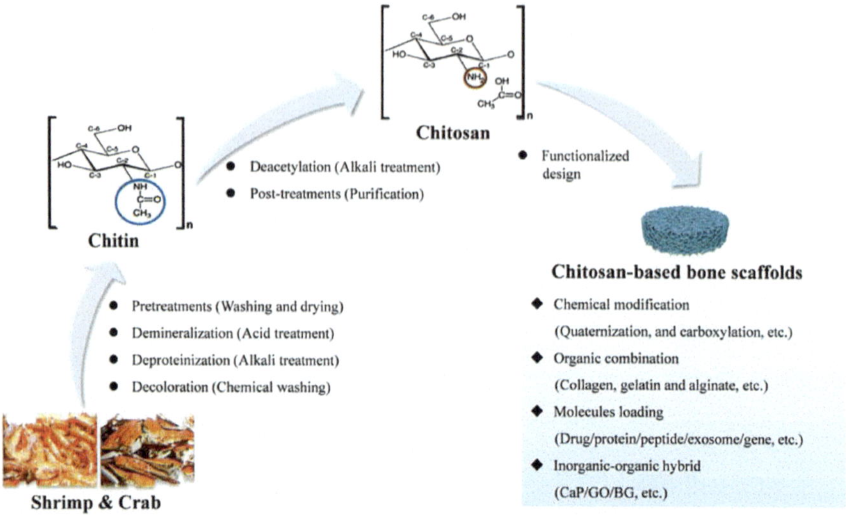

Fig. 16.3 CS is extracted from crustacean shells and applied to the design of bone repair scaffolds through various functionalization strategies. Reproduced with permission from MDPI [3]

the natural process of repairing the damaged tissue (Fig. 16.2). The scaffold operates as a "force of gravity" that draws in and favourably promotes natural stem cells and growth regulators to the injured location. This, in turn, stimulates stem cell proliferation and osseointegration, which ultimately results in the healing of bone defects [14]. In light of this, the FEBTE technique has been able to garner a lot of interest in the realm of scientific research [54, 55]. This is primarily attributable to the fact that it is both practical and economical. Constructing a bone healing scaffold that has outstanding osteoinductivity is the element that stands out as being the most critical and crucial among them.

16.4 Structure and Properties of Chitosan

The potential of CS as a scaffolding material in BTE has generated a great deal of considerable interest. It can produce an intricate pore structure and has a negligible response to foreign substances. CS is a naturally occurring polymer of crystalline amino polysaccharides. It is produced by enzymatic N-deacetylation of chitin, a component of the exoskeleton of crustaceans (Fig. 16.3). The varied biological features of CS and its derivatives are determined by their molecular weight as well as the degree to which they have been deacetylated. The CS crystal structure may be broken down into three distinct sorts, such as α, β and γ types. The most prevalent of these three varieties is α type. A primary hydroxyl group, an amino/acetamido group, and a secondary hydroxyl group are located at C-2, C-3, and C-6, respectively, in CS. The make-up of the amino group is the primary determinant that decides the structures and physicochemical characteristics of these compounds.

16.4.1 Physicochemical Properties

Due to the presence of hydrophobic (–NH_2) groups and hydrophilic (–OH) groups, the cationic character of CS has aroused substantial attention as a biomedical material in BTE. These functional groups help in enhancing degradation rate, biocompatibility, and bioadhesion capability. CS has various physical, chemical and morphological features due to the diverse degree of acetylation (ranging from 50 to 95%) along with their varying molecular mass. In the process of deacetylation, the acetyl groups that are found on the side chains of chitin are removed. This allows the amino groups to be exposed by subjecting the chitin to an alkaline treatment. It is vital to have an exact estimation of the degree of acetylation for the swelling behaviour of the matrix and the release of protein molecules that are caused by interactions between hydrophobic and hydrophilic groups. This also has an impact on the crystallinity and degradation rates of the substance. Therefore, high crystallinity in nature is expressed by having chitin that is 0% deacetylated and CS that is 100% deacetylated. In the same breath, the nature of the other varied intermediate degrees of acetylation reveals itself to be semi-crystalline. However, there are drawbacks, such as insolubility in water, high viscosity, and a propensity to coagulate with proteins when the pH is high. The alteration of the side chains of CS results in the production of a wide variety of derivatives that have uses in biomedicine. A self-cross-linked biomaterial attached on a CS backbone that carries amino and aldehyde moieties has been developed by Fouad et al. [56].

16.4.2 Biological Properties

CS has a broad variety of biological effects and has anti-cancer properties which can be used to treat a variety of medical conditions, including bone formation, wound repair. In the log phase as well as the stationary phase of bacterial growth, it has potent antibacterial action. CS is associated with exceptional haemostatic properties, an increase in the infiltration and movement of neutrophils and macrophages, and assists in the production of granulation tissue, which makes it possible for re-epithelization and fibrous tissue to occur [57]. It has the ability to break through the membranes of cancer cells and exhibits anticancer actions through immunological, antiangiogenic, enzymatic, antioxidant defence, and apoptotic pathways inside the cell [58]. Antioxidant activity is one of the best-known characteristics of CS, which is a characteristic that demonstrates redox regulation action in biological systems. The use of CS in BTE has a number of inherent drawbacks, the most significant of which are its decreased mechanical stability and osteoconductivity. As a result, it is critical to either modify CS or build bio-composites based on CS using one or more biocompatible materials such as polymers or ceramic. In a similar vein, in order to circumvent the problem of CS low solubility and boost the efficacy and efficiency of drug delivery, a variety of scaffolds are currently being developed and researched to achieve therapeutic agents by merely modifying their morphological and functional components [28, 56]. The degree to which CS has been deacetylated not only affects the physical and chemical characteristics of the material, such as degradation rate, crystallinity, hydrophilicity, surface charge, and tensile strength, but it also impacts the physiological reaction to biomaterials [59, 60].

16.5 Fabrication of Chitosan-Based Scaffolds

CS is an inexpensive, biocompatible, osteoconductive polymer that has been used in the development of therapeutic agents-loaded scaffolds for the treatment of bone tissue regeneration [61, 62]. However, the physiochemical and biological properties of scaffold, such as porosity, mechanical strength, biocompatibility, interaction with bone tissue, and delivery of therapeutic agents throughout the post-implantation healing period (sustained release) have been dramatically affected by their method of preparation [63, 64]. As a result, the use of an appropriate fabrication approach is absolutely necessary in order to produce ideal CS-based scaffolds that have a high capacity for the regeneration of bone tissue. The following sections will provide more information on some of the synthesis methods that are utilised most frequently in the process of developing CS-based scaffolds. These methods include phase separation, lyophilization, particulate leaching, electrospinning and 3D printing/rapid prototyping, respectively.

16.5.1 3D-Printing/Rapid Prototyping

3D printing is an emerging strategy that has taken part in numerous biomedical applications. It is also called rapid prototyping and additive manufacturing technique [65]. Unlike conventional techniques, 3D printing provides tailorable CS-based scaffolds with desired porosity and mechanical strength, which could encourage personalized medicine with relatively less cost. In addition, this technique can print scaffolds with micron-scale precision [63]. Coming to fabrication process, the starting point of 3D printing is a computer-aided design (CAD), which directs the printing machine to fabricate the desired scaffold layer-by-layer. Photopolymerization-based technique, i.e., stereolithography (SLG), is a type of 3D printing that involves the fabrication of scaffolds using photon-curable liquid polymers. These polymers are systematically irradiated according to the CAD model to initiate crosslinking, leading to the development of photocured material. However, the photoinitiators and photoabsorbers in the liquid polymer determine the degree of polymerization [66, 67]. Fused deposition modelling (FDM), another kind of 3D printing based on extrusion, is widely utilised for the fabrication of customised scaffolding. This technique involves melt extrusion of beads, sheets, filaments, or wires into geometrical structures according to the command of the CAD model [68–70]. However, the main drawback of this technique is that it can print well using thermoplastic polymers like polylactic acid (PLA). Therefore, a study was developed by PLA-CS composite-based 3D-printed scaffold with good porosity and mechanical strength for efficient bone tissue generation [71]. Another study conducted by Cabral et al. who prepared a composition of gelatin, tricalcium phosphate, graphene oxide, and CS, was further utilized to print 3D scaffolds of the greatest mechanical strength. The prepared CS-composite scaffold exhibited excellent bone regeneration [72]. In this regard, 3D printing is a well-known technique for developing robust CS-based bone tissue regeneration scaffolds.

16.5.2 Electrospinning

Electrospinning is a low-cost and efficient strategy to fabricate micro/nanofibrous structures by applying an electrostatic force to the polymer solution. Electric current draws charged threads from polymer solution to a fibrous state [73, 74]. The electrospinning setup is comprised of three basic parts, i.e., grounded collector, spinneret, and DC voltage. Initially, CS is dissolved in a suitable aqueous solution (acetic acid, lactic acid, etc.) to yield CS viscous solution. Further, the CS solution is loaded into a syringe pump. Upon application of high voltage, the CS solution creates a repulsive force, making CS droplet form a conical structure called a Taylor cone. Further, it has been extruded from the Taylor cone to the grounded collector. Meanwhile, depositing in a grounded collector, the solvent evaporates, making the CS stretch and form a fibrous mat/scaffold [64]. However, the overall electrospinning process can be significantly affected by specific parameters such as applied voltage, distance between the collector and the needle and CS concentration. In order to improve the mechanical properties of CS scaffold, Singh et al. developed monetite nanoparticles (MNPs) imbibed CS scaffolds via electrospinning technique [75]. It was found that the addition of 7% MNPs enhanced the tensile strength of CS scaffold from 6 to 12 MPa. In addition, the MNPs imbibed scaffolds improved the cell attachment, distribution, and mobility of cells in the MG-63 osteoblast cell line, followed by enhancing their proliferation compared to pure CS scaffold. Toullec et al. investigated the curdlan and chitosancomposite-based electrospun scaffold for bone regeneration [76]. Curdlan is a polysaccharide with a demonstrated bone regeneration history by helping mesenchymal stem cell adhesion. The scratch assay endowed that the developed curdlan/chitosan composite scaffold expressed immunomodulatory features by increasing cell migration. Overall, electrospinning is a simple and low-costmethod to fabricate CS-based scaffolds for bone regeneration.

16.5.3 Particulate Leaching

Particulate leaching is the widely used technique for producing polymeric scaffolds with optimum pore size. In this technique, the porogen is homogeneously mixed in CS solution and cast onto the mould. Further, the solvent from CS solution is evaporated or lyophilized, leaving behind the CS matrix and porogen. Finally, the CS matrix containing porogen is washed with ethanol to leach out porogen, resulting in porous CS scaffolds. Some commonly used porogens are sodium chloride, sodium acetate, and sodium citrate. The CS scaffold pore size may also be changed by altering the concentration of the porogen. Studies have reported that CS scaffold fabricated using 90% salt concentration resulted in enhanced pore interconnectivity, cell proliferation, and superior mechanical properties [77, 78]. The major drawback of the particulate leaching technique is that most of the loaded drug is eliminated from the scaffold during porogen removal process. However, to overcome this disadvantage, many studies have reported combining the particulate leaching technique with other techniques such as freeze-drying, compression moulding, and gas foaming [64].

16.5.4 Phase Separation and Lyophilization/Freeze-Drying

Phase separation and lyophilization is a well-known scaffold fabrication technique. This technique utilizes different solubilizing abilities of polymers in the solvent. Firstly, CS solution is prepared using an acetic acid solution, followed by casting it on the respective mould. Further, CS solution is freeze-dried to yield ice crystals. The formed crystals are phase separated from CS acetate salt, followed by undergoing sublimation during the freeze-drying process to obtain CS scaffold. Finally, ethanol or sodium hydroxide is used to hydrate CS scaffold [79]. Some common parameters that affect the pore homogeneity and porosity of the scaffold are CS concentration, solvent composition, solvent type, freezing rate, temperature, etc. Reduced freezing temperature has been proven in studies to lower the pore size while increasing pore homogeneity and interconnectivity in scaffolds. A notable benefit of this technology is that it may be done at extremely low temperatures; due to this, many biomacromolecules and other thermolabile drugs can be easily loaded into CS scaffolds without affecting therapeutic activity [63].

16.6 Mechanism of Drug Release from Chitosan-Based Scaffolds

The scaffolds prepared using CS release drugs in four different ways, such as by diffusion, erosion, swelling and stimulation of the entrenched system. Diffusion-controlled release refers to the process in which a drug is discharged from a drug carrier by permeation from the system's interior to the environment around it. Hydrogels containing polymer chains, release medicines by diffusion when subjected to a concentration gradient. The CS-graft-poly (acrylic acid-co-acrylamide)/Hap nanocomposite scaffold that was developed does the majority of its releasing of celecoxib via the processes of diffusion and erosion. The term "swelling" refers to the process through which the polymer system absorbs substances, which ultimately results in increase in volume. Because the drug-loaded scaffold is in close interaction with the bodily fluid, swelling may sometimes be a prelude to polymer disintegration. This can influence the rate at which drugs are released from the scaffold. Regeneration of craniofacial bone abnormalities is an important topic in the area of bone tissue regeneration, according to Yousefiasl et al. Due to this, innovative methods of therapy have been developed, such as tissue engineering that makes use of porous scaffolds [80].

The degradation of the drug-loaded scaffold is linked to the loss of material that occurs as a result of the monomers and oligomers being dissociated from the polymer. When the polymer hydrogels in the marginal areas begin to deteriorate, erosion takes place, which ultimately leads to the release of the medication that was entangled. Gentile et al. integrated simvastatin-loaded poly-(dl-lactide-co-glycolide) acid microparticles onto CS-gelatin scaffolds. They were capable of achieving a slower localised release of the medication simvastatin by gradual erosion of microspheres. Stimuli-responsive drug delivery systems are sensitive to their surrounding environments. These systems can react to physiological stimuli such as ATP, pH, enzyme, glucose, redox and H_2O_2 as well as to physical stimuli such as heat, light, magnetization, ultrasound, and electricity (Fig. 16.4) [81]. According to El-Husseiny et al. the process of repairing bone abnormalities is a lengthy, self-healing procedure that is dependent on tissue modelling and remodelling. However, the capability of this treatment is restricted, particularly for treating critical-sized bone deficiencies without use of bone augmentation. As a result, BTE, which involves the implementation of therapeutic procedures that combine biomolecules, bioinspired scaffolds, and cells, plays a pivotal role in this line of research. SSRHs, also known as smart and stimuli-responsive hydrogels, are excellent examples of 3D biocompatible polymeric materials that may be used for tissue regeneration and other applications in the medical field. They are able to simulate the natural tissues in a variety of biological, mechanical and physicochemical aspects. In addition to this, they provide a 3D design, an abundance of aqueous conditions, and mechanical stability which are essential for cell development [82]. CS-based pH-sensitive nanocomposites developed by Amiryaghoubi et al., for the targeted administration of doxorubicin (DOX) to osteosarcoma cells. The CS was functionalized using succinic anhydride (SA) so that it could be used in the preparation of the nanocomposite (CS-SA). The formation of CS-folic acid (FA) conjugates was achieved by the formation of an amide link between CS and FA during the conjugation process. Next, Fe_3O_4 magnetic nanoparticles ferrofluid was produced, and the nanocomposite was made utilising magnetic nanoparticles. Additionally, CS–SA/CS-FA and CS–SA were synthesised by the formation of an inclusion between the hydroxyl groups of Fe_3O_4 and –COOH

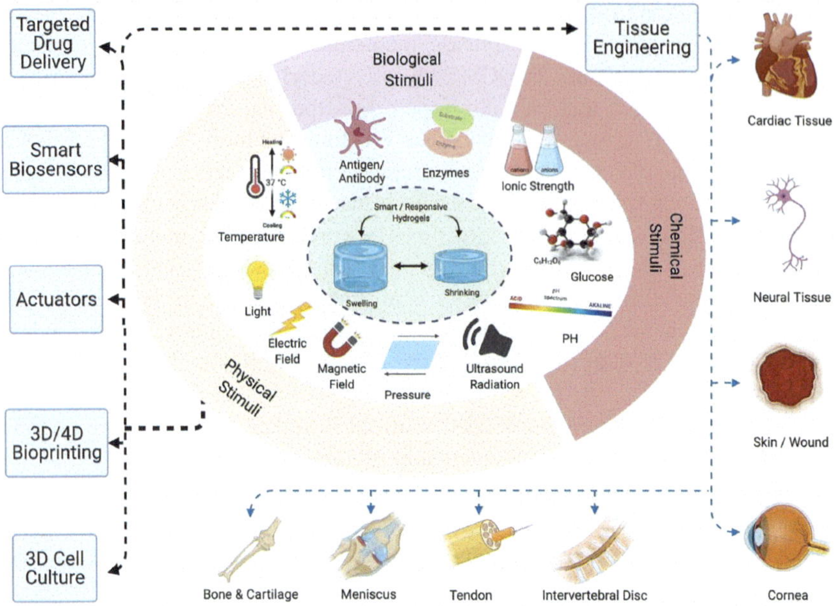

Fig. 16.4 Schematic representation of different smart/stimuli-reactive hydrogels used for different biomedical applications. Reproduced with permission from Elsevier [84]

groups of CS-SA. Lastly, DOX molecules were included into the nanocomposites [83].

16.7 Chitosan Based Scaffolds as Drug Delivery Systems

The purpose of drug delivery systems is to transport a wide range of medications to the appropriate sites of absorption in order to enhance the physiological effects of the medication [60]. In addition to this, they circumvent problems associated with hydrophobicity, poor solubility, reduced bioavailability, drug agglomeration, and lack of specificity. Since CS is cationic, it makes regulated drug release, adhesion, in situ gel formation, transfection, improved permeability, and efflux pump inhibitory actions [64, 85]. All of these features make CS-based scaffolds a promising choice to transport therapeutic substances to the defect location in a targeted and regulated way. The method of loading, the extent to which the scaffold can be broken down, as well as the amount of drug present decide the effectiveness of drug release profile. BTE employs both scaffold-based and scaffold-independent drug delivery strategies [85]. The adverse effects and danger of overdose associated with systemic administration are avoided when a medication is supplied locally through scaffolds, and an appropriate concentration of the medicine reaches the target location more efficiently and effectively during the healing phase. Various ways for delivering particular medicines and tiny bioactive compounds using CS-based scaffolds to enhance bone regeneration have been explored [86].

16.8 Cytotoxicity Assessments on Chitosan Based Scaffolds

The toxicity of CS-based scaffolds to use for osteogenesis must be evaluated, and numerous characteristics, including membrane stability and mitochondrial function, must be examined. In addition, the cellular microenvironment present on the scaffold is a significant contributor to the pro-regenerative capacity of the biocompatible material. The cytotoxicity and biocompatibility of CS and CS-based scaffolds have been investigated in a number of research studies, and the findings of these studies point to the non-toxicity and biocompatibility of these scaffolds under both in vitro and in vivo settings. Patel et al. developed a versatile CS/CNCs hydrogel scaffold by infusing varying concentrations of CNCs into a CS hydrogel [87]. When compared to pure CS hydrogel scaffolds, CS/CNCs shown a significant enhancement in mechanical strength. In the presence of bone marrow-derived mesenchymal stem cells, the cytocompatibility of the scaffolds was assessed. CS/CNC hydrogel scaffolds improved cell survival and nutrient uptake over pure CS hydrogel scaffolds.

16.9 Conclusion

In recent years, there has been a lot of research directed toward CS-based scaffolds that are used in drug delivery systems. Scaffolds made from CS have the potential to distribute drugs in a way that is maintained, extended, and regulated. Small biologically active molecules, such as nutrients, genetic material, polyphenols, antibiotics, statins, bisphosphates, and anti-inflammatory molecules, could be infused, entrapped, and adsorbed onto the desired CS scaffolds to deliver synergistic impacts. These effects will increase bioactivity, and bone regeneration properties. In addition, scaffolds made entirely of CS have certain weak mechanical qualities and a quick rate of deterioration. Therefore, CS might be encapsulated in bioactive ceramics in order to improve its compressive strength, successfully managing its porosity, water-holding capacity, biodegradation rate, and increasing its mechanical characteristics. In the future, the resolution of all of these obstacles might make it possible for SC-based scaffolds to play a major role in drug delivery and bone regeneration and could make it easier for this technology to establish itself in the area of biomedical applications.

References

1. U.G.K. Wegst, H. Bai, E. Saiz et al., Bioinspired structural materials. Nat. Mater. **14**, 23–36 (2015). https://doi.org/10.1038/nmat4089
2. M. Farokhi, F. Mottaghitalab, S. Samani et al., Silk fibroin/hydroxyapatite composites for bone tissue engineering. Biotechnol. Adv. **36**, 68–91 (2018). https://doi.org/10.1016/j.biotechadv.2017.10.001
3. Y. Zhao, S. Zhao, Z. Ma et al., Chitosan-based scaffolds for facilitated endogenous bone re-generation. Pharmaceuticals **15**, 1023 (2022). https://doi.org/10.3390/ph15081023
4. X. Li, L. Wei, J. Li et al., Multifunctional SDF-1-loaded hydroxyapatite/polylactic acid membranes promote cell recruitment, immunomodulation, angiogenesis, and osteogenesis for biomimetic bone regeneration. Appl. Mater. Today **22**, 100942 (2021). https://doi.org/10.1016/j.apmt.2021.100942
5. G.L. Koons, M. Diba, A.G. Mikos, Materials design for bone-tissue engineering. Nat. Rev. Mater. **5**, 584–603 (2020). https://doi.org/10.1038/s41578-020-0204-2
6. H. Qu, H. Fu, Z. Han, Y. Sun, Biomaterials for bone tissue engineering scaffolds: a review. RSC Adv. **9**, 26252–26262 (2019). https://doi.org/10.1039/C9RA05214C
7. A. Oryan, S. Alidadi, A. Moshiri, N. Maffulli, Bone regenerative medicine: classic options, novel strategies, and future directions. J. Orthop. Surg. Res. **9**, 18 (2014). https://doi.org/10.1186/1749-799X-9-18
8. S. Bose, M. Roy, A. Bandyopadhyay, Recent advances in bone tissue engineering scaffolds. Trends Biotechnol. **30**, 546–554 (2012). https://doi.org/10.1016/j.tibtech.2012.07.005
9. C.H. Evans, G.D. Palmer, A. Pascher et al., Facilitated endogenous repair: making tissue engineering simple, practical, and economical. Tissue Eng. **13**, 1987–1993 (2007). https://doi.org/10.1089/ten.2006.0302
10. A.R. Amini, C.T. Laurencin, S.P. Nukavarapu, Bone tissue engineering: recent advances and challenges. Crit. Rev. Biomed. Eng. **40**, 363–408 (2012). https://doi.org/10.1615/CritRevBiomedEng.v40.i5.10
11. Q. Yao, Y. Liu, J. Tao et al., Hypoxia-mimicking nanofibrous scaffolds promote endogenous bone regeneration. ACS Appl. Mater. Interfaces **8**, 32450–32459 (2016). https://doi.org/10.1021/acsami.6b10538
12. J. Chen, Y. Zhang, P. Pan et al., In situ strategy for bone repair by facilitated endogenous tissue engineering. Colloids Surf. B Biointerfaces **135**, 581–587 (2015). https://doi.org/10.1016/j.colsurfb.2015.08.019
13. Q. Wang, Y. Tang, Q. Ke et al., Magnetic lanthanum-doped hydroxyapatite/chitosan scaffolds with endogenous stem cell-recruiting and immunomodulatory properties for bone regeneration. J. Mater. Chem. B **8**, 5280–5292 (2020). https://doi.org/10.1039/D0TB00342E
14. M.N. Collins, G. Ren, K. Young et al., Scaffold fabrication technologies and structure/function properties in bone tissue engineering. Adv. Funct. Mater. **31**, 2010609 (2021). https://doi.org/10.1002/adfm.202010609
15. G. Turnbull, J. Clarke, F. Picard et al., 3D bioactive composite scaffolds for bone tissue engineering. Bioact. Mater. **3**, 278–314 (2018). https://doi.org/10.1016/j.bioactmat.2017.10.001
16. S. Wang, Y. Yang, Z. Zhao et al., Mineralized collagen-based composite bone materials for cranial bone regeneration in developing sheep. ACS Biomater. Sci. Eng. **3**, 1092–1099 (2017). https://doi.org/10.1021/acsbiomaterials.7b00159
17. L. Gambari, E. Amore, R. Raggio et al., Hydrogen sulfide-releasing silk fibroin scaffold for bone tissue engineering. Mater. Sci. Eng. C **102**, 471–482 (2019). https://doi.org/10.1016/j.msec.2019.04.039
18. S.H. Park, J.Y. Park, Y.B. Ji et al., An injectable click-crosslinked hyaluronic acid hydrogel modified with a BMP-2 mimetic peptide as a bone tissue engineering scaffold. Acta Biomater. **117**, 108–120 (2020). https://doi.org/10.1016/j.actbio.2020.09.013

19. R. Deshpande, S. Shukla, R. Sayyad et al., Silk fibroin and ceramic scaffolds: comparative in vitro studies for bone regeneration. Bioeng. Transl. Med. **6**(2021). https://doi.org/10.1002/btm2.10221
20. M.M. Abdallah, N. Fernández, A.A. Matias, M.R. Bronze, Hyaluronic acid and chondroitin sulfate from marine and terrestrial sources: extraction and purification methods. Carbohydr. Polym. **243**, 116441 (2020). https://doi.org/10.1016/j.carbpol.2020.116441
21. M. Pakizeh, A. Moradi, T. Ghassemi, Chemical extraction and modification of chitin and chitosan from shrimp shells. Eur. Polym. J. **159**, 110709 (2021). https://doi.org/10.1016/j.eurpolymj.2021.110709
22. A. Farazin, M. Mohammadimehr, A.H. Ghasemi, H. Naeimi, Design, preparation, and characterization of CS/PVA/SA hydrogels modified with mesoporous Ag_2O/SiO_2 and curcumin nanoparticles for green, biocompatible, and antibacterial biopolymer film. RSC Adv. **11**, 32775–32791 (2021). https://doi.org/10.1039/D1RA05153A
23. X. Zheng, P. Zhang, Z. Fu et al., Applications of nanomaterials in tissue engineering. RSC Adv. **11**, 19041–19058 (2021). https://doi.org/10.1039/D1RA01849C
24. S.L. Levengood, A.E. Erickson, F. Chang, M. Zhang, Chitosan–poly(caprolactone) nanofibers for skin repair. J. Mater. Chem. B **5**, 1822–1833 (2017). https://doi.org/10.1039/C6TB03223K
25. D. Jhala, H. Rather, R. Vasita, Polycaprolactone–chitosan nanofibers influence cell morphology to induce early osteogenic differentiation. Biomater. Sci. **4**, 1584–1595 (2016). https://doi.org/10.1039/C6BM00492J
26. J. Yang, T. Long, N.-F. He et al., Fabrication of a chitosan/bioglass three-dimensional porous scaffold for bone tissue engineering applications. J. Mater. Chem. B **2**, 6611–6618 (2014). https://doi.org/10.1039/C4TB00940A
27. N.T.T. Dang, T.T.L. Chau, D.H. Van et al., Water-soluble chitosan-derived sustainable materials: towards filaments, aerogels, microspheres, and plastics. Soft Matter. **13**, 7292–7299 (2017). https://doi.org/10.1039/C7SM01292F
28. F. Damiri, Y. Bachra, C. Bounacir et al., Synthesis and characterization of lyophilized chitosan-based hydrogels cross-linked with benzaldehyde for controlled drug release. J. Chem. **2020**(2020). https://doi.org/10.1155/2020/8747639
29. L. Keller, A. Regiel-Futyra, M. Gimeno et al., Chitosan-based nanocomposites for the repair of bone defects. Nanomed. Nanotechnol., Biol. Med. **13**, 2231–2240 (2017). https://doi.org/10.1016/j.nano.2017.06.007
30. F. Damiri, Y. Bachra, M. Berrada, Synthesis and characterization of 4-formylphenylboronic acid cross-linked chitosan hydrogel with dual action: glucose-sensitivity and controlled insulin release. Chinese J. Anal. Chem. **50**, 100092 (2022). https://doi.org/10.1016/j.cjac.2022.100092
31. T. Xu, H. Yang, D. Yang, Z.-Z. Yu, Polylactic acid nanofiber scaffold decorated with chitosan island like topography for bone tissue engineering. ACS Appl. Mater. Interfaces **9**, 21094–21104 (2017). https://doi.org/10.1021/acsami.7b01176
32. X. Zhao, Y. Han, J. Li et al., BMP-2 immobilized PLGA/hydroxyapatite fibrous scaffold via polydopamine stimulates osteoblast growth. Mater. Sci. Eng. C **78**, 658–666 (2017). https://doi.org/10.1016/j.msec.2017.03.186
33. F. Tao, Y. Cheng, H. Tao et al., Carboxymethyl chitosan/sodium alginate-based micron-fibers fabricated by emulsion electrospinning for periosteal tissue engineering. Mater. Des. **194**, 108849 (2020). https://doi.org/10.1016/j.matdes.2020.108849
34. Y. Yang, H. Ao, Y. Wang et al., Cytocompatibility with osteogenic cells and enhanced in vivo anti-infection potential of quaternized chitosan-loaded titania nanotubes. Bone Res. **4**, 16027 (2016). https://doi.org/10.1038/boneres.2016.27
35. A.H. Menon, S.P. Soundarya, V. Sanjay et al., Sustained release of chrysin from chitosan-based scaffolds promotes mesenchymal stem cell proliferation and osteoblast differentiation. Carbohydr. Polym. **195**, 356–367 (2018). https://doi.org/10.1016/j.carbpol.2018.04.115
36. A. Soriente, I. Fasolino, A. Gomez-Sánchez et al., Chitosan/hydroxyapatite nanocomposite scaffolds to modulate osteogenic and inflammatory response. J. Biomed. Mater. Res. A **110**, 266–272 (2022). https://doi.org/10.1002/jbm.a.37283

37. M. Stevanović, M. Djošić, A. Janković et al., Antibacterial graphene-based hydroxyapatite/chitosan coating with gentamicin for potential applications in bone tissue engineering. J. Biomed. Mater. Res. A **108**, 2175–2189 (2020). https://doi.org/10.1002/jbm.a.36974
38. T. Ding, W. Kang, J. Li et al., An in situ tissue engineering scaffold with growth factors combining angiogenesis and osteoimmunomodulatory functions for advanced periodontal bone regeneration. J. Nanobiotechnol. **19**, 247 (2021). https://doi.org/10.1186/s12951-021-00992-4
39. P. Brun, A. Zamuner, C. Battocchio et al., Bio-functionalized chitosan for bone tissue engineering. Int. J. Mol. Sci. **22**, 5916 (2021). https://doi.org/10.3390/ijms22115916
40. Y. Zhao, T. Fan, J. Chen et al., Magnetic bioinspired micro/nanostructured composite scaffold for bone regeneration. Colloids Surf. B Biointerfaces **174**, 70–79 (2019). https://doi.org/10.1016/j.colsurfb.2018.11.003
41. R. Zeng, M. Tu, H. Liu et al., Preparation, structure, drug release and bioinspired mineralization of chitosan-based nanocomplexes for bone tissue engineering. Carbohydr. Polym. **78**, 107–111 (2009). https://doi.org/10.1016/j.carbpol.2009.04.035
42. S. Gorantla, N. Dabholkar, S. Sharma et al., Chitosan-based microneedles as a potential platform for drug delivery through the skin: trends and regulatory aspects. Int. J. Biol. Macromol. **184**, 438–453 (2021). https://doi.org/10.1016/j.ijbiomac.2021.06.059
43. F. Damiri, M.H. Rahman, M. Zehravi et al., MXene (Ti3C2Tx)-embedded nanocomposite hydrogels for biomedical applications: a review. Materials (Basel) **15**, 1666 (2022). https://doi.org/10.3390/ma15051666
44. D. Kulkarni, F. Damiri, S. Rojekar et al., Recent advancements in microneedle technology for multifaceted biomedical applications. Pharmaceutics **14**, 1097 (2022). https://doi.org/10.3390/pharmaceutics14051097
45. S. Pardeshi, F. Damiri, M. Zehravi et al., Functional thermoresponsive hydrogel molecule to material design for biomedical applications. Polymers (Basel) **14**, 3126 (2022). https://doi.org/10.3390/polym14153126
46. Y. Liu, J. Lim, S.-H. Teoh, Review: development of clinically relevant scaffolds for vascularised bone tissue engineering. Biotechnol. Adv. **31**, 688–705 (2013). https://doi.org/10.1016/j.biotechadv.2012.10.003
47. T. Sukpaita, S. Chirachanchai, A. Pimkhaokham, R.S. Ampornaramveth, Chitosan-based scaffold for mineralized tissues regeneration. Mar. Drugs **19**, 551 (2021). https://doi.org/10.3390/md19100551
48. S. Cao, Y. Zhao, Y. Hu et al., New perspectives: in-situ tissue engineering for bone repair scaffold. Compos. B Eng. **202**, 108445 (2020). https://doi.org/10.1016/j.compositesb.2020.108445
49. S. Wang, Y. Yang, G.L. Koons et al., Tuning pore features of mineralized collagen/PCL scaffolds for cranial bone regeneration in a rat model. Mater. Sci. Eng. C **106**, 110186 (2020). https://doi.org/10.1016/j.msec.2019.110186
50. G.M. Crane, S.L. Ishaug, A.G. Mikos, Bone tissue engineering. Nat. Med. **1**, 1322–1324 (1995). https://doi.org/10.1038/nm1295-1322
51. J.L. Brown, C.T. Laurencin, Bone tissue engineering. in *Biomaterials Science*, (Elsevier, 2020), pp. 1373–1388
52. S. Datta, A.P. Rameshbabu, K. Bankoti et al., Decellularized bone matrix/oleoyl chitosan derived supramolecular injectable hydrogel promotes efficient bone integration. Mater. Sci. Eng. C **119**, 111604 (2021). https://doi.org/10.1016/j.msec.2020.111604
53. A. Bouchie, Tissue engineering firms go under. Nat. Biotechnol. **20**, 1178–1179 (2002). https://doi.org/10.1038/nbt1202-1178
54. Y. Zhao, J. Chen, L. Zou et al., Facile one-step bioinspired mineralization by chitosan functionalized with graphene oxide to activate bone endogenous regeneration. Chem. Eng. J. **378**, 122174 (2019). https://doi.org/10.1016/j.cej.2019.122174
55. R.M. Raftery, B. Woods, A.L.P. Marques et al., Multifunctional biomaterials from the sea: assessing the effects of chitosan incorporation into collagen scaffolds on mechanical and biological functionality. Acta Biomater. **43**, 160–169 (2016). https://doi.org/10.1016/j.actbio.2016.07.009

56. D. Fouad, Y. Bachra, G. Ayoub, et al. A novel drug delivery system based on nanoparticles of magnetite Fe_3O_4 embedded in an auto cross-linked chitosan. in *Chitin and Chitosan – Physicochemical Properties and Industrial Applications* [Working Title]. IntechOpen
57. P. Feng, Y. Luo, C. Ke et al., Chitosan-based functional materials for skin wound repair: mechanisms and applications. Front. Bioeng. Biotechnol. **9**(2021). https://doi.org/10.3389/fbioe.2021.650598
58. H.S. Adhikari, P.N. Yadav, Anticancer activity of chitosan, chitosan derivatives, and their mechanism of action. Int. J. Biomater. **2018**, 1–29 (2018). https://doi.org/10.1155/2018/2952085
59. I. Aranaz, A.R. Alcántara, M.C. Civera et al., Chitosan: an overview of its properties and applications. Polymers (Basel) **13**, 3256 (2021). https://doi.org/10.3390/polym13193256
60. F. Croisier, C. Jérôme, Chitosan-based biomaterials for tissue engineering. Eur. Polym. J. **49**, 780–792 (2013). https://doi.org/10.1016/j.eurpolymj.2012.12.009
61. S.K.L. Levengood, M. Zhang, Chitosan-based scaffolds for bone tissue engineering. J. Mater. Chem. B **2**, 3161 (2014). https://doi.org/10.1039/c4tb00027g
62. S. Saravanan, R.S. Leena, N. Selvamurugan, Chitosan based biocomposite scaffolds for bone tissue engineering. Int. J. Biol. Macromol. **93**, 1354–1365 (2016). https://doi.org/10.1016/j.ijbiomac.2016.01.112
63. L.R. Yadav, S.V. Chandran, K. Lavanya, N. Selvamurugan, Chitosan-based 3D-printed scaffolds for bone tissue engineering. Int. J. Biol. Macromol. **183**, 1925–1938 (2021). https://doi.org/10.1016/j.ijbiomac.2021.05.215
64. R. Bharathi, S.S. Ganesh, G. Harini et al., Chitosan-based scaffolds as drug delivery systems in bone tissue engineering. Int. J. Biol. Macromol. **222**, 132–153 (2022). https://doi.org/10.1016/j.ijbiomac.2022.09.058
65. A.-V. Do, B. Khorsand, S.M. Geary, A.K. Salem, 3D printing of scaffolds for tissue regeneration applications. Adv. Healthc. Mater. **4**, 1742–1762 (2015). https://doi.org/10.1002/adhm.201500168
66. K. Kim, A. Yeatts, D. Dean, J.P. Fisher, Stereolithographic bone scaffold design parameters: osteogenic differentiation and signal expression. Tissue Eng. B Rev. **16**, 523–539 (2010). https://doi.org/10.1089/ten.teb.2010.0171
67. S.A. Skoog, P.L. Goering, R.J. Narayan, Stereolithography in tissue engineering. J. Mater. Sci. Mater. Med. **25**, 845–856 (2014). https://doi.org/10.1007/s10856-013-5107-y
68. A. Alafaghani, A. Qattawi, M.A. Ablat, Design consideration for additive manufacturing: fused deposition modelling. Open J. Appl. Sci. **07**, 291–318 (2017). https://doi.org/10.4236/ojapps.2017.76024
69. D. Han, H. Lee, Recent advances in multi-material additive manufacturing: methods and applications. Curr. Opin. Chem. Eng. **28**, 158–166 (2020). https://doi.org/10.1016/j.coche.2020.03.004
70. M. Khodaei, K. Amini, A. Valanezhad, Fabrication and characterization of poly lactic acid scaffolds by fused deposition modeling for bone tissue engineering. J. Wuhan Univ. Technol. Sci. Ed. **35**, 248–251 (2020). https://doi.org/10.1007/s11595-020-2250-4
71. B. Ashwin, B. Abinaya, T.P. Prasith et al., 3D-poly (lactic acid) scaffolds coated with gelatin and mucic acid for bone tissue engineering. Int. J. Biol. Macromol. **162**, 523–532 (2020). https://doi.org/10.1016/j.ijbiomac.2020.06.157
72. C.S.D. Cabral, S.P. Miguel, D. de Melo-Diogo et al., Green reduced graphene oxide functionalized 3D printed scaffolds for bone tissue regeneration. Carbon N Y **146**, 513–523 (2019). https://doi.org/10.1016/j.carbon.2019.01.100
73. S. Eap, A. Ferrand, C. Mendoza et al., Electrospun nanofibrous 3D scaffold for bone tissue engineering. Biomed. Mater. Eng. **22**, 137–141 (2012). https://doi.org/10.3233/BME-2012-0699
74. I. Jun, H.-S. Han, J. Edwards, H. Jeon, Electrospun fibrous scaffolds for tissue engineering: viewpoints on architecture and fabrication. Int. J. Mol. Sci. **19**, 745 (2018). https://doi.org/10.3390/ijms19030745

75. Y.P. Singh, B. Mishra, M.K. Gupta et al., Enhancing physicochemical, mechanical, and bioactive performances of monetite nanoparticles reinforced chitosan-PEO electrospun scaffold for bone tissue engineering. J. Appl. Polym. Sci. **139**(2022). https://doi.org/10.1002/app.52844
76. C. Toullec, J. Le Bideau, V. Geoffroy et al., Curdlan-chitosan electrospun fibers as potential scaffolds for bone regeneration. Polymers (Basel) **13**, 526 (2021). https://doi.org/10.3390/polym13040526
77. S. Jana, S.J. Florczyk, M. Leung, M. Zhang, High-strength pristine porous chitosan scaffolds for tissue engineering. J. Mater. Chem. **22**, 6291 (2012). https://doi.org/10.1039/c2jm16676c
78. J.I. Lim, Y.-K. Lee, J.-S. Shin, K.-J. Lim, Preparation of interconnected porous chitosan scaffolds by sodium acetate particulate leaching. J. Biomater. Sci. Polym. Ed. **22**, 1319–1329 (2011). https://doi.org/10.1163/092050610X504783
79. J. Venkatesan, S.-K. Kim, T.W. Wong, Chitosan and its application as tissue engineering scaffolds. in *Nanotechnology Applications for Tissue Engineering* (Elsevier, 2015), pp 133–147
80. S. Yousefiasl, E. Sharifi, E. Salahinejad et al., Bioactive 3D-printed chitosan-based scaffolds for personalized craniofacial bone tissue engineering. Eng. Regen. **4**, 1–11 (2023). https://doi.org/10.1016/j.engreg.2022.09.005
81. Y. Song, Y. Li, Q. Xu, Z. Liu, Mesoporous silica nanoparticles for stimuli-responsive controlled drug delivery: advances, challenges, and outlook. Int. J. Nanomed. **12**, 87–110 (2016). https://doi.org/10.2147/IJN.S117495
82. H.M. El-, E.A. Mady, W.A. El- et al., Smart/stimuli-responsive hydrogels: state-of-the-art platforms for bone tissue engineering. Appl. Mater. Today **101560**(2022). https://doi.org/10.1016/j.apmt.2022.101560
83. N. Amiryaghoubi, E.D. Abdolahinia, A. Nakhlband et al., Smart chitosan–folate hybrid magnetic nanoparticles for targeted delivery of doxorubicin to osteosarcoma cells. Colloids Surf. B Biointerfaces **220**, 112911 (2022). https://doi.org/10.1016/j.colsurfb.2022.112911
84. H.M. El-, E.A. Mady, L. Hamabe et al., Smart/stimuli-responsive hydrogels: cutting-edge platforms for tissue engineering and other biomedical applications. Mater Today Bio **13**, 100186 (2022). https://doi.org/10.1016/j.mtbio.2021.100186
85. F. Dabbarh, N. Elbakali-Kassimi, M. Berrada, Chitosan based biocomposites for hard tissue engineering. in *Chitin and Chitosan – Physicochemical Properties and Industrial Applications* (2021) [Working Title]. IntechOpen
86. M. Lazaridou, D.N. Bikiaris, D.A. Lamprou, 3D Bioprinted chitosan-based hydrogel scaffolds in tissue engineering and localised drug delivery. Pharmaceutics **14**, 1978 (2022). https://doi.org/10.3390/pharmaceutics14091978
87. D.K. Patel, S.D. Dutta, K. Ganguly, K.-T. Lim, Multifunctional bioactive chitosan/cellulose nanocrystal scaffolds eradicate bacterial growth and sustain drug delivery. Int. J. Biol. Macromol. **170**, 178–188 (2021). https://doi.org/10.1016/j.ijbiomac.2020.12.145

Chapter 17
Challenges and Future Perspectives of Chitosan Nanocomposites for Bionanomechanical Applications

Sarat Kumar Swain

Abstract Despite easy availability, biocompatibility, biodegradability and nontoxic nature of chitosan it is not fully accepted for designing of smart devices because of the presence of lots of critical challenges in its strength, stability, processability, etc. Although, the stability is somehow achieved by making its nanocomposite with reinforcement of nanoparticles still numerous constraints need to be addressed to make it in the way of sustainability. The present chapter reveals different challenges of chitosan-based nanocomposites specifically for bionanomechanical applications. Future perspectives regarding different applications of chitosan bionanocomposites are also established in this chapter.

17.1 Challenges in Chitosan Nanocomposite

Chitosan is not getting its deserved attention to meet the global slogan of a green, economic and sustainable world. Although, chitosan is the second easily available material but its area of research is limited to laboratory with a big gap from prospective industrial applications. It is because of lots of unsolved challenges in generating chitosan in the front range of future sustainable materials as chitosan nanocomposites, which are depicted in Fig. 17.1. These challenges can be overcome by interacting chitosan with nanoscale materials as the potential filler in the composite. However, the transfer of these chitosan-based nanocomposites from laboratories to pilot scale production and its utility to community is still a huge challenge. Nowadays industries are afraid of the production of chitosan and its nanocomposite in term of risk in development of products and acceptability in market with viable cost-effectiveness and processability. Therefore, researchers have focused on the development of designing more industrially acceptable and economically synthesized chitosan-based materials. Its biocompatibility, biodegradability, non-toxicity, antibacterial and antifungal

S. K. Swain (✉)
Department of Chemistry, Veer Surendra Sai University of Technology, Burla, Sambalpur, Odisha 768018, India
e-mail: skswain_chem@vssut.ac.in

S. K. Swain and A. Biswal (eds.), *Chitosan Nanocomposites*,
Biological and Medical Physics, Biomedical Engineering,
https://doi.org/10.1007/978-981-19-9646-7_17

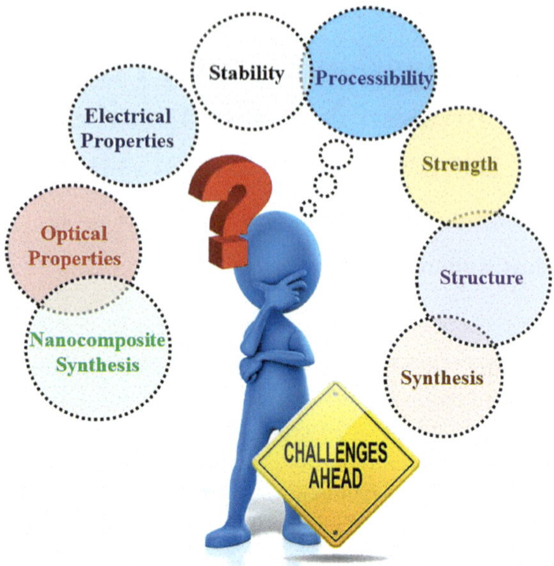

Fig. 17.1 Challenges in nanocomposite

properties have given significant research attentions towards producing chitosan-based nanocomposites for biomedical applications such as tissue engineering, drug delivery, wound dressing, gene therapy, and cancer treatment. It is interesting that chitosan contains primary amine groups which are responsible for its versatile properties such as controlled release, cationic nature, antimicrobial, in situ gelation, micro adhesion etc. The present section mainly targets the challenges of chitosan-based nanocomposites specially for bionanomechanical applications.

Chitosan has excellent film forming properties, but it is a big challenge for food packaging industries to use it as packaging material because chitosan is insoluble in water and highly pH specific. This is due to its high molecular weight and typical structural arrangement. Hence, chitosan has low mechanical strength as individual but achieves somehow its mechanical strength in form of its nanocomposites. There is a big challenge in its stability which can be overcome by applying different environmental factors such as pressure and temperature. Sometimes it requires a proper stabilizing agent and blending agent or ionic agent to acquire proper stability. Numerous internal and external factors such as degree of deacetylation, purity, molecular weight, humidity and processing methods including thermal, physical, and sterilization factors are responsible to determine the stability of chitosan. The external and internal factors affecting the stability of chitosan are schematically represented in Fig. 17.2. Chitosan cannot resist high temperature even till 200 °C. Hence, thermal stability of chitosan is a big challenge. The dispersion of nanostructured material in chitosan polymeric backbone creates strong interfacial adhesion between the surface of chitosan and filler by which the thermal stability somehow increased which still needs to be improved by reinforcement of other potential materials.

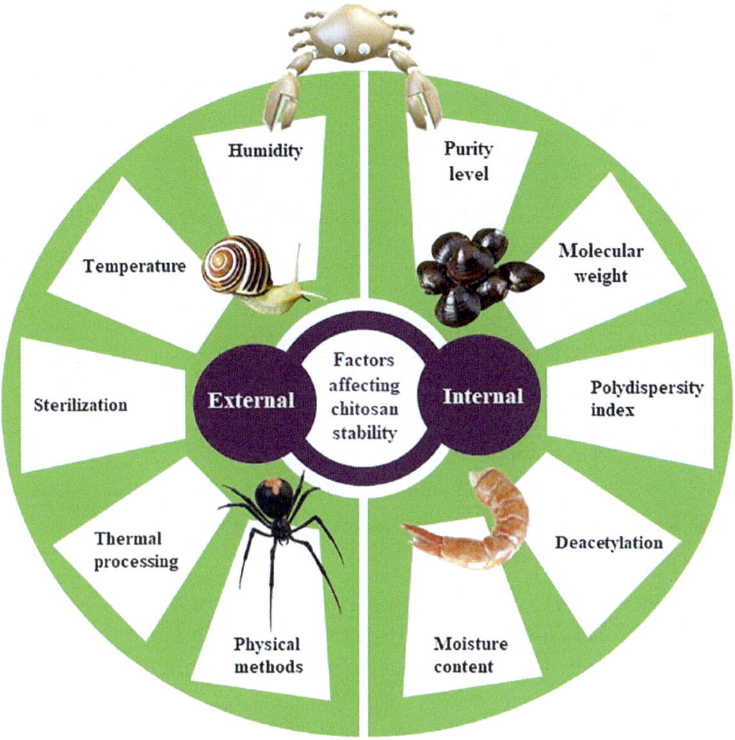

Fig. 17.2 Factors affecting the stability of chitosan

It is critically analysed that most of the synthetic processes related to chitosan-based materials is hardly reproducible which may be due to versatile change in the properties of chitosan and its high dependence upon the processability conditions. Further, the storage of chitosan needs low temperature under 5 °C because of significant chain hydrolysis at room temperature. On other hand, storage of nano chitosan is even more difficult as it requires less than 2 °C, or else it may cause possible agglomeration. Besides this optical, and electrical properties of chitosan-based nanocomposites are scanty in the literature.

17.2 Future Perspectives

Chitosan nanocomposites play a vital role in production of future sustainable materials. Perspectives of chitosan nanocomposites for different bionanomechanical applications are summarized as Fig. 17.3.

Fig. 17.3 Future perspectives for chitosan nanocomposites for bionanomechanical applications

17.2.1 Chitosan Nanocomposites as Future Smart Materials

As the definition of smart materials suggests that these should be controllable as per the requirement, chitosan presents itself as a viable choice under this category. Its characteristic properties like aqueous solubility, biocompatibility, non-toxicity and biodegradable nature makes it a better capping agent for nanocomposites. Various applications of chitosan nanocomposites so far include drug delivery for treatment of various diseases, dye removal, antibacterial and antioxidant activities, packaging material, wound healing properties and many more. Going out of the conventional way instead of having limited applications, the chitosan nanocomposite material has shown multiple properties and uses due to which it is one of most prominent smart materials that has captured the researchers' attention. It is abundantly available and its compatibility is a key factor for its incorporation with various novel nanocomposites.

17.2.2 Perspectives for Chitosan Nanocomposites as Biological Materials

Traditionally, chitosan, a biobased polymer has been used in agricultural sectors for high yield of crops. However, as its biocompatible and biodegradable properties came to light, extensive research is going on for its biomedical applications. Chitosan by itself presents a challenge for broad applications because of its limited mechanical properties. But after addition of various polymers and fillers, it has opened numerous pathways as chitosan nanocomposites. It is being used for bioimaging, biosensors, drug delivery, wound healing and many more. Besides its antifungal, antibacterial, mucoadhesive and haemostatic properties, it has also shown promising potential for non-protein matrix for the application of tissue growth and tissue organization.

These wide variety of applications of chitosan nanocomposites make it a distinctive biological material for the future.

17.2.3 Perspectives for Chitosan Nanocomposites as Antibacterial Materials

The evolution of chitosan and chitosan-based nanocomposite being low cost and multipurpose polymer-based material has attracted a lot of interest in the field of antibacterial materials due to its polycationic nature. Chitosan-based nanocomposite were synthesized by incorporation of different types of nanoparticles including metal oxides, graphene oxide, clay etc. and this leads to improvement in their antibacterial properties. However, the inferior mechanical strength and moderate antibacterial properties are still the drawbacks that restrict their further clinical applications. Despite of these faults, chitosan-based nanocomposite has been the most promising candidate for antibacterial materials.

17.2.4 Perspectives for Chitosan Nanocomposites as Drug Delivery Vehicles

Chitosan being biodegradable and biocompatible are widely used in drug delivery applications including cancer or osteoarthritis. Many nanoparticles including layered double hydroxide, reduced graphene oxide, and metal oxide nanoparticles are utilized for the synthesis of different types of chitosan-based nanocomposites as drug delivery vehicles. These are fabricated by different methods including solution casting, solution blending, in-situ approach, freeze-drying etc. Chitosan-based nanocomposites with tailored properties show targeted drug delivery while it comes to combining with inorganic nanomaterials, they may aggregate under physiological conditions due to their difficulty in modification, side effects for non-targeting and poor stability. However, some faults are there but chitosan-based nanocomposite still is the potential candidate for drugs delivery vehicle.

17.2.5 Perspectives for Chitosan Nanocomposites as Wound Dressing Materials

Numerous numbers of publications presenting various chitosan-based nanocomposites for wound healing application proclaim the excellent properties and significance of these composites in this field. This popularity of chitosan nanocomposites in wound treatment application is due to many advantages like reduced treatment cost, both intrinsic and extrinsic antimicrobial ability, anti-inflammatory property, enhanced blood clotting, effective cell proliferation, promotion in the action of growth factors etc. Also, chitosan nanocomposites help in the reduction of inflammatory phase through the stimulation of inflammatory cells, fibroblasts and macrophases resulting in acceleration of wound healing process. Recently, these wound healing chitosan nanocomposite agents are combined with controlled release drug delivery vehicles leading to improved healing of damages. However, the nanocomposites of chitosan can better be prepared by incorporation with some natural wound healing materials in future to acquire better control to cytotoxicity. Furthermore, the designing of smart wound healing devices is limited, which could get more focus of researchers to configure devices so that we can control and regulate the wound healing process more effectively.

17.2.6 Perspectives for Chitosan Nanocomposites as Tissue Engineering Scaffolds

Chitosan-based nanocomposites have diversified tissue engineering applications including bone tissue regeneration, skin tissue regeneration, cartilage tissue regeneration and nerve tissue regeneration. The main advantage behind choosing chitosan nanocomposites in tissue engineering applications is their high resemblance to the biological cellular environment, fast cell proliferation, and promotion of cell differentiation regulating growth factors. This organ-specific tissue engineering application is also accompanied by biodegradability and biocompatibility of these materials. However, the future study should emphasise on tolerating the disadvantages of using these materials in tissue engineering like uncontrolled degradation, lower mechanical strength in terms of stability and possible infection, by combining chitosan with other inorganic and organic additives like carbon materials, various synthetic polymers such as polylactic acid, polycaprolactone and poly(ethylene oxide) etc.

17.2.7 Perspectives for Chitosan Nanocomposites as Future Materials for Biosensing

Recently chitosan-based nanocomposites were implemented in widespread biosensing applications for the detection of various bio-analytes such as proteins, amino acids, DNA, virus, cancer cells, bacteria, hormones, cholesterol as well as in the detection of various small biomolecules like ascorbic acid, uric acid, dopamine and tryptophan. Chitosan is combined with different metal nanoparticles like gold and silver nanoparticles, metal oxide nanoparticles, carbon quantum dots, graphene quantum dots, etc. showing excellent biosensing applications with high sensitivity, and selectivity, owing to its ability to form an amazing immobilization matrix on the biosensor surface. However, from some analysis carried out in the present book, specific disadvantages in the stability of chitosan nanocomposites in some cases are noticed. Keeping an eye upon this, as "a disadvantage is often a seed of an advantage" we can bind chitosan nanocomposites with several binding polymers like polyvinyl alcohol, starch, polyurethane, and epoxy polymers etc. to increase the stability of chitosan nanocomposites and at the same time these new stabilizer additives can emanate more sensing properties to the nanocomposite. Moreover, the structure, stability and most importantly the optical and electrical properties can be regulated more efficiently for enhanced applications in fluorescence, electrical as well as colorimetric sensing applications of this biomolecule.

17.2.8 Perspectives for Chitosan Nanocomposites as Smart Cancer Therapeutic Materials

Cancer is one of the fatal diseases that immensely causes death around the globe and the number of lives lost due to cancer is increasing day by day. To minimize the death rate caused by cancer, anti-cancer therapeutic agents such as 5-fluorouracil, doxorubicin etc. should be administrated through a proper drug delivery system with target specific and controlled release features. For this purpose, a variety of chitosan nanocomposites have drawn the attention of researchers in the area of cancer treatment. Chitosan is endowed with inflammatory characteristics which affect the polarization effect of immune cells that consequently make chitosan a promising candidate for cancer treatment. In addition to this, stimuli responsive chitosan nanocomposite with graphene, metal, metal oxide and polymer and numerous formulations of chitosan, namely film, solution, gel, micro-needle and nanoparticle with immunomodulatory potential have been implementing in in vitro and in vivo cancer therapy. Regardless of such advancement of chitosan nanocomposites, there are still scopes for the improvement of the synthesis process, employment route of the particular system for targeted specific delivery and enhancement of biocompatibility and cytotoxicity effects.

17.2.9 Perspectives for Chitosan Nanocomposites as Future Materials for Dentistry

Starting from preventive to restorative dentistry chitosan-based nanocomposites stands as a real blessing in this field in view of their remarkable properties like antimicrobial, non-toxicity, bioadhesive, biocompatibility, biodegradability, antiplaque, osteoconduction and anti-inflammatory responses etc. Chitosan is modified with various metal and metal oxide nanoparticles, clay, graphene and graphene oxide as well as other synthetic polymers for outstanding applications in dentistry. However, from all the reviews in the present book, it is noticed that the accurate reproducibility of the designing process and molecular weight is challenging and demands further improvement in terms of reproducibility of the process. There is no doubt that chitosan has expanding future in dentistry application but the development of their clinical application in actual practical field still needs much attention. The future outlook should be high quality research leading to translation of dentistry application of chitosan-based nanocomposites from research to clinical trials by means of in vivo studies.

17.2.10 Perspectives for Chitosan Nanocomposites as Nanomedicine

Nanomedicine is a burgeoning area of nanotechnology which has the capability to transfigure the entire healthcare and medicine field with the assistance of therapeutic and diagnostic tools. So, the designing and development of new and unique nanomedicines are the absolute necessity. Chitosan nanocomposite emerged as a potential nanocarrier in sundry kinds of biomedical applications such as drug administration, antibacterial activity, gene therapy, cancer treatment, wound healing etc. This bio-nanocomposite holds the ability to encapsulate medications and conveys them to the requisite sites leading to administration in a sustainable and controlled manner. Being non-toxic and of suitable size, the chitosan nanocrystal is also useful in anticancer action. Furthermore, chitosan nanocomposites can enter the nucleus to interrupt protein synthesis inside the bacterial cell and cause the inhibition of bacteria production. In case of wound dressing, haemostatic properties of this material play a major role in plugging the bleeding wound surface.

17.3 Outlook and Remark

Chitosan nanocomposites can be chosen as one of the promising candidates for designing of advanced smart materials and devices keeping an eye upon the efficient translation from research to practical applications. Now it is an opportunity to explore new direction of research in the way of chitosan-based nanocomposites with a big solution to different challenges of individual chitosan. The cost effectiveness, safe processability, environmental friendly nature of chitosan nanocomposites need to be considered for device applications in the area of bionanomechanical applications. This chapter in particular and the book as a whole may discover a novel path in solving the problems of existing challenges towards proposing chitosan-based bionanocomposites towards biomechanical applications.

Index

S. K. Swain and A. Biswal (eds.), *Chitosan Nanocomposites*, Biological and Medical Physics, Biomedical Engineering,
https://doi.org/10.1007/978-981-19-9646-7

C

D

E

The manufacturer's authorised representative in the EU is Springer Nature Customer Service Centre GmbH, Europaplatz 3, 69115 Heidelberg, Germany. If you have any concerns regarding our products, please contact ProductSafety@springernature.com

Printed and bound by CPI Group (UK) Ltd, Croydon, CR0 4YY
30/09/2026
02225299-0008